AF677097

The Stochastic Perturbation Method for Computational Mechanics

The Stochastic Perturbation Method for Computational Mechanics

Marcin Kamiński

Department of Structural Mechanics
Technical University of Łódź, Poland

WILEY

A John Wiley & Sons, Ltd., Publication

This edition first published 2013 © 2013, John Wiley & Sons Ltd

Registered office
John Wiley & Sons Ltd, The Atrium, Southern Gate, Chichester, West Sussex, PO19 8SQ, United Kingdom

For details of our global editorial offices, for customer services and for information about how to apply for permission to reuse the copyright material in this book please see our website at www.wiley.com.

The right of the author to be identified as the author of this work has been asserted in accordance with the Copyright, Designs and Patents Act 1988.

All rights reserved. No part of this publication may be reproduced, stored in a retrieval system, or transmitted, in any form or by any means, electronic, mechanical, photocopying, recording or otherwise, except as permitted by the UK Copyright, Designs and Patents Act 1988, without the prior permission of the publisher.

Wiley also publishes its books in a variety of electronic formats. Some content that appears in print may not be available in electronic books.

Designations used by companies to distinguish their products are often claimed as trademarks. All brand names and product names used in this book are trade names, service marks, trademarks or registered trademarks of their respective owners. The publisher is not associated with any product or vendor mentioned in this book. This publication is designed to provide accurate and authoritative information in regard to the subject matter covered. It is sold on the understanding that the publisher is not engaged in rendering professional services. If professional advice or other expert assistance is required, the services of a competent professional should be sought.

Computations in this book were performed by using Maple™. Maple is a trademark of Waterloo Maple Inc.

Library of Congress Cataloging-in-Publication Data

Kaminski, M. M. (Marcin M.), 1969-
The stochastic perturbation method for computational mechanics / Marcin Kaminski.
pages cm
Includes bibliographical references and index.
ISBN 978-0-470-77082-5 (hardback)
1. Engineering–Statistical methods. 2. Perturbation (Mathematics) I. Title.
TA340.K36 2013
620.001′51922–dc23

2012029897

A catalogue record for this book is available from the British Library.

Print ISBN: 978-0-470-77082-5

Typeset in 10/12.5 Palatino by Laserwords Private Limited, Chennai, India from an electronic version supplied by the Author

Printed and bound in Singapore by Markono Print Media Pte Ltd

Contents

Acknowledgments

The author would like to acknowledge the financial support of the Polish Ministry of Science and Higher Education in Warsaw under Research Grant No. 519-386-636 entitled "Computer modeling of the aging processes using stochastic perturbation method" transferred recently to the Polish National Science Center in Cracow, Poland. This grant enabled me to make most of the research findings contained in this book. Its final shape is thanks to a professor's grant from the Rector of the Technical University of Łódź during the year 2011. Undoubtedly, my PhD students – with their curiosity, engagement in computer work, and research questions – helped me to prepare the numerical illustrations provided in the chapter focused on the stochastic finite element method.

Introduction

Uncertainty and stochasticity accompany our life from the very beginning and are still a matter of interest, guesses, and predictions made by mathematicians, economists, and fortune tellers. Their results may be as dramatic as car or airplane accidents, sudden weather changes, stock price fluctuations, diseases, and mortality in larger populations. All these phenomena and processes, although completely unpredictable for most people, have mathematical models to explain some trends and limited prognosis. There is a philosophical issue undertaken by various famous scientists whether the universe has a deterministic nature and some marginal stochastic noise – some kind of chaos or, in contrast, everything is uncertain – more, less, or fully.

In civil engineering we may observe the most dangerous aspects resulting from earthquakes, tornadoes, ice covers, and extensive rainfalls. These are the cases when stochastic fluctuations may also be treated as fully unpredictable, usually having no mean (expected) value and quantified coefficient of variation, so we are unable to provide any specific computer simulation. Let us recall that engineering codes usually apply the Poisson process to model huge catastrophic failures but they need extended and reliable statistics unavailable in many countries and sometimes even non-existent due to the enormous technological progress required. On a smaller scale (counting economic disasters and their possible consequences) we notice almost every day wind-blow variations and their results [158], accidental loading of cars and railways on bridges during rush hours, statistical strength properties of building materials, corrosion, interface cracks, volumetric structural defects, and a number of geometrical imperfections in structural engineering [142]. These are all included in mathematical and computational models with basic statistics coming from observations, engineering experience and, first of all, experimental verification. We need to assume that our design parameters have some distribution function and the most practical assumption is that they have Gaussian distributions. This reflects the Central Limit Theorem, stating that the mixture of different random variables tends to this particular distribution when their total number tends to infinity.

We are not interested in analyses and predictions without expectations in this book; computational analysis is strictly addressed to engineering and scientific problems having perfectly known expected values as well as standard deviations and to the case where the initial random dispersion is Gaussian or may be approximated by a Gaussian distribution with relatively small modeling error. In exceptional circumstances it is possible to consider lognormal distributions as they have recursive equations for higher-order probabilistic moments. From the probabilistic point of view we provide up to a fourth central probabilistic moments analysis of state functions like deformations, stresses, temperatures, and eigenfrequencies, because then it is possible to verify whether these functions really may have Gaussian distributions or not. The stochastic perturbation technique of course has a non-statistical character so we cannot engage any statistical hypothesis and we are interested in quantification of the resulting skewness and kurtosis. Recognition of the Gaussian output probability density function (PDF) will simplify further numerical experiments of similar character since these PDFs are uniquely defined by their first two moments and then the numerical determination of higher moments may be postponed.

From a historical point of view the first contribution to probability theory was made by the Italian mathematician Hieronimus Cardanus in the first part of his book entitled *Philologica, Logica, Moralia* published more than 100 years after he finished it in seventeenth century. As many later elaborations, it was devoted to the probability of winning in random games and had some continuation and extension in the work of Christian Huygens. It was summarized and published in London, in 1714, under the self-explanatory title *The Value of All Chances in Games of Fortune; Cards, Dice, Wagers, Lotteries & C. Mathematically Demonstrated*. The main objective at that time was to study the discrete nature of random events and combinatorics, as also documented by the pioneering works of Blaise Pascal and Pierre de Fermat. One of the most amazing facts joining probability theory with the world of analytical continuous functions is that the widely known PDF named after the German mathematician Karl Friedrich Gauss was nevertheless elaborated by Abraham de Moivre, most famous for his formula in complex number theory. The beginnings of modern probability theory date to the 1930s and are connected with the axioms proposed by Andriei Kolmogorov (exactly 200 years after the normal distribution introduced by de Moivre). However, the main engine of this branch of mathematics was, as in the previous century, just mechanics and, particularly, quantum mechanics based on the statistical and unpredictable nature noticed on the molecular scale, especially for gases. Studies slowly expanded to other media exhibiting strong statistical aspects in laboratory experiments performed in long repeatable series. There is no doubt today that a second milestone was the technical development in computer machinery and sciences, enabling large statistical simulations.

Probabilistic methods in engineering and applied sciences follow mathematical equations and methods [158], however the recent fast progress of computers and relevant numerical techniques has brought about some new perspectives, a little bit unavailable for broader audience because of mathematical complexity. Historically, it is necessary to mention a variety of mathematical methods, where undoubtedly the oldest one is based on straightforward evaluation of the probabilistic moments of the resulting analytical functions on the basis of moments of some input parameters. This can be done using integral definitions of moments or using specific algebraic properties of probabilistic moments themselves; similar considerations may be provided for the time series defining some random time fluctuations of engineering systems and populations as well as related simple stochastic processes. It is possible, of course, to provide analytical calculations and justification that some structure or system gives a stationary (or not) stochastic response. According to the progress of mathematical disciplines after classical probability theory, at the beginning of the twentieth century we noticed an elaboration of the theory of stochastic differential equations and their solutions for specific cases having applications in non-stationary technical processes like structural vibrations and signal analysis [158].

Nowadays these methods have brand new applications with the enormous expansion of computer algebra systems, where analytical and visualization tools give new opportunities in conjunction with old, well-established mathematical theories. Since these systems work as neural networks, we are able to perform statistical reasoning and decision-making based on the verification of various statistical hypotheses implemented. The successive expansion of uncertainty analysis continued thanks to computers, important for large data set analysis and, naturally, additional statistical estimators. The first of the computer-based methods, following traditional observation and laboratory experiments, is of course the Monte Carlo simulation technique [5, 25, 53, 71], where a large set of computational realizations of the original deterministic problem on the generated population returns through statistical estimation the desired probabilistic moments and coefficients. The pros and cons of this technique result from the quality and subprocedures of the internal random number generator (generation itself and shuffling routines) as well as the estimators (especially important for higher-order moments) implemented in the computer program. Usually, precise information about these estimator types is not included in commercial software guides. An application of this method needs an *a priori* definition of both basic moments and the PDF of the random or stochastic input, however, we usually restrict ourselves to the Gaussian, truncated Gaussian, or lognormal PDF because of a difficulty in recovering and analytical processing of the probabilistic moments. The next technique that evolved was fuzzy analysis [132], where an engineer needs precise information about the maximum and minimum values of a given random parameter, which also naturally comes from observation or experiments. Then, this method operates using interval analysis to show the admissible intervals for the resulting state functions on the basis of the intervals for given input parameters. A separate direction is represented by the spectral methods widely implemented in the finite element method (FEM), with commercial software like ABAQUS or ANSYS, for instance. These are closely related to vibration analysis, where a structure with deterministic characteristics is subjected to some random excitation with the first two probabilistic moments given [117, 153]. Application of the FEM system makes it possible to determine the power spectral density (PSD) function for the nodal response. General stochastic vibration analysis is still the subject of many works [30, 143], and many problems in that area remain unsolved.

We also have the family of perturbation methods of first, second, and general order applied in computational mechanics and, also, the Karhunen–Loeve expansion techniques [38, 39] as well as some mixed hybrid techniques, popular especially for multiscale models [176]. These expansion techniques are provided using the eigenfunctions and eigenvectors of the covariance kernel for the input random fields or processes, both Gaussian and non-Gaussian [168, 174]. They need more assumptions and mathematical effort to randomize the given physical problem than the perturbation methods and, further, determination of higher moments is not so straightforward. Moreover, there is no commercial implementation in any of the popular existing FEM systems in this case. There are some new theoretical ideas in random analysis for both discrete [55] and continuous variables or processes [33, 52, 173], but they have no widely available computational realizations or general applications in engineering. The reader is advised to study [41, 154] for a comprehensive review of modern probabilistic methods in structural mechanics.

Restricting our overview to the perturbation method we need to mention that the first-order technique is useful for the very small random dispersion of input random variables (with coefficient of variation smaller than $\alpha < 0.10$) to replace Monte Carlo simulations in simplified first-two-moments analysis. The second-order techniques [112, 118] are applicable for $\alpha < 0.15$ in second-moment analysis also for both symmetrical distributions (second-order second-moment analysis – SOSM) and for some non-symmetrical probability functions like the Weibull distribution (the so-called Weibull second-order third-moment approach – WSOTM). The main idea of the generalized stochastic perturbation method proposed here is to calculate higher-order moments and coefficients to recognize the resulting distributions of structural response. The second purpose is to allow for larger input coefficients of variation, but higher moments were initially derived in many numerical experiments contained in this book using fourth- and sixth-order expansions only. Implementation of the given general-order stochastic perturbation technique was elaborated first of all to minimize the modeling error [139] and now is based on polynomials of uncertain input variable with deterministic coefficients. It needs to be mentioned that random or stochastic polynomials appeared in probabilistic analysis before [50, 147], but were never connected with the perturbation method and deterministic structural response determination.

It should be emphasized further that the perturbation method was neither strictly connected with the stochastic or probabilistic analysis nor developed for these problems [135]. The main idea of this method is to make an analytical expansion of some input parameter or phenomenon around its mean value thanks to some series representation, where Taylor series expansions are traditionally the most popular. Deterministic applications of this technique are known first of all from dynamical problems, where system vibrations are frequently found thanks to such an expansion in more complex situations. One interesting application is the homogenization method, where effective material properties tensors of some multi-material systems are found from the solution of the so-called homogenization problem including initial perturbation-based expansions of these effective tensor components with respect to various separate geometrical scales [6, 56, 151]. Further, as also demonstrated in this book, such a deterministic expansion may be linked with probabilistic analysis, where many materials constituting such a structure are separately statistically homogeneous (finite and constant expectations and deviations of physical properties) and results in a statistically heterogeneous global system (partially constant expectations and deviations of physical properties). This is the case when the geometry is perfectly periodic and the physical nature of the composite exhibits some random fluctuation. Then such a homogenization procedure returns statistical homogeneity using some mixing procedure and remains clearly deterministic, because expansion deals with geometric scales that show no uncertainty.

Let us note that the very attractive aspect of the perturbation method is that it includes sensitivity analysis [35, 44, 83, 91] since first-, second-, and higher-order partial derivatives of the objective function with respect to the design parameter(s) must be known before the expansions are provided. Therefore, before we start uncertainty analysis of some state function in the given boundary value problem, we should perform first-order sensitivity analysis and randomize only these parameters whose gradients (after normalization) have dominating and significant values. Further, the stochastic perturbation method is not really associated with any discrete computational technique available [111, 152] like FEM, the Finite Difference Method (FDM), the Finite Volume Method (FVM), the Boundary Element Method (BEM), various meshless techniques, or even molecular dynamics simulations. We can use it first of all to make additional probabilistic expansions of the given analytical solutions exhibiting some parametric randomness or even to solve analytically some algebraic or differential equations using explicit, implicit, and even symbolic techniques.

The stochastic perturbation technique is shown here in two different realizations – with use of the Diret Differentiation Method (DDM) and in conjunction with the Response Function Method (RFM). First of them is based on the straightforward differentiation of the basic deterministic counterpart of the stochastic problem, so that we obtain for a numerical solution a system of hierarchical equations with increasing order. The zeroth-order solution is computed from the first equation and inserted into the second equation, where first-order approximation is obtained and so on, until the highest-order solution is completed. Computational implementation of the DDM proceeds through direct implementation with the deterministic source code or, alternatively, with use of some of the automatic differentiation tools available widely as shareware. Although higher-order partial derivatives are calculated analytically at the mean values of input parameters, and so that are determined exactly, the final solution of the system of algebraic equations of increasing order enlarges the final error in probabilistic moments – the higher order of the solution, the larger possible numerical error. The complexity of the general-order implementation, as well as this aspect, usually results in DDM implementations of lowest order – as first or the second. Contrary to numerous previous models, now full tenth-order stochastic expansions are used to recover all the probabilistic moments and coefficients; this significantly increases the accuracy of the final results.

We employ the RFM consecutively, where we carry out numerical determination of the analytical function for a given structural response like displacement or temperature as the polynomial representation of the chosen random input design parameter (to determine its deterministic coefficients). Generally, it can be implemented in a global sense, where a single function connects the probabilistic output and input and, in a more delicate manner – in the local formulation, where the approximating polynomial form varies from the mesh or grid node to another node in the discrete model. It is apparent that global approximation is much faster but may show a larger modeling error; such a numerical error [139] in the local formulation is partially connected with the discretization procedure and may need some special adaptivity tools similar to these worked out in deterministic analyses. The main advantages of RFM over DDM are that (i) error analysis issues deal with the deterministic approximation problems and (ii) there is an opportunity for a relatively easy interoperability with commercial (or any) packages for discrete computational techniques. The RFM procedures do not need any symbolic algebra system because we differentiate well-known polynomials of random variables, so this differentiation is also of deterministic character. The RFM is used here in the few different realizations starting from classical polynomial interpolation with the given order, some interval spline approximations, through the non-weighted least-squares method until more sophisticated weighted optimized least-squares methods. This aspect is now closely related to the computer algebra system and this choice also follows enriched visualization procedures, but may be implemented in classical programming language. The RFM is somewhat similar to the response surface method (RSM) applicable in reliability analysis [175] or the response function technique known from vibration analysis. The major and very important difference is that the RFM uses a higher-order polynomial response relating a single input random variable with the structural output, whereas the RSM is based on first- or second-order approximations of this output with respect to multiple random structural parameters. An application of the RSM is impossible in the current context because the second-order truncation of the response eliminates all higher-order terms necessary for reliable computation of the probabilistic structural response. Furthermore, the RSM has some statistical aspects and issues, while the RFM has a purely deterministic character and exhibits some errors typical for mathematical approximation theory methods only.

Finally, let us note that the generalized stochastic perturbation technique was initially worked out for a single input random variable but we have some helpful comments in this book concerning how to complete its realization in case of a vector of correlated or not random input sources. The uncorrelated situation is a simple extension of the initial single-variable case, while non-zero cross-correlations, especially of higher order, will introduce a large number of new components into the perturbation-based equations for the probabilistic moments, even for expectations.

It is clear that stochastic analysis in various branches of engineering does not result from a fascination with random dispersion and stochastic fluctuations in civil or aerospace structures, mechanical as well as electronic systems – it is directly connected with reliability assessment and durability predictions [1]. Recently we noticed a number of probabilistic numerical studies in non-linear problems in mechanics dealing particularly with the design of experiments [45], gradient plasticity [177], and viscoelastic structures [42], summarized for multiscale random media in [140]. Even the simplest model of the first-order reliability method is based on the reliability index giving quantified information about the safety margin computed using the expected values and standard deviations for two or more components of the limit function. According to various numerical illustrations presented here, the tenth-order stochastic perturbation technique is as efficient for this purpose as the MCS method and frequently does not need further comparative studies. It is also independent of the input random dispersion of the given variable of the problem and should be checked for correlated variables also. As is known, the second-order reliability methods [128] include some correction factors and/or multipliers like the curvature of the limit functions usually expressed by the second partial derivatives of the objective function with respect to the random input. The generalized perturbation technique serves in a straightforward manner in this situation, because these derivatives are included in the Taylor expansions themselves, so there is no need for an additional numerical procedure. As has been documented, this stochastic perturbation-based finite element method (SFEM) implemented using the RFM idea may be useful at least for civil engineers following Eurocode 0 statements and making simulations on commercial FEM software. It is worth emphasizing that the stochastic perturbation method may be efficient in time-dependent reliability analysis, where time series having Gaussian coefficients approximate time fluctuations of the given design parameters. There are some further issues not discussed in this book, like the adaptivity method related to the stochastic finite elements [171], which may need some new approaches to the computational implementation of the perturbation technique.

This book is organized into five main chapters – Chapter 1 is devoted to the mathematical aspects of the stochastic perturbation technique, necessary definitions and properties of the probability theory. It is also full of computational examples showing implementations of various engineering problems with uncertainty into the computer algebra system Maple™ [17] supporting all further examples and solutions. Some of these are shown directly as scripts with screenshots, especially once some analytical derivations have been provided. The remaining case studies, where numerical data has been processed, are focused on a discussion of the results visualized as the parametric plots of probabilistic moments and characteristics, mostly with respect to the input random dispersion coefficient. They are also illustrated with the Maple™ scripts accompanying the book, which are still being expanded by the author and may be obtained by special request in the most recent versions. Special attention is given to the RFM here, various-order approximations of the moments in the stochastic perturbation technique, some comparisons against the Monte Carlo technique and computerized analytical methods, as well as simple time-series analysis with the perturbation technique.

Chapter 2 is the largest in the book and is devoted entirely to the SFEM. It starts with the statements of various more important boundary-value or boundary-initial problems in engineering with random parameters, which are then transformed into additional variational statements, also convenient for general nth-order stochastic formulations. According to the above considerations, these stochastic variational principles and the resulting systems of algebraic equations are expanded using both DDM and RFM approaches to enable alternative implementations depending on the source code and automatic differentiation routines availability; there are multiple Maple™ source codes for most of the numerical illustrations here, as also in the preceding chapter. Theoretical developments start from the FEM for the uncoupled equilibrium problems with scalar and vector state functions and are continued until the thermo-electro-elastic couplings as well as Navier–Stokes equations for incompressible and non-turbulent Newtonian fluid flows. The particular key computational experiments obey Newtonian viscous unidirectional and 2D fluid flows, linear elastic response and buckling of a spatial elastic system, elasto-plastic behavior of a simple 2D truss, eigenvibrations analysis of a 3D steel tower, non-stationary heat transfer in a unidirectional rod, as well as forced vibrations in a 2 DOF system, all with randomized material parameters. It is demonstrated that the Maple™ system may be used efficiently as the FEM postprocessor, making a visualization of the mesh together with the desired probabilistic characteristics in vector form; three-dimensional graphics are not so complicated in this environment, but physical interpretation of higher-order moments does not require such sophisticated tools right now. The discussion is restricted each time to the first four probabilistic moments and coefficients for the structural response shown as functions of the input coefficient of variation and, sometimes, the stochastic perturbation order. Usually, we (i) check the probabilistic convergence of the SFEM together with its order, (ii) detect the influence of an initial uncertainty source, and (iii) verify the output PDF.

Chapter 3 describes the basic equilibrium equations and computational implementation of the Stochastic Perturbation-based Boundary Element Method (SBEM) related to the linear isotropic elasticity of the statistically homogeneous and multi-component domains; numerical work has been completed using the open-source academic BEM code [4]. The basic equations have all been rewritten in the response functions language with numerical illustrations showing uncertain elastic behavior of a steel plane panel, an analogous composite layered element with perfect interface, as well as a composite with some interface defects between the constituents. A comparison of the SBEM implemented using triangular and Dirac distributions of the weights in Least Squares Method is also given here using the example of the first four probabilistic characteristics presented as functions of the input coefficient of variation for the last problem.

Chapter 4 is addressed to anyone who is interested in Stochastic analysis using the specially adopted Finite Difference Method (SFDM) and additional source codes. According to the main philosophy of the method we rewrite the particular differential equations in the difference forms and introduce first of all their DDM versions to carry out computational modeling directly using the Maple™ program. The example problem with random parameters is the linear elastic equilibrium of the Euler–Bernoulli beam with constant and linearly varying cross-sectional area; further, this structure is analyzed numerically on an elastic single parameter random foundation. Let us note that stochastic analysis of beams with random stiffness in civil and mechanical engineering is of significant practical importance and has been many times studied theoretically and numerically [31, 112]. Other models include non-stationary heat transfer in a homogeneous rod with Gaussian physical parameters, eigenvibration analysis of a simply supported beam and a thin plate, as well as the unidirectional diffusion equation. Some examples show the behavior of the probabilistic moments computed together with increasing density of the grid, others are shown to make a comparison with the results obtained from the analytical predictions.

Chapter 5 is particularly and entirely devoted to the homogenization procedure presented as the unique application of the double perturbation method, where deterministic expansion with respect to the scale parameter is used in conjunction with stochastic expansions of the basic elastic parameters. Homogenization of the perfectly periodic two-component composite is the main objective in this chapter, and its effective elasticity tensor in a probabilistic and stochastic version is studied for material parameters of fiber and matrix defined as Gaussian random variables or time series with Gaussian coefficients. The main purpose is to verify the stochastic perturbation technique and its FEM realization against the Monte Carlo simulation, as well as some novel computational techniques using the RFM based on analytical integration implemented in the Maple™ system. The examples are used to confirm the Gaussian character of the resulting homogenized tensor components, check the perturbation technique convergence for various approximation orders, show the probabilistic entropy fluctuations in the homogenization procedure, and provide some perspectives for further development of both SFEM and RFM techniques.

The last part of this book is given as the Appendix, where all more popular probability distributions are contrasted. Particularly, their up to the tenth central probabilistic moments are derived symbolically to serve the Readers in their own stochastic implementations.

The major conclusion of this book is that the stochastic perturbation technique is a universal numerical method useful with any discrete or symbolic, academic or commercial computer programs, and environments. The applicability range for expectations is practically unbounded, for second moments – extremely large (much larger than before) but for third- and fourth-order statistics – limited (but may be given precisely in terms of an input random dispersion). Mathematical simplicity and time savings are attractive for engineers, but we need to remember that this is not a computational hammer to randomize everything. Special attention is necessary in case of coupled problems with huge random fluctuations, where output coefficients of variation at some iteration step (even the first one) can make it practically useless. The local and global response functions are usually matched very well by the polynomial forms proposed here, and, sometimes, resulting moments show no singularities with respect to the input coefficient of variation. This situation, however, may change in systems with state-dependent physical and mechanical properties (for example, with respect to large temperature variations).

The book in its present shape took me almost 20 years of extensive work, from the very beginning of my career with the second order version of the SFEM at the Institute of Fundamental Technological Research in Warsaw, Poland [112]. Slowly my interest in the finite elements domain evolved towards other discrete computational techniques and, after that, an idea of any-order Taylor expansion appeared around 10 years ago. I would like to express special thanks to my PhD students at the Technical University of Łódź for their help in reworking and reorganizing many numerical examples for this book, but also for their never-ending questions – pushing me to carefully check many times the same issues. I appreciate the comments of many colleagues from all around the world who are interested in my work, as well as the anonymous reviewers who took care over the precision of my formulations.

1

Mathematical Considerations

1.1 Stochastic Perturbation Technique Basis

The input random variable of a problem is denoted here consecutively by $b(\omega)$ and its probability density by $g_b(x)$. The expected value of this variable is expressed by Feller [34] and Vanmarcke [165] as

$$E[b] = \int_{-\infty}^{+\infty} b \; g_b(x)\mathrm{d}x, \tag{1.1}$$

while its mth central probabilistic moment is

$$\mu_m(b) = \int_{-\infty}^{+\infty} (b - E[b])^m \, g_b(x)\mathrm{d}x. \tag{1.2}$$

Since we are mostly focused on the Gaussian distribution application, we recall now its probability density function:

$$g_b(x) = \frac{1}{\sqrt{2\pi \; \mathrm{Var}(b)}} \exp\left(-\frac{(x - E[b])^2}{2\mathrm{Var}(b)}\right). \tag{1.3}$$

The coefficient of variation, skewness, flatness and kurtosis are introduced in the form

$$\alpha(b) = \frac{\sqrt{\mu_2(b)}}{E[b]} = \frac{\sqrt{\mathrm{Var}(b)}}{E[b]} = \frac{\sigma(b)}{E[b]}, \quad \beta(b) = \frac{\mu_3(b)}{\sigma^3(b)}, \quad \gamma(b) = \frac{\mu_4(b)}{\sigma^4(b)} = \kappa(b) + 3. \tag{1.4}$$

Nowadays, computer algebra software is usually employed to provide analytical formulas following these statements. A symbolic solution provided in the system Maple™ for the well-known case of two Gaussian random variables X and Y having

The Stochastic Perturbation Method for Computational Mechanics, First Edition. Marcin Kamiński.
© 2013 John Wiley & Sons, Ltd. Published 2013 by John Wiley & Sons, Ltd.

defined expectations and standard deviations equal to *EX*, *EY* and *SIGX*, *SIGY* is given below. As is supposed, we can have more variables, combined in all algebraic forms implemented into this system and, finally, random variables not necessarily Gaussian.

```
>restart: with(plots): with(plottools): with(Statistics):
>X:=RandomVariable(Normal(EX,SX)): Y:=RandomVariable(Normal(EY,SY)):
>G:=X*Y:
>EG:=ExpectedValue(G);
```

$$EG := EX\,EY$$

```
>VarG:=Variance(G);
```

$$VarG := EY^2\,SX^2 + SY^2\,EX^2 + SY^2\,SX^2$$

```
>StdG:=StandardDeviation(G);
```

$$StdG := \sqrt{EY^2\,SX^2 + SY^2\,EX^2 + SY^2\,SX^2}$$

```
>alfaG:=Variation(G);
```

$$alfaG := \frac{\sqrt{EY^2\,SX^2 + SY^2\,EX^2 + SY^2\,SX^2}}{EX\,EY}$$

```
>skewG:=Skewness(G);
```

$$skewG := \frac{6\,EX\,EY\,SX^2\,SY^2}{(EY^2\,SX^2 + SY^2\,EX^2 + SY^2\,SX^2)^{3/2}}$$

```
>kappaG:=Kurtosis(G);
```

$$kappaG := (3(EY^4\,SX^4 + 2\,EY^2\,SX^2\,SY^2\,EX^2 + 6\,EY^2\,SX^4\,SY^2 + SY^4\,EX^4 + 6\,SY^4\,EX^2\,SX^2 + 3\,SY^4\,SX^4))/(EY^2\,SX^2 + SY^2\,EX^2 + SY^2\,SX^2)^2$$

```
>m3G:=CentralMoment(G,3);
```

$$m3G := 6\,EX\,EY\,SX^2\,SY^2$$

```
>m4G:=CentralMoment(G,4);
```

$$m4G := 3\,EY^4\,SX^4 + 6\,EY^2\,SX^2\,SY^2\,EX^2 + 18\,EY^2\,SX^4\,SY^2 + 3\,SY^4\,EX^4 + 18\,SY^4\,EX^2\,SY^2 + 9\,SY^4\,SX^4$$

The second, less trivial opportunity with this program is recovery of the probabilistic moments for the other probability distributions widely applied in engineering, whose formulas are not available in the literature or are hard to find (contained in the

Appendix). The cases of lognormal and Gumbel distributions serve as an example below – one can use more sophisticated algebraic combinations of course.

```
>restart; with(Statistics): a::real, 0 < b: X1:=RandomVariable(Gumbel(a,b)):
>EX1:=ExpectedValue(X1); VX1:=Variance(X1); MX1:=Median(X1); KX1:=Kurtosis(X1):
SKX1:=Skewness(X1): COVX1:=Variation(X1); CM3X1:=CentralMoment(X1,3):
CM4X1:=CentralMoment(X1,4):
```

$$EX1 := a + \gamma b$$

$$VX1 := \frac{1}{6} b^2 \pi^2$$

$$MX1 := a - b \ln(\ln(2))$$

$$COVX1 := \frac{1}{6} \frac{\sqrt{6} b \pi}{a + \gamma b}$$

```
>X2:=RandomVariable(LogNormal(a,b)):
>EX2:=ExpectedValue(X2); VX2:=Variance(X2); MX2:=Median(X2); KX2:=Kurtosis(X2);
SKX2:=Skewness(X2); COVX2:=Variation(X2); CM3X2:=CentralMoment(X2,3);
CM4X2:=CentralMoment(X2,4);
```

$$EX2 := e^{a+\frac{1}{2}b^2} VX2 := e^{2a+b^2}\left(e^{b^2} - 1\right)$$

$$MX2 := e^a$$

$$KX2 := -\frac{-e^{4a+8b^2} + 4e^{4a+5b^2} - 6e^{4a+3b^2} + 3e^{2b^2+4a}}{\left(e^{2a+b^2}\right)^2 \left(e^{b^2-1}\right)^2}$$

$$SKX2 := \frac{e^{3a+\frac{9}{2}b^2} - 3e^{3a+\frac{5}{2}b^2} + 2e^{\frac{3}{2}b^2+3a}}{\left(e^{2a+b^2}\left(e^{b^2} - 1\right)\right)^{3/2}}$$

$$COVX2 := \frac{\sqrt{e^{2a+b^2}\left(e^{b^2} - 1\right)}}{e^{a+\frac{1}{2}b^2}}$$

$$CM3X2 := e^{3a+\frac{9}{2}b^2} - 3e^{3a+\frac{5}{2}b^2} + 2e^{\frac{3}{2}b^2+3a}$$

$$CM4X2 := e^{4a+8b^2} - 4e^{4a+5b^2} + 6e^{4a+3b^2} - 3e^{2b^2+4a}$$

Besides the probabilistic moments and coefficients, the entropy of random variables and processes is also sometimes considered. Probabilistic entropy [155, 156] (contrary to that popular in thermodynamics) illustrates an uncertainty of occurrence of some event in the next moment, so that entropy equal to 0 accompanies a probability equal to 1 (or 0) for any random experiment showing no randomness at all. If the countable set of random events has n elements associated with the probabilities p_i for $i = 1, \ldots, n$, then the entropy in this space is defined uniquely by the following sum [155, 156]:

$$H(x) = -\sum_{i=1}^{n} p_i \log_r (p_i), \tag{1.5}$$

where the logarithm basis r is the entropy unit; computational information theory is naturally based on bits, where $r=2$. This discrete definition restricts the values to the non-negative real numbers, where $H(x)$ reaches maximum for two elements' random space with both events having the same probability -1 (like a bit of entropy per single throw with a geometrically regular coin). Its generalization to continuous variables in case of the Gaussian distribution is

$$h(x) = -\int_{-\infty}^{+\infty} \frac{1}{\sigma\sqrt{2\pi}} \exp\left(-\frac{(x-m)^2}{2\sigma^2}\right) \log\left(\frac{1}{\sigma\sqrt{2\pi}} \exp\left(-\frac{(x-m)^2}{2\sigma^2}\right)\right) dx. \tag{1.6}$$

where m, σ denote traditionally its expectation and standard deviation.

An integration process is carried out using classical normalization:

$$t = \frac{x-m}{\sqrt{2}\sigma}; \qquad dx = \sqrt{2}\sigma \ dt \tag{1.7}$$

and therefore

$$\begin{aligned}
h(t) &= -\int_{-\infty}^{+\infty} \frac{1}{\sigma\sqrt{2\pi}} \exp(-t^2) \log\left(\frac{1}{\sigma\sqrt{2\pi}} \exp(-t^2)\right) \sqrt{2}\sigma dt \\
&= -\frac{1}{\sqrt{\pi}} \int_{-\infty}^{+\infty} \exp(-t^2) \log\left(\frac{1}{\sigma\sqrt{2\pi}} \exp(-t^2)\right) dt \\
&= -\frac{1}{\sqrt{\pi}} \log\left(\frac{1}{\sigma\sqrt{2\pi}}\right) \int_{-\infty}^{+\infty} \exp(-t^2) dt - \frac{1}{\sqrt{\pi}} \int_{-\infty}^{+\infty} (-t^2) \exp(-t^2) dt \\
&= -\frac{\sqrt{\pi}}{\sqrt{\pi}} \log\left(\frac{1}{\sigma\sqrt{2\pi}}\right) + \frac{1}{\sqrt{\pi}} \int_{-\infty}^{+\infty} t^2 \exp(-t^2) dt \\
&= -\log\left(\frac{1}{\sigma\sqrt{2\pi}}\right) + \frac{1}{\sqrt{\pi}} \frac{\sqrt{\pi}}{2} = \frac{1}{2} \log(2\pi e \sigma^2).
\end{aligned} \tag{1.8}$$

The entropy formula remains unimplemented in most computer algebra systems, so this integral definition may appear useful in some engineering applications, especially with time series or stochastic processes. As could be expected in the case of Gaussian variables it is entirely affected by the standard deviation, so that the proposed stochastic perturbation technique – with its perfect agreement with the other numerical techniques in determination of the second-order probabilistic moments – is a reliable computational tool to determine entropy also.

A very interesting problem for any state function and its uncertainty source would be the entropy variation, and this can be defined through initial and final values as

$$\Delta h = h(f(b)) - h(b) = -\int_{-\infty}^{+\infty} g_{f(b)}(x) \log\left(g_{f(b)}(x)\right) dx$$

$$+ \int_{-\infty}^{+\infty} g_b(x) \log\left(g_b(x)\right) dx = \log\left(\sqrt{2\pi e}\sigma\left(f(b)\right)\right) - \log\left(\sqrt{2\pi e}\sigma(b)\right)$$

$$= \log\left(\frac{\sigma\left(f(b)\right)}{\sigma(b)}\right). \tag{1.9}$$

This entropy change shows whether the uncertainty may be amplified by the given boundary value problem, preserved or damped. The most interesting case, from a probabilistic point of view, would be $\Delta h = 0$, which can be interpreted as no influence of the problem solution method or the problem itself on the initial random dispersion.

Now let us focus on the generalized stochastic perturbation technique – the main philosophy of this method is to expand all state parameters and the response functions in an initially deterministic problem (heat conductivity, heat capacity, temperature, and its gradient as well as the material density) using a given-order Taylor series with random coefficients. It is provided by the following representation of the random function $u(b)$ with respect to its parameter b around its mean value [74, 81]:

$$\mathbf{u}(b) = \mathbf{u}^0\left(b^0\right) + \varepsilon \left.\frac{\partial \mathbf{u}(b)}{\partial b}\right|_{b=b^0} \Delta b + \cdots + \frac{\varepsilon^n}{n!} \left.\frac{\partial^n \mathbf{u}(b)}{\partial b^n}\right|_{b=b^0} \Delta b^n, \tag{1.10}$$

where ε is a given small perturbation (usually taken equal to 1), while the nth-order variation is given as follows:

$$\varepsilon^n \Delta b^n = (\delta b)^n = \varepsilon^n \left(b - b^0\right)^n. \tag{1.11}$$

The expected values can be derived exactly with use of the tenth-order expansion for Gaussian variables as

$$E[\mathbf{u}(b)] = \mathbf{u}^0\left(b^0\right) + \frac{\varepsilon^2}{2} \frac{\partial^2 \mathbf{u}(b)}{\partial b^2} \mu_2(b)$$

$$+ \frac{\varepsilon^4}{4!} \frac{\partial^4 \mathbf{u}(b)}{\partial b^4} \mu_4(b) + \frac{\varepsilon^6}{6!} \frac{\partial^6 \mathbf{u}(b)}{\partial b^6} \mu_6(b) + \frac{\varepsilon^8}{8!} \frac{\partial^8 \mathbf{u}(b)}{\partial b^8} \mu_8(b) + \frac{\varepsilon^{10}}{10!} \frac{\partial^{10} \mathbf{u}(b)}{\partial b^{10}} \mu_{10}(b) \tag{1.12}$$

for any natural m with μ_{2m} being central probabilistic moment of 2mth order. It is obtained via substitution of an expansion (1.10) into the definition (1.1), by dropping off all odd order terms and integration of all the remaining order variations. It returns even order central probabilistic moments of variable b as well as deterministic odd order partial derivatives with respect to this b at its mean value. Usually, according to some previous computational convergence studies, we may limit this expansion to tenth order but consecutively for all moments of interest here. Quite similar considerations lead to the expressions for higher moments, like the variance, for instance:

$$\begin{aligned}
\mathrm{Var}\,(\mathbf{u}(b)) &= \mu_2\,(\mathbf{u}(b)) = \int_{-\infty}^{+\infty} (\mathbf{u}(b) - E\,[\mathbf{u}(b)])^2\, g_b(x)\mathrm{d}x \\
&= \mu_2(b)\left(\frac{\partial \mathbf{u}(b)}{\partial b}\right)^2 + \mu_4(b)\left\{\frac{1}{4}\left(\frac{\partial^2 \mathbf{u}(b)}{\partial b^2}\right)^2 + \frac{1}{3}\frac{\partial^3 \mathbf{u}(b)}{\partial b^3}\frac{\partial \mathbf{u}(b)}{\partial b}\right\} \\
&+ \mu_6(b)\left\{\frac{1}{36}\left(\frac{\partial^3 \mathbf{u}(b)}{\partial b^3}\right)^2 + \frac{1}{24}\frac{\partial^4 \mathbf{u}(b)}{\partial b^4}\frac{\partial^2 \mathbf{u}(b)}{\partial b^2} + \frac{1}{60}\frac{\partial^5 \mathbf{u}(b)}{\partial b^5}\frac{\partial \mathbf{u}(b)}{\partial b}\right\} \\
&+ \mu_8(b)\left\{\frac{1}{576}\left(\frac{\partial^4 \mathbf{u}(b)}{\partial b^4}\right)^2 + \frac{1}{360}\frac{\partial^5 \mathbf{u}(b)}{\partial b^5}\frac{\partial^3 \mathbf{u}(b)}{\partial b^3} + \frac{1}{2520}\frac{\partial^7 \mathbf{u}(b)}{\partial b^7}\frac{\partial \mathbf{u}(b)}{\partial b}\right. \\
&\qquad \left. + \frac{1}{720}\frac{\partial^6 \mathbf{u}(b)}{\partial b^6}\frac{\partial^2 \mathbf{u}(b)}{\partial b^2}\right\} \\
&+ \mu_{10}(b)\left\{\frac{1}{14400}\left(\frac{\partial^5 \mathbf{u}(b)}{\partial b^5}\right)^2 + \frac{1}{40320}\frac{\partial^8 \mathbf{u}(b)}{\partial b^8}\frac{\partial^2 \mathbf{u}(b)}{\partial b^2} + \frac{1}{8640}\frac{\partial^6 \mathbf{u}(b)}{\partial b^6}\frac{\partial^4 \mathbf{u}(b)}{\partial b^4}\right\} \\
&+ \mu_{10}(b)\left\{\frac{1}{15120}\frac{\partial^7 \mathbf{u}(b)}{\partial b^7}\frac{\partial^3 \mathbf{u}(b)}{\partial b^3} + \frac{1}{181440}\frac{\partial^9 \mathbf{u}(b)}{\partial b^9}\frac{\partial \mathbf{u}(b)}{\partial b}\right\}.
\end{aligned} \tag{1.13}$$

One may notice that each component corresponds to the next consecutive order in Equation (1.12), while a linear increase of the components is noticed from each order to the next one for the variance.

The third probabilistic moment may be recovered from this scheme as

$$\begin{aligned}
\mu_3\,(\mathbf{u}(b)) &= \int_{-\infty}^{+\infty} (\mathbf{u}(b) - E\,[\mathbf{u}(b)])^3\, g_b(x)\mathrm{d}x = \frac{3}{2}\mu_4(b)\left(\frac{\partial \mathbf{u}(b)}{\partial b}\right)^2 \frac{\partial^2 \mathbf{u}(b)}{\partial b^2} \\
&+ \mu_6(b)\left\{\frac{1}{8}\left(\frac{\partial^2 \mathbf{u}(b)}{\partial b^2}\right)^3 + \frac{1}{2}\frac{\partial \mathbf{u}(b)}{\partial b}\frac{\partial^2 \mathbf{u}(b)}{\partial b^2}\frac{\partial^3 \mathbf{u}(b)}{\partial b^3} + \frac{1}{8}\left(\frac{\partial \mathbf{u}(b)}{\partial b}\right)^2\frac{\partial^4 \mathbf{u}(b)}{\partial b^4}\right\} \\
&+ \mu_8(b)\left\{\frac{1}{24}\frac{\partial \mathbf{u}(b)}{\partial b}\frac{\partial^3 \mathbf{u}(b)}{\partial b^3}\frac{\partial^4 \mathbf{u}(b)}{\partial b^4} + \frac{1}{40}\frac{\partial \mathbf{u}(b)}{\partial b}\frac{\partial^2 \mathbf{u}(b)}{\partial b^2}\frac{\partial^5 \mathbf{u}(b)}{\partial b^5}\right. \\
&\qquad \left. + \frac{1}{240}\left(\frac{\partial \mathbf{u}(b)}{\partial b}\right)^2\frac{\partial^6 \mathbf{u}(b)}{\partial b^6}\right\}
\end{aligned}$$

$$+\mu_8(b)\left\{\frac{1}{32}\left(\frac{\partial^2\mathbf{u}(b)}{\partial b^2}\right)^2\frac{\partial^4\mathbf{u}(b)}{\partial b^4}+\frac{1}{24}\left(\frac{\partial^3\mathbf{u}(b)}{\partial b^3}\right)^2\frac{\partial^2\mathbf{u}(b)}{\partial b^2}\right\}$$

$$+\mu_{10}(b)\left\{\frac{1}{480}\frac{\partial\mathbf{u}(b)}{\partial b}\frac{\partial^4\mathbf{u}(b)}{\partial b^4}\frac{\partial^5\mathbf{u}(b)}{\partial b^5}+\frac{1}{1680}\frac{\partial\mathbf{u}(b)}{\partial b}\frac{\partial^2\mathbf{u}(b)}{\partial b^2}\frac{\partial^7\mathbf{u}(b)}{\partial b^7}\right.$$

$$\left.+\frac{1}{720}\frac{\partial\mathbf{u}(b)}{\partial b}\frac{\partial^3\mathbf{u}(b)}{\partial b^3}\frac{\partial^6\mathbf{u}(b)}{\partial b^6}\right\}$$

$$+\mu_{10}(b)\left\{\frac{1}{240}\frac{\partial^2\mathbf{u}(b)}{\partial b^2}\frac{\partial^3\mathbf{u}(b)}{\partial b^3}\frac{\partial^5\mathbf{u}(b)}{\partial b^5}+\frac{1}{13440}\left(\frac{\partial\mathbf{u}(b)}{\partial b}\right)^2\frac{\partial^8\mathbf{u}(b)}{\partial b^8}\right.$$

$$\left.+\frac{1}{960}\left(\frac{\partial^2\mathbf{u}(b)}{\partial b^2}\right)^2\frac{\partial^6\mathbf{u}(b)}{\partial b^6}\right\}$$

$$+\mu_{10}(b)\left\{\frac{1}{384}\left(\frac{\partial^4\mathbf{u}(b)}{\partial b^4}\right)^2\frac{\partial^2\mathbf{u}(b)}{\partial b^2}+\frac{1}{288}\left(\frac{\partial^3\mathbf{u}(b)}{\partial b^3}\right)^2\frac{\partial^4\mathbf{u}(b)}{\partial b^4}\right\} \tag{1.14}$$

while the fourth probabilistic moment computation proceeds with use of the following formula:

$$\mu_4(\mathbf{u}(b))=\int_{-\infty}^{+\infty}(\mathbf{u}(b)-E[\mathbf{u}(b)])^4 g_b(x)\mathrm{d}x$$

$$=\mu_4(b)\left(\frac{\partial\mathbf{u}(b)}{\partial b}\right)^4+\mu_6(b)\left\{\frac{3}{2}\left(\frac{\partial\mathbf{u}(b)}{\partial b}\right)^2\left(\frac{\partial^2\mathbf{u}(b)}{\partial b^2}\right)^2+\frac{2}{3}\left(\frac{\partial\mathbf{u}(b)}{\partial b}\right)^3\frac{\partial^3\mathbf{u}(b)}{\partial b^3}\right\}$$

$$+\mu_8(b)\left\{\frac{1}{16}\left(\frac{\partial^2\mathbf{u}(b)}{\partial b^2}\right)^4+\frac{1}{30}\left(\frac{\partial\mathbf{u}(b)}{\partial b}\right)^3\frac{\partial^5\mathbf{u}(b)}{\partial b^5}+\frac{1}{6}\left(\frac{\partial\mathbf{u}(b)}{\partial b}\right)^2\left(\frac{\partial^3\mathbf{u}(b)}{\partial b^3}\right)^2\right\}$$

$$+\mu_8(b)\left\{\frac{1}{4}\left(\frac{\partial\mathbf{u}(b)}{\partial b}\right)^2\frac{\partial^2\mathbf{u}(b)}{\partial b^2}\frac{\partial^4\mathbf{u}(b)}{\partial b^4}+\frac{1}{2}\frac{\partial\mathbf{u}(b)}{\partial b}\left(\frac{\partial^2\mathbf{u}(b)}{\partial b^2}\right)^2\frac{\partial^3\mathbf{u}(b)}{\partial b^3}\right\}$$

$$+\mu_{10}(b)\left\{\frac{1}{1260}\left(\frac{\partial\mathbf{u}(b)}{\partial b}\right)^3\frac{\partial^7\mathbf{u}(b)}{\partial b^7}+\frac{1}{96}\left(\frac{\partial\mathbf{u}(b)}{\partial b}\right)^2\left(\frac{\partial^4\mathbf{u}(b)}{\partial b^4}\right)^2+\frac{1}{54}\frac{\partial\mathbf{u}(b)}{\partial b}\left(\frac{\partial^3\mathbf{u}(b)}{\partial b^3}\right)^3\right\}$$

$$+\mu_{10}(b)\left\{\frac{1}{48}\frac{\partial^4\mathbf{u}(b)}{\partial b^4}\left(\frac{\partial^2\mathbf{u}(b)}{\partial b^2}\right)^3+\frac{1}{24}\left(\frac{\partial^2\mathbf{u}(b)}{\partial b^2}\right)^2\left(\frac{\partial^3\mathbf{u}(b)}{\partial b^3}\right)^2\right.$$

$$\left.+\frac{1}{12}\frac{\partial\mathbf{u}(b)}{\partial b}\frac{\partial^2\mathbf{u}(b)}{\partial b^2}\frac{\partial^3\mathbf{u}(b)}{\partial b^3}\frac{\partial^4\mathbf{u}(b)}{\partial b^4}\right\}$$

$$+\mu_{10}(b)\left\{\frac{1}{60}\left(\frac{\partial\mathbf{u}(b)}{\partial b}\right)^2\frac{\partial^3\mathbf{u}(b)}{\partial b^3}\frac{\partial^5\mathbf{u}(b)}{\partial b^5}+\frac{1}{120}\left(\frac{\partial\mathbf{u}(b)}{\partial b}\right)^2\frac{\partial^2\mathbf{u}(b)}{\partial b^2}\frac{\partial^6\mathbf{u}(b)}{\partial b^6}\right\}$$

$$+\mu_{10}(b)\frac{1}{40}\frac{\partial\mathbf{u}(b)}{\partial b}\left(\frac{\partial^2\mathbf{u}(b)}{\partial b^2}\right)^2\frac{\partial^5\mathbf{u}(b)}{\partial b^5}. \tag{1.15}$$

Of course, the higher probabilistic moment, the larger Taylor expansion and the faster increase of the components number corresponding to the neighboring order central moments.

The central moments of the Gaussian variable b may obviously be simply recovered here as

$$\mu_p(b) = \begin{cases} 0; & p = 2k+1 \\ \{\sigma(b)\}^p (p-1)!!; & p = 2k \end{cases} \tag{1.16}$$

for any natural $k \leq 1$. As one may suppose, the higher-order moments we need to compute the higher-order perturbations need to be included in all formulas, so that the complexity of the computational model grows non-proportionally together with the precision and size of the output information needed. Once we take the polynomial $f(b) = \sum_{i=1}^{10} c_i b^i$, then its general perturbation-based formula for the tenth-order expectation equals

$$\begin{aligned} E[f(b)] &= \frac{1}{2}c_1 + \frac{1}{4}c_2 + \frac{1}{8}c_3 + \frac{1}{16}c_4 + \frac{1}{32}c_5 + \frac{1}{64}c_6 + \frac{1}{128}c_7 + \frac{1}{256}c_8 + \frac{1}{512}c_9 + \frac{1}{1024}c_{10} \\ &+ \alpha^2(b)\left(\frac{1}{4}c_2 + \frac{3}{8}c_3 + \frac{3}{8}c_4 + \frac{5}{16}c_5 + \frac{15}{64}c_6 + \frac{21}{128}c_7 + \frac{7}{64}c_8 + \frac{9}{128}c_9 + \frac{45}{1024}c_{10}\right) \\ &+ \alpha^4(b)\left(\frac{3}{16}c_4 + \frac{15}{32}c_5 + \frac{45}{64}c_6 + \frac{105}{128}c_7 + \frac{105}{128}c_8 + \frac{189}{256}c_9 + \frac{315}{512}c_{10}\right) \\ &+ \alpha^6(b)\left(\frac{105}{128}c_7 + \frac{105}{64}c_8 + \frac{315}{128}c_9 + \frac{1575}{512}c_{10}\right) \\ &+ \alpha^8(b)\left(\frac{105}{256}c_8 + \frac{945}{512}c_9 + \frac{4725}{1024}c_{10}\right) + \alpha^{10}(b)\frac{945}{1024}c_{10}. \end{aligned} \tag{1.17}$$

The variance, third and fourth probabilistic moments of this function, considering their lengths, are omitted here and may be found in the Maple™ source files located on the book's website.

It is obvious that the symmetric probability density functions do not require full expansions, but for the general distribution and specifically non-symmetric distributions such as the lognormal, we need to complete them with odd-order terms. These additional terms are specified below:

$$\begin{aligned} E'[\mathbf{u}(b)] &= \varepsilon\frac{\partial \mathbf{u}(b)}{\partial b}\mu_1(b) + \frac{\varepsilon^3}{3!}\frac{\partial^3 \mathbf{u}(b)}{\partial b^3}\mu_3(b) \\ &+ \frac{\varepsilon^5}{5!}\frac{\partial^5 \mathbf{u}(b)}{\partial b^5}\mu_5(b) + \frac{\varepsilon^7}{7!}\frac{\partial^7 \mathbf{u}(b)}{\partial b^7}\mu_7(b) + \frac{\varepsilon^9}{9!}\frac{\partial^9 \mathbf{u}(b)}{\partial b^9}\mu_9(b) \end{aligned} \tag{1.18}$$

for the variances:

$$\begin{aligned} \mathrm{Var}'(\mathbf{u}(b)) &= \mu_2'(\mathbf{u}(b)) \\ &= \mu_3(b)\frac{\partial \mathbf{u}(b)}{\partial b}\frac{\partial^2 \mathbf{u}(b)}{\partial b^2} + \mu_5(b)\left\{\frac{1}{12}\frac{\partial^4 \mathbf{u}(b)}{\partial b^4}\frac{\partial \mathbf{u}(b)}{\partial b} + \frac{1}{6}\frac{\partial^3 \mathbf{u}(b)}{\partial b^3}\frac{\partial^2 \mathbf{u}(b)}{\partial b^2}\right\} \end{aligned}$$

$$
+\mu_7(b)\left\{\frac{1}{72}\frac{\partial^4\mathbf{u}(b)}{\partial b^4}\frac{\partial^3\mathbf{u}(b)}{\partial b^3}+\frac{1}{120}\frac{\partial^5\mathbf{u}(b)}{\partial b^5}\frac{\partial^2\mathbf{u}(b)}{\partial b^2}+\frac{1}{360}\frac{\partial^6\mathbf{u}(b)}{\partial b^6}\frac{\partial\mathbf{u}(b)}{\partial b}\right\}
$$

$$
+\mu_9(b)\left\{\frac{1}{5040}\frac{\partial^7\mathbf{u}(b)}{\partial b^7}\frac{\partial^2\mathbf{u}(b)}{\partial b^2}+\frac{1}{20160}\frac{\partial^8\mathbf{u}(b)}{\partial b^8}\frac{\partial\mathbf{u}(b)}{\partial b}\right\}
$$

$$
+\mu_9(b)\left\{\frac{1}{1440}\frac{\partial^5\mathbf{u}(b)}{\partial b^5}\frac{\partial^4\mathbf{u}(b)}{\partial b^4}+\frac{1}{2160}\frac{\partial^6\mathbf{u}(b)}{\partial b^6}\frac{\partial^3\mathbf{u}(b)}{\partial b^3}\right\} \tag{1.19}
$$

in case of the third central probabilistic moment:

$$
\mu_3'(\mathbf{u}(b))=\frac{1}{3}\mu_3(b)\left(\frac{\partial\mathbf{u}(b)}{\partial b}\right)^3
$$

$$
+\mu_5(b)\left\{\frac{3}{4}\left(\frac{\partial^2\mathbf{u}(b)}{\partial b^2}\right)^2\frac{\partial\mathbf{u}(b)}{\partial b}+\frac{1}{2}\left(\frac{\partial\mathbf{u}(b)}{\partial b}\right)^2\frac{\partial^3\mathbf{u}(b)}{\partial b^3}\right\}
$$

$$
+\mu_7(b)\left\{\frac{1}{40}\left(\frac{\partial\mathbf{u}(b)}{\partial b}\right)^2\frac{\partial^5\mathbf{u}(b)}{\partial b^5}+\frac{1}{12}\frac{\partial\mathbf{u}(b)}{\partial b}\left(\frac{\partial^3\mathbf{u}(b)}{\partial b^3}\right)^2\right\}
$$

$$
+\mu_7(b)\left\{\frac{1}{8}\frac{\partial^3\mathbf{u}(b)}{\partial b^3}\left(\frac{\partial^2\mathbf{u}(b)}{\partial b^2}\right)^2+\frac{1}{8}\frac{\partial\mathbf{u}(b)}{\partial b}\frac{\partial^2\mathbf{u}(b)}{\partial b^2}\frac{\partial^4\mathbf{u}(b)}{\partial b^4}\right\}
$$

$$
+\mu_9(b)\left\{\frac{1}{216}\left(\frac{\partial^3\mathbf{u}(b)}{\partial b^3}\right)^3+\frac{1}{1680}\left(\frac{\partial\mathbf{u}(b)}{\partial b}\right)^2\frac{\partial^7\mathbf{u}(b)}{\partial b^7}+\frac{1}{192}\frac{\partial\mathbf{u}(b)}{\partial b}\left(\frac{\partial^4\mathbf{u}(b)}{\partial b^4}\right)^2\right\}
$$

$$
+\mu_9(b)\left\{\frac{1}{160}\left(\frac{\partial^2\mathbf{u}(b)}{\partial b^2}\right)^2\frac{\partial^5\mathbf{u}(b)}{\partial b^5}+\frac{1}{48}\frac{\partial^2\mathbf{u}(b)}{\partial b^2}\frac{\partial^3\mathbf{u}(b)}{\partial b^3}\frac{\partial^4\mathbf{u}(b)}{\partial b^4}\right\}
$$

$$
+\mu_9(b)\left\{\frac{1}{120}\frac{\partial\mathbf{u}(b)}{\partial b}\frac{\partial^3\mathbf{u}(b)}{\partial b^3}\frac{\partial^5\mathbf{u}(b)}{\partial b^5}+\frac{1}{240}\frac{\partial\mathbf{u}(b)}{\partial b}\frac{\partial^2\mathbf{u}(b)}{\partial b^2}\frac{\partial^6\mathbf{u}(b)}{\partial b^6}\right\} \tag{1.20}
$$

as well as the fourth one:

$$
\mu_4'(\mathbf{u}(b))=2\mu_5(b)\left(\frac{\partial\mathbf{u}(b)}{\partial b}\right)^3\frac{\partial^2\mathbf{u}(b)}{\partial b^2}
$$

$$
+\mu_7(b)\left\{\frac{1}{6}\frac{\partial^4\mathbf{u}(b)}{\partial b^4}\left(\frac{\partial\mathbf{u}(b)}{\partial b}\right)^3+\frac{1}{2}\frac{\partial\mathbf{u}(b)}{\partial b}\left(\frac{\partial^2\mathbf{u}(b)}{\partial b^2}\right)^3+\left(\frac{\partial\mathbf{u}(b)}{\partial b}\right)^2\frac{\partial^2\mathbf{u}(b)}{\partial b^2}\frac{\partial^3\mathbf{u}(b)}{\partial b^3}\right\}
$$

$$
+\mu_9(b)\left\{\frac{1}{180}\left(\frac{\partial\mathbf{u}(b)}{\partial b}\right)^3\frac{\partial^6\mathbf{u}(b)}{\partial b^6}+\frac{1}{12}\left(\frac{\partial^2\mathbf{u}(b)}{\partial b^2}\right)^3\left(\frac{\partial^3\mathbf{u}(b)}{\partial b^3}\right)^2\right.
$$

$$
\left.+\frac{1}{12}\left(\frac{\partial\mathbf{u}(b)}{\partial b}\right)^2\frac{\partial^3\mathbf{u}(b)}{\partial b^3}\frac{\partial^4\mathbf{u}(b)}{\partial b^4}\right\}+\mu_9(b)\frac{\partial\mathbf{u}(b)}{\partial b}\left\{\frac{1}{20}\frac{\partial\mathbf{u}(b)}{\partial b}\frac{\partial^2\mathbf{u}(b)}{\partial b^2}\frac{\partial^5\mathbf{u}(b)}{\partial b^5}\right.
$$

$$
\left.+\frac{1}{8}\frac{\partial^4\mathbf{u}(b)}{\partial b^4}\left(\frac{\partial^2\mathbf{u}(b)}{\partial b^2}\right)^2+\frac{1}{6}\frac{\partial^2\mathbf{u}(b)}{\partial b^2}\left(\frac{\partial^3\mathbf{u}(b)}{\partial b^3}\right)^2\right\}. \tag{1.21}
$$

The situation becomes definitely more complicated when we consider a problem with multiple random variables, let's say p random variables being totally uncorrelated – we vectorize these variables here as b_r for $r = 1, \ldots, p$. Then the Taylor expansion with random coefficients proposed in Equation (1.10) is provided for all these components as

$$\mathbf{u}(b_r) = \mathbf{u}^0(b_r^0) + \varepsilon \left.\frac{\partial \mathbf{u}(b_r)}{\partial b_r}\right|_{b_r = b_r^0} \Delta b_r + \cdots + \frac{\varepsilon^n}{n!} \left.\frac{\partial^n \mathbf{u}(b_r)}{\partial b_r^n}\right|_{b_r = b_r^0} \Delta b_r^n. \tag{1.22}$$

The most fundamental difference is that the zeroth-order component is calculated only once – for the mean values of the design vector components, but higher-order terms include partial derivatives of the response function with respect to all these p components separately. So, the tenth-order expansion, instead of 11 components for the single input random variable, will contain $10p + 1$ independent terms. In view of the above, the expectation for the structural response is calculated as (where the summation convention is replaced for brevity with a classical sum)

$$\begin{aligned} E[\mathbf{u}(b_r)] = {} & \mathbf{u}^0(b_r^0) + \sum_{p=1}^{r} \frac{\varepsilon^2}{2} \frac{\partial^2 \mathbf{u}(b_p)}{\partial b_p^2} \mu_2(b_p) + \sum_{p=1}^{r} \frac{\varepsilon^4}{4!} \frac{\partial^4 \mathbf{u}(b_p)}{\partial b_p^4} \mu_4(b_p) \\ & + \sum_{p=1}^{r} \frac{\varepsilon^6}{6!} \frac{\partial^6 \mathbf{u}(b_p)}{\partial b_p^6} \mu_6(b_p) + \sum_{p=1}^{r} \frac{\varepsilon^8}{8!} \frac{\partial^8 \mathbf{u}(b_p)}{\partial b_p^8} \mu_8(b_p) \\ & + \sum_{p=1}^{r} \frac{\varepsilon^{10}}{10!} \frac{\partial^{10} \mathbf{u}(b_p)}{\partial b_p^{10}} \mu_{10}(b_p). \end{aligned} \tag{1.23}$$

Therefore, following this idea it is relatively easy to extend Equations (1.13)–(1.15) with the additional summation procedure over the independent components of the input random variables vector to get multi-parametric equations for the variances as well as the third and fourth central probabilistic moments. As one can realize, the correlation effect in these expansions will result in cross-correlations (of higher order also) between all the components of the vector b_r. It yields, for the second-order expansion of three random variables after Equation (1.23),

$$\begin{aligned} E[\mathbf{u}(b_1, b_2, b_3)] = {} & \mathbf{u}^0(b_1^0, b_2^0, b_3^0) + \frac{\partial^2 \mathbf{u}(b_1)}{\partial b_1^2} \mu_2(b_1) + \frac{\partial^2 \mathbf{u}(b_2)}{\partial b_2^2} \mu_2(b_2) + \frac{\partial^2 \mathbf{u}(b_3)}{\partial b_3^2} \mu_2(b_3) \\ & + \frac{\partial \mathbf{u}(b_1)}{\partial b_1} \frac{\partial \mathbf{u}(b_2)}{\partial b_2} \mathrm{Cov}(b_1, b_2) + \frac{\partial \mathbf{u}(b_1)}{\partial b_1} \frac{\partial \mathbf{u}(b_3)}{\partial b_3} \mathrm{Cov}(b_1, b_3) \\ & + \frac{\partial \mathbf{u}(b_2)}{\partial b_2} \frac{\partial \mathbf{u}(b_3)}{\partial b_3} \mathrm{Cov}(b_2, b_3) \end{aligned} \tag{1.24}$$

where $\mathrm{Cov}(b_1, b_2)$ stands for the covariance matrix of two random quantities defined classically as [34]

$$\mathrm{Cov}\left(b_1, b_2\right) = \int_{-\infty}^{+\infty} \left(b_1 - E\left[b_1\right]\right)\left(b_2 - E\left[b_2\right]\right) g_{b_1 b_2}(x_1, x_2)\mathrm{d}x, \tag{1.25}$$

replaced frequently with the non-dimensional and normalized correlation coefficient introduced as

$$\rho\left(b_1, b_2\right) = \frac{\mathrm{Cov}\left(b_1, b_2\right)}{\sigma\left(b_1\right)\sigma\left(b_1\right)} = \frac{\int_{-\infty}^{+\infty} \left(b_1 - E\left[b_1\right]\right)\left(b_2 - E\left[b_2\right]\right) g_{b_1 b_2}(x_1, x_2)\mathrm{d}x}{\sqrt{\int_{-\infty}^{+\infty} \left(b_1 - E\left[b_1\right]\right)^2 g_{b_1}(x_1)\mathrm{d}x \times \int_{-\infty}^{+\infty} \left(b_2 - E\left[b_2\right]\right)^2 g_{b_2}(x_2)\mathrm{d}x}} \tag{1.26}$$

taking values $-1 \leq \rho(b_1, b_2) \leq 1$ only. Of course, $g_{b_1 b_2}\left(x_1, x_2\right)$ denotes here the joint probability density function of the variables b_1 and b_2. The basic problem with higher-order perturbation terms is the necessity of including higher-order cross-correlations between all input random variables. This is not due to the mathematical and/or numerical level of complexity of equations for all basic random characteristics, but the lack of practical engineering knowledge about these correlations. Usually, this knowledge reduces to the ordinary covariance of two or more random parameters, which can be a subject or the result of some statistical regression models.

As seen above, the most important numerical issue is determination of the partial derivatives of the state function $\mathbf{u}(b)$ with respect to the input parameter b and, depending on the case study, we can apply an analytical technique – fully implemented in the computer algebra system, a semi-analytical approach – a combination of the symbolic calculus with some other discrete technique implementation software (like the finite element method (FEM) solver) or, finally, just a typical numerical solution provided entirely by this solver itself. Nevertheless, we need to consider the following equation system:

$$\mathbf{L}(b)\mathbf{u}(b) = \mathbf{f}(b), \tag{1.27}$$

where $\mathbf{L}(b)$ usually represents the main system matrix, $\mathbf{f}(b)$ includes the boundary conditions imposed on the system, while $\mathbf{u}(b)$ is the structural response. More advanced problems, like transient heat transfer or the dynamical equilibrium equations, are considered further in this context – see Chapter 2. According to the main philosophy proposed above, we provide the nth-order expansion of both sides of this statement to get

$$\left\{\varepsilon^0 \mathbf{L}^0\left(b^0\right)+\sum_{j=1}^n \frac{\varepsilon^j}{j!} \frac{\partial^j \mathbf{L}(b)}{\partial b^j}(\Delta b)^j\right\}\left\{\varepsilon^0 \mathbf{u}^0\left(b^0\right)+\sum_{j=1}^n \frac{\varepsilon^j}{j!} \frac{\partial^j \mathbf{u}(b)}{\partial b^j}(\Delta b)^j\right\}$$
$$=\mathbf{f}^0\left(b^0\right)+\sum_{j=1}^n \frac{\varepsilon^j}{j!} \frac{\partial^j \mathbf{f}(b)}{\partial b^j}(\Delta b)^j . \tag{1.28}$$

After multiplication of the left-hand side (LHS) in Equation (1.28), we collect components of same order of the perturbation parameter ε to arrive at the increasing-order hierarchical equilibrium equations

$$\begin{cases}\varepsilon^0: \mathbf{L}^0\left(b^0\right) \mathbf{u}^0\left(b^0\right)=\mathbf{f}^0\left(b^0\right) \\ \varepsilon^1: \dfrac{\partial \mathbf{L}(b)}{\partial b} \mathbf{u}^0(b)+\mathbf{L}^0(b) \dfrac{\partial \mathbf{u}(b)}{\partial b}=\dfrac{\partial \mathbf{f}(b)}{\partial b} \\ \varepsilon^2: \dfrac{\partial^2 \mathbf{L}(b)}{\partial b^2} \mathbf{u}^0(b)+2 \dfrac{\partial \mathbf{L}(b)}{\partial b} \dfrac{\partial \mathbf{u}(b)}{\partial b}+\mathbf{L}^0(b) \dfrac{\partial^2 \mathbf{u}(b)}{\partial b^2}=\dfrac{\partial^2 \mathbf{f}(b)}{\partial b^2} \\ \ldots \\ \varepsilon^n: \sum_{j=0}^n \dbinom{n}{j} \dfrac{\partial^{n-j} \mathbf{L}(b)}{\partial b^{n-j}} \dfrac{\partial^j \mathbf{u}(b)}{\partial b^j}=\dfrac{\partial^n \mathbf{f}(b)}{\partial b^n} .\end{cases} \tag{1.29}$$

Further, leaving the highest-order derivative of the solution $\mathbf{u}(b)$ on the LHS only and using simple algebra, one gets zeroth-, first-, second-, and finally nth-order equations

$$\begin{cases}\mathbf{u}^0\left(b^0\right)=\left\{\mathbf{L}^0\left(b^0\right)\right\}^{-1} \mathbf{f}^0\left(b^0\right) \\ \dfrac{\partial \mathbf{u}(b)}{\partial b}=\left\{\mathbf{L}^0\left(b^0\right)\right\}^{-1}\left\{\dfrac{\partial \mathbf{f}(b)}{\partial b}-\dfrac{\partial \mathbf{L}(b)}{\partial b} \mathbf{u}^0(b)\right\} \\ \dfrac{\partial^2 \mathbf{u}(b)}{\partial b^2}=\left\{\mathbf{L}^0\left(b^0\right)\right\}^{-1}\left\{\dfrac{\partial^2 \mathbf{f}(b)}{\partial b^2}-\dfrac{\partial^2 \mathbf{L}(b)}{\partial b^2} \mathbf{u}^0(b)-2 \dfrac{\partial \mathbf{L}(b)}{\partial b} \dfrac{\partial \mathbf{u}(b)}{\partial b}\right\} \\ \ldots \\ \dfrac{\partial^n \mathbf{u}(b)}{\partial b^n}=\left\{\mathbf{L}^0\left(b^0\right)\right\}^{-1}\left\{\dfrac{\partial^n \mathbf{f}(b)}{\partial b^n}-\sum_{j-1}^n \dbinom{n}{j} \dfrac{\partial^{n-j} \mathbf{L}(b)}{\partial b^{n-j}} \dfrac{\partial^j \mathbf{u}(b)}{\partial b^j}\right\} .\end{cases} \tag{1.30}$$

Hence it is apparent that the zeroth-order equation returns a zeroth-order solution, as in a deterministic problem, which, inserted into the first-order equation next gives the first-order partial derivative of the function $\mathbf{u}(b)$. Then, both-order terms for this function inserted into the second-order equation result in $\frac{\partial^2 \mathbf{u}(b)}{\partial b^2}$, and so on, until the highest-order partial derivative of this function is determined. It is characteristic that during computational implementation only the main matrix of the system (i.e., stiffness or heat conduction) needs to be inverted and only once during the entire solution procedure. A second observation is that small modifications of the existing solvers are necessary since the right-hand side (RHS) vector is modified, while the LHS procedures typical for a solution of the linear algebraic equation system (like various decompositions of the main system matrix) remain the same. Usually, when $\mathbf{L}=\mathbf{L}(b)$, $\mathbf{f}$ may remain deterministic and vice versa, and then we can make some further simplifications. The following hold:

1. For $\mathbf{f}=\mathbf{f}(b)$ *and* $\mathbf{L}\neq\mathbf{L}(b)$

$$\begin{cases} \mathbf{u}^0\left(b^0\right)=\left\{\mathbf{L}^0\left(b^0\right)\right\}^{-1}\mathbf{f}^0\left(b^0\right) \\ \dfrac{\partial \mathbf{u}(b)}{\partial b}=\left\{\mathbf{L}^0\left(b^0\right)\right\}^{-1}\dfrac{\partial \mathbf{f}(b)}{\partial b} \\ \dfrac{\partial^2 \mathbf{u}(b)}{\partial b^2}=\left\{\mathbf{L}^0\left(b^0\right)\right\}^{-1}\dfrac{\partial^2 \mathbf{f}(b)}{\partial b^2} \\ \ldots \\ \dfrac{\partial^n \mathbf{u}(b)}{\partial b^n}=\left\{\mathbf{L}^0\left(b^0\right)\right\}^{-1}\dfrac{\partial^n \mathbf{f}(b)}{\partial b^n}. \end{cases} \tag{1.31}$$

Calculation of the expectation for $\mathbf{u}(b)$ in this case seems to be very straightforward and

$$E[\mathbf{u}(b)]=\left\{\mathbf{L}^0\left(b^0\right)\right\}^{-1}\left\{\mathbf{f}^0(b)+\sum_{j=1}^{n}\frac{1}{j!}\frac{\partial^j \mathbf{f}(b)}{\partial b^j}\mu_j(b)\right\} \tag{1.32}$$

where further modifications result from Equation (1.16) – zeroing of the odd central probabilistic moments $\mu_j(b)$ on the RHS if b is Gaussian.

2. For $\mathbf{f}\neq\mathbf{f}(b)$ *and* $\mathbf{L}=\mathbf{L}(b)$, Equation (1.21) reduces to

$$\begin{cases} \mathbf{u}^0\left(b^0\right)=\left\{\mathbf{L}^0\left(b^0\right)\right\}^{-1}\mathbf{f}^0\left(b^0\right) \\ \dfrac{\partial \mathbf{u}(b)}{\partial b}=-\left\{\mathbf{L}^0\left(b^0\right)\right\}^{-1}\dfrac{\partial \mathbf{L}(b)}{\partial b}\mathbf{u}^0(b) \\ \dfrac{\partial^2 \mathbf{u}(b)}{\partial b^2}=-\left\{\mathbf{L}^0\left(b^0\right)\right\}^{-1}\left\{\dfrac{\partial^2 \mathbf{L}(b)}{\partial b^2}\mathbf{u}^0(b)+2\dfrac{\partial \mathbf{L}(b)}{\partial b}\dfrac{\partial \mathbf{u}(b)}{\partial b}\right\} \\ \ldots \\ \dfrac{\partial^n \mathbf{u}(b)}{\partial b^n}=-\left\{\mathbf{L}^0\left(b^0\right)\right\}^{-1}\displaystyle\sum_{j=1}^{n}\binom{n}{j}\dfrac{\partial^{n-j}\mathbf{L}(b)}{\partial b^{n-j}}\dfrac{\partial^j \mathbf{u}(b)}{\partial b^j}. \end{cases} \tag{1.33}$$

The very special case appearing relatively frequently in a number of engineering problems is a linear dependence $\mathbf{L}(b)$ from b and then we may drop some further terms from these equations, for example

$$\begin{cases} \mathbf{u}^0\left(b^0\right) = \left\{\mathbf{L}^0\left(b^0\right)\right\}^{-1}\mathbf{f}^0\left(b^0\right) \\ \dfrac{\partial \mathbf{u}(b)}{\partial b} = -\left\{\mathbf{L}^0\left(b^0\right)\right\}^{-1}\dfrac{\partial \mathbf{L}(b)}{\partial b}\mathbf{u}^0(b) \\ \dfrac{\partial^2 \mathbf{u}(b)}{\partial b^2} = -2\left\{\mathbf{L}^0\left(b^0\right)\right\}^{-1}\dfrac{\partial \mathbf{L}(b)}{\partial b}\dfrac{\partial \mathbf{u}(b)}{\partial b} \\ \dots \\ \dfrac{\partial^n \mathbf{u}(b)}{\partial b^n} = -n\left\{\mathbf{L}^0\left(b^0\right)\right\}^{-1}\dfrac{\partial \mathbf{L}(b)}{\partial b}\dfrac{\partial^{n-1}\mathbf{u}(b)}{\partial b^{n-1}}. \end{cases} \tag{1.34}$$

Embedding the partial derivatives into Equations (1.12)–(1.15) we recover all necessary probabilistic characteristics of the structural response. This is usually verified against the Monte Carlo simulation, which, as is well known, is treated as the exact solution and is based on a series of computational experiments with randomly generated random spaces of additional input parameters. We need to remember that independently of the implementation type of this method (crude simulation or stratified sampling, for instance), the estimators given below are convergent for number of experiments tending to infinity and that the random number generators also have their deficiencies. Therefore, it is very reasonable to compare first the given simulation tool with the probabilistic analytical results for various types of expected output distributions. The computations according to the Monte Carlo simulations follow the well-known, most reliable estimators of the expected value, variance, as well as the nth-order central probabilistic moment. The following hold [5]:

$$E\left[\mathbf{u}(b)\right] = \frac{1}{M}\sum_{i=1}^{M}\mathbf{u}^{(i)}(b), \tag{1.35}$$

$$\mathrm{Var}\left(\mathbf{u}(b)\right) = \frac{1}{M-1}\sum_{i=1}^{M}\left(\mathbf{u}^{(i)}(b) - E\left[\mathbf{u}(b)\right]\right)^2, \tag{1.36}$$

$$\mu_n\left(\mathbf{u}(b)\right) = \frac{1}{M}\sum_{i=1}^{M}\left(\mathbf{u}^{i}(b) - E\left[\mathbf{u}(b)\right]\right)^n, \tag{1.37}$$

and having estimated these moments we employ Equation (1.4) to calculate the coefficient of variation, skewness, and kurtosis shown in the next two sections (M is the total number of random trials). Since most of the verifications are provided via the simulation tool implemented in the computer algebra system Maple™, further discussion on statistical aspects may be found in the literature. Very simple illustrations are given only to enable readers to carry out a simple verification of their own statistical simulations in this package – it is confirmed below that the standard error of higher-moments determination for the same number of random trials is definitely larger than for the first two, even for a Gaussian distribution.

```
>restart: with(Statistics):
>X:=RandomVariable(Normal(10,1)): A:=Sample(X,10000):
>Bootstrap(Mean,X,replications=10000,output=['value','standarderror']);

                    [10.00033977, 0.03184353072]

>Bootstrap(Mean,A,replications=10000,output=['value','standarderror']);

                    [9.991094312, 0.01001230197]

>Mean(X);
                                10

>Bootstrap(Variance,X,replications=10000,output=['value','standarderror']);

                    [0.9997266603, 0.0451189936]

>Bootstrap(Variance,A,replications=10000,output=['value','standarderror']);

                    [0.9992068113, 0.01406359292]

>Variance(X);
                                1

>Bootstrap(Skewness,X,replications=10000,output=['value','standarderror']);

                    [-0.0001756788883, 0.07814788973]

>Bootstrap(Skewness,A,replications=10000,output=['value','standarderror']);

                    [0.01192136730, 0.02285198363]

>Skewness(X);
                                0

>Bootstrap(Kurtosis,X,replications=10000,output=['value','standarderror']);

                    [2.992500847, 0.1545298225]

>Bootstrap(Kurtosis,A,replications=10000,output=['value','standarderror']);

                    [2.981760768, 0.04047992704]

>Kurtosis(X);
                                3
```

We study an interrelation of the probabilistic techniques described briefly above for the two computational examples below – for the relatively simple, well-known engineering problems where an analytical expression for $\mathbf{u}(b)$ does exist, but not always can be simply integrated according to classical definitions.

Example 1.1: Simple tension of an elastic bar

We consider first a linear elastic rod with constant cross-sectional area A, Young's modulus e, length l under tension on the RHS with force P. Extension of this bar is the state function dependent on the Gaussian variable e. This problem has a well-known solution from the strength of materials, that is, $u\,(e) = \frac{Pl}{eA}$ and its second-order second-moment (SOSM) solution may be found in [112]. It is clear from the very beginning that $u(e)$ has an inverse Gaussian distribution, but since usually output distributions cannot be predicted straightforwardly without such an analytical solution, we focus on the expectations, coefficients of variation, skewness, and kurtosis all computed as functions of the input coefficient of variation. The numerical data in this test are taken as $E[e]=210\,\text{GPa}$, $A=0.0001\,\text{m}^2$, $l=10\,\text{m}$, $P=10\,\text{kN}$, while the results of stochastic perturbation technique (SPT)-based (full tenth-order method) and Monte Carlo simulation (MCS)-based ($M=5\times10^5$ random trials) computations are presented in Figures 1.1 and 1.2. We notice a perfect agreement of the first two probabilistic characteristics in the entire range of the input coefficient of variation. The worst coincidence is obtained in case of skewness, while kurtosis agrees well up to $\alpha(e)<0.15$.

Example 1.2: Simply supported elastic beam under uniform transverse load

The second example concerns a deterministically homogeneous and linear elastic stainless steel beam simply supported at both ends and loaded with a constant distributed load. Its parameters are defined as $e=210\,\text{GPa}$, $J=0.0001\,\text{m}^4$, $E[l]=10\,\text{m}$,

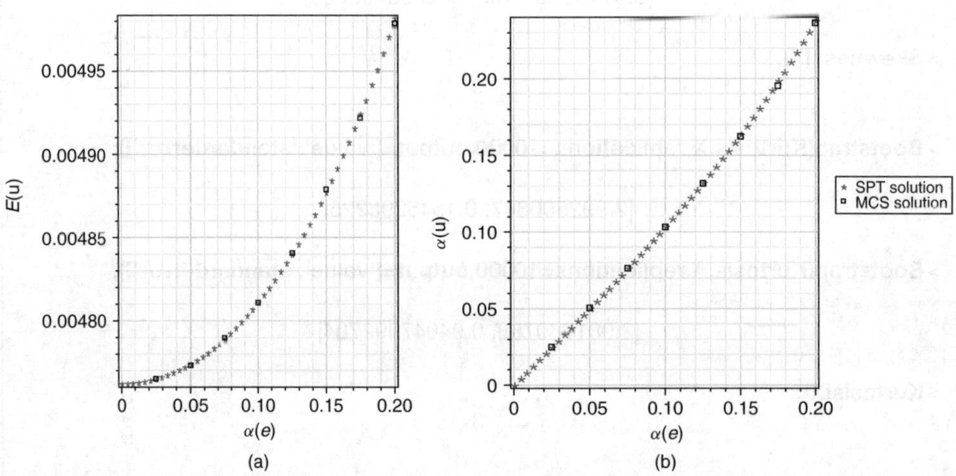

Figure 1.1 Expected values (a) and coefficients of variation (b) for $u\,(e\,(\omega)) = \frac{Pl}{e(\omega)A}$

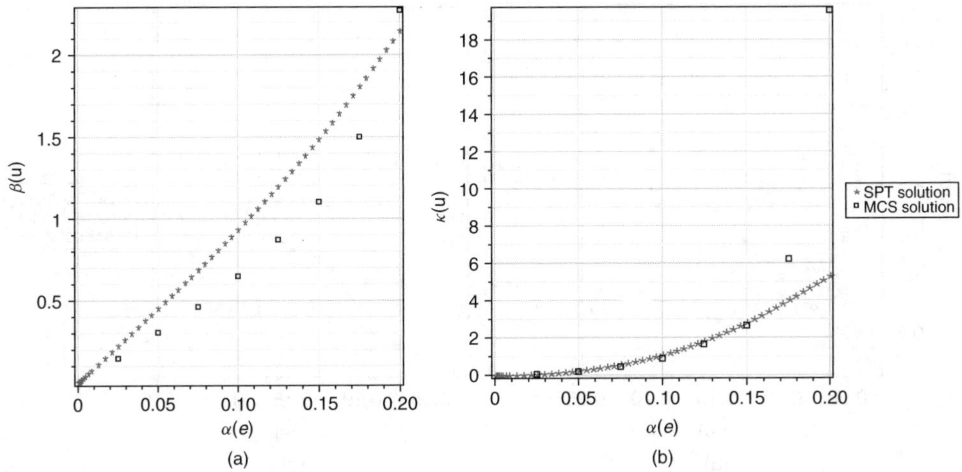

Figure 1.2 Skewness (a) and kurtosis (b) for $u\left(e\left(\omega\right)\right)=\frac{Pl}{e(\omega)A}$

and $q=10\,\mathrm{kN/m}$. Once more, we provide a visualization of the expectations, coefficients of variation (Figure 1.3), skewness, and kurtosis (Figure 1.4) – all computed as functions of the input coefficient of variation for the beam length. Naturally, our state function is the maximum deflection at the mid-span of this beam widely used in civil engineering as the basis for reliability analysis, $u\left(l\left(\omega\right)\right)=\frac{5}{384}\frac{q(l(\omega))^4}{eJ}$. Now, analytical calculations are relatively easy, so thanks to the additional internal functions of the

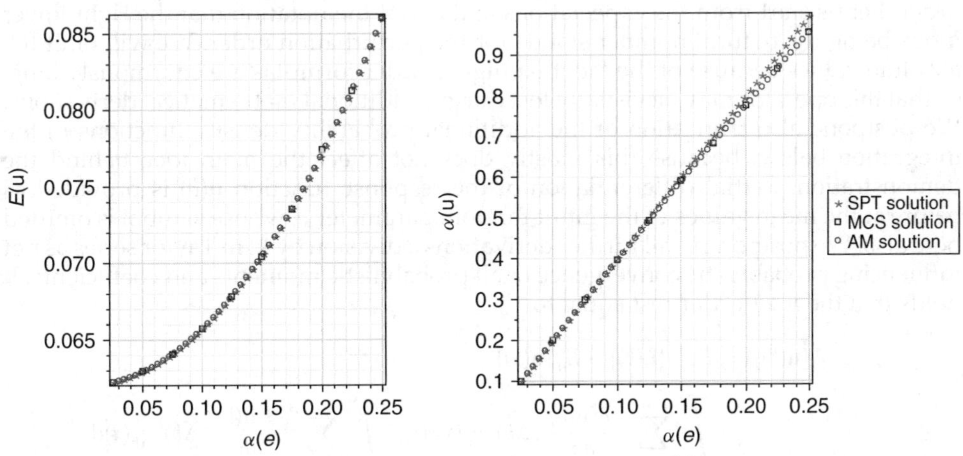

Figure 1.3 Expectations (a) and coefficients of variation (b) for $u\left(l\left(\omega\right)\right)=\frac{5}{384}\frac{q\left(l(\omega)\right)^4}{eJ}$

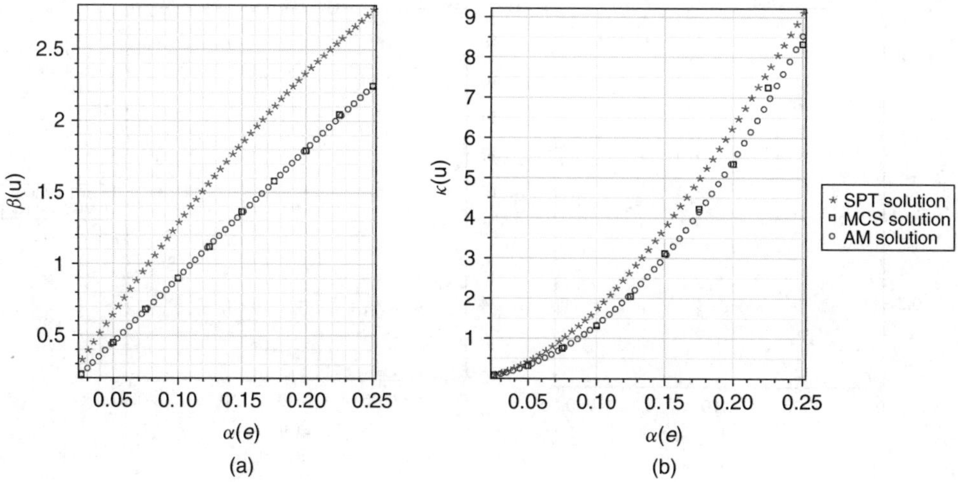

Figure 1.4 Skewness (a) and kurtosis (b) for $u\left(l\left(\omega\right)\right)=\frac{5}{384}\frac{q\left(l(\omega)\right)^{4}}{eJ}$

Maple™ system, we compare the triples (analytical (AM) versus perturbation (SPT) versus simulation (MCS) methods) of basic probabilistic moments and coefficients.

In this particular case all the methods coincide perfectly for all probabilistic characteristics, which leads to the conclusion that the perturbation method seems to be very efficient in polynomial transforms of random variables; this efficiency is much worse once the input random variable is to be inversed.

Let us focus now on the convergence of the generalized stochastic perturbation technique implemented in conjunction with the polynomial-based response function method. Using classical theorems on mathematical analysis we will show that the first four probabilistic moments and coefficients converge together with the analysis order. Let us start from the expectations and adopt the notation that the right lower index being the natural number stands for the perturbation order. We will omit the absolute values because of the fact that higher-order formulas are enormously long, so that this operator may be visually lost during additional mathematical derivations. We postpone also truncation of the additional probability density functions in the integration below, because this mostly does not affect the main idea behind the demonstration. Partial differentiation of the response function $\mathbf{u}(b)$ is provided as before at the mean values of the input random parameter b, whose symbol is omitted here also. We postpone in all further derivations a division by k! in Taylor series as not influencing probabilistic convergence of all probabilistic moments and coefficients. It holds that the $n+1$ order rest equal to

$$\begin{aligned}\Delta_{n+1}\left(E\left[\mathbf{u}(b)\right]\right) &= E_{n+1}\left[\mathbf{u}(b)\right]-E_{n}\left[\mathbf{u}(x)\right]\\ &= \int_{-\infty}^{+\infty}\sum_{k=0}^{n+1}\varepsilon^{k}\frac{\partial^{k}\mathbf{u}(b)}{\partial b^{k}}\left(\Delta b\right)^{k}g_{b}(x)\mathrm{d}x-\int_{-\infty}^{+\infty}\sum_{k=0}^{n}\varepsilon^{k}\frac{\partial^{k}\mathbf{u}(b)}{\partial b^{k}}\left(\Delta b\right)^{k}g_{b}(x)\mathrm{d}x\\ &= \int_{-\infty}^{+\infty}\varepsilon^{n+1}\frac{\partial^{n+1}\mathbf{u}(b)}{\partial b^{n+1}}\left(\Delta b\right)^{n+1}g_{b}(x). \end{aligned} \tag{1.38}$$

Using as above a unitary perturbation parameter and inserting a definition of the nth central probabilistic moment, one obtains

$$\Delta_{n+1}\left(E\left[\mathbf{u}(b)\right]\right)=\frac{\partial^{n+1}\mathbf{u}(b)}{\partial b^{n+1}}\mu_{n+1}(b). \tag{1.39}$$

because $n+1$ partial derivative of $\mathbf{u}(b)$ with respect to b is evaluated at $b=b^0$ in traditional deterministic way and may be excluded from an integrations process over the given P.D.F. as simply constant.

So, it remains clear that unconditional convergence of the expected value approximation is equivalent to zeroing of the $(n+1)$th-order partial derivative of the function $\mathbf{u}(b)$ (with finite value of the n+1 central moment of b itself), which means that it is quite enough to adopt it as the nth-order polynomial, as is exploited in further computational experiments. A similar derivation for the variance is a little more complex – we try to reorganize the first integral to group the nth-order terms and the rest by reducing the zeroth-order components in both integrals with their expectations, so that we have

$$\begin{aligned}
&\Delta_{n+1}\left(\mathrm{Var}\left(\mathbf{u}(b)\right)\right)=\mathrm{Var}_{n+1}\left(\mathbf{u}(b)\right)-\mathrm{Var}_{n}\left(\mathbf{u}(b)\right)\\
&=\int_{-\infty}^{+\infty}\left\{\sum_{k=0}^{n+1}\varepsilon^{k}\frac{\partial^{k}\mathbf{u}(b)}{\partial b^{k}}\left(\Delta b\right)^{k}-E\left[\mathbf{u}(b)\right]\right\}^{2}g_{b}(x)dx\\
&-\int_{-\infty}^{+\infty}\left\{\sum_{k=0}^{n}\varepsilon^{k}\frac{\partial^{k}\mathbf{u}(b)}{\partial b^{k}}\left(\Delta b\right)^{k}-E\left[\mathbf{u}(b)\right]\right\}^{2}g_{b}(x)dx\\
&=\int_{-\infty}^{+\infty}\left\{\sum_{k=1}^{n+1}\varepsilon^{k}\frac{\partial^{k}\mathbf{u}(b)}{\partial b^{k}}\left(\Delta b\right)^{k}\right\}^{2}g_{b}(x)dx-\int_{-\infty}^{+\infty}\left\{\sum_{k=1}^{n}\varepsilon^{k}\frac{\partial^{k}\mathbf{u}(b)}{\partial b^{k}}\left(\Delta b\right)^{k}\right\}^{2}g_{b}(x)dx\\
&=\int_{-\infty}^{+\infty}\left\{\sum_{k=1}^{n}\varepsilon^{k}\frac{\partial^{k}\mathbf{u}(b)}{\partial b^{k}}\left(\Delta b\right)^{k}\right\}^{2}g_{b}(x)dx-\int_{-\infty}^{+\infty}\left\{\sum_{k=1}^{n}\varepsilon^{k}\frac{\partial^{k}\mathbf{u}(b)}{\partial b^{k}}\left(\Delta b\right)^{k}\right\}^{2}g_{b}(x)dx\\
&+2\int_{-\infty}^{+\infty}\sum_{k=1}^{n}\varepsilon^{k}\frac{\partial^{k}\mathbf{u}(b)}{\partial b^{k}}\left(\Delta b\right)^{k}\varepsilon^{n+1}\frac{\partial^{n+1}\mathbf{u}(b)}{\partial b^{n+1}}\left(\Delta b\right)^{n+1}g_{b}(x)dx\\
&+\int_{-\infty}^{+\infty}\left\{\varepsilon^{n+1}\frac{\partial^{n+1}\mathbf{u}(b)}{\partial b^{n+1}}\left(\Delta b\right)^{n+1}\right\}^{2}g_{b}(x)dx\\
&=\frac{\partial^{n+1}\mathbf{u}(b)}{\partial b^{n+1}}\left(2\int_{-\infty}^{+\infty}\sum_{k=1}^{n}\varepsilon^{k}\frac{\partial^{k}\mathbf{u}(b)}{\partial b^{k}}\left(\Delta b\right)^{k}\varepsilon^{n+1}\left(\Delta b\right)^{n+1}g_{b}(x)dx\right.\\
&\qquad\left.+\int_{-\infty}^{+\infty}\varepsilon^{2(n+1)}\frac{\partial^{n+1}\mathbf{u}(b)}{\partial b^{n+1}}\left(\Delta b\right)^{2(n+1)}g_{b}(x)dx\right).
\end{aligned} \tag{1.40}$$

Using $\varepsilon = 1$ and a definition of the central probabilistic moment, we arrive at the final conclusion that

$$\Delta_{n+1}(\mathrm{Var}(\mathbf{u}(b))) = \frac{\partial^{n+1}\mathbf{u}(b)}{\partial b^{n+1}}\left(2\int_{-\infty}^{+\infty}\sum_{k=1}^{n}\frac{\partial^{k}\mathbf{u}(b)}{\partial b^{k}}(\Delta b)^{k}(\Delta b)^{n+1}g_b(x)\mathrm{d}x + \frac{\partial^{n+1}\mathbf{u}(b)}{\partial b^{n+1}}\mu_{2(n+1)}(b)\right)$$

$$= \frac{\partial^{n+1}\mathbf{u}(b)}{\partial b^{n+1}}\left(2\sum_{k=1}^{n}\frac{\partial^{k}\mathbf{u}(b)}{\partial b^{k}}\mu_{k+n+1}(b) + \frac{\partial^{n+1}\mathbf{u}(b)}{\partial b^{n+1}}\mu_{2(n+1)}(b)\right) \quad (1.41)$$

which has analogous properties as the error for the expected values – since higher central probabilistic moments exist and are bounded, the governing factor for the series convergence is a higher-order partial derivative of the state function with respect to the random input parameter. Now we proceed with the third central probabilistic moment, also from its definition. Denoting by $\mu_{3,n}(\mathbf{u}(b))$ the nth-order stochastic perturbation-based approximation of this moment yields

$$\Delta_{n+1}(\mu_3(\mathbf{u}(b))) = \mu_{3,n+1}(\mathbf{u}(b)) - \mu_{3,n}(\mathbf{u}(b))$$

$$= \int_{-\infty}^{+\infty}\left\{\sum_{k=0}^{n+1}\varepsilon^{k}\frac{\partial^{k}\mathbf{u}(b)}{\partial b^{k}}(\Delta b)^{k} - E[\mathbf{u}(b)]\right\}^{3}g_b(x)\mathrm{d}x$$

$$-\int_{-\infty}^{+\infty}\left\{\sum_{k=0}^{n}\varepsilon^{k}\frac{\partial^{k}\mathbf{u}(b)}{\partial b^{k}}(\Delta b)^{k} - E[\mathbf{u}(b)]\right\}^{3}g_b(x)\mathrm{d}x$$

$$= \int_{-\infty}^{+\infty}\left\{\sum_{k=1}^{n+1}\varepsilon^{k}\frac{\partial^{k}\mathbf{u}(b)}{\partial b^{k}}(\Delta b)^{k}\right\}^{3}g_b(x)\mathrm{d}x - \int_{-\infty}^{+\infty}\left\{\sum_{k=1}^{n}\varepsilon^{k}\frac{\partial^{k}\mathbf{u}(b)}{\partial b^{k}}(\Delta b)^{k}\right\}^{3}g_b(x)\mathrm{d}x. \quad (1.42)$$

Further, we employ the Pascal triangle to divide the first series integral into n components and the last one. Substituting for $\varepsilon = 1$ we obtain

$$\Delta_{n+1}(\mu_3(\mathbf{u}(b)))$$

$$= \int_{-\infty}^{+\infty}\left\{\sum_{k=1}^{n}\frac{\partial^{k}\mathbf{u}(b)}{\partial b^{k}}(\Delta b)^{k}\right\}^{3}g_b(x)\mathrm{d}x - \int_{-\infty}^{+\infty}\left\{\frac{\partial^{k}\mathbf{u}(b)}{\partial b^{k}}(\Delta b)^{k}\right\}^{3}g_b(x)\mathrm{d}x$$

$$+3\int_{-\infty}^{+\infty}\left\{\sum_{k=1}^{n}\frac{\partial^{k}\mathbf{u}(b)}{\partial b^{k}}(\Delta b)^{k}\right\}^{2}\frac{\partial^{n+1}\mathbf{u}(b)}{\partial b^{n+1}}(\Delta b)^{n+1}g_b(x)\mathrm{d}x$$

$$+3\int_{-\infty}^{+\infty}\sum_{k=1}^{n}\frac{\partial^{k}\mathbf{u}(b)}{\partial b^{k}}(\Delta b)^{k}\left\{\frac{\partial^{n+1}\mathbf{u}(b)}{\partial b^{n+1}}(\Delta b)^{n+1}\right\}^{2}g_b(x)\mathrm{d}x$$

$$+\int_{-\infty}^{+\infty}\sum_{k=1}^{n}\left\{\frac{\partial^{n+1}\mathbf{u}(b)}{\partial b^{n+1}}(\Delta b)^{n+1}\right\}^3 g_b(x)\mathrm{d}x$$

$$=\frac{\partial^{n+1}\mathbf{u}(b)}{\partial b^{n+1}}\left(3\int_{-\infty}^{+\infty}\left\{\sum_{k=1}^{n}\frac{\partial^k\mathbf{u}(b)}{\partial b^k}(\Delta b)^k\right\}^2(\Delta b)^{n+1}g_b(x)\mathrm{d}x\right.$$

$$\left.+3\frac{\partial^{n+1}\mathbf{u}(b)}{\partial b^{n+1}}\sum_{k=1}^{n}\frac{\partial^k\mathbf{u}(b)}{\partial b^k}\mu_{k+2(n+1)}(b)\right)$$

$$+\frac{\partial^{n+1}\mathbf{u}(b)}{\partial b^{n+1}}\left(\frac{\partial^{n+1}\mathbf{u}(b)}{\partial b^{n+1}}\right)^2\mu_{3(n+1)}(b). \tag{1.43}$$

Once more, independently of the summation and bounded moments included in the brackets, the zeroing of the n $(z+1)$th-order rest may be guaranteed by the polynomial of order equal or less than the nth one. We need to realize that an integration of the series consisting of higher-order partial derivatives as well as perturbations of b itself around the mean value also returns almost the same neighboring odd- and even-order moments, so that at least for a Gaussian distribution we will obtain some zero-valued components there. Finally, we consider the fourth-order moment and it holds that

$$\Delta_{n+1}\left(\mu_4\left(\mathbf{u}(b)\right)\right)=\mu_{4,n+1}\left(\mathbf{u}(b)\right)-\mu_{4,n}\left(\mathbf{u}(b)\right)$$

$$=\int_{-\infty}^{+\infty}\left\{\sum_{k=0}^{n+1}\varepsilon^k\frac{\partial^k\mathbf{u}(b)}{\partial b^k}(\Delta b)^k-E\left[\mathbf{u}(b)\right]\right\}^4 g_b(x)\mathrm{d}x$$

$$-\int_{-\infty}^{+\infty}\left\{\sum_{k=0}^{n}\varepsilon^k\frac{\partial^k\mathbf{u}(b)}{\partial b^k}(\Delta b)^k-E\left[\mathbf{u}(b)\right]\right\}^4 g_b(x)\mathrm{d}x$$

$$=\int_{-\infty}^{+\infty}\left\{\sum_{k=1}^{n+1}\varepsilon^k\frac{\partial^k\mathbf{u}(b)}{\partial b^k}(\Delta b)^k\right\}^4 g_b(x)\mathrm{d}x-\int_{-\infty}^{+\infty}\left\{\sum_{k=1}^{n}\varepsilon^k\frac{\partial^k\mathbf{u}(b)}{\partial b^k}(\Delta b)^k\right\}^4 g_b(x)\mathrm{d}x. \tag{1.44}$$

Almost identically as before, taking $\varepsilon=1$ and reorganizing the components from the fourth power of the $(n+1)$th order we obtain

$$\Delta_{n+1}\left(\mu_4\left(\mathbf{u}(b)\right)\right)$$

$$=\int_{-\infty}^{+\infty}\left\{\sum_{k=1}^{n}\frac{\partial^k\mathbf{u}(b)}{\partial b^k}(\Delta b)^k\right\}^4 g_b(x)\mathrm{d}x-\int_{-\infty}^{+\infty}\left\{\sum_{k=1}^{n}\frac{\partial^k\mathbf{u}(b)}{\partial b^k}(\Delta b)^k\right\}^4 g_b(x)\mathrm{d}x$$

$$+4\int_{-\infty}^{+\infty}\left\{\sum_{k=1}^{n}\frac{\partial^k\mathbf{u}(b)}{\partial b^k}(\Delta b)^k\right\}^3\frac{\partial^{n+1}\mathbf{u}(b)}{\partial b^{n+1}}(\Delta b)^{n+1}g_b(x)\mathrm{d}x$$

$$+6\int_{-\infty}^{+\infty}\left\{\sum_{k=1}^{n}\frac{\partial^k \mathbf{u}(b)}{\partial b^k}(\Delta b)^k\right\}^2\left\{\frac{\partial^{n+1}\mathbf{u}(b)}{\partial b^{n+1}}(\Delta b)^{n+1}\right\}^2 g_b(x)\mathrm{d}x$$

$$+4\int_{-\infty}^{+\infty}\sum_{k=1}^{n}\frac{\partial^k \mathbf{u}(b)}{\partial b^k}(\Delta b)^k\left\{\frac{\partial^{n+1}\mathbf{u}(b)}{\partial b^{n+1}}(\Delta b)^{n+1}\right\}^3 g_b(x)\mathrm{d}x$$

$$+\int_{-\infty}^{+\infty}\left\{\frac{\partial^{n+1}\mathbf{u}(b)}{\partial b^{n+1}}(\Delta b)^{n+1}\right\}^4 g_b(x)\mathrm{d}x. \tag{1.45}$$

Finally, we eliminate the perturbation parameter as not influencing the overall convergence and try to get the total multiplier whose zeroing may give a desired result. It holds that

$$\begin{aligned}
\Delta_{n+1}\left(\mu_4\left(\mathbf{u}(b)\right)\right) &= 4\frac{\partial^{n+1}\mathbf{u}(b)}{\partial b^{n+1}}\int_{-\infty}^{+\infty}\left\{\sum_{k=1}^{n}\frac{\partial^k \mathbf{u}(b)}{\partial b^k}(\Delta b)^k\right\}^3(\Delta b)^{n+1} g_b(x)\mathrm{d}x \\
&+4\frac{\partial^{n+1}\mathbf{u}(b)}{\partial b^{n+1}}\left\{\frac{\partial^{n+1}\mathbf{u}(b)}{\partial b^{n+1}}\right\}^2\int_{-\infty}^{+\infty}\sum_{k=1}^{n}\frac{\partial^k \mathbf{u}(b)}{\partial b^k}(\Delta b)^k(\Delta b)^{3(n+1)} g_b(x)\mathrm{d}x \\
&+6\frac{\partial^{n+1}\mathbf{u}(b)}{\partial b^{n+1}}\frac{\partial^{n+1}\mathbf{u}(b)}{\partial b^{n+1}}\int_{-\infty}^{+\infty}\left\{\sum_{k=1}^{n}\frac{\partial^k \mathbf{u}(b)}{\partial b^k}(\Delta b)^k\right\}^2(\Delta b)^{2(n+1)} g_b(x)\mathrm{d}x \\
&+\frac{\partial^{n+1}\mathbf{u}(b)}{\partial b^{n+1}}\left\{\frac{\partial^{n+1}\mathbf{u}(b)}{\partial b^{n+1}}\right\}^3\int_{-\infty}^{+\infty}(\Delta b)^{4(n+1)} g_b(x)\mathrm{d}x \\
&=\frac{\partial^{n+1}\mathbf{u}(b)}{\partial b^{n+1}}\left(4\int_{-\infty}^{+\infty}\left\{\sum_{k=1}^{n}\frac{\partial^k \mathbf{u}(b)}{\partial b^k}(\Delta b)^k\right\}^3(\Delta b)^{n+1} g_b(x)\mathrm{d}x\right. \\
&\qquad\left.+4\left\{\frac{\partial^{n+1}\mathbf{u}(b)}{\partial b^{n+1}}\right\}^2\sum_{k=1}^{n}\frac{\partial^k \mathbf{u}(b)}{\partial b^k}\mu_{k+3(n+1)}(b)\right) \\
&+\frac{\partial^{n+1}\mathbf{u}(b)}{\partial b^{n+1}}\left(6\frac{\partial^{n+1}\mathbf{u}(b)}{\partial b^{n+1}}\int_{-\infty}^{+\infty}\left\{\sum_{k=1}^{n}\frac{\partial^k \mathbf{u}(b)}{\partial b^k}(\Delta b)^{k+n+1}\right\}^2 g_b(x)\mathrm{d}x\right. \\
&\qquad\left.+\left\{\frac{\partial^{n+1}\mathbf{u}(b)}{\partial b^{n+1}}\right\}^3\mu_{4(n+1)}(b)\right).
\end{aligned} \tag{1.46}$$

Concluding these derivations we see that having the least-squares approximation of a certain order for the state function $\mathbf{u}(b)$ it is just enough to apply the perturbation technique only a single order higher, quite independent of the probabilistic moment under consideration. Analysis of the convergence for the probabilistic coefficients is carried out using the other well-known theorem in mathematical analysis. Let us consider now the coefficient of variation, its $(n+1)$th- combined with its nth-order approximation, and the following ratio:

$$\delta_{n+1}\left(\alpha\left(\mathbf{u}(b)\right)\right)=\frac{\alpha_{n+1}\left(\mathbf{u}(b)\right)}{\alpha_n\left(\mathbf{u}(b)\right)}=\frac{\sqrt{\mathrm{Var}_{n+1}\left(\mathbf{u}(b)\right)}}{E_{n+1}\left[\mathbf{u}(b)\right]}\frac{E_n\left[\mathbf{u}(b)\right]}{\sqrt{\mathrm{Var}_n\left(\mathbf{u}(b)\right)}}. \tag{1.47}$$

Now we add an artificial zero to the counter of the second fraction in the following form (according to the order n of the response polynomial):

$$0=\int_{-\infty}^{+\infty}\varepsilon^{n+1}\frac{\partial^{n+1}\mathbf{u}(b)}{\partial b^{n+1}}\left(\Delta b\right)^{n+1}g_b(x)\mathrm{d}x \tag{1.48}$$

to get simply

$$\begin{aligned}\delta_{n+1}\left(\alpha\left(\mathbf{u}(b)\right)\right)&=\frac{\alpha_{n+1}\left(\mathbf{u}(b)\right)}{\alpha_n\left(\mathbf{u}(b)\right)}\\&=\frac{\sqrt{\mathrm{Var}_{n+1}\left(\mathbf{u}(b)\right)}}{E_{n+1}\left[\mathbf{u}(b)\right]}\frac{E_n\left[\mathbf{u}(b)\right]+\int_{-\infty}^{+\infty}\varepsilon^{n+1}\frac{\partial^{n+1}\mathbf{u}(b)}{\partial b^{n+1}}\left(\Delta b\right)^{n+1}g_b(x)\mathrm{d}x}{\sqrt{\mathrm{Var}_n\left(\mathbf{u}(b)\right)}}\\&=\frac{\sqrt{\mathrm{Var}_{n+1}\left(\mathbf{u}(b)\right)}}{E_{n+1}\left[\mathbf{u}(b)\right]}\frac{E_{n+1}\left[\mathbf{u}(b)\right]}{\sqrt{\mathrm{Var}_n\left(\mathbf{u}(b)\right)}}=\sqrt{\frac{\mathrm{Var}_{n+1}\left(\mathbf{u}(b)\right)}{\mathrm{Var}_n\left(\mathbf{u}(b)\right)}}.\end{aligned} \tag{1.49}$$

This means that the coefficient of variation is convergent, together with a variance of the random function $\mathbf{u}(b)$. Analogously we can proceed to study the skewness coefficient; we have

$$\begin{aligned}\delta_{n+1}\left(\beta\left(\mathbf{u}(b)\right)\right)&=\frac{\beta_{n+1}\left(\mathbf{u}(b)\right)}{\beta_n\left(\mathbf{u}(b)\right)}=\frac{\mu_{3,n+1}\left(\mathbf{u}(b)\right)}{\left(\mathrm{Var}_{n+1}\left(\mathbf{u}(b)\right)\right)^{\frac{3}{2}}}\frac{\left(\mathrm{Var}_n\left(\mathbf{u}(b)\right)\right)^{\frac{3}{2}}}{\mu_{3,n}\left(\mathbf{u}(b)\right)}\\&=\frac{\mu_{3,n+1}\left(\mathbf{u}(b)\right)}{\mu_{3,n}\left(\mathbf{u}(b)\right)}\left\{\frac{\left(\mathrm{Var}_n\left(\mathbf{u}(b)\right)\right)}{\left(\mathrm{Var}_{n+1}\left(\mathbf{u}(b)\right)\right)}\right\}^{\frac{3}{2}}\xrightarrow{n\to\infty}1\end{aligned} \tag{1.50}$$

since the limits of both components in this product also tend together to 1, as documented above. The situation is quite similar for the coefficient of flatness and kurtosis, because

$$\begin{aligned}\delta_{n+1}\left(\gamma\left(\mathbf{u}(b)\right)\right)&=\frac{\beta_{n+1}\left(\mathbf{u}(b)\right)}{\beta_n\left(\mathbf{u}(b)\right)}=\frac{\mu_{4,n+1}\left(\mathbf{u}(b)\right)}{\left(\mathrm{Var}_{n+1}\left(\mathbf{u}(b)\right)\right)^2}\frac{\left(\mathrm{Var}_n\left(\mathbf{u}(b)\right)\right)^2}{\mu_{4,n}\left(\mathbf{u}(b)\right)}\\&=\frac{\mu_{4,n+1}\left(\mathbf{u}(b)\right)}{\mu_{4,n}\left(\mathbf{u}(b)\right)}\left\{\frac{\left(\mathrm{Var}_n\left(\mathbf{u}(b)\right)\right)}{\left(\mathrm{Var}_{n+1}\left(\mathbf{u}(b)\right)\right)}\right\}^2\xrightarrow{n\to\infty}1\end{aligned} \tag{1.51}$$

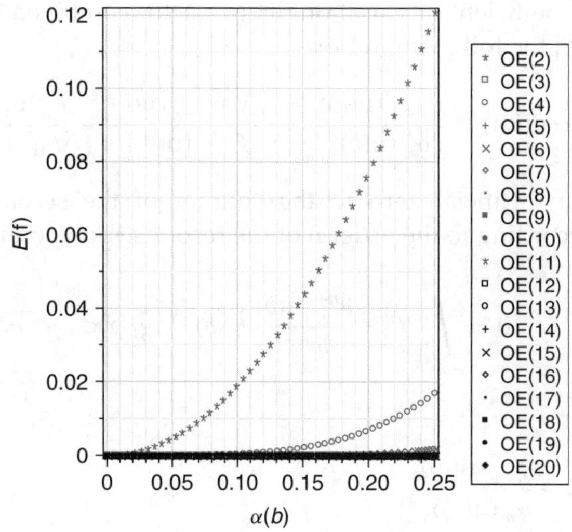

Figure 1.5 Probabilistic convergence of the expected values for $f(b) = \sum_{i=1}^{10} b^i$, $E(b) = 0.5$

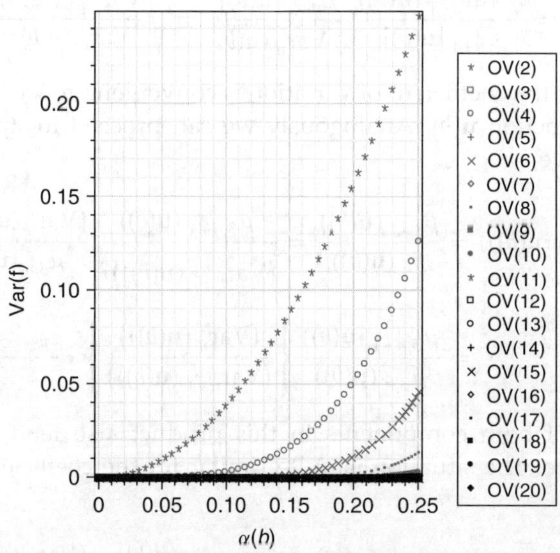

Figure 1.6 Probabilistic convergence of the variances for $f(b) = \sum_{i=1}^{10} b^i$, $E(b) = 0.5$

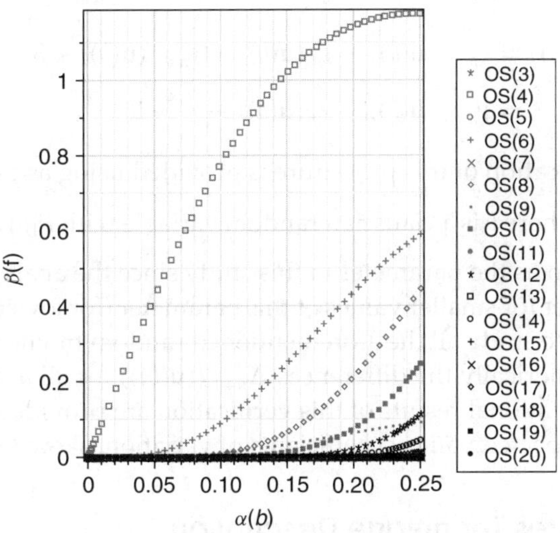

Figure 1.7 Probabilistic convergence of the skewness for $f(b) = \sum_{i=1}^{10} b^i$, $E(b) = 0.5$

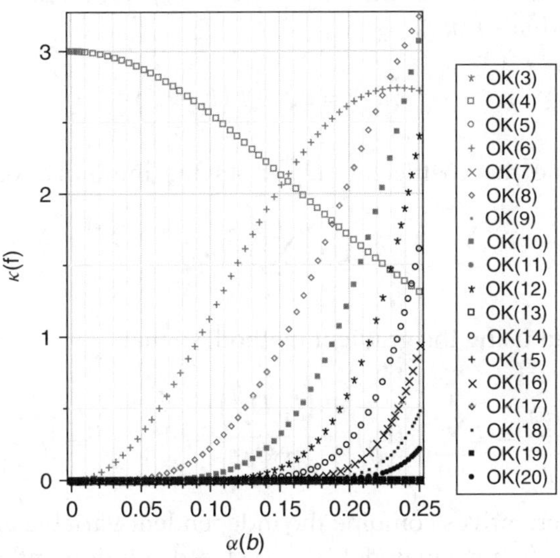

Figure 1.8 Probabilistic convergence of the kurtosis for $f(b) = \sum_{i=1}^{10} b^i$, $E(b) = 0.5$

and, therefore, the kurtosis

$$\Delta_{n+1}(\gamma(\mathbf{u}(b))) = \kappa_{n+1}(\mathbf{u}(b)) - \kappa_n(\mathbf{u}(b)) = \gamma_{n+1}(\mathbf{u}(b)) + 3 - \gamma_n(\mathbf{u}(b)) - 3$$
$$= \gamma_{n+1}(\mathbf{u}(b)) - \gamma_n(\mathbf{u}(b)) \xrightarrow{n\to\infty} 0. \tag{1.52}$$

Numerical verification of this procedure is provided using as example the polynomial $f(b) = \sum_{i=1}^{10} b^i$, where b is a Gaussian random variable with $E[b] = 0.5$. Its coefficient of variation is taken as the parameter in this study since the analyzed convergence is expected to be faster for smaller values of this parameter (for lower-order approximations, of course). We study all the aforementioned random moments and characteristics and verify numerically the differences $\Delta_{n+1}((\mathbf{u}(b)))$ for all $n \leq 20$ using the 20th-order SPT. Computational results of this verification are provided in Figures 1.5–1.8 in turn – for the expected values, coefficients of variation, skewness, and kurtosis.

1.2 Least-Squares Technique Description

Consider a set of m data points $\left(b_i, u_\beta^{(i)}\right)$ for $\beta = 1, \ldots, N$, the non-linear continuous function to be found: $u_\beta = f(b)$ and a curve (approximating function) $u_\beta = f(b, D_\beta)$, which additionally depends on the parameters $D_\beta^{(j)}, j = 1, \ldots, n$, where $m \geq n$ [9]. We define also the residuals $r_i(u_\beta)$:

$$r_i(u_\beta) = u_\beta^{(i)} - f\left(b_i, D_\beta^{(i)}\right) \tag{1.53}$$

to finally determine the coefficients $D_\beta^{(j)}$ from the minimization of the following functional:

$$S(u) = \sum_{i=1}^{m} r_i^2(u_\beta). \tag{1.54}$$

This is carried out using the gradient method, so that

$$\frac{\partial S(u)}{\partial D_\beta^{(j)}} = 2\sum_{i=1}^{m} r_i(u_\beta) \frac{\partial f\left(b_i, D_\beta^{(i)}\right)}{\partial D_\beta^{(j)}} = 0; \qquad j = 1, \ldots, n. \tag{1.55}$$

Since the LHS derivatives combine the independent variable and the parameters, so these gradient equations may not have a closed solution and some initial values must be adopted for these parameters. They are finally determined via some iterative approximation as

$$D_\beta^{(j)} \cong {}^{(k+1)}D_\beta^{(j)} = {}^{(k)}D_\beta^{(j)} + \Delta D_\beta^{(j)} \tag{1.56}$$

where k denotes an iteration number and $\Delta D_\beta^{(j)}$ is the so-called shift vector. The following linearization based on a first-order Taylor series representation is applied at each iteration:

$$f\left(b_i, D_\beta^{(i)}\right) \cong f\left(b_i, {}^{(k)}D_\beta^{(i)}\right) + \sum_{j=1}^{n} \frac{\partial f\left(b_i, D_\beta^{(i)}\right)}{\partial D_\beta^{(j)}} \left(D_\beta^{(j)} - {}^{(k)}D_\beta^{(j)}\right)$$

$$\cong f\left(b_i, {}^{(k)}D_\beta^{(i)}\right) + \sum_{j=1}^{n} J_{ij}^{\beta} \Delta D_\beta^{(j)}. \tag{1.57}$$

Since $\frac{\partial r_i}{\partial D_\beta^{(j)}} = -J_{ij}^{\beta}$ it holds that

$$r_i\left(u_\beta\right) = \Delta u_\beta^{(i)} - \sum_{s=1}^{n} J_{is}^{\beta} \Delta D_\beta^{(s)}; \quad \Delta u_\beta^{(i)} = u_\beta^{(i)} - f\left(b_i, {}^{(k)}D_\beta^{(i)}\right). \tag{1.58}$$

Substituting these expressions into the gradient Equation (1.55), one gets

$$-2\sum_{i=1}^{m} J_{ij}^{\beta} \left(\Delta u_\beta^{(i)} - \sum_{s=1}^{n} J_{is}^{\beta} \Delta D_\beta^{(s)}\right) = 0, \tag{1.59}$$

which effectively becomes a system of linear equations

$$\sum_{i=1}^{m}\sum_{s=1}^{n} J_{ij}^{\beta} J_{is}^{\beta} \Delta D_\beta^{(s)} = \sum_{i=1}^{n} J_{ij}^{\beta} \Delta u_\beta^{(i)}, \qquad j = 1, \ldots, n. \tag{1.60}$$

Finally, we obtain

$$(\mathbf{J}^T\mathbf{J})\Delta\mathbf{D} = \mathbf{J}^T \Delta\mathbf{u}. \tag{1.61}$$

Further determination of the probabilistic moments and characteristics is based on determination of the partial derivatives of the response function with respect to the input random parameter b using the analytical form

$$\frac{\mathrm{d}^k f}{\mathrm{d}b^k} = \prod_{i=1}^{k} (n-i)\, a_n b^{n-k} + \prod_{i-2}^{k} (n-i)\, a_{n-1} b^{n-(k+1)} + \cdots + a_{n-k}. \tag{1.62}$$

Application of this method to the tensors needs small extensions only; observe the basic equations set on the example of the fourth-order tensor serving further in Chapter 5 for a solution of the homogenization problem. Let us consider once more a data set of m trial points $\left(b_i, C_{\alpha\beta\gamma\delta}^{(eff)(i)}\right)$ with $\alpha, \beta, \gamma, \delta = 1, 2, 3$, some non-linear continuous function $C_{\alpha\beta\gamma\delta}^{(eff)} = f(b)$ and a curve (approximating function) $C_{\alpha\beta\gamma\delta}^{(eff)} = f\left(b, D_{\alpha\beta\gamma\delta}\right)$, which additionally depends on n parameters $D_{\alpha\beta\gamma\delta}^{(j)}$, $j = 1, \ldots, n$, where $m \geq n$. The upper index "*eff*" relates the tensor under consideration to the homogenization problem only. We define the residuals $r_i\left(C_{\alpha\beta\gamma\delta}^{(eff)}\right)$:

$$r_i\left(C_{\alpha\beta\gamma\delta}^{(eff)}\right) = C_{\alpha\beta\gamma\delta}^{(eff)(i)} - f\left(b_i, D_{\alpha\beta\gamma\delta}^{(i)}\right) \tag{1.63}$$

to finally determine the components of the tensor $D_{\alpha\beta\gamma\delta}$ from the minimization of

$$S = \sum_{i=1}^{m} r_i^2 \left(C_{\alpha\beta\gamma\delta}^{(eff)} \right). \tag{1.64}$$

We employ the formula

$$\frac{\partial S}{\partial D_{\alpha\beta\gamma\delta}^{(j)}} = 2 \sum_{i=1}^{m} r_i \left(C_{\alpha\beta\gamma\delta}^{(eff)} \right) \frac{\partial f\left(b_i, D_{\alpha\beta\gamma\delta}^{(i)}\right)}{\partial D_{\alpha\beta\gamma\delta}^{(j)}} = 0; \quad j = 1, \ldots, n. \tag{1.65}$$

Analogously to the above, an iterative approximation is proposed as

$$D_{\alpha\beta\gamma\delta}^{(j)} \cong {}^{(k+1)}D_{\alpha\beta\gamma\delta}^{(j)} = {}^{(k)}D_{\alpha\beta\gamma\delta}^{(j)} + \Delta D_{\alpha\beta\gamma\delta}^{(j)} \tag{1.66}$$

where k denotes an iteration number and $\Delta D_{\alpha\beta\gamma\delta}^{(j)}$ stands for the shift vector. The following linearization for the tensor is applied:

$$f\left(b_i, D_{\alpha\beta\gamma\delta}^{(i)}\right) \cong f\left(b_i, {}^{(k)}D_{\alpha\beta\gamma\delta}^{(i)}\right) + \sum_{j=1}^{n} \frac{\partial f\left(b_i, D_{\alpha\beta\gamma\delta}^{(i)}\right)}{\partial D_{\alpha\beta\gamma\delta}^{(j)}} \left(D_{\alpha\beta\gamma\delta}^{(j)} - {}^{(k)}D_{\alpha\beta\gamma\delta}^{(i)} \right)$$
$$\cong f\left(b_i, {}^{(k)}D_{\alpha\beta\gamma\delta}^{(i)}\right) + \sum_{j=1}^{n} J_{ij}^{\beta} \Delta D_{\alpha\beta\gamma\delta}^{(j)}. \tag{1.67}$$

because $\frac{\partial r_i}{\partial D_{\alpha\beta\gamma\delta}^{(j)}} = -J_{ij}^{\alpha\beta\gamma\delta}$ and with no summation on α,β,γ,δ, it holds that

$$r_i \left(C_{\alpha\beta\gamma\delta}^{(eff)} \right) = \Delta C_{\alpha\beta\gamma\delta}^{(eff)(i)} - \sum_{s=1}^{n} J_{is}^{\alpha\beta\gamma\delta} \Delta D_{\alpha\beta\gamma\delta}^{(s)}; \quad \Delta C_{\alpha\beta\gamma\delta}^{(eff)(i)} = C_{\alpha\beta\gamma\delta}^{(eff)(i)} - f\left(b_i, {}^{(k)}D_{\alpha\beta\gamma\delta}^{(i)}\right). \tag{1.68}$$

Substituting these expressions for the gradient Equation (1.65), they become

$$-2 \sum_{i=1}^{m} J_{ij}^{\alpha\beta\gamma\delta} \left(\Delta C_{\alpha\beta\gamma\delta}^{(eff)(i)} - \sum_{s=1}^{n} J_{is}^{\alpha\beta\gamma\delta} \Delta D_{\alpha\beta\gamma\delta}^{(s)} \right) = 0, \tag{1.69}$$

which finally equals to

$$\sum_{i=1}^{m} \sum_{s=1}^{n} J_{ij}^{\alpha\beta\gamma\delta} J_{is}^{\alpha\beta\gamma\delta} \Delta D_{\alpha\beta\gamma\delta}^{(s)} = \sum_{i=1}^{n} J_{ij}^{\alpha\beta\gamma\delta} \Delta C_{\alpha\beta\gamma\delta}^{(eff)(i)}, \quad j = 1, \ldots, n. \tag{1.70}$$

We obtain in matrix notation as

$$(\mathbf{J}^T\mathbf{J}) \ \Delta\mathbf{D} = \mathbf{J}^T \Delta\mathbf{C}^{(eff)}. \tag{1.71}$$

The observations may also be minimized using the weighted sum for the tensor as follows:

$$\tilde{S} = \sum_{i=1}^{m} W_{ii}\tilde{r}_i^2\left(C^{(eff)}_{\alpha\beta\gamma\delta}\right). \tag{1.72}$$

Each element of the diagonal weights matrix $\mathbf{W}$ equals the reciprocal of the measurement variance (completely independent of the variance of the input random parameter in this problem). Then, the equation system Equation (1.71) becomes

$$(\mathbf{J}^T\mathbf{W}\mathbf{J})\Delta\mathbf{D} = \mathbf{J}^T\mathbf{W}\Delta\mathbf{C}^{(eff)}. \tag{1.73}$$

Of course, some further numerical improvements of this algorithm are available now, but according to the numerical results obtained, they do not seem to be necessary. The sense and value of the least-square method approximation behind other, well-described numerical methods can be displayed with the following simple computational example.

Example 1.3: Comparison of various approximation techniques

Let us consider an approximation problem consisting of the set of trial points with values equal to $l=$[3.5E−8, 3.6E−8, 3.7E−8, 3.8E−8, 3.9E−8, 4.0E−8, 4.1E−8, 4.2E−8, 4.3E−8] resulting in some structural response given by the values vector $f(l)=$[7.4568E11, 7.2497E11, 7.0537E11, 6.8681E11, 6.6910E11, 6.5247E11, 6.3656E11, 6.2140E11, 6.0695E11]. The goal is to provide a curvilinear continuous approximation of this function and to compare the first- and second-order partial derivatives with respect to the independent parameter l [95].

The solution to this problem with Maple™ is visualized in Figures 1.9 and 1.10, showing in turn various approximations for the function $f(l)$, its first and second derivatives computed on the basis of the functions recovered via Newton, Lagrange, monomial, power, and least-squares methods (using a quadratic form only). The trial points set is given in Figure 1.9 using large dots, while the remaining data corresponds to the comments in the legend.

As is apparent, practically all the methods within the numerical error range return the same functions, however the differences are significant for both first- and second-order derivatives. All the methods except the least-squares method (LSM) return functions whose both derivatives demonstrate inacceptable fluctuations at both ends of the given computational domain. Even if the second-order partial derivatives computed at the mean value of the design parameter l all have almost the same value, the first-order derivatives are totally different at this point; monomial and power methods return a few percentage points higher value at this point, while Newton and Lagrange algorithms agree rather well with the results obtained thanks to the least-squares algorithm. This last technique enables a determination of two very stable derivatives, which furthermore have the same values (for the second-order derivative) or the values with relatively small fluctuations (for the first-order derivative) over the entire variability interval of l, quite contrary to the first four methods. Considering this result, we consequently use the LSM technique to determine all the response

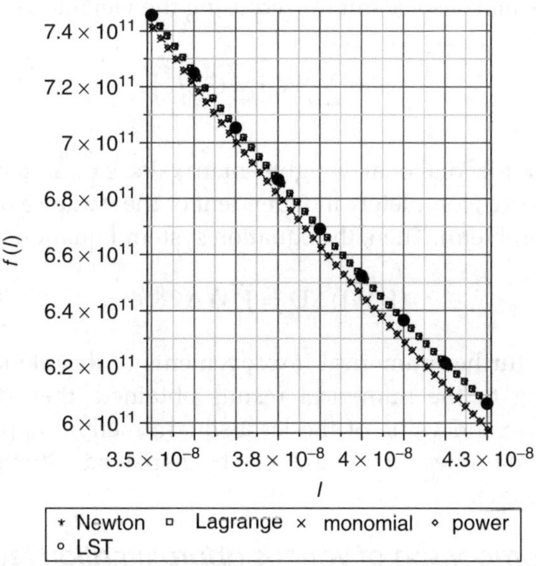

Figure 1.9 Various approximations for the response function $f(l)$. Reproduced with permission from Springer

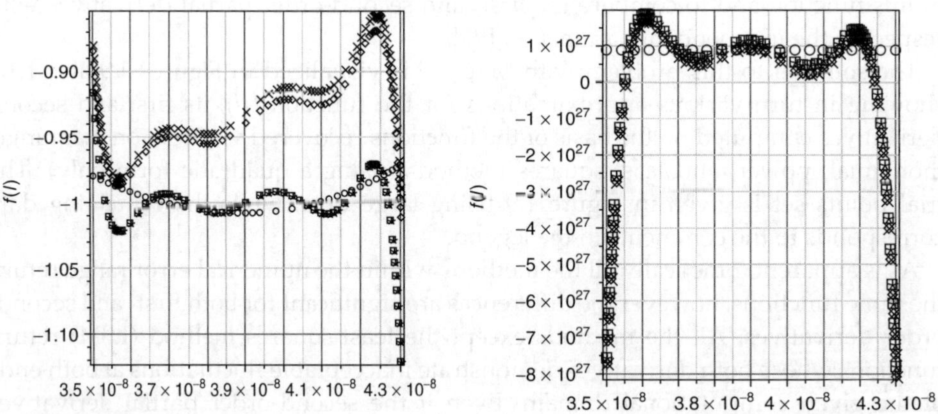

Figure 1.10 Various approximations for the first and second partial derivatives of the response function $f(l)$ with respect to l. Reproduced with permission from Springer

functions – this is especially advised in case when the general form of structural response cannot be predicted from previous examples or theoretical considerations, otherwise polynomial interpolation implemented in Maple™ may be recommended. The second issue, shown in Equation (1.73), is the weighting procedure and its general influence on the LSM approximation results for the stochastic perturbation method and the particular moments. This is explained well within the next example, where our main goal is to find such a weights distribution over a computational domain of the design parameter to assure the best stability of these moments and coefficients as well as their fastest probabilistic convergence. We expect that a systematic increase of the perturbation order applied should result in a decrease of the differences between the results obtained for these increasing order terms.

Example 1.4: Probabilistic moments efficiency in weighted least-squares technique

The main aim of this computational study is to look at the influence of the weighting procedure on the probabilistic moments computed via the least-squares approximation of a certain set of trial data points; this is done for the homogenization problem described in detail in the last chapter of this book [85]. Let us consider a derivation of the expected values, coefficients of variation, skewness, and coefficients of flatness for the set of trial points $b = [0.29, 0.30, 0.31, 0.32, 0.33, 0.34, 0.35, 0.36, 0.37, 0.38, 0.39]$ representing an input random variable, which results in some structural response given by the first components of the fourth-order tensor as $C_{1111}(b) = [$9.4956E9, 9.7022E9, 9.9404E9, 1.0204E10, 1.0500E10, 1.0841E10, 1.1221E10, 1.1664E10, 1.2165E10, 1.2744E10, 1.3442E10]. The primary goal is to provide from the first- up to the ninth-order polynomial LSM-based continuous approximation of the function relating these two data sets for the unweighted LSM (ULSM) procedure, with triangular weights distribution according to the pattern [1,2,3,4,5,6,5,4,3,2,1,] and, finally, for the Dirac-type distribution of these weights given as [1,1,1,1,1,6,1,1,1,1,1]. Secondly, we are going to compare the first four probabilistic characteristics to check the influence of the weighting procedure and find an optimal perturbation procedure order in this context. Since the LSM implemented in Maple™ has an opportunity to propose an approximating polynomial degree, one may use this capability to calibrate an optimal order and this degree first, and then solve for the probabilistic response instead of automatic usage of this procedure for all structural problems.

The response functions are given in Figures 1.11 and 1.12 for the three methods with respect to the input design parameter b, while the remaining functions – expectations (Figures 1.13 and 1.14), coefficients of variation (Figures 1.15 and 1.16), skewness (Figures 1.17 and 1.18), as well as flatness coefficient (Figures 1.19 and 1.20) – are shown with respect to the input coefficient of variation $\alpha \in [0.0, 0.2]$.

It is apparent that all order approximations (except for the first) exhibit really very small differences and it is almost impossible to distinguish between higher-order functions at all. One may notice that the distance between the first-order approximation and the final function taken at the expectation of input variable is greatest for the non-weighted technique and significantly smaller for the weighted

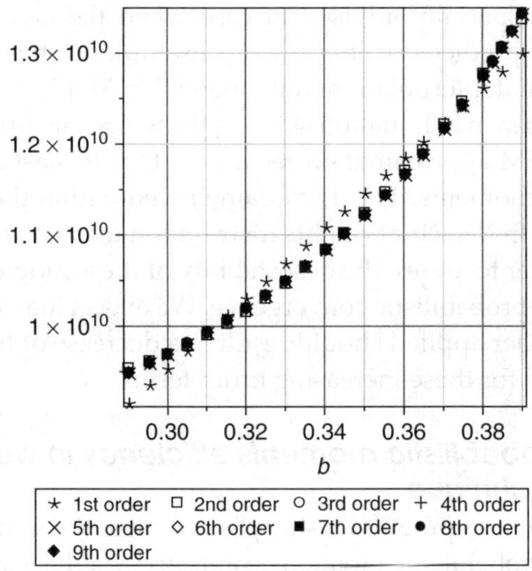

Figure 1.11 Various approximations for the response function $C_{1111}(b)$, ULSM. Reproduced with permission from IFTR PAS

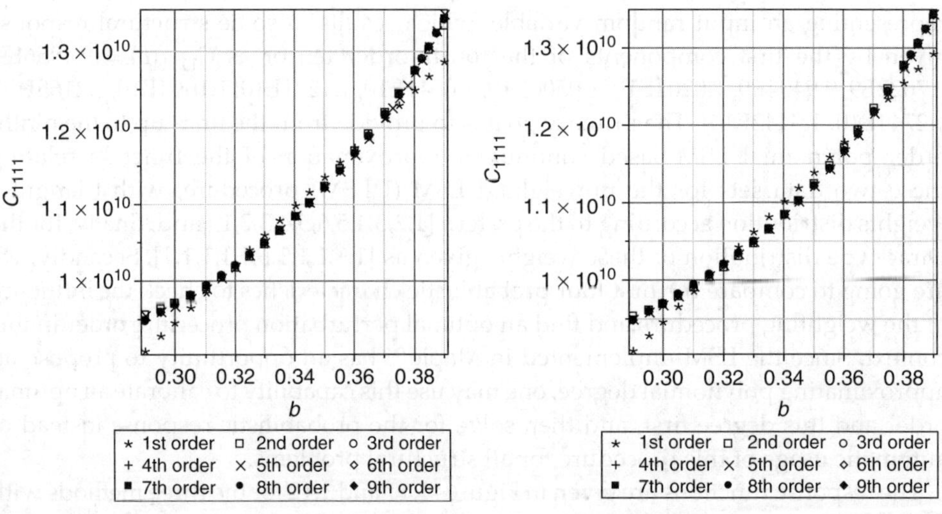

Figure 1.12 Various approximations for the response function $C_{1111}(b)$, triangular versus Dirac weights. Reproduced with permission from IFTR PAS

methods. The general conclusion from this example is that almost independently of the perturbation order and the type of probabilistic characteristic being studied, all values increase together with input $\alpha(b)$, with some small exception for the kurtosis; nevertheless, they all strongly depend on $\alpha(b)$ unlike for the Monte Carlo simulations or in analytical calculus. These fluctuations may not look so large for significant

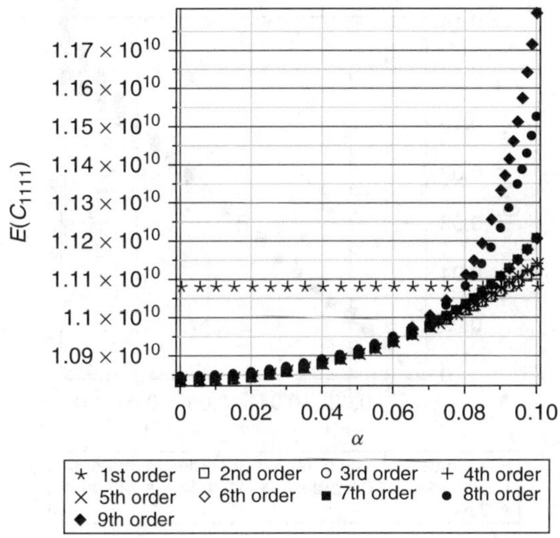

Figure 1.13 The expected values of the component $C_{1111}(b)$, ULSM. Reproduced with permission from IFTR PAS

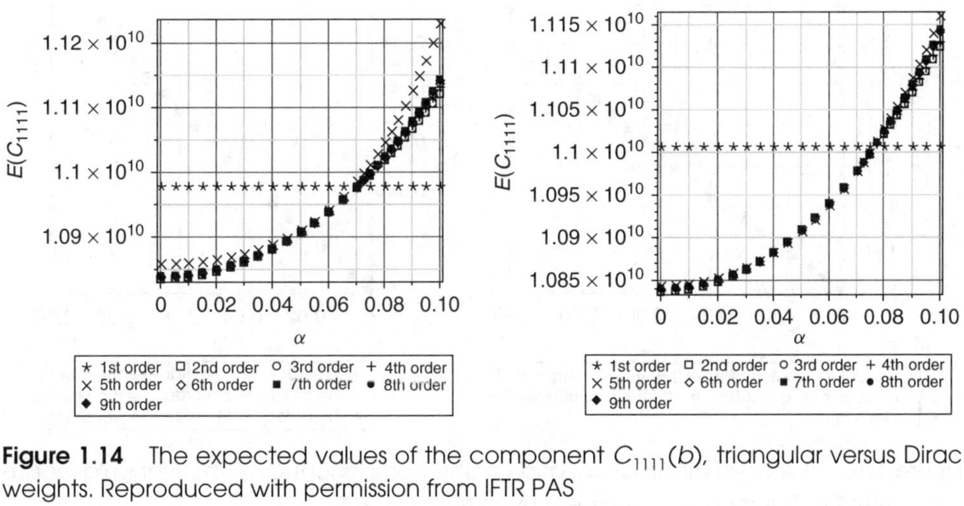

Figure 1.14 The expected values of the component $C_{1111}(b)$, triangular versus Dirac weights. Reproduced with permission from IFTR PAS

numbers of given responses, however, they cannot be simply postponed. The next set of figures (Figures 1.13 and 1.14) show essentially different expectations computed and presented as functions of the input coefficient of variation. It should be mentioned that determination of these interrelations without the usage of some computer algebra system would be extremely difficult. The expectations computed even for the first-order response polynomial are different for the three models compared, but

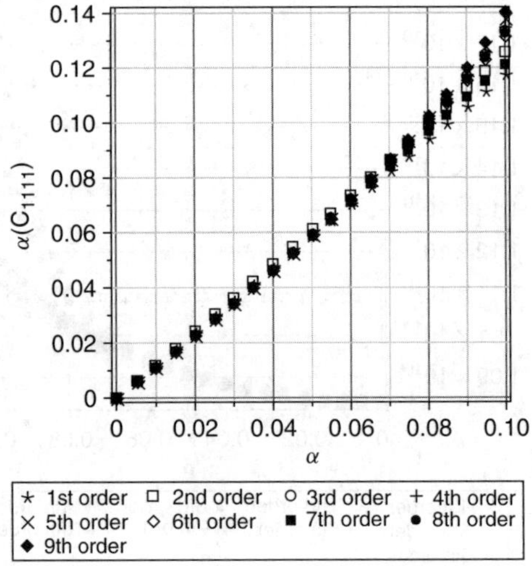

Figure 1.15 The coefficients of variation of the component $C_{1111}(b)$, ULSM. Reproduced with permission from IFTR PAS

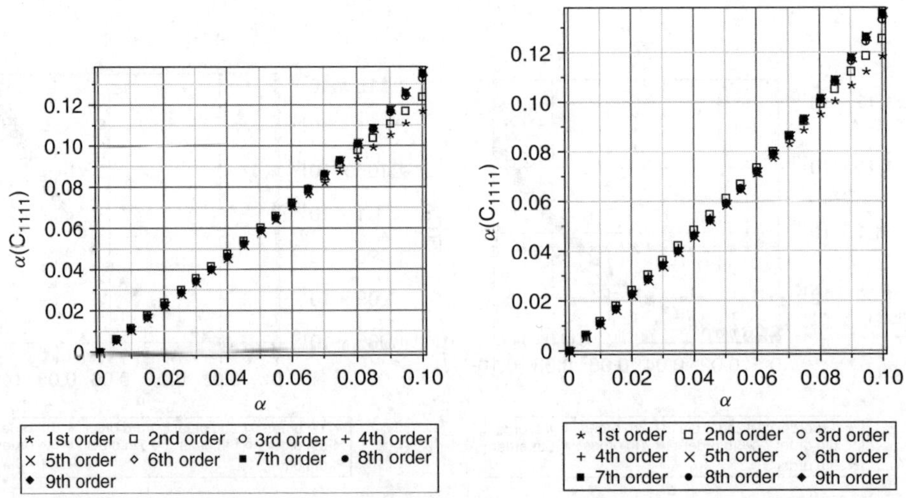

Figure 1.16 The coefficients of variation of the component $C_{1111}(b)$, triangular versus Dirac weights. Reproduced with permission from IFTR PAS

each time are quite insensitive to input random dispersion. Obviously, the higher-order response functions result each time in non-linear, fully convex interrelations between $C_{1111}(b)$ and b; apparently, the ULSM appears to be almost divergent for the maximum value of the parameter $\alpha(b)$, whereas both WLSM approaches show good convergence together with both additional increases of the approximating polynomial order and input random dispersion. Comparing carefully the left and right diagrams in Figure 1.14, one can see that the application of the Dirac weights

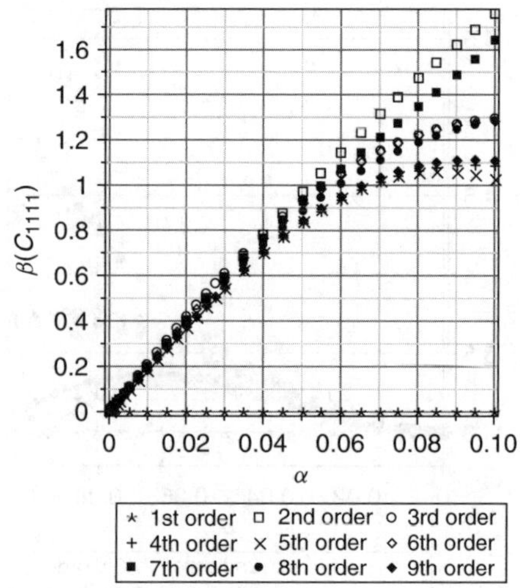

Figure 1.17 The skewness of the component $C_{1111}(b)$, ULSM. Reproduced with permission from IFTR PAS

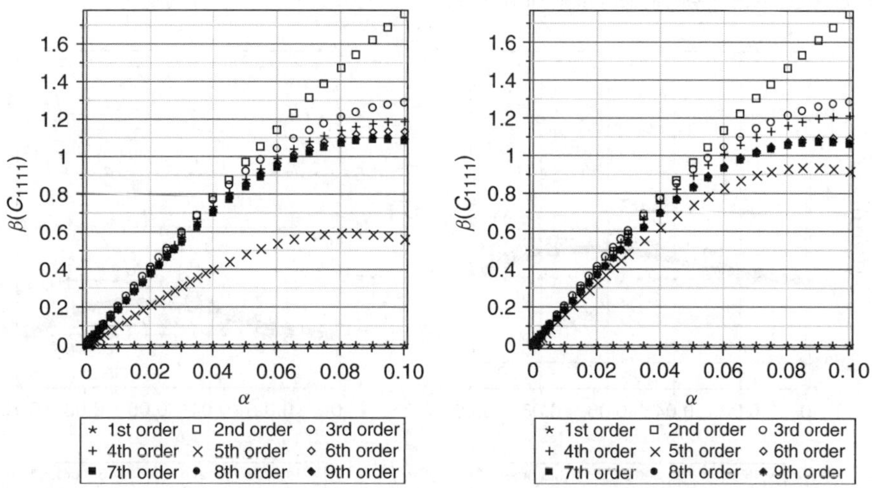

Figure 1.18 The skewness of the component $C_{1111}(b)$, triangular versus Dirac weights. Reproduced with permission from IFTR PAS

definitely eliminates some numerical discrepancies that can be obtained for the middle-order approximants (such as the fifth one here) – some subtle differences in the response functions weighted using these two types of distributions are found. Finally, the maximum expected values obtained for $\alpha(b) = 0.10$ are also different – the largest are computed for ULSM, then for triangular WLSM, and the minimum is detected for Dirac WLSM, so that the conclusion is that the last approach guarantees the most stable results for the expectations. Figures 1.15 and 1.16 contain the coefficient of variation of $C_{1111}(b)$ shown as a function of $\alpha(b)$ and the tendency of

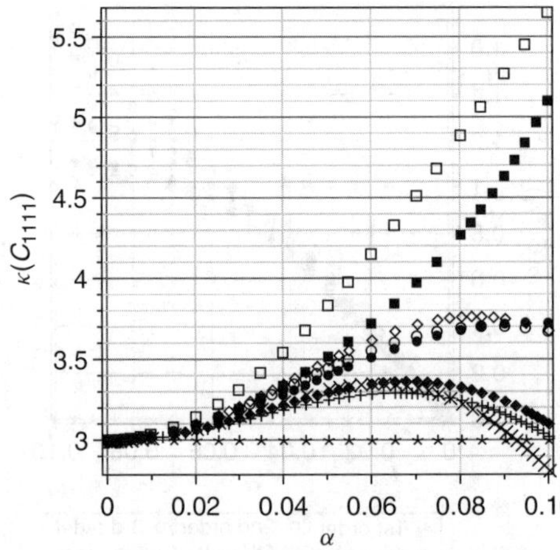

Figure 1.19 The coefficients of flatness of the component $C_{1111}(b)$, ULSM. Reproduced with permission from IFTR PAS

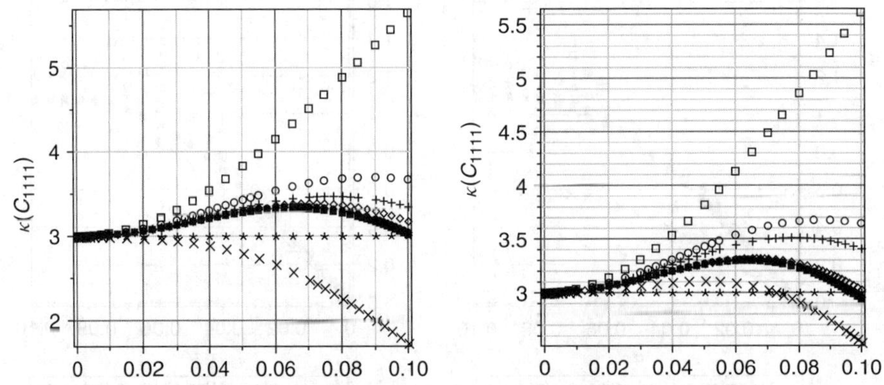

Figure 1.20 The coefficients of flatness of the component $C_{1111}(b)$, triangular versus Dirac weights. Reproduced with permission from IFTR PAS

the ULSM approach to overestimate this output characteristic is quite clear – this method returns a stable and convergent result for smaller values of the input random variable only. The additional interrelations between the input and output coefficient of variation presented in both parts of Figure 1.16 as the function of the response polynomial order show that we need to exclude the first two orders from the WLSM approach results, while the remaining higher orders return very stable values which are very close to each other. The skewnesses given in Figures 1.17 and 1.18 better stabilize in the weighted methods for the maximum value of the input $\alpha(b)$, while for the unweighted method they still increase together with this input characteristic. In all cases they are really different from 0 typically for Gaussian variables and processes, somewhat contrary to the coefficients of flatness given in Figures 1.19 and 1.20. This coefficient increases for smaller perturbation orders but together with an increase of the perturbation order it becomes closer and closer to the value 3, known from the previous formulas. To conclude this example it is necessary to point out that the weighted version of the LSM is recommended, preferably with the Dirac-type distribution of the weights, but some order optimization of the stochastic perturbation technique may reduce a total computational effort since higher-order terms do not always introduce significant differences to the existing results.

A discussion of the response functions and their approximation methods cannot be completed without a longer remark on their direct usage (without the stochastic perturbation technique) in the integrals defining the basic probabilistic moments. Now the objective is the real function f of the following polynomial form [88]:

$$f(X;n) = \sum_{i=0}^{n} A_i X^i \tag{1.74}$$

where A_i, $i=1, \ldots, n$ are some real-valued parameters and X is the given Gaussian random variable. The key issue is a determination of the basic probabilistic moments and characteristics of this power series expansion with respect to the expectation and standard deviation of X itself. As is known, analytical derivation results in very long expansions, especially for higher probabilistic moments like the third or fourth, and alternatively may be carried out using the complex exponential characteristic function of X. Even if computer algebra systems, like Maple™, are employed to derive the additional formulas describing the basic probabilistic characteristics for a general polynomial form, it appears to be extremely difficult. We consider only the difference between the nth and $(n-1)$th orders in this polynomial to demonstrate the complexity level of the derivation. Taking this as $O(f(X);n) = A_n X^n$ one can show that the following hold true for the standardized Gaussian variable X [88]:

- expected value

$$E\left[A_n X^n\right] = \frac{(-1)^n 2^{-1+\frac{1}{2}n} A_n \Gamma\left(\frac{1}{2}n+\frac{1}{2}\right)}{\sqrt{\pi}}, \tag{1.75}$$

- variances

$$\mathrm{Var}\left(A_n X^n\right) = \frac{1}{4}\frac{A_n^2\left(4(-1)^{2n}2^{n-1}\Gamma\left(\frac{1}{2}+n\right)\pi + (-1)^{2n+1}2^n\Gamma\left(\frac{1}{2}+\frac{1}{2}n\right)^2\sqrt{\pi}\right)}{\pi^{\frac{3}{2}}}, \tag{1.76}$$

- standard deviations

$$\sigma\left(A_n X^n\right) = \frac{A_n}{2\pi^{\frac{3}{4}}}\sqrt{4(-1)^{2n}2^{n-1}\Gamma\left(\frac{1}{2}+n\right)\pi + (-1)^{2n+1}2^n\Gamma\left(\frac{1}{2}+\frac{1}{2}n\right)^2\sqrt{\pi}}, \tag{1.77}$$

- coefficient of variation

$$\alpha\left(A_nX^n\right)=\frac{1}{(-1)^n\,2^{\frac{1}{2}n}\Gamma\left(\frac{1}{2}n+\frac{1}{2}\right)}\times\sqrt{\frac{4\,(-1)^{2n}\,2^{n-1}\Gamma\left(\frac{1}{2}+n\right)\pi+(-1)^{2n+1}\,2^n\Gamma\left(\frac{1}{2}+\frac{1}{2}n\right)^2\sqrt{\pi}}{\pi^{\frac{1}{2}}}},\tag{1.78}$$

- third central probabilistic moment

$$\mu_3\left(A_nX^n\right)=\frac{A_n^3}{4\pi^{\frac{3}{2}}}\left(3\,(-1)^{3n+1}\,2^{\frac{3}{2}n}\Gamma\left(\frac{1}{2}+\frac{1}{2}n\right)\Gamma\left(\frac{1}{2}+n\right)\sqrt{\pi}+4\,(-1)^{3n}\,2^{-1+\frac{3}{2}n}\Gamma\left(\frac{1}{2}+\frac{3}{2}n\right)\pi\right)+\frac{A_n^3}{4\pi^{\frac{3}{2}}}\left((-1)^{3n}\,2^{\frac{3}{2}n}\Gamma\left(\frac{1}{2}+\frac{1}{2}n\right)^3\right),\tag{1.79}$$

- fourth central probabilistic moment

$$\mu_4\left(A_nX^n\right)=\frac{A_n^4}{16\pi^2}\left(48\sqrt{\pi}\,(-1)^{4n}\,2^{2n-2}\Gamma\left(\frac{1}{2}n+\frac{1}{2}\right)^2\Gamma\left(\frac{1}{2}+n\right)+8\pi^{\frac{3}{2}}\,(-1)^{4n}\,4^n\Gamma\left(\frac{1}{2}n+\frac{1}{2}\right)\right)+\frac{A_n^4}{16\pi^2}\left(-16\,(-1)^{4n}\,4^n\Gamma\left(\frac{1}{2}n+\frac{1}{2}\right)\Gamma\left(\frac{1}{2}+\frac{3}{2}n\right)\pi+3\,(-1)^{4n+1}\,4^n\Gamma\left(\frac{1}{2}n+\frac{1}{2}\right)^4\right),\tag{1.80}$$

- skewness

$$\beta\left(A_nX^n\right)=\frac{(-1)^{n+1}\,2^{\frac{1}{2}n+1}\left(-3\Gamma\left(\frac{1}{2}n+\frac{1}{2}\right)\Gamma\left(\frac{1}{2}+n\right)\sqrt{\pi}+2\Gamma\left(\frac{3}{2}n+\frac{1}{2}\right)\pi+\Gamma\left(\frac{1}{2}n+\frac{1}{2}\right)^3\right)}{\left(-2\sqrt{\pi}\Gamma\left(\frac{1}{2}+n\right)+\Gamma\left(\frac{1}{2}+\frac{1}{2}n\right)^2\right)\sqrt{(-1)^{2n}\,2^n\left(2\sqrt{\pi}\Gamma\left(\frac{1}{2}+n\right)-\Gamma\left(\frac{1}{2}n+\frac{1}{2}\right)^2\right)}},\tag{1.81}$$

- kurtosis

$$\kappa\left(A_nX^n\right)=\frac{12\pi^{\frac{3}{2}}\Gamma\left(\frac{1}{2}n+\frac{1}{2}\right)^2\Gamma\left(\frac{1}{2}+n\right)+8\pi^{\frac{5}{2}}\Gamma\left(\frac{1}{2}+2n\right)}{\left(2\Gamma\left(\frac{1}{2}+n\right)\pi-\Gamma\left(\frac{1}{2}n+\frac{1}{2}\right)^2\sqrt{\pi}\right)^2}-\frac{16\pi^2\Gamma\left(\frac{1}{2}n+\frac{1}{2}\right)\Gamma\left(\frac{1}{2}+\frac{3}{2}n\right)-3\pi\Gamma\left(\frac{1}{2}+\frac{1}{2}n\right)^4}{\left(2\Gamma\left(\frac{1}{2}+n\right)\pi-\Gamma\left(\frac{1}{2}n+\frac{1}{2}\right)^2\sqrt{\pi}\right)^2}.\tag{1.82}$$

According to the Legendre definition, the gamma function is identified as

$$\Gamma(z) = \int_0^\infty t^{z-1} e^{-t} dt, \tag{1.83}$$

leading through a simple induction to its well-known properties

$$\Gamma(n) = (n-1)!, \quad \Gamma\left(n + \frac{1}{2}\right) = \frac{(2n-1)!!}{2^n} \sqrt{\pi}. \tag{1.84}$$

As is intuitively clear, the expected values and standard deviations are linear functions of the approximating polynomial coefficients A_i, the variances – parabolic function, third and fourth central probabilistic moments are expressed as the third and fourth powers of A_i, respectively, whereas the probabilistic coefficients are entirely independent of them; obviously they all depend on n. Since the ninth order of polynomial approximation is most frequently used in numerical analysis, it is recommended to notice that in this case the rest equals

$$\begin{aligned} O\left(E\left[A_n X^n\right], 9\right) &= 945 C_9 \sigma^8 (X) E[X] + 1260 C_9 \sigma^6 (X) E^3 [X] + 378 C_9 \sigma^4 (X) E^5 [X] \\ &+ 36 C_9 \sigma^2 (X) E^7 [X] + C_9 E^9 [X]. \end{aligned} \tag{1.85}$$

When one studies the analogous rest for the standardized X (with zero mean and unitary standard deviation), then automatically this rest is 0, while the equation adjacent to the rest in variances yields

$$\begin{aligned} O\left(\mathrm{Var}\left(A_n X^n\right), 9\right) &= 34459425 C_9^2 + 4054050 C_9 C_7 + 270270 C_9 C_5 + 20790 C_9 C_3 \\ &+ 1890 C_9 C_1. \end{aligned} \tag{1.86}$$

It is natural that the procedure described by Equations (1.75)–(1.82) depends strongly on the initial probability density function, so that an application of the lognormal variable with mean log parameter $\mu = 0$ and scale parameter $\sigma = 1$ returns these moments as

$$\begin{cases} E\left[A_n X^n\right] = C_n \exp\left(\frac{1}{2} n^2\right), \\ \mathrm{Var}(A_n X^n) = C_n^2 \left[\exp\left(2n^2\right) - \exp\left(n^2\right)\right], \\ \mu_3\left(A_n X^n\right) = C_n^3 \left[\exp\left(\frac{9}{2} n^2\right) - 3\exp\left(\frac{5}{2} n^2\right) + 2\exp\left(\frac{3}{2} n^2\right)\right], \\ \mu_4\left(A_n X^n\right) = C_n^4 \left[\exp\left(8n^2\right) - 4\exp\left(5n^2\right) + 6\exp\left(3n^2\right) - 3\exp\left(2n^2\right)\right]. \end{cases} \tag{1.87}$$

It is shown that assuming X as the lognormal variable (instead of Gaussian) returns decisively simpler equations for the probabilistic moments of the nth-order rests in polynomial approximation and, since we derive automatically full expansions, then one can expect shorter equations for this type of distribution at least. It is also clear at this point that the general recursive procedure for any type of density function and given nth-order rest and kth probabilistic central (or ordinary) moment may fail because the analytical formulas simply may not exist, even for linear transforms of most probability distributions (except the Gaussian and lognormal). It should be underlined at this point that the main objection behind the wide usage of this technique is the necessity of generating all the probabilistic moment equations by the computer system itself – a control is not always possible, while for the Gaussian variables this reduces to the values of the key parameters only. Nevertheless, there is no doubt that this method may become popular with more powerful computers and new versions of the computer algebra systems.

1.3 Time Series Analysis

Let us consider a specific class of the non-stationary random process [34], that is,

$$X(\omega,t) = \sum_{i=0}^{n} C_i(\omega)\, t^i; \qquad t \in [0,\infty),\ i \in N, \tag{1.88}$$

where $C_i(\omega)$ are all uncorrelated Gaussian random variables with specified and bounded first two probabilistic moments. Then, the expected value of this process for the time moment $\tilde{t}$ equals

$$E\left[X(\omega,\tilde{t})\right] = \sum_{i=0}^{n} E\left[C_i(\omega)\right] \tilde{t}^i. \tag{1.89}$$

The variance for this process is defined here as

$$\mathrm{Var}\left(X(\omega,\tilde{t})\right) = \sum_{i=0}^{n} \mathrm{Var}\left(C_i(\omega)\right) \tilde{t}^{2i}. \tag{1.90}$$

To develop analogous formulas for the higher probabilistic moments we notice that if

$$Y(\omega) = aX(\omega), \qquad a \in \mathcal{R} \tag{1.91}$$

then the kth central probabilistic moment is found from its definition [34] as

$$\begin{aligned}\mu_k(Y(\omega)) &= \int_{-\infty}^{+\infty} (Y(\omega) - E[Y(\omega)])^k p(\omega)\mathrm{d}\omega \\ &= \int_{-\infty}^{+\infty} (aX(\omega) - aE[X(\omega)])^k p(\omega)\mathrm{d}\omega = a^k \mu_k(X).\end{aligned} \tag{1.92}$$

Turning back to the initial time series, one obtains

$$\begin{cases} \mu_3\left(X\left(\omega,\tilde{t}\right)\right) = \sum_{i=0}^{n} \mu_3(C_i(\omega))\tilde{t}^{3i}, \\ \mu_4\left(X\left(\omega,\tilde{t}\right)\right) = \sum_{i=0}^{n} \mu_4(C_i(\omega))\tilde{t}^{4i} \end{cases} \tag{1.93}$$

and the general expression yields

$$\mu_k\left(X\left(\omega,\tilde{t}\right)\right) = \sum_{i=0}^{n} \mu_k(C_i(\omega))\tilde{t}^{ki}, \qquad k \in N. \tag{1.94}$$

The basic coefficients of variation, asymmetry, and flatness are simply determined as

$$\alpha\left(X\left(\omega,\tilde{t}\right)\right) = \frac{\sqrt{\sum_{i=0}^{n} \mathrm{Var}\left(C_i\left(\omega\right)\right)\tilde{t}^{2i}}}{\sum_{i=0}^{n} E\left[C_i\left(\omega\right)\right]\tilde{t}^{i}},$$

$$\beta\left(X\left(\omega,\tilde{t}\right)\right) = \frac{\sum_{i=0}^{n} \mu_3(C_i(\omega))\tilde{t}^{3i}}{\left(\sum_{i=0}^{n} \mathrm{Var}\left(C_i\left(\omega\right)\right)\tilde{t}^{2i}\right)^{\frac{3}{2}}},$$

$$\gamma\left(X\left(\omega,\tilde{t}\right)\right) = \frac{\sum_{i=0}^{n} \mu_4(C_i(\omega))\tilde{t}^{4i}}{\left(\sum_{i=0}^{n} \mathrm{Var}\left(C_i\left(\omega\right)\right)\tilde{t}^{2i}\right)^{2}}. \tag{1.95}$$

Since $C_i(\omega)$ are all Gaussian, only half of these moments have non-zero values, that is,

$$\mu_{2l}\left(X\left(\omega,\tilde{t}\right)\right) = \sum_{i=0}^{n} \mu_{2l}(C_i(\omega))\tilde{t}^{2li}, \qquad l \in N. \tag{1.96}$$

It can be relatively easy to prove that the random process given by Equation (1.88) is non-stationary, have continuous realizations continuous in time.

Now, we can also provide the required computational process via the Maple™ numerics similarly to the analytical calculus for the basic probabilistic moments and characteristics. This is done by considering a simple case of two linear forms analogous to Equation (1.88), namely $X = X_1 t + X_0$ and $Y = Y_1 t + Y_0$, where X_1, X_0, Y_1, Y_0 are Gaussian random variables with specified and bounded first two moments. The procedure is relatively easy, however it returns extremely long fourth-order moments and kurtosis analytical formulas and they are omitted in the expanded forms in page 42.

```
>restart: with(plots): with(plottools): with(Statistics):
>X0:=RandomVariable(Normal(EX0,SX0)): X1:=RandomVariable(Normal(EX1,SX1)):
>Y0:=RandomVariable(Normal(EY0,SY0)): Y1:=RandomVariable(Normal(EY1,SY1)):
>X:=X1*t+X0: Y:=Y1*t+Y0:
>G:=X*Y: EG:=ExpectedValue(G);
```

$$EG := (EX1\,t + EX0)(EY1\,t + EY0)$$

```
>VarG:=Variance(G);
```

$$\begin{aligned}VarG :=\ & SY0^2\,SX0^2 + SY0^2\,EX1^2\,t^2 + SY0^2\,EX0^2 + SY0^2\,t^2\,SX1^2\\ & + 2\,SY0^2\,EX1\,EX0\,t + EY0^2\,t^2\,SX1^2 + EY0^2\,SX0^2\\ & + 2\,EY1\,EX0\,SX0^2\,t + 2\,EY1\,EY0\,t^3\,SX1^2 + EY1^2\,SX0^2\,t^2\\ & + EY1^2\,t^4\,SX1^2 + t^2\,SY1^2\,SX0^2 + t^4\,SY1^2\,EX1^2 + t^2\,SY1^2\,EX0^2\\ & + t^4\,SY1^2\,SX1^2 + 2\,t^3\,SY1^2\,EX1\,EX0\end{aligned}$$

```
>StdG:=StandardDeviation(G);
```

$$\begin{aligned}StdG :=\ &\\ & (SY0^2\,SX0^2 + SY0^2\,EX1^2\,t^2 + SY0^2\,EX0^2\\ & + SY0^2\,t^2\,SX1^2 + 2\,SY0^2\,EX1\,EX0\,t + EY0^2\,t^2\,SX1^2\\ & + EY0^2\,SX0^2 + 2\,EY1\,EY0\,SX0^2\,t + 2\,EY1\,EY0\,t^3\,SX1^2\\ & + EY1^2\,SX0^2\,t^2 + EY1^2\,t^4\,SX1^2 + t^2\,SY1^2\,SX0^2 + t^4\,SY1^2\,EX1^2\\ & + t^2\,SY1^2\,EX0^2 + t^4\,SY1^2\,SX1^2 + 2\,t^3\,SY1^2\,EX1\,EX0)^{1/2}\end{aligned}$$

```
>alfaG:=Variation(G);
```

$$\begin{aligned}alfaG :=\ &\\ & \frac{1}{(EX1\,t + EX0)(EY1\,t + EY0)}(SY0^2\,SX0^2\\ & + SY0^2\,EX1^2\,t^2 + SY0^2\,EX0^2 + SY0^2\,t^2\,SX1^2\\ & + 2\,SY0^2\,EX1\,EX0\,t + EY0^2\,t^2\,SX1^2 + EY0^2\,SX0^2\\ & + 2\,EY1\,EY0\,SX0^2\,t + 2\,EY1\,EY0\,t^3\,SX1^2 + EY1^2\,SX0^2\,t^2\\ & + EY1^2\,t^4\,SX1^2 + t^2\,SY1^2\,SX0^2 + t^4\,SY1^2\,EX1^2 + t^2\,SY1^2\,EX0^2\\ & + t^4\,SY1^2\,SX1^2 + 2\,t^3\,SY1^2\,EX1\,EX0)^{1/2}\end{aligned}$$

```
>skewG:=Skewness(G);
```

$$\begin{aligned}skewG := (6(&SY0^2\,SX0^2\,EX0\,EY1\,t + SY0^2\,SX0^2\,EXt\,t^2\,EY1\\ & + SY0^2\,SX0^2\,EXt\,t\,EY0 + SY0^2\,SXt^2\,t^3\,EX0\,EY1\end{aligned}$$

$$+SY0^2 SX0^2 t^2 EX0\,EY0+SY0^2 SX1^2 t^4 EX1\,EY1$$
$$+SY0^2 SX1^2 t^3 EX1\,EY0+SY0^2 SX0^2 EX0\,EY0$$
$$+t^3 SY1^2 SX0^2 EX0\,EY1+t^4 SY1^2 SX0^2 EX1\,EY1$$
$$+t^3 SY1^2 SX0^2 EX1\,EY0+t^5 SY1^2 SX1^2 EX0\,EY1$$
$$+t^4 SY1^2 SX1^2 EX0\,EY0+t^6 SY1^2 SX1^2 EX1\,EY1$$
$$+t^5 SY1^2 SX1^2 EX1\,EY0+t^2 SY1^2 SX0^2 EX0\,EY0))/$$
$$(SY0^2 SX0^2+SY0^2 EX1^2 t^2+SY0^2 EX0^2+SY0^2 t^2 SX1^2$$
$$+2SY0^2 EX1\,EX0\,t+EY0^2 t^2 SX1^2+EY0^2 SX0^2$$
$$+2EY1\,EY0\,SX0^2 t+2EY1\,EY0\,t^3 SX1^2+EY1^2 SX0^2 t^2$$
$$+EY1^2 t^4 SX1^2+t^2 SY1^2 SX0^2+t^4 SY1^2 EX1^2+t^2 SY0^2 EX0^2$$
$$+t^4 SY1^2 SX1^2+2t^3 SY1^2 EX1\,EX0)^{3/2}$$

```
>kappaG:=Kurtosis(G):
>m3G:=CentralMoment(G,3);
```

$$m3G:=6SY0^2 SX0^2 EX0\,EY0+6SY0^2 SX0^2 EX0\,EY1\,t$$
$$+6SY0^2 SX0^2 EX1\,t^2 EY1+6SY0^2 SX0^2 EX1\,t\,EY0$$
$$+6SY0^2 SX1^2 t^2 EX0\,EY0+6SY0^2 SX1^2 t^3 EX1\,EY0$$
$$+6SY0^2 SX1^2 t^3 EX0\,EY1+6SY0^2 SX1^2 t^4 EX1\,EY1$$
$$+6t^2 SY1^2 SX0^2 EX0\,EY0+6t^3 SY1^2 SX0^2 EX0\,EY1$$
$$+6t^4 SY1^2 SX0^2 EX1\,EY1+6t^3 SY1^2 SX0^2 EX1\,EY0$$
$$+6t^4 SY1^2 SX1^2 EX0\,EY0+6t^5 SY1^2 SX1^2 EX1\,EY0$$
$$+6t^5 SY1^2 SX1^2 EX0\,EY1+6t^6 SY1^2 SX1^2 EX1\,EY1$$

```
>m4G:=CentralMoment(G,4):
```

Example 1.5: Probabilistic characteristics for the power law ageing scheme

We consider a specific class of time series with Gaussian random coefficients of special importance in reliability engineering, where $v(t)=ct^b$ describes the total number of failures appearing in some system (detailed information on applications of time series and time-dependent reliability may be found in [1, 166]). This function is tested in civil engineering at least with a few different powers b resulting from experimental verification, so our computational example focuses on $E[b]=\{2, 1, 0.5, 0.4, 0.125\}$ having all 10% standard deviations; additionally, the coefficient c takes expectation equal to 10 and the same level of standard deviation. We compare, as usual, the first four probabilistic moments and coefficients to monitor numerically not only the first two statistics necessary in the reliability

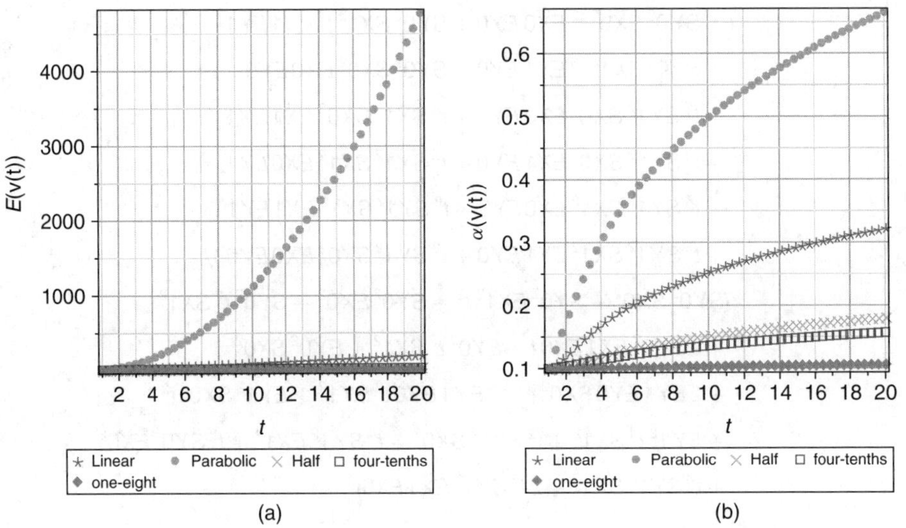

Figure 1.21 The expected values (a) and coefficients of variation (b) for $\nu(t)=ct^b$

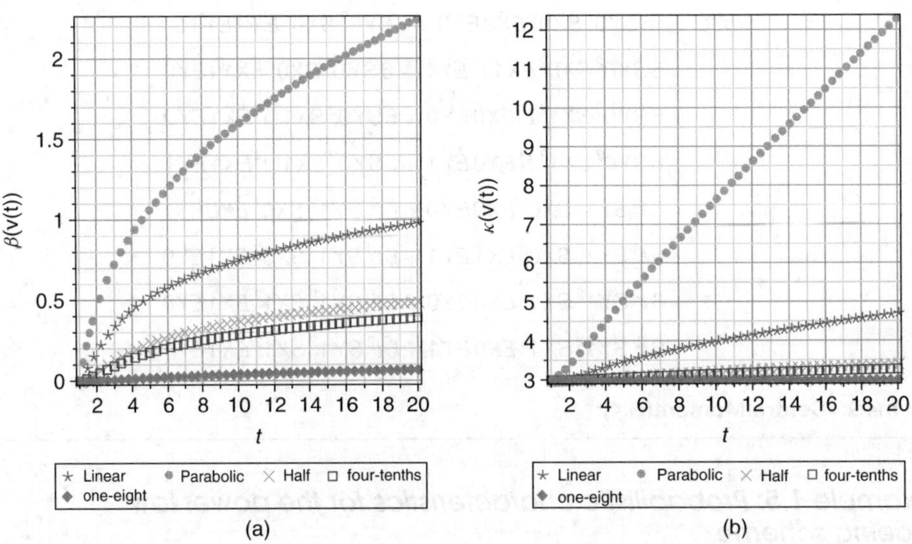

Figure 1.22 The skewness (a) and kurtosis (b) for $\nu(t)=ct^b$

index determination but also fluctuations of the probability distribution. The data are contained in Figures 1.21 and 1.22 as expectations, coefficients of variation, skewness, and kurtosis – all being functions of time.

Essentially, the largest values and the most apparent variations are noticed for the parabolic function here, where all the moments increase in time. The linear time series with $E[b]=1.0$ shows also that time increases everywhere, but this is not so evident in the remaining models. None of the models tested is Gaussian, since there is positive kurtosis significantly larger than 0. Let us note that the resulting coefficient of variation is always higher than the input one depending on the model considered. The skewness is positive in all cases, giving a PDF close to the symmetric one for $E[b] = 0.125$ only. In conclusion, we observe that higher-order analysis will usually be necessary in engineering computations with such time series, which is partially confirmed in the next numerical example.

Example 1.6: Probabilistic and stochastic elastoplastic stability analysis

Let us consider for the simple bar a stability loss above the proportionality limit according to Hooke's law, where inelastic (non-linear) strains are observed. We assume that the compressive force has an entirely static character. The non-linearity in the constitutive relation diagram is replaced with a linear relation uniquely characterized by the tangent modulus e_t such that

$$e_t = \frac{d\sigma}{d\varepsilon}. \tag{1.97}$$

We also assume that the deformed cross-section is planar and then, the neutral axis remains straight, which means

$$\varepsilon' = \frac{y}{\rho}, \tag{1.98}$$

where y is an axis perpendicular to the neutral one, while ρ stands for the curvature radius. The constitutive relations for the compressed (A_1, $y < 0$) and tensioned (A_2, $y > 0$) parts of the considered cross-section are given respectively as

$$\sigma'_1 = e_t \frac{y}{\rho}; \qquad \sigma'_2 = e\frac{y}{\rho}. \tag{1.99}$$

Then, we provide the equilibrium equations in this system and it holds that

$$\begin{cases} \int\limits_{A_2} \sigma_1 \; dA + \int\limits_{A_2} \sigma_2 \; dA = P, \\ \int\limits_{A_1} \sigma_1 y \; dA + \int\limits_{A_2} \sigma_2 y \; dA = M \end{cases} \tag{1.100}$$

with P being a compressive force and M standing for the resulting bending moment. The next trivial substitutions

$$P = \sigma_0 A, \quad \sigma_1 = \sigma_0 + \sigma'_1, \qquad \sigma_2 = \sigma_0 + \sigma'_2 \tag{1.101}$$

together with the constitutive relations (1.99) lead to the following equation system:

$$\begin{cases} e_t \int\limits_{A_1} y \; dA + e \int\limits_{A_2} y \; dA = 0, \\ e_t \frac{1}{\rho} \int\limits_{A_1} y^2 \; dA + e \frac{1}{\rho} \int\limits_{A_2} y^2 \; dA = M. \end{cases} \tag{1.102}$$

Recognizing the inertia moments in Equation (1.102) and denoting them as J_1 and J_2, correspondingly, we simply obtain the additional modulus accompanying this phenomenon e' as

$$e' = \frac{e_t J_1 + e J_2}{J}. \tag{1.103}$$

Using further a classical equation for the Euler stability limit

$$e' J \frac{d^2 y}{dx^2} + P y = 0, \tag{1.104}$$

one obtains (similarly to the well-known formula) the modified equation for the critical stresses

$$\sigma_{cr} = \frac{\pi^2 e'}{\lambda^2}, \tag{1.105}$$

which is frequently contrasted with the Tetmajer–Jasiński formula

$$\sigma_{cr}^{TJ} = a - b\lambda \tag{1.106}$$

with experimentally given parameters a, b as well as with the Johnson–Ostenfeld theory, where

$$\sigma_{cr}^{JO} = A - B\lambda^2; \tag{1.107}$$

with A and B some material parameters determined experimentally in the form of engineering statistics. It is apparent after Equation (1.103) that the modulus e' is strongly affected by the initial shape of the cross-section and is different for the rectangular prismatic beams and for the I-profiles. A derivation of e' for the last type of cross-section is provided below, because a probabilistic Taylor expansion is based on its partial derivatives with respect to the web and flange thicknesses calculated in the traditional deterministic way. We assume that they are Gaussian variables having the expectations equal to $E[t_f]$ and $E[t_w]$, respectively, and the same coefficient of variation α. The aforementioned elastoplastic models are also contrasted with three-parameter models, where the constitutive relation is proposed as

$$\sigma = e\varepsilon - \frac{1}{2}\frac{\left(4e\varepsilon_0 - 5f_y\right)\varepsilon^3}{\varepsilon_0^3} + \frac{1}{2}\frac{\left(2e\varepsilon_0 - 3f_y\right)\varepsilon^5}{\varepsilon_0^5} \tag{1.108}$$

with tangent modulus defined as

$$e_t = e - \frac{3}{2}\frac{\left(4e\varepsilon_0 - 5f_y\right)\varepsilon^2}{\varepsilon_0^3} + \frac{5}{2}\frac{\left(2e\varepsilon_0 - 3f_y\right)\varepsilon^4}{\varepsilon_0^5}. \tag{1.109}$$

where f_y is of course plastic stress.

We examine three values of plastic deformation – in turn $\varepsilon_0 = \left\{\frac{3f_y}{2e}, \frac{5f_y}{4e}, \frac{15f_y}{8e}\right\}$ resulting in various orders of stress functions and different final tangent moduli. We derive the tenth-order approximation for the expected values of the critical force in the Engesser model as

$$E\left(P_{cr}\right)=\frac{1}{(e+\beta e)\,l^2}\left(\pi^2\beta e^2\left(\frac{1}{3}b_fE^3\left[t_f\right]+b_fE\left[t_f\right]\left(E\left[t_f\right]+b_w\right)^2\right)+\frac{1}{12}E\left[t_w\right]b_w^3\right)$$
$$+\frac{\pi^2\beta e^2\left(2b_fe\left[t_f\right]+2b_f\left(E\left[t_f\right]+b_w\right)\right)E^2\left[t_f\right]}{(e+\beta e)\,l^2}\alpha^2. \tag{1.110}$$

We denote by l the length of the structural element, while b_f and b_w stand for the widths of the flange and the web.

Consequently, we obtain the additional sixth-order formula for its variance. It holds that

$$\mathrm{Var}\left(P_{cr}\right)=\frac{\pi^4e^2\alpha^2E^2\left[t_f\right]}{l^4}\left(\frac{1}{2}b_fE^2\left[t_f\right]+\frac{1}{2}b_f\left(E\left[t_f\right]+b_w\right)^2\right.$$
$$\left.+E\left[t_f\right]b_f\left(E\left[t_f\right]+b_w\right)\right)^2$$
$$+\frac{\pi^4e^2b_w^2E^2\left[t_w\right]}{144\ l^4}\alpha^2+\frac{3\alpha^4E^4\left[t_f\right]}{4l^4}\left(\pi^4e^2\left(2b_fE\left[t_f\right]+b_f\left(E\left[t_f\right]+b_w\right)\right)^2\right)$$
$$+\frac{3\alpha^4E^4\left[t_f\right]}{3l^4}\left(\pi^4e^2\left(\frac{1}{2}b_fE^2\left[t_f\right]+\frac{1}{2}b_f\left(E\left[t_f\right]+b_w\right)^2+b_fE\left[t_f\right]\left(E\left[t_f\right]+b_w\right)\right)b_f\right)$$
$$+\frac{20}{3}\frac{\alpha^6E^6\left[t_f\right]\pi^4e^2b_f^2}{l^4}. \tag{1.111}$$

Quite analogously one finds the third central probabilistic moment for the stability limit as

$$\mu_3\left(P_{cr}\right)=\frac{960\alpha^6\pi^6E^6\left[t_f\right]e^6\beta^3\left(b_fE\left[t_f\right]++b_fE\left[t_f\right]+b_fb_w\right)^3}{(e+\beta e)^3\,l^6}$$
$$+\left(b_fE\left[t_f\right]+b_fE\left[t_f\right]+b_fb_w\right)$$
$$\times\frac{72\alpha^4\pi^6E^4\left[t_f\right]e^6\beta^3\left(\frac{1}{2}b_fE^2\left[t_f\right]+\frac{1}{2}b_f\left(E\left[t_f\right]+b_w\right)^2+b_fE\left[t_f\right]\left(E\left[t_f\right]+b_w\right)\right)^2}{(e+\beta e)^3\,l^6} \tag{1.112}$$

and, finally, the fourth central probabilistic moment

$$\mu_4\left(P_{cr}\right)=\frac{48\alpha^4E^4\left[t_f\right]\pi^8e^8\beta^4}{(e+\beta e)^4\,l^8}$$

$$\times\left(\frac{1}{2}b_fE^2\left[t_f\right]+\frac{1}{2}b_f\left(E\left[t_f\right]+b_w\right)^2+b_fE\left[t_f\right]\left(E\left[t_f\right]+b_w\right)\right)^4$$

$$+\frac{360\alpha^6E^6\left[t_f\right]\pi^8e^8\beta^4}{(e+\beta e)^4l^8}\left(\frac{1}{2}b_fE^2\left[t_f\right]+\frac{1}{2}b_f\left(E\left[t_f\right]+b_w\right)^2\right.$$

$$\left.+b_fE\left[t_f\right]\left(E\left[t_f\right]+b_w\right)^2\left(2b_fE\left[t_f\right]+b_f\left(E\left[t_f\right]+b_w\right)\right)^2\right). \quad (1.113)$$

Finally, having two Gaussian random variables with specified first two probabilistic moments, we derive the additional central probabilistic moments of the power time series with non-integer power, that is,

$$d\left(t\right)=A\left(\omega\right)t^{B(\omega)}. \quad (1.114)$$

We would like to use this expression to model stochastic time fluctuations of the flange and the web's thicknesses due to the corrosion, for instance.

Provided that $A(\omega)$ and $B(\omega)$ are uniquely defined by the moments $E[A]$ and $E[B]$ as well as $\sigma(A)$ and $\sigma(B)$, one gets time fluctuations of the expectation as

$$E\left[d\left(t\right)\right]=E\left[A\right]\exp\left(-\frac{1}{2}\frac{E^2\left[B\right]}{\sigma^2(b)}\right)\exp\left(\frac{1}{2}\frac{\left(E\left[B\right]+\ln\left(t\right)\sigma^2(b)\right)^2}{\sigma^2(b)}\right) \quad (1.115)$$

and the variances in the following form:

$$\mathrm{Var}\left(d\left(t\right)\right)=E^2\left[A\right]t^{2E[B]}\left\{\exp\left(2\ln^2\left(t\right)\sigma^2(b)\right)-\exp\left(\ln^2\left(t\right)\sigma^2(b)\right)\right\}$$

$$+\sigma^2\left(A\right)t^{2E[B]}\exp\left(2\ln^2\left(t\right)\sigma^2(b)\right). \quad (1.116)$$

The third and fourth central probabilistic moments are derived as

$$\mu_3\left(d\left(t\right)\right)=E^3\left[A\right]t^{3E[B]}\left\{\exp\left(\frac{9}{2}\ln^2\left(t\right)\sigma^2(b)\right)-3\exp\left(\frac{5}{2}\ln^2\left(t\right)\sigma^2(b)\right)\right\}$$

$$+2E^3\left[A\right]t^{3E[B]}\exp\left(\frac{3}{2}\ln^2\left(t\right)\sigma^2(b)\right)$$

$$+3E\left[A\right]\sigma^2\left(A\right)t^{3E[B]}\left\{\exp\left(\frac{9}{2}\ln^2\left(t\right)\sigma^2(b)\right)-\exp\left(\frac{5}{2}\ln^2\left(t\right)\sigma^2(b)\right)\right\} \quad (1.117)$$

and

$$\mu_4\left(d\left(t\right)\right)=E^4\left[A\right]t^{4E[B]}\exp\left(8\ln^2\left(t\right)\sigma^2(b)\right)$$

$$+6E^2\left[A\right]\sigma^2\left(A\right)t^{4E[B]}\exp\left(3\ln^2\left(t\right)\sigma^2(b)\right)-4E^4\left[A\right]t^{4E[B]}\exp\left(5\ln^2\left(t\right)\sigma^2(b)\right)$$

$$+3\sigma^4\left(A\right)t^{4E[B]}\exp\left(8\ln^2\left(t\right)\sigma^2(b)\right)+6E^4\left[A\right]t^{4E[B]}\exp\left(3\ln^2\left(t\right)\sigma^2(b)\right)$$

$$- 3E^4 [A] t^{4E[B]} \exp\left(2\ln^2(t)\,\sigma^2(b)\right) + 6E^2 [A]\,\sigma^2(A)\, t^{4E[B]} \exp\left(8\ln^2(t)\,\sigma^2(b)\right)$$
$$- 12E^2 [A]\,\sigma^2(A)\, t^{4E[B]} \exp\left(5\ln^2(t)\,\sigma^2(b)\right). \tag{1.118}$$

Having derived these characteristics, the statistical data from the Table 1.1 are applied to verify the influence of the stochastic corrosion on the critical force of this profile. Probabilistic numerical analysis was carried out entirely using the system Maple™, where an elastoplastic critical force determination algorithm for I-profiles has been implemented. Symbolic derivation enabled us to obtain this critical force in the deterministic case as

$$P_{cr} = \frac{\pi^2 T J}{l^2}, \qquad T = \frac{2\beta e}{1+\beta} \tag{1.119}$$

where $\beta = 0.01$ is substituted. This result clearly differs from a formula valid for the rectangular cross-section. Furthermore, it has been enriched with symbolic differentiation procedures enabling a final determination of the probabilistic moments and characteristics of the critical force according to the Engesser theory, of course. Such a randomized computational procedure tested on all HEB profiles is extended then to stochastic analysis of a corrosion process decreasing the effective cross-sectional area for all profiles, where time-dependent stochastic fluctuations are analyzed in the stability limit. Let us note that this stability problem may be generalized in a deterministic context with the simultaneous solution of three coupled partial differential equations of double deflection in two perpendicular planes and with a single twisting angle.

Probabilistic analysis deals with the engineering case, where the probabilistic moments and coefficients of the critical force in axial compression for the steel HEB 6.0 m long bar are computed. Because this is a preliminary study just before full stochastic analysis, we examine the situation where the thicknesses of both flange and webs are input truncated Gaussian random variables with specified expectations (following engineering tables for the HEBs). The random dispersion expressed here uniquely by the coefficient of variation α has exactly the same value and is taken as the additional parameter in this study; shown on the horizontal axis of all diagrams. The expected values (Figure 1.23), standard deviations (Figure 1.24), coefficients of variation (Figure 1.25), skewness coefficients (Figure 1.26), and kurtosis (Figure 1.27) are computed with respect to $\alpha(t) \in [0.0, 0.20]$. All the results are subdivided into two separate data for brevity of presentation – for HEB100 until HEB300 and, separately, for HEB300 until HEB1000. The profile HEB300 is taken specifically since the web width remains the same and the height increases only for larger HEBs, so that the critical behavior should be somewhat different for this second group than for smaller HEBs.

As one could expect on the basis of the previous results, the expected values are totally independent of the coefficient $\alpha(t)$ when compared with each other (on the scale of all profiles taken together). They usually exhibit some variations when a single profile is analyzed, but these fluctuations are about the second or

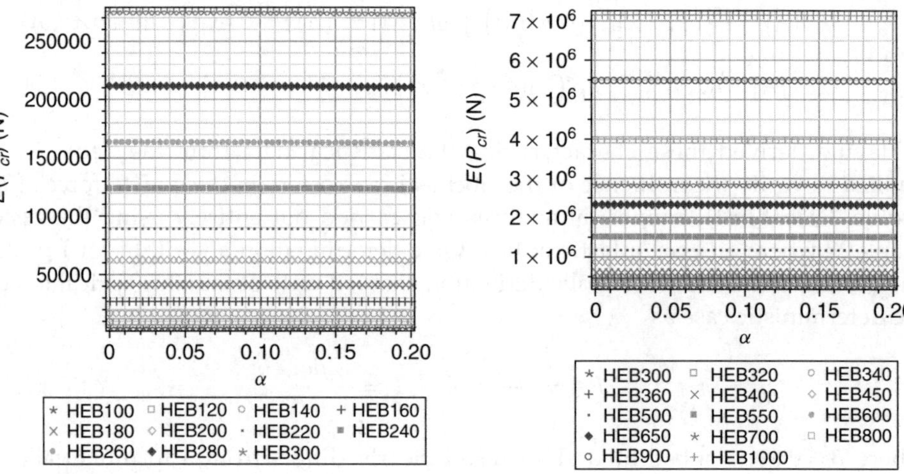

Figure 1.23 The expected values of the critical forces for HEB profiles

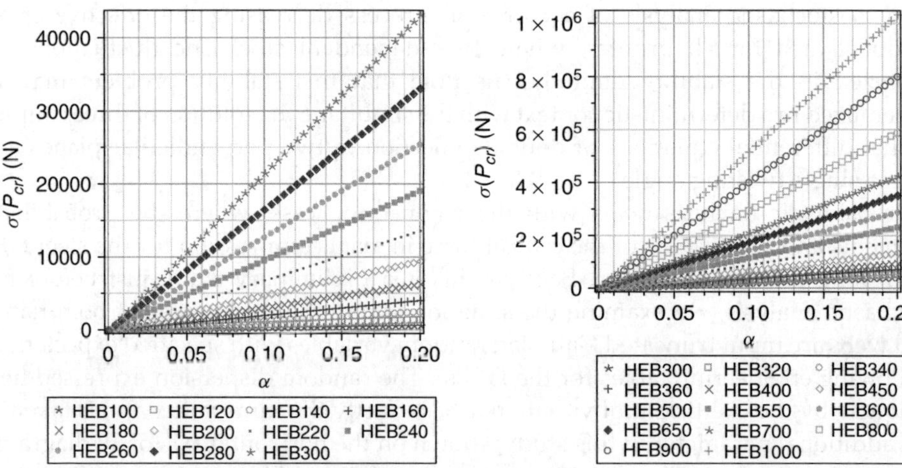

Figure 1.24 Standard deviations of the critical forces for HEB profiles

even third order of importance. Generally, as one may expect, the larger the HEB profile, the greater the value of the critical force expectation; they increase non-proportionally together with the given HEB identification number. The standard deviations (Figure 1.24) are plotted next, also as the functions of $\alpha(t)$, because they are necessary in all the reliability index models. These deviations increase linearly together with an input coefficient of variation and, as a rule, the larger the HEB, the greater the value of the computed deviation. Of course, they all start from 0 for $\alpha(t) = 0$ as this corresponds to the deterministic situation, where no fluctuation is

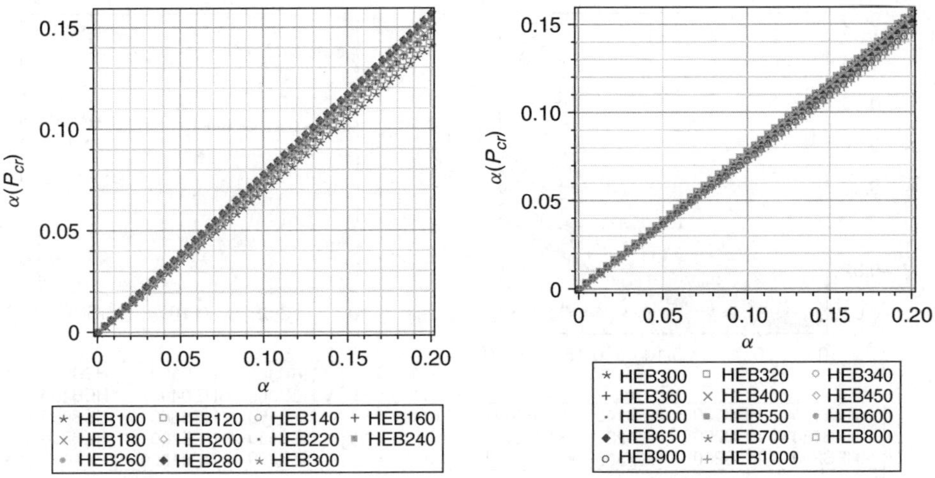

Figure 1.25 Coefficients of variation of the critical forces for HEB profiles

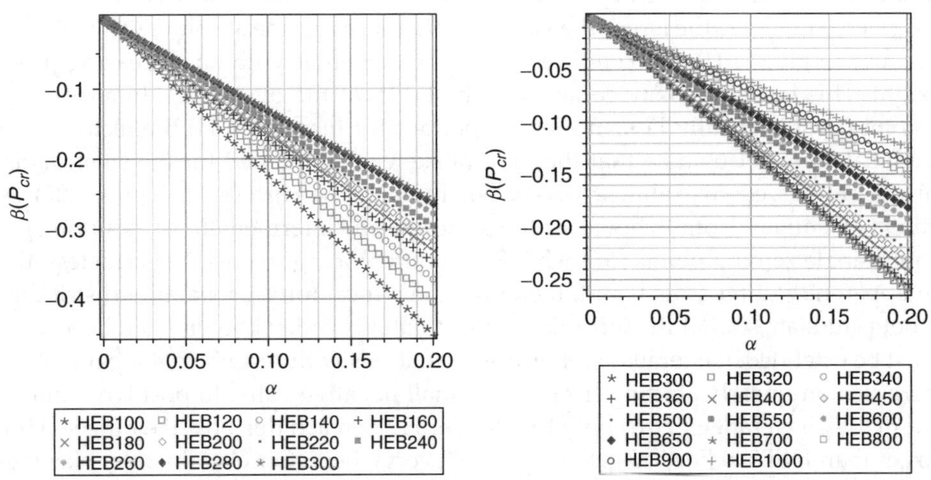

Figure 1.26 Coefficients of skewness of the critical forces for HEB profiles

noticed. Even more informative is Figure 1.25, where the $\alpha(P_{cr})$ values are collected. They generally must show analogous tendencies to the standard deviations, but the functions obtained for neighboring HEBs are definitely closer than before. It is clear on the basis of the detailed numerical results that the critical force model has some probabilistic damping properties since the ratio of the output to input coefficients of variation is decisively smaller than unity. Nevertheless, a linear interrelation is noticed everywhere starting from 0 and increasing together with $\alpha(t)$ and the HEB size. The coefficients of skewness given in Figure 1.26 demonstrate almost opposite

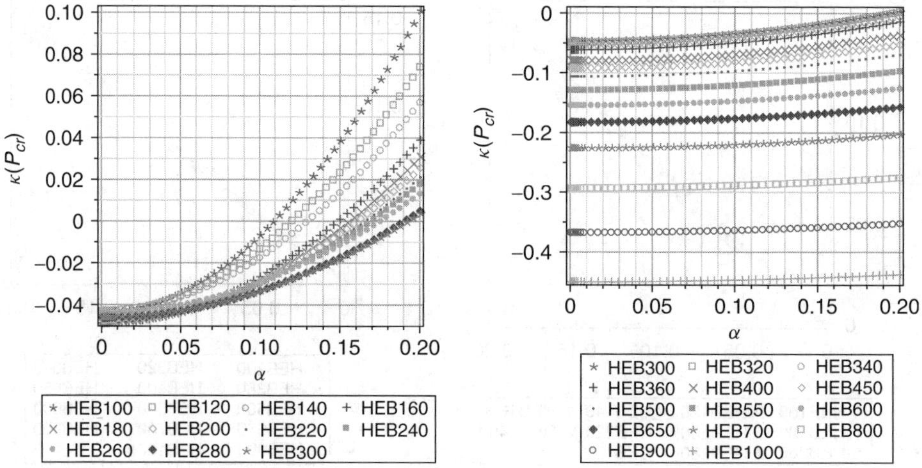

Figure 1.27 Kurtosis of the critical forces for HEB profiles

behavior – they are all negative, while also linearly dependent on $\alpha(t)$. According to the given equations, the larger the value of $\alpha(t)$, the larger the absolute value of $\beta(P_{cr})$ but for the given $\alpha(t)$ this coefficient decreases together with an increasing profile height. This last effect may come from the fact that the overall randomness effect becomes systematically less and less important for increasing HEB effective cross-sectional area. Let us note that the ratio of skewness obtained for the smallest and largest profiles, equal to almost 4, is extremely large. The kurtosis (cf. Figure 1.27) also differs essentially from 0, however for some specific intermediate values of $\alpha(t)$ we notice single zeroing points for each HEB for different arguments. Nevertheless, they all apparently differ from 0 and, therefore, the critical force probability distribution is not Gaussian, so that the formula for the reliability index known from Eurocode 0 must be extended. Generally, all kurtosis' variations with respect to the parameter α increase non-linearly starting from some small negative value to positive numbers, but in the given range of $\alpha(t)$ only the smaller HEBs reach this positive range – HEBs larger than HEB300 keep negative kurtosis everywhere. Once again, after skewness collection, we notice that the larger the HEB, the smaller the kurtosis. The data collected here may be used directly in probabilistic engineering reliability of any order for steel columns under axial compression and, after small modifications of the input data, for all metallic and composite bisymmetric and thin-walled bars. The limit function, contrary to the ultimate strength and allowable deflection limit states, would consist of a difference between the acting force(s) and the critical one and may appear crucial for overall column safety.

The second group of computational experiments is devoted to stochastic analysis of the corrosion process and its influence on the critical force probabilistic characteristics. It is assumed that both webs and flange are subjected to uniform loss of thickness d treated as a truncated Gaussian variable with expectations and standard deviations as functions of time given in Table 1.1; other data have not been considered since they contain efficiently smaller standard deviations, so that the final random dispersion would also not be so influential. The corrosion reduces the I-profile thicknesses on both sides of the webs and flange, and hence

$$\begin{cases} t_w(\omega; t) = t_w^0 - 2d(\omega; t), \\ t_f(\omega; t) = t_f^0 - 2d(\omega; t) \end{cases} \tag{1.120}$$

so that we have their first two moments as

$$\begin{cases} E\left[t_w(\omega; t)\right] = t_w^0 - 2E\left[d(\omega; t)\right], \\ \mathrm{Var}\left(t_w(\omega; t)\right) = 4\ \mathrm{Var}\left(d(\omega; t)\right) \end{cases} \tag{1.121}$$

and analogously for the flange corrosion. The data from Table 1.1 form the basis of the spline approximation of the first two probabilistic moments for the penetration depth. The remaining derivations with use of the stochastic perturbation technique are also carried out in the computer algebra system Maple™, and all probabilistic parameters of the critical force are determined as continuous functions of the exploitation time t; these time fluctuations are computed for $t \in [0.0, 5.0\ \mathrm{yr}]$ to correspond precisely with the experimental data on corrosion [128].

The expected values (Figure 1.28), standard deviations (Figure 1.29), coefficients of variation (Figure 1.30), coefficients of skewness (Figure 1.31), and kurtosis (Figure 1.32) as well as, finally, entropy variations (Figure 1.33) are all calculated for the entire family of HEB profiles and, once more, divided into two groups – HEB100 to HEB300 (left-hand graphs) as well as HEB300 to HEB1000 (right-hand graphs). As is intuitively clear, the expected values of the critical force for all the profiles decrease over time and even for the first five years of their exploitation by the few percentage points. The standard deviations and coefficients of variation are strongly non-linear and, starting from a zero value at $t=0$, reach some maximum for about $t=1.3$, then once more a minimum at $t=3.7$ to monotonously increase further. Similarly to the expectations, the larger the height of the HEB profile, the greater standard deviation for all t within a time domain; this relation is quite opposite to that for the coefficient of variation. It is apparent that both standard deviations and coefficients of variation are significantly smaller than in the previous computational example. The basic reason is that only a very small part of the cross-sectional thickness is really random during

Table 1.1 Experimental statistics on the first two moments of the corrosion rate

Observation time, t (years)	Mean corrosion depth, d (mm)	Standard deviation of the corrosion depth at time t (mm)	Coefficient of variation about the mean in time t
0.5	0.075	0.0373	0.50
1.0	0.144	0.0674	0.47
3.0	0.292	0.0614	0.21
5.0	0.421	0.0470	0.12

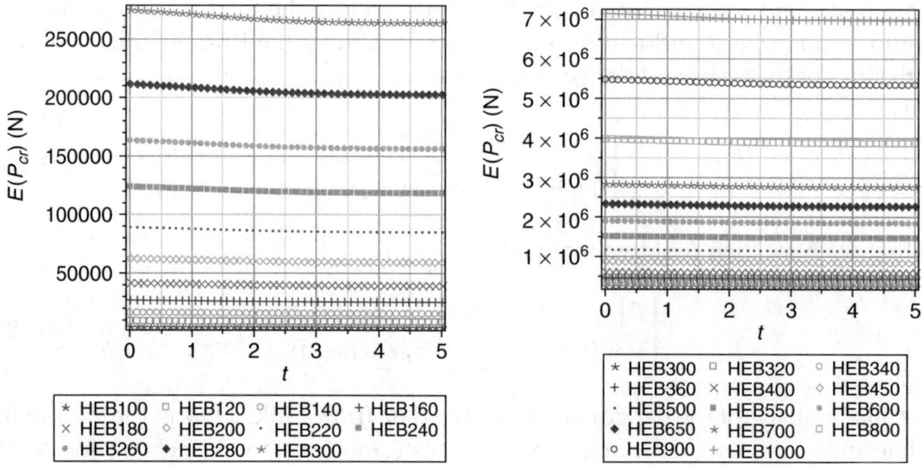

Figure 1.28 The expected values of the critical forces for HEB profiles

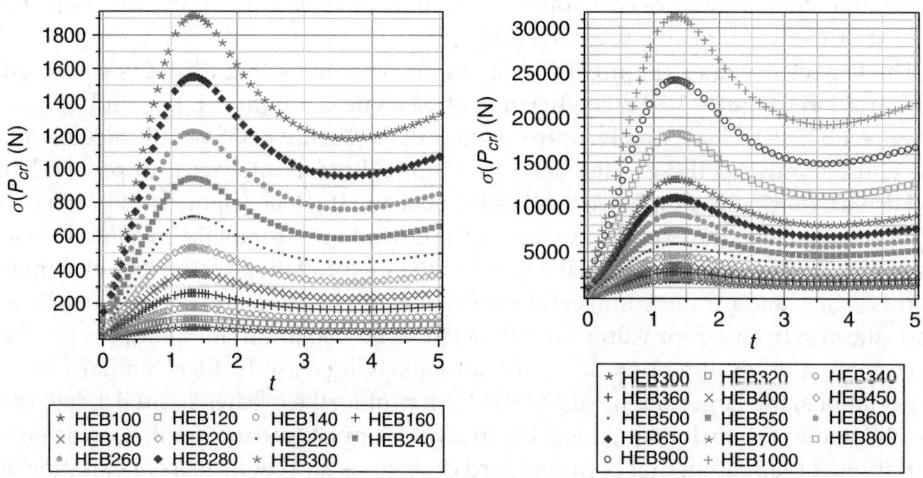

Figure 1.29 Standard deviations of the critical forces for HEB profiles

simulation of the corrosion, while the entire cross-section was a random variable previously. Then, the smaller the size of the HEB, the larger coefficient of variation of the critical force for the bars subjected to such corrosion – almost three times larger for HEB100 than for HEB1000.

The coefficients of skewness and kurtosis show time fluctuations very similar in shape to the coefficient of variation, with negative values of the skewness as the main exception (and some short interval of negative kurtosis also). Now, the absolute values of all probabilistic coefficients increase together with an additional decrease of the HEB profile size. The local extrema of the standard deviations, coefficients

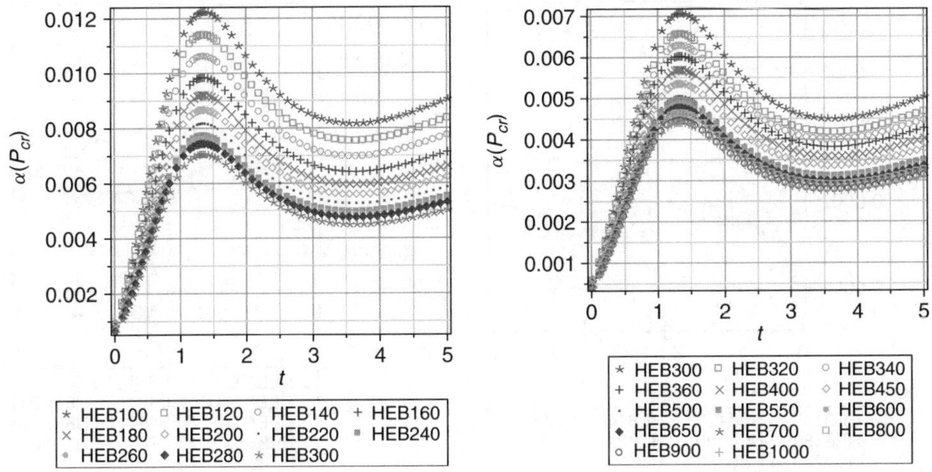

Figure 1.30 Coefficients of variation of the critical forces for HEB profiles

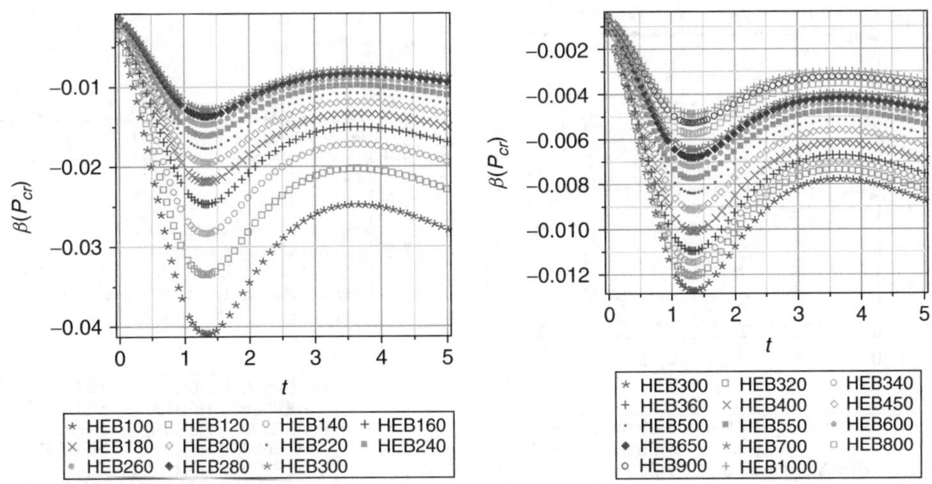

Figure 1.31 Coefficients of skewness of the critical forces for HEB profiles

α, β, and κ for the critical force coincide at the arguments specified above; their shape variations also look really similar. Now, contrary to the previous example, one can conclude that the critical force for the structural elements under uniform stochastic corrosion has Gaussian realizations for different exploitation moments t at the very beginning of this process. So, determination of the time-dependent reliability index may proceed efficiently using the first two probabilistic moments and simplified formula from Eurocode 0. It is necessary, however, to point out that uniform corrosion is not a common mechanism in all steel engineering structures,

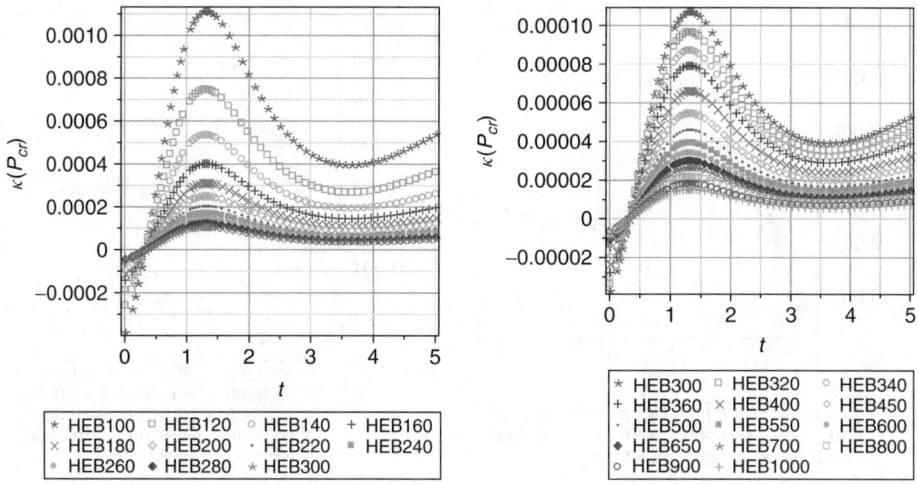

Figure 1.32 Kurtosis of the critical forces for HEB profiles

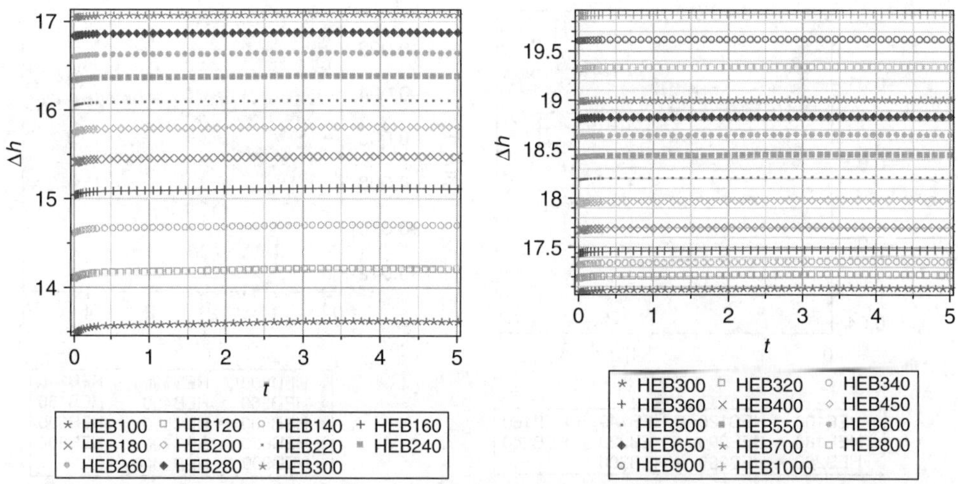

Figure 1.33 Entropy of the critical forces for HEB profiles

where sometimes the lower web and lower part of the flange are corroded most (in bridge decks especially) and may even result in local holes close to the supports. A conclusion on the normality of $P_{cr}(\omega; t)$ is necessary for the final entropy variations determination following the statement (1.9), whose computations are presented in Figure 1.33, also in the form of time fluctuations. These are almost linear and have constant positive values. It is evident that the larger the size of the HEB profile, the greater entropy change in the perturbation-based calculations of the critical force under corrosion. Effectively, the entropy accompanying $P_{cr}(\omega; t)$ is larger than that associated with the stochastic corrosion process itself, so that the uncertainty of

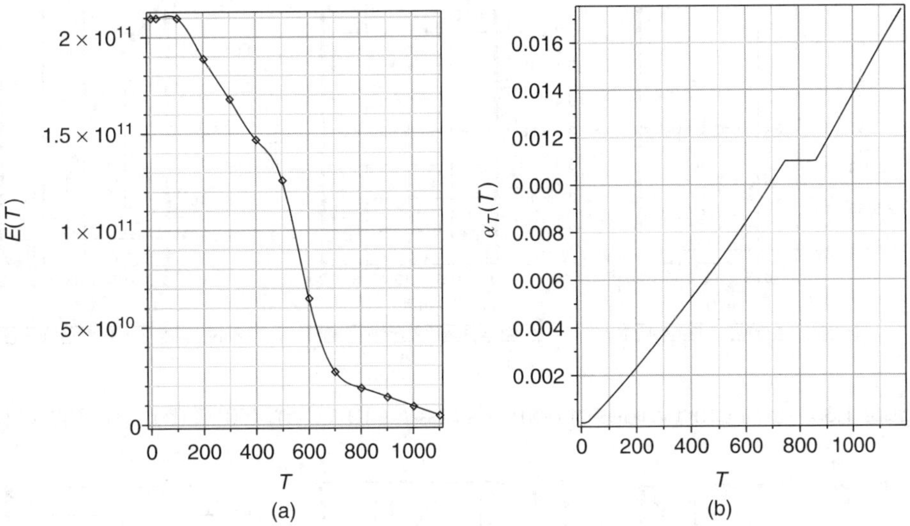

Figure 1.34 Temperature variations of structural steel Young's modulus (a) and coefficient of thermal expansion (b)

notification of the specific critical force value is larger than the analogous uncertainty for the corrosion penetration depth.

Example 1.7: Probabilistic thermal stresses in the heated bar

This numerical example deals with a simple steel bar fixed at both ends subjected along its length to a constant temperature with expectations specified by Eurocode 3 and belonging to the range of fire. The normal force within the bar is equal to $N(T)=e(T)\ A\ \alpha_T(T)\ \Delta T$; we may assume with relatively good accuracy that the temperature is constant in the system and this is the basis to treat it as a random variable. Using the spline method applied to some experimental data on $e=e(T)$, we provide the following second-order polynomial approximation: $e(T)=aT^2+bT+c$ according to the data included in Eurocode 3 for the structural steel; this also affects some perturbation-based equations. Material properties of the steel in the fire temperature range, like Young's modulus as well as the coefficient of thermal expansion, are given in Eurocode 3; the cross-sectional area of the bar is assumed to be $A=0.00391\ \mathrm{m}^2$. According to this code, we need to provide a linear interpolation, however, considering the need for an efficient partial differentiation some parabolic functions between the trial points are provided. The coefficient of linear expansion is defined using interval functions for the same input temperature range (see Figure 1.34). It is necessary to emphasize here the very fast loss of Young's modulus value for the temperature range $T=500$–$700\,^\circ$C, which strongly affects all further probabilistic computations.

First of all we compute probabilistic analytical solutions to the classical formulas for thermal stresses and strains known from the strength of materials via symbolic integration with use of the Gaussian probability density function of the temperature (marked with diamonds in Figures 1.35–1.42). Its coefficient of variation $\alpha(T)$ marked

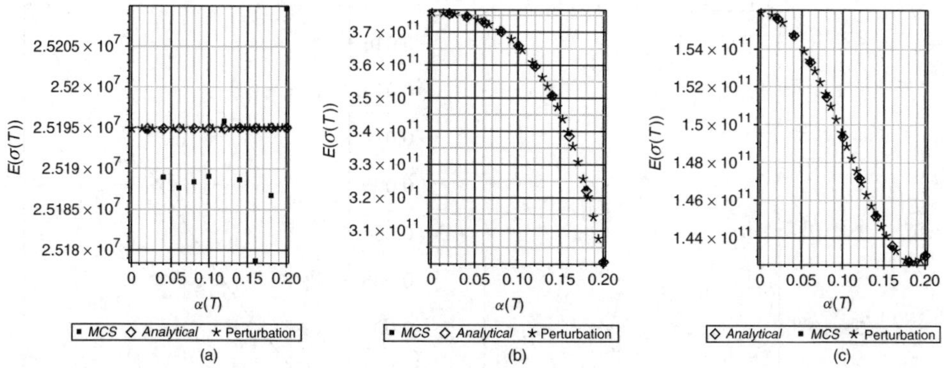

Figure 1.35 Expected values of normal stress $T = 10\,^\circ\mathrm{C}$ (a), $450\,^\circ\mathrm{C}$ (b), and $850\,^\circ\mathrm{C}$ (c)

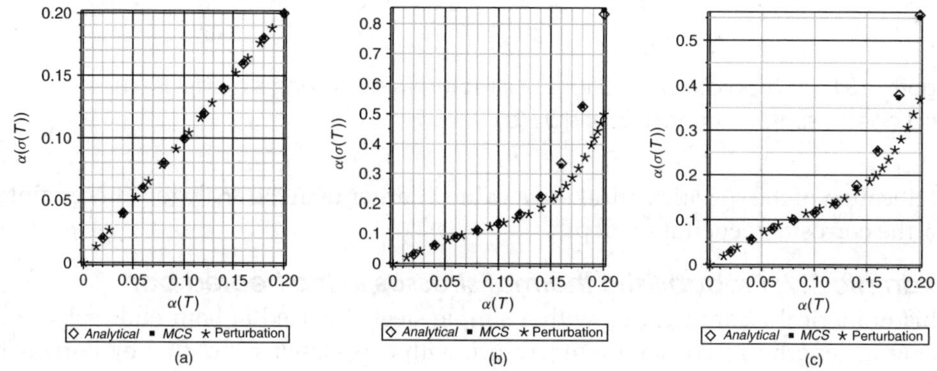

Figure 1.36 Coefficient of variation of normal stress $T = 10\,^\circ\mathrm{C}$ (a), $450\,^\circ\mathrm{C}$ (b), and $850\,^\circ\mathrm{C}$ (c)

on the horizontal graphs is the additional input variable in this study. Further, we compare these results against the Monte Carlo simulation experiments also obtained in the package Maple™ for the set of coefficients of variation $\alpha = [0.02, 0.04, \ldots, 0.20]$. Finally, we provide stochastic perturbation-based computations using the tenth order for expectations, and sixth order for coefficients of variation as well as the least-order approximants for the third and fourth central moments needed for skewness and kurtosis determination (the series marked with stars).

A comparison of the data summarized in Figure 1.35 allows us to notice that the thermal stress expectations increase from a minimum value for $T = 10\,^\circ\mathrm{C}$ to a maximum for about $T = 450\,^\circ\mathrm{C}$ (around four orders larger) and then slowly decrease through the values computed for $T = 850\,^\circ\mathrm{C}$ (only two times smaller than the maximum). Initially the expectations are quite insensitive to the input coefficients of variation – some small fluctuations are almost in the range of numerical error. For larger temperatures the higher the random dispersion of the temperature, the smaller the expectation; which is more important – all the probabilistic methods return the same results for $\alpha \in [0.0, 0.2]$. The coefficient of variation of thermal stresses

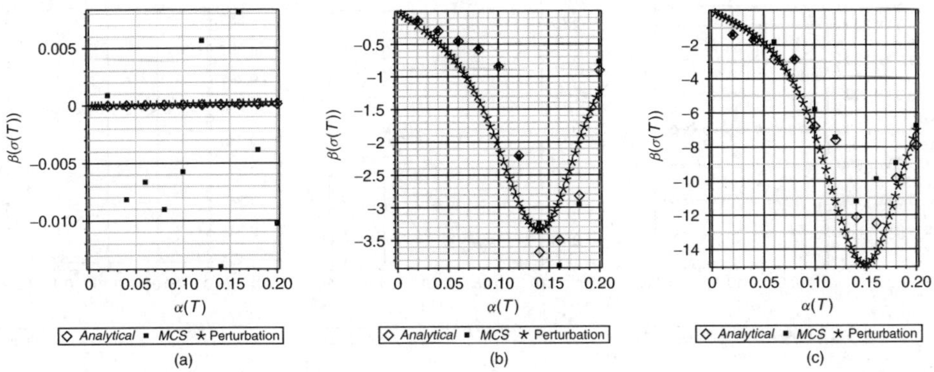

Figure 1.37 Skewness of normal stress $T = 10\,°\mathrm{C}$ (a), $450\,°\mathrm{C}$ (b), and $850\,°\mathrm{C}$ (c)

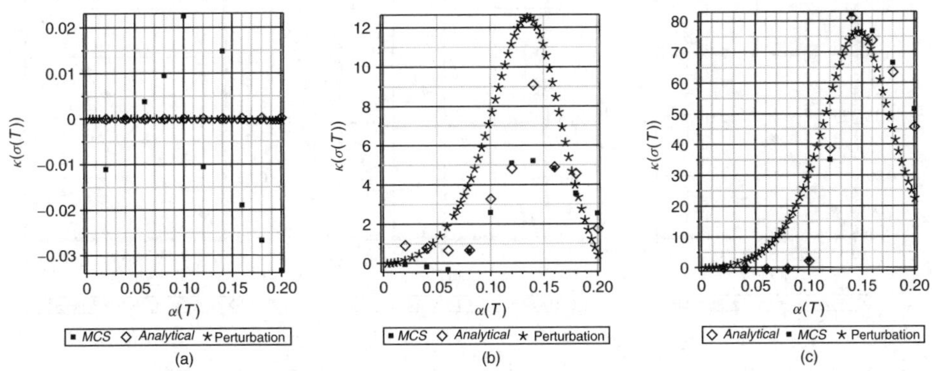

Figure 1.38 Kurtosis of normal stress $T = 10\,°\mathrm{C}$ (a), $450\,°\mathrm{C}$ (b), and $850\,°\mathrm{C}$ (c)

demonstrates different behavior – it increases four times for $T = [10\,°\mathrm{C}, 450\,°\mathrm{C}]$ and then starts to decrease until $T = 850\,°\mathrm{C}$, but now a perfect agreement is obtained for the initial temperatures only. For higher temperatures such a perfect coincidence is between analytical and simulation methods, while the stochastic perturbation technique underestimates this coefficient for $\alpha > 0.12$. The interrelation for $\alpha(\sigma(T))$ and $\alpha(T)$ is non-linear for larger temperatures but for the entire domain the larger $\alpha(T)$ the larger $\alpha(\sigma(T))$. It is indeed interesting that random dispersion has some maximum for the intermediate temperatures, which is adjacent to the beginning of the steepest descent of the $E[T]$ function given in Figure 1.34. The skewnesses presented in Figure 1.37 decrease systematically from the value 0 obtained initially to some significant negative values. Three methods coincide practically for $T = 10\,°\mathrm{C}$ only but all have the same tendencies and orders – a decrease from $\alpha = 0$ until $\alpha = 0.15$ to exhibit some small increase afterwards. Contrasting the skewness and kurtosis (Figure 1.38), one can notice that the thermal stresses undoubtedly have a Gaussian distribution only for smaller environmental temperatures. Then, they both start from

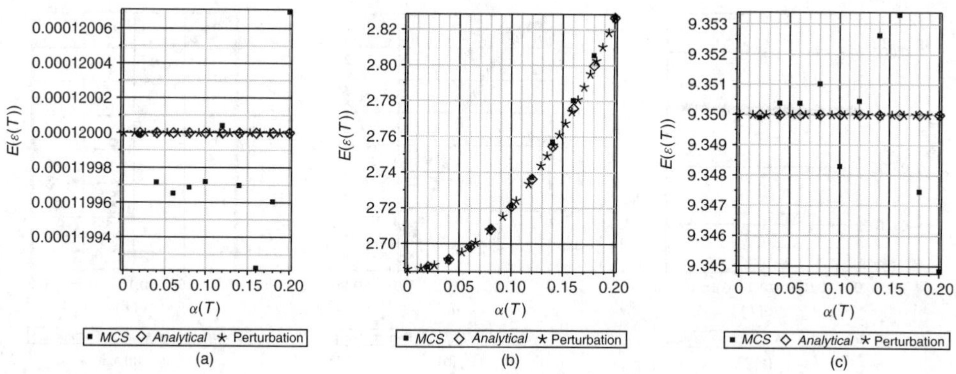

Figure 1.39 Expected values of normal strains $T = 10\,^\circ\text{C}$ (a), $450\,^\circ\text{C}$ (b), and $850\,^\circ\text{C}$ (c)

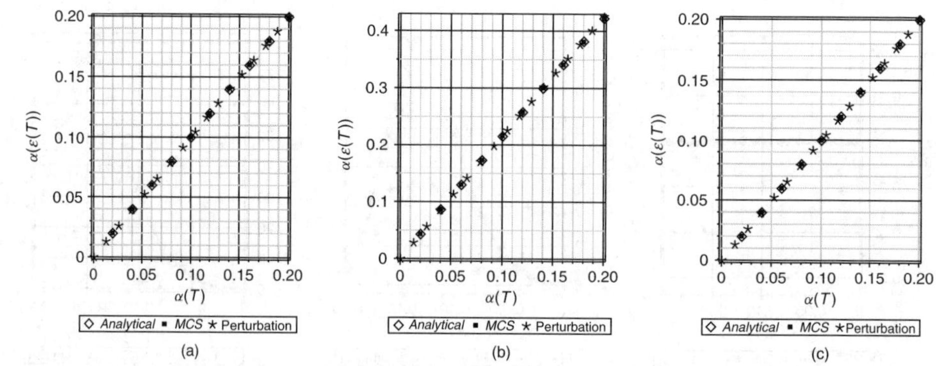

Figure 1.40 Coefficient of variation of normal strains $T = 10\,^\circ\text{C}$ (a), $450\,^\circ\text{C}$ (b), and $850\,^\circ\text{C}$ (c)

0 (for $\alpha(T) = 0$ adequate to the deterministic situation, of course) and have extremum values for almost the same input coefficients of variation – the skewness minimum coincides with the kurtosis maximum.

The probabilistic moments of the thermal strain (Figures 1.39–1.42) show rather good coincidence of the three numerical methods for $\alpha \in [0.0,0.2]$; some exceptions follow computational errors in the Monte Carlo simulation. It is interesting that for minimum and maximum temperatures these expectations are quite insensitive to the input coefficient of variation, while for $T = 450\,^\circ\text{C}$ we notice that $E[\varepsilon(T)]$ increases non-linearly together with $\alpha(T)$ according to the convex curve – the higher the temperature in the system, the larger the expectation of course. The coefficients of variation, similarly to those computed for stresses, increase from the initial temperature to $T = 450\,^\circ\text{C}$ and then show a reversed tendency, nevertheless for all

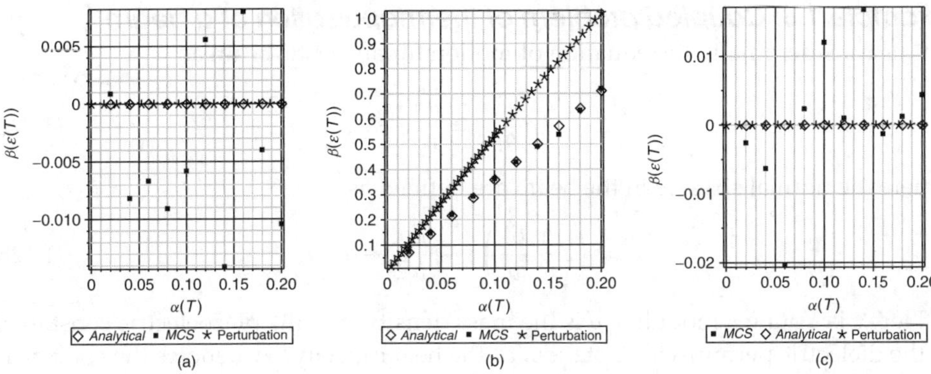

Figure 1.41 Skewness of normal strains $T = 10\,^{\circ}\mathrm{C}$ (a), $450\,^{\circ}\mathrm{C}$ (b), and $850\,^{\circ}\mathrm{C}$ (c)

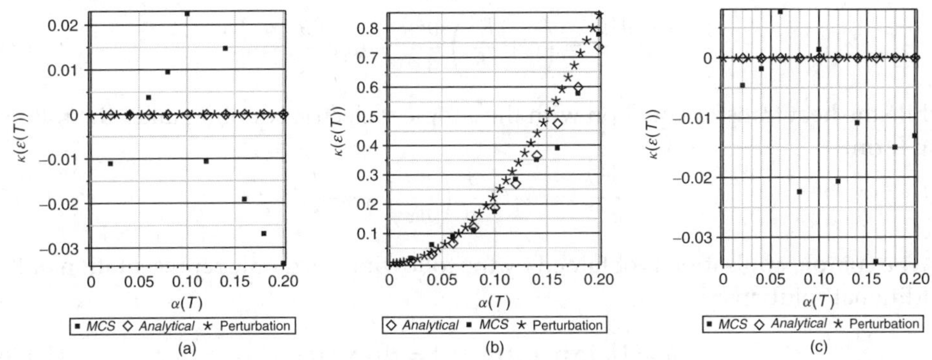

Figure 1.42 Kurtosis of normal strains $T = 10\,^{\circ}\mathrm{C}$ (a), $450\,^{\circ}\mathrm{C}$ (b), and $850\,^{\circ}\mathrm{C}$ (c)

temperatures the output and input coefficients are linearly proportional to each other. The skewnesses (Figure 1.41) and the kurtosis (Figure 1.42) demonstrate fluctuations very close to the expectations – perfectly constant and equal to 0 for minimum and maximum temperature, showing some slightly non-linear increase together with $\alpha(T)$ for $T = 450\,^{\circ}\mathrm{C}$ and, therefore, becoming non-Gaussian. Now, the stochastic perturbation technique shows some overestimation with respect to the other methods applied.

Let us underline that the computational time consumption is still the smallest for the stochastic perturbation scheme, then a little bit larger for the analytical approach, while, as always, the largest one is in the case of Monte Carlo simulation. It needs to be mentioned that this is valid for Gaussian variables only, since other probability density functions may cause larger time consumption for the symbolic integration and, furthermore, for some specific cases it may be impossible.

Example 1.8: Coupled problem of thermopulsation of a beam

Let us consider finally the equation of motion for an elastic beam as

$$\rho \frac{\partial^2 u}{\partial t^2} = \frac{\partial \sigma}{\partial x}, \tag{1.122}$$

where the normal stresses in the beam are defined as

$$\sigma = \left(e + \frac{E^2}{b} \right) \varepsilon - \frac{E}{b} D - \alpha_T e \theta. \tag{1.123}$$

Here e is Young's modulus, ρ is the mass density, E is the piezoelectric constant, b is the dielectric permittivity, c stands for the heat capacity, α_T denotes the coefficient of thermal expansion, k is the heat conductivity coefficient, T_0 stands for the reference temperature, and l is the length of the vibrating structural element.

A substitution of the stresses into the equation of motion yields

$$\rho \frac{\partial^2 u}{\partial t^2} = \left(e + \frac{E^2}{b} \right) \frac{\partial^2 u}{\partial x^2} - \alpha_T e \frac{\partial \theta}{\partial x}, \tag{1.124}$$

while the heat transfer equation with thermopiezoelectric coupling takes the following form:

$$k \frac{\partial^2 \theta}{\partial x^2} = e \alpha_T \theta_0 \frac{\partial}{\partial t} \left(\frac{\partial u}{\partial x} \right) + c \frac{\partial \theta}{\partial t}. \tag{1.125}$$

The complex pulsation is obtained by the following harmonic representation of the additional solutions:

$$u = U_0 \exp(i\omega t); \quad \theta = \theta_0 \exp(i\omega t), \tag{1.126}$$

where the first mechanical and thermal modes are given by

$$U_0 = A \sin\left(\frac{\pi x}{l} \right); \quad \theta_0 = B \cos\left(\frac{\pi x}{l} \right), \tag{1.127}$$

where A and B depend on each other. Substituting these representations into the relation (1.124), one obtains

$$B = A \left(\frac{\pi}{l} \frac{1}{\alpha_T} \left(1 + \frac{E^2}{eb} \right) - \omega^2 \frac{l}{\pi} \frac{\rho}{e \alpha_T} \right). \tag{1.128}$$

Therefore, the thermopiezoelectric pulsation ω of this bar is the real eigenvalue fulfilling the following equation [95, 116]:

$$\frac{\rho}{e} \left(\frac{l}{\pi} \right)^2 i\omega^3 - \frac{\rho}{e} \frac{k}{c} \omega^2 + \left(1 + \frac{E^2}{eb} + \frac{\theta_0 \alpha_T^2 e}{c} \right) i\omega + \frac{k}{c} \left(\frac{\pi}{l} \right)^2 \left(1 + \frac{E^2}{eb} \right) = 0. \tag{1.129}$$

All computational experiments are devoted to the silicon beam having, in a deterministic case, the following properties: $e = 1.58 \times 10^{11} \frac{\mathrm{N}}{\mathrm{m}^2}$, $\rho = 2300 \frac{\mathrm{kg}}{\mathrm{m}^3}$, $E = 0.1711 \frac{\mathrm{C}}{\mathrm{m}^2}$, $b = 3.992 \times 10^{-11} \frac{\mathrm{F}}{\mathrm{m}^2}$, $c = 711 \frac{\mathrm{J}}{\mathrm{kg\,K}}$, $\alpha_T = 2.5 \times 10^{-6} \frac{1}{\mathrm{K}}$, $k = 170 \frac{\mathrm{W}}{\mathrm{mK}}$, $T_0 = 298\,\mathrm{K}$,

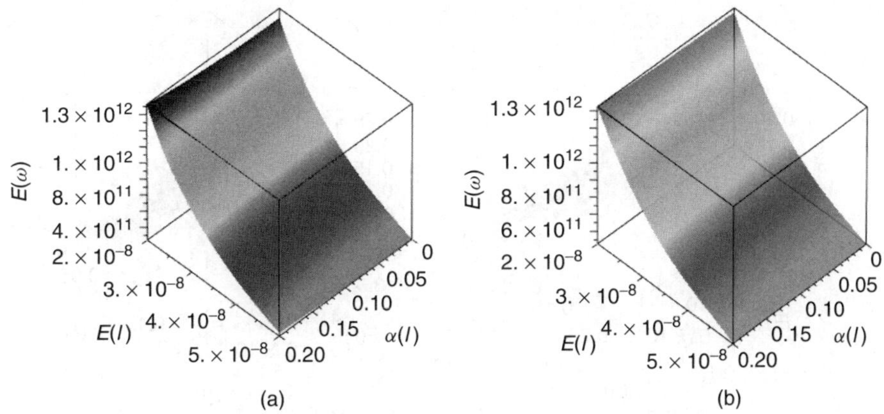

Figure 1.43 Expected values from polynomial interpolation (a) and least-squares method approximation (b) of the beam thermopulsation eigenfrequency

and $l = 3.9 \times 10^{-8}$m. The unique real solution to Equation (1.129) is unavailable in analytical form even in the symbolic environment of Maple™, so numerical experiments are based alternatively on (i) polynomial interpolation with 13 experimental points (left-hand graphs) and (ii) least-squares approximation of the response function (right-hand graphs) provided on the basis of the same points with the use of a third-order polynomial expression; all output moments and characteristics determined using second (2), fourth (4), sixth (6), eight (8) and tenth (10) order expansions. It is assumed now that the beam length is the truncated Gaussian random variable and the parametric computational analysis and visualization is provided with respect to its expected value $E[l] \in [2.0 \times 10^{-8}, 5.0 \times 10^{-8}]$ and coefficient of variation $\alpha(l) \in [0.0, 0.20]$ identifying uniquely the given Gaussian distribution. We analyze the parametric variability of the expectations (Figure 1.43), coefficients of variation (Figure 1.44), skewness (Figure 1.45), as well as the kurtosis (Figure 1.46) with respect to both $E[l]$ and $\alpha(l)$. A general impression is that the expected values only are very similar to each other in both numerical methods (where additionally all the perturbation orders give invisible differences), the remaining characteristics have very close results, but the corresponding surfaces are slightly different, especially for the largest input random dispersion. These expectations, unlike the remaining characteristics, have the same vertical axis ranges and the only apparent difference is that the right surface is almost horizontal for a minimum value of the beam length, while it shows some small increase together with the input value of $\alpha(l)$. Now, the surfaces corresponding to the particular orders can be identified, however the major difference is between the second and the remaining orders resulting in almost the same values. It is interesting that while polynomial interpolation results here in almost a straight-line increase of the computed $\alpha(\omega)$ with respect to $E[l]$ and $\alpha(l)$, the least-squares results show some waviness, especially for maximum input random dispersion. Incidentally, there are slightly larger extremum values for a maximum combination of both inputs. Skewness and kurtosis computed using two different

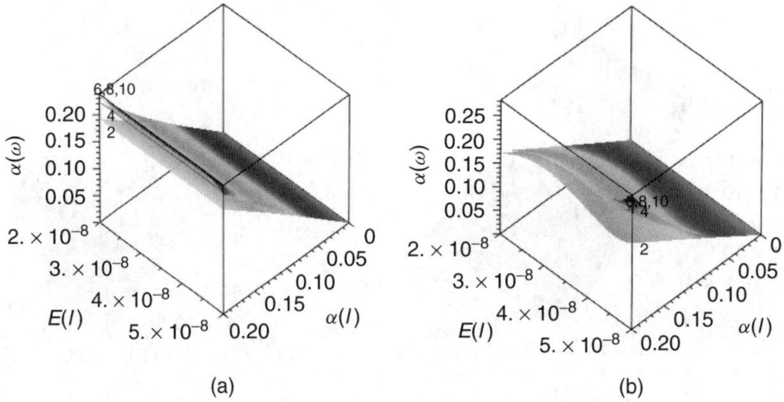

Figure 1.44 Coefficients of variation from polynomial interpolation (a) and least--squares method approximation (b) of the beam thermopulsation eigenfrequency

methods have only close values but their parametric variations are totally different and the only common point is that they start from 0 for $\alpha(l)=0$, which is trivial. Polynomial interpolation is mainly affected by the input coefficient of variation, whereas fluctuations with respect to E[l] are negligible here. They are both almost expectation's insensitive except for the maximum values of both input parameters, where some local extremum is obtained. The least-squares approach results in an almost opposite effect – less sensitivity to the input randomness than to the beam length expectation. Maximum and positive values of both higher-order characteristics are determined for the initial mean value of this length, where the differences between particular perturbation analysis orders are more transparent. Skewness and kurtosis of the thermopulsation eigenfrequency shows that, quite independent of the computer method applied, the final distribution cannot be Gaussian.

Now, let us consider the stochastic variations in the piezoelectric beam length using the following linear representation for a decay in its length: $l(\omega;t)=l^0(\omega)-\dot{l}(\omega)\,t$. The Gaussian random variable $l^0(\omega)$ is equivalent to the initial random fluctuations of the beam length, whereas $\dot{l}(\omega)$ represents the aging rate for this length. From a mathematical point of view, this process represented as the simplest time series is continuous with continuous time and, which can be relatively easily proven, non-stationary (owing to the unboundedness of the variance). For the purposes of computer simulation, we analyze once more the same elastic beam and the following additional data: $E\left[l^0(\mathbf{x};\omega)\right]=3.9\times10^{-8}$m, $E[\dot{l}(\mathbf{x};\omega)]=3.9\times10^{-10}$m/yr, which is equivalent to 1% loss of the initial length per year. As documented in Figures 1.47–1.50, the expectations are replaced with the time domain for $t\in[0,50$ years], and the second independent parameter is the coefficient of variation taken as $\alpha(l)\in[0.0,0.20]$. We analyze in turn time fluctuations of the expected values (Figure 1.47), the coefficients of variation (Figure 1.48), skewness (Figure 1.49), as well as kurtosis (cf. Figure 1.50), all related to the thermopulsation eigenfrequency of this device. All the left-hand figures are, as before, adjacent to the traditional polynomial

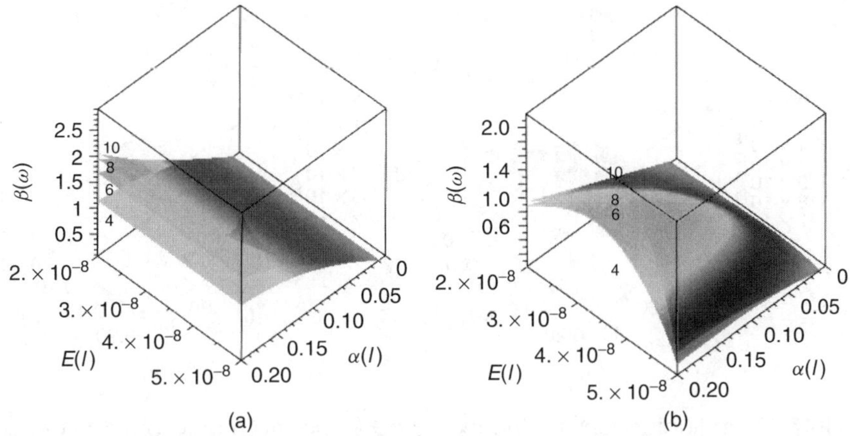

Figure 1.45 Skewness from polynomial interpolation (a) and least-squares method approximation (b) of the beam thermopulsation eigenfrequency

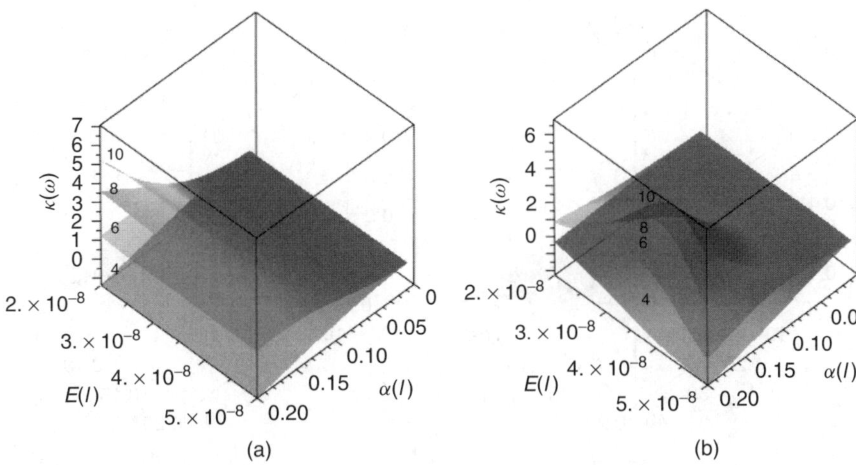

Figure 1.46 Kurtosis from polynomial interpolation (a) and least-squares method approximation (b) of the beam thermopulsation eigenfrequency

interpolation of the response function, while the right-hand graphs correspond to the third-order polynomial least-squares approximation.

As is clear from this set of numerical results, all stochastic properties are strongly affected by the input coefficient of variation even for initial time and aging velocity for the device length. As one could expect from the results obtained in the previous subsection, both expectations and standard deviations increase together with the lifetime of the device, while skewness and kurtosis demonstrate a more complex tendency showing rather their final decrease together with time. A contrast of the left-hand graphs with the right-hand set leads to the conclusion that the polynomial interpolation usually significantly overestimates the absolute values of all up to

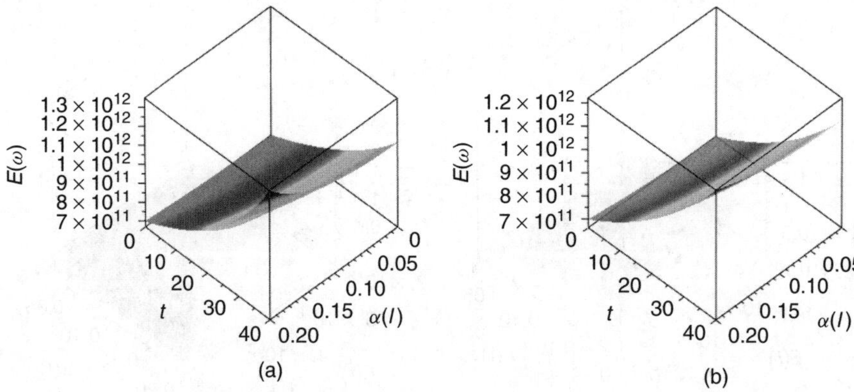

Figure 1.47 Time fluctuations for the expected values from polynomial interpolation (a) and least-squares method approximation (b) of the thermopulsation eigenfrequency

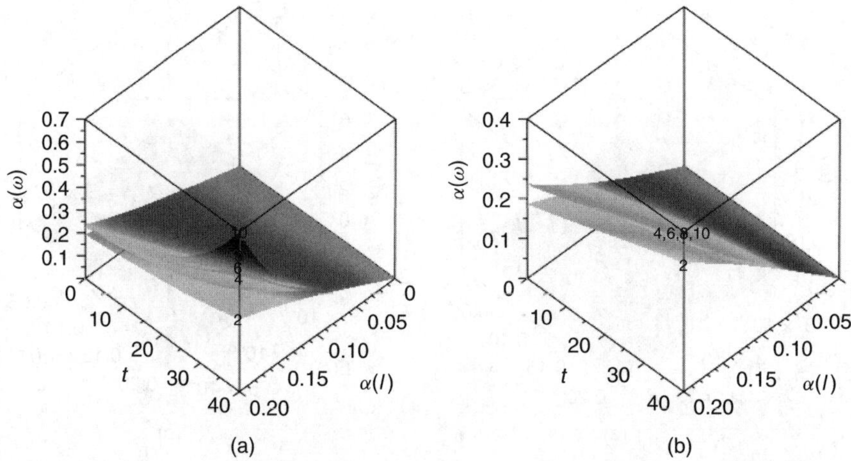

Figure 1.48 Time fluctuations for the coefficients of variation from polynomial interpolation (a) and least-squares method approximation (b) of the thermopulsation eigenfrequency

fourth-order probabilistic characteristics in comparison to the lower-order least-squares-based response function, having also polynomial form. The parametric variability of the expectations in both methods is almost the same – it is impossible to distinguish on the given scale between the results corresponding to the particular perturbation orders. The surfaces computed for the coefficients of variation also have similar shapes for both methods, however now there are clear differences for the given order results and, further, the extremum value of $\alpha(\omega)$ and the surface curvatures obtained for the polynomial interpolation are larger for the ordinary polynomial

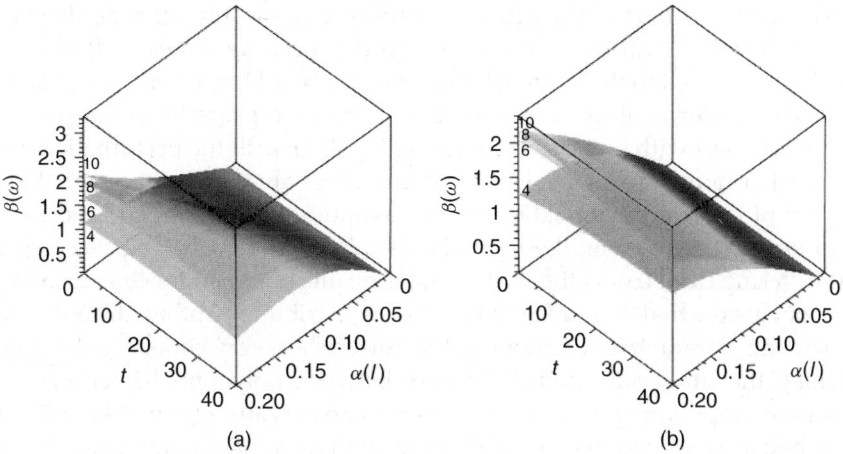

Figure 1.49 Time fluctuations for the skewness from polynomial interpolation (a) and least-squares method approximation (b) of the thermopulsation eigenfrequency

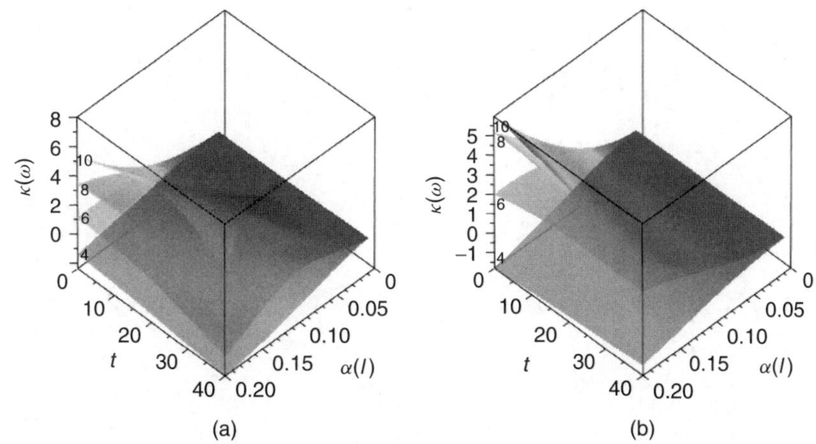

Figure 1.50 Time fluctuations for the kurtosis from polynomial interpolation (a) and least-squares method approximation (b) of the thermopulsation eigenfrequency

interpolation. Therefore, we see that the first method may result in large gradients of probabilistic characteristics, especially for increasing values of an input coefficient of variation; this local phenomenon is also noticeable for the expectations, but is a few times smaller. It should be mentioned that, contrary to the previous probabilistic analysis, now the output variation coefficients are almost two or even more than three times larger at the end of the process with respect to the initial stage and for the maximum value of the input parameter $\alpha(l)$; this effect follows an influence of time, because their initial values agree with the results presented in Figure 1.44. Skewness time fluctuations exhibit slightly different parametric variations for both approximation methods – least squares results in an almost linear time change, non-linear with small curvature – with respect to the input parameter $\alpha(l)$. Polynomial interpolation applied to this parameter returns an almost linear time variation in case of the lowest-order analysis and then, this function appears to be non-linear with respect to $\alpha(l)$ and with increasing curvature together with the perturbation analysis order. Furthermore, it has some local extremum for values of input parameters close to the end of the computational domain (maximum $\alpha(l)$ and large t) and for larger time values it decreases, unlike in the least-squares approach. The clearly positive values differing from 0, together with similar results obtained for the kurtosis, leads to the conclusion that this uncertainty has a distribution rather distant from the Gaussian one. Parametric variations of the kurtosis are very similar to the skewness concerning the similarities and differences between polynomial interpolation and least-squares approximation. Some tendencies like curvature for higher orders in the first method as well as local extremum are even more apparent in Figure 1.50, where we can see that the skewness and kurtosis increase slightly together with input $\alpha(l)$ except for lower-order surfaces in the least-squares analysis; time variations in right Figs. 1.49 and 1.50 are not so remarkable.

2

The Stochastic Finite Element Method

2.1 Governing Equations and Variational Formulations

2.1.1 Linear Potential Problems

Let us consider the following partial differential equation:

$$\sum_i \frac{\partial}{\partial x_i}\left(k_i \frac{\partial \phi}{\partial x_i}\right) + g = 0; \;\; x_i \in \Omega, \tag{2.1}$$

equivalent to the field problem in a certain anisotropic domain, where ϕ denotes the potential function, q is equivalent to the source, while k_x, k_y, and k_z denote physical parameters of the domain governing its behavior for the particular scalar field considered. This equation is traditionally solved with the following essential

$$\phi = \hat{\phi}; \;\; x_i \in \partial\Omega_\phi \tag{2.2}$$

and natural boundary conditions

$$\sum_i k_i \frac{\partial \phi}{\partial x_i} n_i = \hat{q}; \;\; x_i \in \partial\Omega_q, \tag{2.3}$$

where $\partial\Omega_\phi \cup \partial\Omega_q = \partial\Omega$ with $\partial\Omega_\phi \cap \partial\Omega_q = 0$. Now let us consider the arbitrary virtual potential distribution $\delta\phi$, which allows us to rewrite Equation (2.1) with constant coefficients in the following integral form:

$$\int_\Omega \left\{ \left(\sum_i k_i \frac{\partial^2 \phi}{\partial x_i^2} + g \right) \delta\phi \right\} d\Omega = 0; \;\; x_i \in \Omega. \tag{2.4}$$

The Stochastic Perturbation Method for Computational Mechanics, First Edition. Marcin Kamiński.
© 2013 John Wiley & Sons, Ltd. Published 2013 by John Wiley & Sons, Ltd.

Application of the classical differentiation rules together with the divergence theorem leads to

$$\int_{\Omega} \left\{ \sum_i \frac{\partial}{\partial x_i} \left(\delta\phi k_i \frac{\partial \phi}{\partial x_i} \right) \right\} d\Omega = \int_{\partial\Omega} \left(\sum_i k_i \frac{\partial \phi}{\partial x_i} n_i \right) \delta\phi \, d\left(\partial\Omega\right). \tag{2.5}$$

Further usage of the boundary conditions with $\delta\phi = 0$ for $x_i \in \partial\Omega_\phi$ results in the variational statement

$$\int_{\Omega} \left(\sum_i \delta\left(\frac{\partial \phi}{\partial x_i} \right) k_i \frac{\partial \phi}{\partial x_i} \right) d\Omega = \int_{\Omega} \left(g\,\delta\phi \right) d\Omega + \int_{\partial\Omega_q} \hat{q}_i n_i \,\delta\hat{\phi}\, d\left(\partial\Omega\right), \;\; x_i \in \Omega. \tag{2.6}$$

The above equation is known as the principle of the virtual potentials, which can be the basis of the FEM standard discretization procedure. It starts from the approximation of the potential function $\phi = \phi(x_i)$ for all nodes in the FEM mesh as

$$\phi\left(x_i\right) = N_\alpha\left(x_i\right) \Phi_\alpha, \; \alpha = 1, \ldots, R, \tag{2.7}$$

where N_α represents the shape functions of the finite elements applied, Φ_α are the nodal values of the potential, and R denotes here the total number of degrees of freedom. Similarly, one discretizes the spatial derivatives of the function ϕ, that is,

$$\frac{\partial \phi}{\partial x} = \frac{\partial N_\alpha}{\partial x} \Phi_\alpha, \; \alpha = 1, \ldots, R. \tag{2.8}$$

Then, the left-hand side of Equation (2.6) is represented as the additional sum of the finite element contributions (denoted by $f = 1, \ldots, F$), so that the following definition is applied:

$$k_{\alpha\beta}^{(f)} = \int_{\Omega_f} \left(\sum_i k_i \frac{\partial N_\alpha}{\partial x_i} \frac{\partial N_\beta}{\partial x_i} \right) d\Omega, \tag{2.9}$$

which constitutes the system elemental matrix equivalent to the stiffness matrix in the elasticity problems. Representing similarly the right-hand side of Equation (2.6), it may be written that

$$Q_\alpha^{(f)} = \int_{\Omega_f} \left(g N_\alpha \right) d\Omega + \int_{\partial\Omega_q} \hat{q}_i n_i N_\alpha \, d\left(\partial\Omega\right). \tag{2.10}$$

The assemblage and condensation procedures for the system matrices and vectors lead to the final well-known algebraic equations system

$$K_{\alpha\beta} \Phi_\beta = Q_\alpha, \tag{2.11}$$

whose solution enables us to compute the nodal solution's vector Φ_β and to approximate the potential values between the nodes. Further numerical illustrations concerning the problems with random coefficients deal with the unidirectional viscous flow described by the differential equation

$$2\mu \frac{\partial^2 \phi}{\partial x^2} + g = 0; \tag{2.12}$$

the coefficient μ characterizes the fluid viscosity, whereas the potential function is equivalent to the fluid velocity.

The second example is devoted to the 2D torsion of the planar domain described by the equation

$$\left(\frac{1}{G}\phi_{,x}\right)_{,x} + \left(\frac{1}{G}\phi_{,y}\right)_{,y} + 2\theta = 0; \quad x_i \in \Omega, \tag{2.13}$$

where G stands for the Kirchhoff modulus, θ is the twisting angle, and where the stresses are given as

$$\tau_{zx} = \phi_{,y}, \quad \tau_{zy} = -\phi_{,x}; \quad x_i \in \Omega. \tag{2.14}$$

Therefore, the stochastic Taylor expansion is applied to the integral formulation of the field problem given by Equation (2.6), so that the following set of increasing-order variational statements has been returned [82]:

$$\begin{aligned} &\int_\Omega \left(\sum_i \delta\left(\frac{\partial \phi}{\partial x_i}\right) \left(\sum_{m=0}^{n} \binom{n}{m} \frac{\partial^m k_i}{\partial b^m} \frac{\partial^{n-m}}{\partial b^{n-m}} \frac{\partial \phi}{\partial x_i} \right) \right) d\Omega \\ &= \int_\Omega \left(\frac{\partial^n g}{\partial b^n} \delta\phi \right) d\Omega + \int_{\partial\Omega_g} \frac{\partial^n \hat{q}_i}{\partial b^n} n_i \, \delta\phi \, d(\partial\Omega) \end{aligned} \tag{2.15}$$

As is known, the FEM does not demand formulation based on the scalar function, like the twisting angle above; we can alternatively use some formulation with scalar function gradients like heat fluxes (or stress Airy functions), for instance [63, 64, 125]. We consider for this purpose torsion of a linear, isotropic, and homogeneous medium given by Equation (2.13) together with the stress tensor components introduced in different notation than for Equation (2.14) as

$$\sigma_{13} = \frac{\partial \phi}{\partial x_2}, \quad \sigma_{23} = -\frac{\partial \phi}{\partial x_1}, \tag{2.16}$$

while the constitutive relation can be rewritten as

$$\frac{\partial \phi}{\partial x_2} = G\left(\frac{\partial u}{\partial x_1} - \theta x_2\right), \quad \frac{\partial \phi}{\partial x_2} = G\left(\frac{\partial u}{\partial x_2} + \theta x_1\right) \tag{2.17}$$

in the case of an externally applied twisting moment $Q(x_i) = -2\theta$. Finally, the complementary energy necessary for finite element discretization can be expressed as

$$\Sigma(\phi) = \int_{\Omega} \frac{1}{2G}\left[\left(\frac{\partial \phi}{\partial x_1}\right)^2 + \left(\frac{\partial \phi}{\partial x_2}\right)^2\right] d\Omega - \int_{\partial \Omega_u} Q\phi d\Omega. \tag{2.18}$$

As is known, the torsion problem is the specific scalar field problem and is equivalent to heat conduction, seepage, and other related physical phenomena described by the same Laplace partial differential equation; solution of this particular problem makes it possible to describe at the same time all equivalent problems in the context of field analogies [3].

2.1.2 Linear Elastostatics

Let us consider a statistically homogeneous and bounded continuum $\Omega \subset \mathcal{R}$ without any initial stresses and strains. The elastic properties and geometry of Ω may be treated as design random parameters and they result in a random displacement field and random stress tensor satisfying the classical boundary value problem of linear elasticity. Let us assume that there are non-empty subsets of external boundaries of Ω, namely $\partial\Omega_\sigma$ and $\partial\Omega_u$, where the stress and displacement boundary conditions are defined. The boundary differential equations system describing this equilibrium problem can be written as follows:

$$\sigma_{ij} = C_{ijkl}\varepsilon_{kl}, \tag{2.19}$$

$$\varepsilon_{ij} = \frac{1}{2}\left(\frac{\partial u_i}{\partial x_j} + \frac{\partial u_j}{\partial x_i}\right), \tag{2.20}$$

$$\sigma_{ij,j} + \rho f_i = 0, \tag{2.21}$$

$$u_i = \hat{u}_i; \; x_i \in \partial\Omega_u, \tag{2.22}$$

$$\sigma_{ij} n_j = \hat{t}_i; \; x_i \in \partial\Omega_\sigma, \tag{2.23}$$

where

$$C_{ijkl} = \delta_{ij}\delta_{kl}\frac{e\nu}{(1+\nu)(1-2\nu)} + \left(\delta_{ik}\delta_{jl} + \delta_{il}\delta_{jk}\right)\frac{e}{2(1+\nu)} \tag{2.24}$$

for $i, j, k, l = 1, 2$; e denotes here Young's modulus, while ν stands for the Poisson ratio. Let us note that theoretical foundations and some numerical aspects of elliptic boundary value problems with random coefficients are proposed in [2]. Generally, the equation system posed above is solved using well-established numerical methods. Usually, it should be transformed first to the variational formulation. This yields

$$\int_{\Omega} C_{ijkl}\, u_{i,j}\delta u_{k,l} d\Omega = \int_{\Omega} \rho f_i \delta u_i d\Omega + \int_{\partial\Omega_\sigma} \hat{t}_i \delta u_i d(\partial\Omega), \tag{2.25}$$

where the left-hand side of Equation (2.25) corresponds to elastic behavior of the structure, the first component on the right-hand side includes the body force effects, while the second is equivalent to the stress boundary conditions applied. Thus, the stochastic version of the minimum potential energy principle has the following form [74, 112]:

- zeroth-order equation

$$\int_{\Omega} \delta u_{k,l} C^0_{ijkl} u^0_{i,j} d\Omega = \int_{\Omega} \rho^0 f^0_i \delta u_i d\Omega + \int_{\partial\Omega_\sigma} \delta u_i \hat{t}^0_i d(\partial\Omega) \tag{2.26}$$

- first-order equation

$$\int_{\Omega} \delta u_{k,l} C^0_{ijkl} \frac{\partial u_{i,j}}{\partial b} d\Omega = \int_{\Omega} \left(\frac{\partial \rho}{\partial b} f^0_i + \rho^0 \frac{\partial f_i}{\partial b} \right) \delta u_i d\Omega + \int_{\partial\Omega_\sigma} \delta u_i \frac{\partial \hat{t}_i}{\partial b} d(\partial\Omega) - \int_{\Omega} \delta u_{k,l} \frac{\partial C_{ijkl}}{\partial b} u^0_{i,j} d\Omega \tag{2.27}$$

- second-order equation

$$\int_{\Omega} \delta u_{k,l} C^0_{ijkl} \frac{\partial^2 u_{i,j}}{\partial b^2} d\Omega = \int_{\partial\Omega_\sigma} \delta u_i \frac{\partial^2 \hat{t}_i}{\partial b^2} d(\partial\Omega) - \int_{\Omega} \delta u_{k,l} \left(2 \frac{\partial C_{ijkl}}{\partial b} \frac{\partial u_{i,j}}{\partial b} + \frac{\partial^2 C_{ijkl}}{\partial b^2} u^0_{i,j} \right) d\Omega$$

$$+ \int_{\Omega} \left(\frac{\partial^2 \rho}{\partial b^2} f^0_i + 2 \frac{\partial \rho}{\partial b} \frac{\partial f_i}{\partial b} + \rho^0 \frac{\partial^2 f_i}{\partial b^2} \right) \delta u_i d\Omega \tag{2.28}$$

- nth-order equation

$$\int_{\Omega} \delta u_{k,l} C^0_{ijkl} \frac{\partial^n u_{i,j}}{\partial b^n} d\Omega = \int_{\partial\Omega_\sigma} \delta u_i \frac{\partial^n \tilde{t}_i}{\partial b^n} d(\partial\Omega) - \int_{\Omega} \delta u_{k,l} \sum_{k=1}^{n} \binom{n}{n-k} \frac{\partial^k C_{ijkl}}{\partial b^k} \frac{\partial^{n-k} u_{i,j}}{\partial b^{n-k}} d\Omega$$

$$+ \int_{\Omega} \sum_{k=0}^{n} \binom{n}{n-k} \frac{\partial^k \rho}{\partial b^k} \frac{\partial^{n-k} f_i}{\partial b^{n-k}} \delta u_i d\Omega \tag{2.29}$$

If, for instance, the Young's modulus of Ω is considered as a random variable of the problem, that is, $b \equiv e$, then the first partial derivatives of the elasticity tensor with respect to this variable are derived as

$$A_{ijkl} = \frac{\partial C_{ijkl}}{\partial b} = \delta_{ij}\delta_{kl} \frac{\nu}{(1+\nu)(1-2\nu)} + (\delta_{ik}\delta_{jl} + \delta_{il}\delta_{jk}) \frac{1}{2(1+\nu)}; \tag{2.30}$$

and they are all deterministic quantities. Further, all higher-order partial derivatives of this tensor with respect to the same variable are equal to 0. Neglecting the body force effects and eliminating all partial derivatives of the stress boundary conditions, we can write down the fundamental equations of the problem as:

- zeroth-order equation

$$\int_{\Omega} \delta u_{k,l} C^0_{ijkl} u^0_{i,j} d\Omega = \int_{\partial\Omega_\sigma} \delta u_i \hat{t}^0_i d(\partial\Omega) \tag{2.31}$$

- first-order equation

$$\int_{\Omega} \delta u_{k,l} C^0_{ijkl} \frac{\partial u_{i,j}}{\partial b} d\Omega = -\int_{\Omega} \delta u_{k,l} A_{ijkl} u^0_{i,j} d\Omega \tag{2.32}$$

- second-order equation

$$\int_{\Omega} \delta u_{k,l} C^0_{ijkl} \frac{\partial^2 u_{i,j}}{\partial b^2} d\Omega = -2\int_{\Omega} \delta u_{k,l} A_{ijkl} \frac{\partial u_{i,j}}{\partial b} d\Omega \tag{2.33}$$

- nth-order equation

$$\int_{\Omega} \delta u_{k,l} C^0_{ijkl} \frac{\partial^n u_{i,j}}{\partial b^n} d\Omega = -(n-1)\int_{\Omega} \delta u_{k,l} A_{ijkl} \frac{\partial^{n-1} u_{i,j}}{\partial b^{n-1}} d\Omega. \tag{2.34}$$

Alternatively, the compliance tensor may be introduced for Ω as

$$c_{ijkl} = \frac{1}{e}\Big((1+\nu)\,\delta_{ik}\delta_{jl} - \nu\,(1+\kappa\nu)\,\delta_{ij}\delta_{kl}\Big), \tag{2.35}$$

where $\kappa = 0, 1$ denotes the plane stress or plane strain problem, respectively. The boundary value problem of linear elasticity is defined using Equations (2.19)–(2.23) together with the constitutive relation

$$\varepsilon_{ij} = c_{ijkl}\sigma_{kl}; \quad x_i \in \Omega. \tag{2.36}$$

A variational formulation for the boundary value problem can be proposed using a statically admissible stress space consisting of the real-defined symmetric tensors fulfilling equilibrium and boundary conditions. Introducing the stress variation $\delta\sigma_{ij}$, one can represent the equivalent variational equation in the region Ω as

$$\int_{\Omega} \left(\varepsilon_{ij} - \frac{1}{2}\left(u_{i,j} + u_{j,i}\right)\right) \delta\sigma_{ij} \mathrm{d}\Omega = 0. \tag{2.37}$$

Applying further the Green formula to displacement field components u_i, the so-called complementary energy principle is obtained for any $\delta\sigma_{ij}$ in the form

$$\int_{\Omega} \varepsilon_{ij}\delta\sigma_{ij} \mathrm{d}\Omega - \int_{\partial\Omega_u} \hat{u}_i \delta\sigma_{ij} n_j \mathrm{d}(\partial\Omega) = 0 \tag{2.38}$$

or, alternatively, with use of the following functional:

$$\Sigma(\sigma) = \frac{1}{2}\int_{\Omega} c_{ijkl}\sigma_{ij}\sigma_{kl}\mathrm{d}\Omega - \int_{\partial\Omega_u} \hat{u}_i\sigma_{ij}n_j\mathrm{d}(\partial\Omega), \tag{2.39}$$

whose minimization leads to the real stress field being a solution to the basic equilibrium problem. We need to clearly emphasize that probabilistic solution to the given elastostatic boundary value problem using displacement function differs significantly from that obtained via stress functions. Despite of classical differences resulting from potential and complementary energies applied in the FEM implementations, the output probability density function for the structural response coming from the same random input may be different. When we randomize Young's modulus for the statistically homogeneous medium, an inversion procedure contained in Equation (2.35) essentially affects higher order perturbation terms, which simply vanish in displacement formulation as a consequence of Equation (2.30).

2.1.3 Non-linear Elasticity Problems

The following boundary value problem is considered now:

$$\Delta\sigma_{kl,l} + \rho\Delta f_k = 0; \quad x_i \in \Omega \tag{2.40}$$

$$\Delta\tilde{\sigma}_{kl} = C_{klmn}\Delta\varepsilon_{mn}; \quad x_i \in \Omega \tag{2.41}$$

$$\Delta\varepsilon_{kl} = \frac{1}{2}[\Delta u_{k,l} + \Delta u_{l,k} + u_{i,k}\Delta u_{i,l} + \Delta u_{i,k}u_{i,l} + \Delta u_{i,k}\Delta u_{i,l}]; \quad x_i \in \Omega \tag{2.42}$$

with the boundary conditions

$$\Delta\sigma_{kl}n_l = \Delta\hat{t}_k; \quad x_i \in \partial\Omega_\sigma, \; k = 1, 2, 3 \tag{2.43}$$

$$\Delta u_k = \Delta\hat{u}_k; \quad x_i \in \partial\Omega_u, \; k = 1, 2, 3 \tag{2.44}$$

This problem is solved for the displacement vector $u_k(\mathbf{x})$, the strain tensor $\varepsilon_{kl}(\mathbf{x})$, and the stress tensor $\sigma_{kl}(\mathbf{x})$ fulfilling the system [Equations (2.40)–(2.44)]. Let us note that the tensor functions $\Delta\sigma_{kl}(\mathbf{x})$, $\Delta\tilde{\sigma}_{kl}(\mathbf{x})$ denote here the first and second Piola–Kirchhoff tensors

$$\Delta\sigma_{kl} = \Delta F_{km}\Delta\tilde{\sigma}_{ml} + F_{km}\Delta\tilde{\sigma}_{ml} + \Delta F_{km}\tilde{\sigma}_{ml}; \quad x_i \in \Omega, \tag{2.45}$$

where

$$\Delta F_{km} = \Delta u_{k,m}; \quad x_i \in \Omega. \tag{2.46}$$

The following functional defined on Δu_k is introduced in order to obtain a numerical solution to this problem:

$$J\left(\Delta u_k\right) = \int_{\Omega} \left(\frac{1}{2} C_{klmn} \Delta \varepsilon_{kl} \Delta \varepsilon_{mn} + \frac{1}{2} \tilde{\sigma}_{kl} \Delta u_{i,k} \Delta u_{i,l} - \rho \Delta f_k \Delta u_k \right) d\Omega - \int_{\partial\Omega} \Delta \hat{t}_k \Delta u_k d\left(\partial\Omega\right). \tag{2.47}$$

This solution is determined from the minimization of the incremental version of the potential energy stationarity principle

$$\delta J\left(\Delta u_k\right) = \frac{\partial J}{\partial \Delta u_k} \delta\left(\Delta u_k\right). \tag{2.48}$$

2.1.4 Variational Equations of Elastodynamics

Let us consider the system of equations representing the elastodynamics problem with deterministic parameters assuming that all state variables are sufficiently smooth and continuous functions of **x** and τ. It is a simple consequence of Equations (2.19)–(2.23) [112]:

- equations of motion

$$\left(C_{ijkl} u_{k,l}\right)_{,j} + \rho f_i = \rho \ddot{u}_i, \quad x_i \in \Omega, \quad \tau \in \left[t_0, \infty\right), \; i, j, k, l = 1, 2, 3 \tag{2.49}$$

- constitutive equations

$$\sigma_{ij} = C_{ijkl} \varepsilon_{kl}, \quad x_i \in \Omega, \quad \tau \in \left[t_0, \infty\right), \quad i,j,k,l = 1,2,3 \tag{2.50}$$

- geometrical equations

$$\varepsilon_{ij} = \frac{1}{2}\left(u_{i,j} + u_{j,i}\right), \quad x_i \in \Omega, \quad \tau \in \left[t_0, \infty\right), \quad i,j,k,l = 1,2,3 \tag{2.51}$$

- displacement boundary conditions

$$u_j = \hat{u}_j, \quad x_i \in \partial\Omega_u, \quad \tau \in \left[t_0, \infty\right), \; j = 1,2,3 \tag{2.52}$$

- stress boundary conditions

$$\sigma_{ij} n_j = \hat{t}_i, \quad x_i \in \partial\Omega_\sigma, \quad \tau \in \left[t_0, \infty\right), \; i,j = 1,2,3 \tag{2.53}$$

- initial conditions

$$u_j = \hat{u}_j^0, \quad \dot{u}_j = \hat{\dot{u}}_j^0, \quad x_i \in \partial\Omega_u, \quad \tau = t_0, \; j = 1,2,3 \tag{2.54}$$

Introducing a variation $\delta u_i\,(\mathbf{x}, t)$ for any time moment we obtain the above equations as

$$-\int\limits_{\Omega}\left(\sigma_{ij,j}+\rho f_i-\rho\ddot{u}_i\right)\ \delta u_i \mathrm{d}\Omega+\int\limits_{\partial\Omega_\sigma}\left(\sigma_{ij}n_j-\hat{t}_i\right)\ \delta u_i \mathrm{d}\left(\partial\Omega\right). \tag{2.55}$$

Assuming known $u\left(\mathbf{x}, t_1\right)=0$ and $u\left(\mathbf{x}, t_2\right)=0$, this yields

$$\delta u\left(\mathbf{x}, t_1\right)=0, \quad \delta u\left(\mathbf{x}, t_2\right)=0. \tag{2.56}$$

Integration by parts with respect to $\mathbf{x}$ and τ gives

$$\int\limits_{t_1}^{t_2}\left[\delta T-\int\limits_{\Omega}\sigma_{ij,j}\,\delta\varepsilon_{ij} d\Omega+\int\limits_{\Omega}\rho f_i\,\delta u_i d\Omega+\int\limits_{\partial\Omega_\sigma}\hat{t}_i\,\delta u_i d\left(\partial\Omega\right)\right]d\tau=0, \tag{2.57}$$

which is used together with

$$\begin{aligned}\delta\varepsilon_{ij}&=\frac{1}{2}\left(\delta u_{i,j}+\delta u_{j,i}\right),\quad x_i\in\Omega,\ \ \tau\in\left[t_1, t_2\right],\\ \delta u_i&=0,\quad x_i\in\partial\Omega_u,\ \ \tau\in\left[t_1, t_2\right].\end{aligned} \tag{2.58}$$

Assuming independence of the vectors f_j and $\hat{t}_j$ from u_i, Equation (2.57) may be written as

$$\delta\int\limits_{t_1}^{t_2}\left(T-J_p\right)\ \mathrm{d}\tau=0 \tag{2.59}$$

and is known as the Hamilton principle, where

$$T=\frac{1}{2}\int\limits_{\Omega}\rho\dot{u}_i\dot{u}_i \mathrm{d}\Omega \tag{2.60}$$

is the kinetic energy stored in Ω and the potential energy is given as

$$J_p=\frac{1}{2}\int\limits_{\Omega}C_{ijkl}\varepsilon_{ij}\varepsilon_{kl}\mathrm{d}\Omega-\int\limits_{\Omega}\rho f_i u_i \mathrm{d}\Omega-\int\limits_{\partial\Omega_\sigma}\hat{t}_i u_i \mathrm{d}\Omega=0. \tag{2.61}$$

2.1.5 Transient Analysis of the Heat Transfer

Generally, the transient heat flow problem consists of determining the temperature field $\theta=\theta(\mathbf{x}, \tau)$ governed by the following differential equation [13, 97]:

$$\rho c\dot{\theta}-\left(k_{ij}\theta_{,j}\right)_{,i}-g=0;\quad x_i\in\Omega;\quad \tau\in[0,\infty), \tag{2.62}$$

where c is the heat capacity characterizing the region Ω, ρ is the density of the material contained in Ω, k_{ij} is the thermal conductivity tensor, while g is the rate of heat generated per unit volume, and the variables θ and τ denote temperature field values and time, respectively. This equation should fulfill the boundary conditions on the additional subsets of the external boundary $\partial\Omega$, given as follows:

1. temperature (essential) boundary conditions

$$\theta = \tilde{\theta};\ x_i \in \partial\Omega_\theta \tag{2.63}$$

2. heat flux (natural) boundary conditions

$$\frac{\partial\theta}{\partial n_i} = \hat{q}_i;\ \ x_i \in \partial\Omega_q, \tag{2.64}$$

where $\partial\Omega_\theta \cup \partial\Omega_q = \partial\Omega$ and $\partial\Omega_\theta \cap \partial\Omega_q = \{\emptyset\}$.

The initial conditions are proposed here as

$$\theta^0 = \theta\left(x_i; 0\right);\ x_i \in \Omega,\ \tau = 0. \tag{2.65}$$

Let us consider further some continuous temperature variations $\delta\theta(x_i)$ defined in the interior of the region Ω and vanishing on $\partial\Omega_\theta$. Multiplying Equation (2.62) by the test function specified and integrating it over the entire Ω, we obtain [96, 97]

$$\int_\Omega \left(\rho c\dot{\theta} - \left(k_{ij}\theta_{,j}\right)_{,i} - g\right)\ \delta\theta\, \mathrm{d}\Omega = 0;\ \ x_i \in \Omega;\ \ \ \tau \in [0,\infty). \tag{2.66}$$

Taking into account that the derivative defined on the temperature variation is in fact a variation of the respective temperature derivative

$$\frac{\partial\left(\delta\theta\right)}{\partial x_i} = \delta\left(\frac{\partial\theta}{\partial x_i}\right) \equiv \delta\theta_{,i}, \tag{2.67}$$

we can arrive at

$$\int_\Omega \left(\rho c\dot{\theta}\delta T - \left(k_{ij}\theta_{,j}\delta\theta\right)_{,i} - \left(k_{ij}\theta_{,j}\right)\delta\theta_{,i} - g\delta\theta\right)\ \mathrm{d}\Omega = 0;\ \ \ x_i \in \Omega;\ \ \ \tau \in [0,\infty). \tag{2.68}$$

Introducing the additional heat transfer boundary conditions

$$\int_\Omega \left(k_{ij}\theta_{,j}\,\delta\theta\right)d\Omega = \int_{\partial\Omega} k_{ij}\,\theta_{,j}\,n_i\,\delta\theta d\left(\delta\Omega\right) = \int_{\partial\Omega_q} \hat{q}_i n_i\,\delta\theta d\left(\partial\Omega\right) \tag{2.69}$$

and integrating by parts, we obtain

$$\int_{\Omega} \left(\rho c \dot{\theta} \delta T + k_{ij} \theta_{,j} \delta \theta_{,i} - g \delta \theta \right) d\Omega - \int_{\partial \Omega_q} \hat{q}_i n_i \, \delta \theta d\,(\partial \Omega) = 0; \ \ x_i \in \Omega; \quad \tau \in [0, \infty). \tag{2.70}$$

This equation is the transient formulation of the principle of virtual temperature and is used to provide its stochastic perturbation technique counterpart. As a result, we obtain a set of algebraic equations of systematically increasing order (from zeroth up to nth). The following hold:

- zeroth-order partial differential equation

$$\int_{\Omega} \left(\rho^0 c^0 \dot{\theta}^0 \delta T + k_{ij}^0 \theta_{,j}^0 \delta \theta_{,i} \right) d\Omega = \int_{\partial \Omega_q} \hat{q}_i^0 n_i \, \delta \theta d\,(\partial \Omega) + \int_{\Omega} g^0 \delta \theta \, d\Omega \,, \tag{2.71}$$

- first-order partial differential equation

$$\begin{aligned} \int_{\Omega} \left(\rho^0 c^0 \frac{\partial \dot{\theta}}{\partial b} \delta T + k_{ij}^0 \frac{\partial \theta_{,j}}{\partial b} \delta \theta_{,i} \right) d\Omega &= \int_{\partial \Omega_q} \frac{\partial \hat{q}_i}{\partial b} n_i \, \delta \theta \, d\,(\partial \Omega) + \int_{\Omega} \frac{\partial g}{\partial b} \delta \theta d\Omega \\ &- \int_{\Omega} \left(\left(\frac{\partial \rho}{\partial b} c^0 + \rho^0 \frac{\partial c}{\partial b} \right) \dot{\theta}^0 \delta \theta + \frac{\partial k_{ij}}{\partial b} \theta_{,j}^0 \delta \theta_{,i} \right) d\Omega, \end{aligned} \tag{2.72}$$

- second-order partial differential equation

$$\begin{aligned} \int_{\Omega} \left(\rho^0 c^0 \frac{\partial^2 \dot{\theta}}{\partial b^2} \delta T + k_{ij}^0 \frac{\partial^2 \theta_{,j}}{\partial b^2} \delta \theta_{,i} \right) d\Omega &= \int_{\partial \Omega_q} \frac{\partial^2 \hat{q}_i}{\partial b^2} n_i \, \delta \theta d\,(\partial \Omega) + \int_{\Omega} \frac{\partial^2 g}{\partial b^2} \delta \theta \, d\Omega + \\ &- \int_{\Omega} \left(\left(\frac{\partial^2 \rho}{\partial b^2} c^0 + 2 \frac{\partial \rho}{\partial b} \frac{\partial c}{\partial b} + \rho^0 \frac{\partial^2 c}{\partial b^2} \right) \dot{\theta}^0 + \left(\frac{\partial \rho}{\partial b} c^0 + \rho^0 \frac{\partial c}{\partial b} \right) \frac{\partial \dot{\theta}}{\partial b} \right) \delta \theta d\Omega + \\ &- \int_{\Omega} \left(\frac{\partial^2 k_{ij}}{\partial b^2} \theta_{,j}^0 + 2 \frac{\partial k_{ij}}{\partial b} \frac{\partial \theta_{,j}}{\partial b} \right) \delta \theta_{,i} d\Omega, \end{aligned} \tag{2.73}$$

- nth-order equation

$$\begin{aligned} &\int_{\Omega} \left(\sum_{k=0}^{n} \binom{n}{k} \left(\sum_{m=0}^{k} \binom{k}{m} \frac{\partial^k \rho}{\partial b^k} \frac{\partial^{k-m} c}{\partial b^{k-m}} \right) \frac{\partial^{n-k} \dot{\theta}}{\partial b^{n-k}} \right) \delta T d\Omega + \int_{\Omega} \left(\sum_{k=0}^{n} \binom{n}{k} \frac{\partial^k k_{ij}}{\partial b^k} \frac{\partial^{n-k} \theta_{,j}}{\partial b^{n-k}} \right) \delta \theta_{,i} d\Omega \\ &= \int_{\partial \Omega_q} \frac{\partial^n \hat{q}_i}{\partial b^n} n_i \, \delta \theta d\,(\partial \Omega) + \int_{\Omega} \frac{\partial^n g}{\partial b^n} \delta \theta d\Omega \end{aligned} \tag{2.74}$$

Having solved these equations for θ^0 up to the nth-order partial derivative $\frac{\partial^n \theta}{\partial b^n}$, respectively, we derive the expressions for the expected values and higher probabilistic moments and the coefficients for the temperature field and its time fluctuations.

2.1.6 Thermopiezoelectricity Governing Equations

Let us consider the free energy of the thermopiezoelectric continuum consisting of thermal, electric, and mechanical field contributions [95, 116, 137]:

$$W\left(\varepsilon_{ij}, E_i, \Theta\right) = \frac{1}{2}C_{ijkl}\varepsilon_{ij}\varepsilon_{kl} - e_{ijk}E_i\varepsilon_{jk} - \frac{1}{2}b_{ij}E_iE_j - \lambda_{ij}\theta\varepsilon_{ij} - d_iE_i\theta - \frac{1}{2}\alpha_T\theta^2. \quad (2.75)$$

Here ε_{ij} are the components of the overall strain tensor, E_i the components of the electric field vector, and θ the actual temperature change from the reference temperature θ_0 defining the state with no initial stresses or strains. The fourth- and third-order tensors C_{ijkl} and e_{ijk} contain the elastic and piezoelectric constants, while b_{ij} stands for the dielectric permittivity. Further, the tensors λ_{ij} and d_i represent thermomechanical and thermopiezoelectric coupling constants. Additionally, the dissipation function F is considered here as

$$F\left(e_i\right) = \frac{1}{2}k_{ij}e_ie_j, \quad (2.76)$$

where k_{ij} means the heat conductivity tensor and e_i is the thermal field vector. As we know, the constitutive equations are derived from Equation (2.75), whereas the Fourier law is a consequence of the dissipation function. The following hold:

$$\sigma_{ij} = \frac{\partial W\left(\varepsilon_{ij}, E_i, \Theta\right)}{\partial \varepsilon_{ij}} = C_{ijkl}\varepsilon_{kl} - e_{ijk}E_k - \lambda_{ij}\theta, \quad (2.77)$$

$$D_i = -\frac{\partial W\left(\varepsilon_{ij}, E_i, \Theta\right)}{\partial E_i} = e_{ijk}\varepsilon_{jk} + b_{ij}E_j + d_i\theta, \quad (2.78)$$

$$S = -\frac{\partial W\left(\varepsilon_{ij}, E_i, \Theta\right)}{\partial \theta} = \lambda_{ij}\varepsilon_{ij} + d_iE_i + \alpha_T\theta, \quad (2.79)$$

$$q_i = \frac{\partial F\left(e_i\right)}{\partial e_i} = k_{ij}e_j. \quad (2.80)$$

The stress tensor σ_{ij}, the components of the electric displacement D_i, the entropy S, and the components of the heat flux vector q_i are determined uniquely from the above equations. The following assumptions accompany this solution:

1. small displacements limitation, where

$$\varepsilon_{kl} = \frac{1}{2}\left(\frac{\partial u_k}{\partial x_l} + \frac{\partial u_l}{\partial x_k}\right) \quad (2.81)$$

2. linear piezoelectricity

$$E_i = -\Phi_{,i} \tag{2.82}$$

where Φ denotes the scalar potential function
3. the following description of the thermal field vector:

$$e_i = -\theta_{,i}. \tag{2.83}$$

Finally, the Hamilton principle is used (see Equation (2.59)):

$$\delta \int_{t_1}^{t_2} (L + W)\, \mathrm{d}t = 0, \tag{2.84}$$

where L and W include all thermal, electric, and mechanical contributions with all variations (of displacement, temperature, etc.) vanishing at the beginning t_1 and at the end t_2 of the considered time period. The Lagrangian L is defined here as follows:

$$L = \int_{\Omega} \left(T + F - U - S\theta - S\theta_0\dot{\theta}\right) \mathrm{d}\Omega, \tag{2.85}$$

where

$$T = \frac{1}{2}\rho\dot{u}_i\dot{u}_i \tag{2.86}$$

as before, is the kinetic energy, while the remaining components become

$$U = \frac{1}{2}\sigma_{ij}\varepsilon_{ij} - \frac{1}{2}D_iE_i - \frac{1}{2}S\theta, \tag{2.87}$$

$$F = \frac{1}{2}q_ie_i. \tag{2.88}$$

Determining all variations with respect to displacement, temperature, and electric potential one can obtain the following set of variational equations having unique solution [137]:

$$\delta_u (L + W) = -\int_{t_1}^{t_2}\int_{\Omega} \left(\rho\ddot{u}_i\delta u_i + \sigma_{ij}\delta\varepsilon_{ij}\right) d\Omega dt + \int_{t_1}^{t_2}\int_{\partial\Omega_\sigma} \hat{t}_i\delta u_i\, d(\partial\Omega)\, dt = 0, \tag{2.89}$$

$$\delta_\Phi (L + W) = -\int_{t_1}^{t_2}\int_{\Omega} D_i\delta\Phi_{,i}\; d\Omega dt + \int_{t_1}^{t_2}\int_{\partial\Omega_e} \hat{q}_e\delta\Phi\; d(\partial\Omega)\, dt = 0, \tag{2.90}$$

$$\delta_\theta (L + W) = \int_{t_1}^{t_2}\int_{\Omega} \left(-q_i\delta\theta_{,i} + \dot{S}\theta_0\delta\theta\right) d\Omega dt + \int_{t_1}^{t_2}\int_{\partial\Omega_S} \hat{q}_S\delta\theta\, d(\partial\Omega)\, dt = 0. \tag{2.91}$$

These equations are valid for some continuum Ω with external boundary $\partial\Omega$, where the boundary conditions expressed by $\hat{t}_i$, $\hat{q}_e$, and $\hat{q}_S$ represent the boundary tractions, the charge density, and the boundary heat fluxes, all uniquely defined. Variational Equations (2.89)–(2.91) are consequently rewritten in the context of the stochastic perturbation technique and, for a computational implementation, they are presented in the recursive form. This starts traditionally from the zeroth-order equations, being in fact almost the same as these given above and that is why we suppress the first- and the second-order relations [95]. The following hold:

$$\int_{t_1}^{t_2}\int_{\Omega}\left(\rho^0\ddot{u}_i^0\delta u_i+\sigma_{ij}^0\delta\varepsilon_{ij}\right)d\Omega dt=\int_{t_1}^{t_2}\int_{\partial\Omega_\sigma}\hat{t}_i^0\delta u_i\, d\left(\partial\Omega\right)dt, \tag{2.92}$$

$$\int_{t_1}^{t_2}\int_{\Omega}D_i^0\delta\Phi_{,i}\,d\Omega dt=\int_{t_1}^{t_2}\int_{\partial\Omega_e}\hat{q}_e^0\delta\Phi\, d\left(\partial\Omega\right)dt, \tag{2.93}$$

$$\int_{t_1}^{t_2}\int_{\Omega}\left(q_i^0\delta\theta_{,i}-\dot{S}^0\theta_0^0\delta\theta\right)d\Omega dt=\int_{t_1}^{t_2}\int_{\partial\Omega_S}\hat{q}_S^0\delta\theta\, d\left(\partial\Omega\right)dt. \tag{2.94}$$

Having solved the zeroth-order quantities, the higher-order responses need to be computed. Therefore, the general nth-order variational equations are introduced as

$$\int_{t_1}^{t_2}\int_{\Omega}\left(\sum_{k=0}^{n}\binom{n}{k}\frac{\partial^k\rho}{\partial b^k}\frac{\partial^{n-k}\ddot{u}}{\partial b^{n-k}}\delta u_i+\frac{\partial^n\sigma_{ij}}{\partial b^n}\delta\varepsilon_{ij}\right)d\Omega dt=\int_{t_1}^{t_2}\int_{\partial\Omega_\sigma}\frac{\partial^n\hat{t}_i}{\partial b^n}\delta u_i\, d\left(\partial\Omega\right)dt, \tag{2.95}$$

$$\int_{t_1}^{t_2}\int_{\Omega}\frac{\partial^n D_i}{\partial b^n}\delta\Phi_{,i}\,d\Omega dt=\int_{t_1}^{t_2}\int_{\partial\Omega_e}\frac{\partial^n\hat{q}_e}{\partial b^n}\delta\Phi\, d\left(\partial\Omega\right)dt, \tag{2.96}$$

$$\int_{t_1}^{t_2}\int_{\Omega}\left(\frac{\partial^n q_i}{\partial b^n}\delta\theta_{,i}-\sum_{k=0}^{n}\binom{n}{k}\frac{\partial^k\dot{S}}{\partial b^k}\frac{\partial^{n-k}\theta_0}{\partial b^{n-k}}\delta\theta\right)d\Omega dt=\int_{t_1}^{t_2}\int_{\partial\Omega_S}\frac{\partial^n\hat{q}_S}{\partial b^n}\delta\theta\, d\left(\partial\Omega\right)dt. \tag{2.97}$$

where $\frac{\partial^n\sigma_{ij}}{\partial b^n}$ is equivalent to the nth-order partial derivative of the stress tensor with respect to the input random variable, and so forth. The remaining part of the solution proceeds relatively easily – zeroth-order solutions are inserted into the first-order versions of Equations (2.95)–(2.97), where the first-order solutions are extracted to be included in the second-order equations, and etc. The choice of highest-order stochastic expansion depends strongly on the desired accuracy of the computational process and needs to be the subject of separate analysis. As we demonstrate further, the straightforward differentiation technique applied to the Hamilton principle may be replaced with some alternative techniques to determine partial derivatives of the structural response with respect to some random variables. Finally, let us mention that some analytical solutions to such coupled problems (even with magnetic field contributions) are available, but they are valid for specific domains or component distributions in space, so that it is impossible to provide any general probabilistic methodology on their basis; one may use in these cases the Monte Carlo simulation only to calculate statistical parameters for the structural response.

2.1.7 Navier–Stokes Equations

Let us consider the phenomenon of a viscous incompressible flow for a homogeneous fluid in the region Ω. The equations of momentum, of continuity, as well as the constitutive relations defining our problem can be listed as follows [3, 94]:

$$\rho\left(\frac{\partial v_i}{\partial t}+v_{i,j}v_j\right)=\sigma_{ij,j}+f_i, \tag{2.98}$$

$$v_{i,i}=0, \tag{2.99}$$

$$\sigma_{ij}=-p\delta_{ij}+2\mu\varepsilon_{ij}, \tag{2.100}$$

$$\rho c\left(\frac{\partial \theta}{\partial t}+\theta_{,i}v_i\right)-\left(k\theta_{,i}\right)_{,i}=g, \tag{2.101}$$

where

$$\varepsilon_{ij}=\frac{1}{2}\left(v_{i,j}+v_{j,i}\right). \tag{2.102}$$

The state variables v_i, p, ε_{ij}, σ_{ij}, and θ denote velocity and pressure of the fluid, components of velocity strain and stress tensor, as well as temperature, respectively, while f_i is a component of the body force vector (force per unit volume of fluid). The variables – that is, ρ, μ, c, and k – denote mass density, fluid (laminar) viscosity, heat capacity, and heat conductivity, respectively. Equations (2.98)–(2.102) represent the standard Navier–Stokes equations governing the motion of a viscous, incompressible fluid in laminar flow coupled with heat transfer. The boundary conditions corresponding to these equations are introduced in the following form:

- fluid velocity boundary conditions

$$v_i=\hat{v}_i;\;\; x_i\in\partial\Omega_v, \tag{2.103}$$

- stress boundary conditions

$$\sigma_{ij}n_j=\hat{t}_i;\;\; x_i\in\partial\Omega_\sigma. \tag{2.104}$$

- temperature boundary conditions

$$\theta = \hat{\theta}; \quad x_i \in \partial\Omega_\theta \tag{2.105}$$

and prescribed heat fluxes

$$k\frac{\partial\theta}{\partial x} = \hat{q}; \quad x_i \in \partial\Omega_q. \tag{2.106}$$

Next, the weak formulation is proposed which makes it possible to introduce the FEM equations of the problem. For this purpose, Equation (2.98) with inserted Equation (2.100) is multiplied by the virtual velocity δv_i and integrated over the fluid domain Ω as follows:

$$\int_\Omega \delta v_i \rho \left(\dot{v}_i + v_{i,j} v_j\right) d\Omega + \int_\Omega \delta\varepsilon_{ij}\left(2\mu\varepsilon_{ij} - p\delta_{ij}\right) d\Omega = \int_\Omega \delta v_i f_i d\Omega + \int_{\partial\Omega_\sigma} \delta v_i \hat{t}_i d\left(\partial\Omega_\sigma\right). \tag{2.107}$$

Equation (2.99) is transformed through multiplication by the virtual pressure δp:

$$\int_\Omega \delta p v_{i,i} \mathrm{d}\Omega = 0, \tag{2.108}$$

whereas Equation (2.101) adjacent to the heat transfer is obtained as

$$\int_\Omega \delta\theta \rho c\left(\dot{\theta} + \theta_{,i} v_i\right) d\Omega + \int_\Omega k\delta\theta_{,i}\theta_{,i} d\Omega = \int_\Omega \delta\theta g d\Omega + \int_{\partial\Omega_q} \delta\hat{\theta}\hat{q}_i n_i d\left(\partial\Omega\right). \tag{2.109}$$

Equations (2)–(2.109) are next extended using the generalized perturbation technique to formulate and solve the corresponding flow problem. Equating terms of equal order in the resulting expressions, the zeroth-, first-, second and n-th order equations for the flow considered we obtained as follows [67, 94]:

- zeroth-order (ε^0 terms, one partial differential equation)

$$\int_\Omega \delta v_i \rho^0 \left(\dot{v}_i^0 + v_{i,j}^0 v_j^0\right) d\Omega + \int_\Omega \delta\varepsilon_{ij}\left(2\mu^0\varepsilon_{ij}^0 - p^0\delta_{ij}\right) d\Omega = $$

$$= \int_\Omega \delta v_i \left(f_i\right)^0 d\Omega + \int_{\partial\Omega_\sigma} \delta v_i \left(\hat{t}_i\right)^0 d\left(\partial\Omega\right), \tag{2.110}$$

$$\int_\Omega \delta p v_{i,i}^0 \mathrm{d}\Omega = 0, \tag{2.111}$$

$$\int_{\Omega} \delta\theta \rho^0 c^0 \left(\dot{\theta}^0 + \theta^0_{,i} v^0_i\right) d\Omega + \int_{\Omega} k^0 \delta\theta_{,i} \theta^0_{,i} d\Omega = \int_{\Omega} \delta\theta \left(g\right)^0 d\Omega + \int_{\partial\Omega_q} \delta\hat{\theta} \left(\hat{q}_i\right)^0 n_i d\left(\partial\Omega\right); \tag{2.112}$$

- first-order equations

$$\int_{\Omega} \delta v_i \left(\frac{\partial\rho}{\partial b} \dot{v}^0_i + \rho^0 \frac{\partial \dot{v}_i}{\partial b} + \frac{\partial\rho}{\partial b} v^0_{i,j} v^0_j + \rho^0 \frac{\partial v_{i,j}}{\partial b} v^0_j + \rho^0 v^0_{i,j} \frac{\partial v_j}{\partial b} \right) d\Omega +$$
$$+ \int_{\Omega} \delta\varepsilon_{ij} \left(2\frac{\partial\mu}{\partial b} \varepsilon^0_{ij} + 2\mu^0 \frac{\partial\varepsilon_{ij}}{\partial b} - \frac{\partial p}{\partial b} \delta_{ij} \right) d\Omega = \int_{\Omega} \delta v_i \frac{\partial f_i}{\partial b} d\Omega + \int_{\partial\Omega_\sigma} \delta v_i \frac{\partial \hat{t}_i}{\partial b} d\left(\partial\Omega\right) \tag{2.113}$$

$$\int_{\Omega} \delta p \; \frac{\partial v_{i,i}}{\partial b} \mathrm{d}\Omega = 0, \tag{2.114}$$

$$\int_{\Omega} \delta\theta \left(\frac{\partial\rho}{\partial b} c^0 \dot{\theta}^0 + \rho^0 \frac{\partial c}{\partial b} \dot{\theta}^0 + \rho^0 c^0 \frac{\partial\dot{\theta}}{\partial b} + \frac{\partial\rho}{\partial b} c^0 \theta^0_{,i} v^0_i + \rho^0 \frac{\partial c}{\partial b} \theta^0_{,i} v^0_i \right) d\Omega +$$
$$+ \int_{\Omega} \delta\theta \left(\rho^0 c^0 \frac{\partial\theta_{,i}}{\partial b} v^0_i + \rho^0 c^0 \theta^0_{,i} \frac{\partial v_i}{\partial b} \right) d\Omega$$
$$+ \int_{\Omega} \delta\theta_{,i} \left(\frac{\partial k}{\partial b} \theta^0_{,i} + k^0 \frac{\partial\theta_{,i}}{\partial b} \right) d\Omega = \int_{\Omega} \delta\theta \frac{\partial g}{\partial b} d\Omega + \int_{\partial\Omega_q} \delta\hat{\theta} \frac{\partial \hat{q}_i}{\partial b} n_i d\left(\partial\Omega\right) \tag{2.115}$$

- second-order equations

$$\int_{\Omega} \delta v_i \rho^0 \left(\frac{\partial^2 \dot{v}_i}{\partial b^2} + \frac{\partial^2 v_{i,j}}{\partial b^2} v^0_j + v^0_{i,j} \frac{\partial^2 v_j}{\partial b^2} \right) d\Omega + 2 \int_{\Omega} \delta\varepsilon_{ij} \mu^0 \frac{\partial^2 \varepsilon_{ij}}{\partial b^2} d\Omega$$
$$= - \int_{\Omega} \delta\varepsilon_{ij} \left(4 \frac{\partial\mu}{\partial b} \frac{\partial\varepsilon_{ij}}{\partial b} - \frac{\partial^2 p}{\partial b^2} \delta_{ij} \right) d\Omega + \int_{\Omega} \delta v_i \frac{\partial^2 f_i}{\partial b^2} d\Omega + \int_{\partial\Omega_\sigma} \delta v_i \frac{\partial^2 \hat{t}_i}{\partial b^2} d\left(\partial\Omega\right) +$$
$$- 2 \int_{\Omega} \delta v_i \left(\frac{\partial\rho}{\partial b} \frac{\partial\dot{v}_i}{\partial b} + \frac{\partial\rho}{\partial b} \frac{\partial v_{i,j}}{\partial b} v^0_j + \rho^0 \frac{\partial v_{i,j}}{\partial b} \frac{\partial v_j}{\partial b} + \frac{\partial\rho}{\partial b} v^0_{i,j} \frac{\partial v_j}{\partial b} \right) d\Omega \tag{2.116}$$

$$\int_{\Omega} \delta p \; \frac{\partial^2 v_{i,i}}{\partial b^2} \mathrm{d}\Omega = 0, \tag{2.117}$$

$$\int_{\Omega} \delta\theta \rho^0 c^0 \left(\frac{\partial^2 \dot{\theta}}{\partial b^2} + \frac{\partial^2 \theta_{,i}}{\partial b^2} v^0_i + \theta^0_{,i} \frac{\partial^2 v_i}{\partial b^2} \right) d\Omega + \int_{\Omega} \delta\theta_{,i} k^0 \frac{\partial^2 \theta_{,i}}{\partial b^2} d\Omega = \int_{\Omega} \delta\theta \frac{\partial^2 g}{\partial b^2} d\Omega$$

$$+\int_{\partial\Omega_q}\delta\hat{\theta}\frac{\partial^2\hat{q}_i}{\partial b^2}n_i d(\partial\Omega)-2\int_{\Omega}\delta\theta\left(\frac{\partial\rho}{\partial b}c^0\frac{\partial\theta_{,i}}{\partial b}v_i^0+\frac{\partial\rho}{\partial b}c^0\theta^0_{,i}\frac{\partial v_i}{\partial b}\right)d\Omega+$$
$$-2\int_{\Omega}\delta\theta\left(\rho^0\frac{\partial c}{\partial b}\frac{\partial\theta_{,i}}{\partial b}v_i^0+\rho^0\frac{\partial c}{\partial b}\theta^0_{,i}\frac{\partial v_i}{\partial b}+\rho^0c^0\frac{\partial\theta_{,i}}{\partial b}\frac{\partial v_i}{\partial b}\right)d\Omega-2\int_{\Omega}\delta\theta_{,i}\frac{\partial k}{\partial b}\frac{\partial\theta_{,i}}{\partial b}d\Omega \tag{2.118}$$

Let us observe that this formulation can be generalized to the nth-order perturbation and then, the following coupled equations are obtained thanks to the Leibniz chain differentiation rule:

$$\int_{\Omega}\delta v_i\left(\sum_{k=0}^{n}\binom{n}{k}\frac{\partial^k\rho}{\partial b^k}\frac{\partial^{n-k}\dot{v}_i}{\partial b^{n-k}}+\sum_{k=0}^{n}\binom{n}{k}\frac{\partial^k\rho}{\partial b^k}\left(\sum_{m=0}^{n-k}\binom{n-k}{m}\frac{\partial^m v_{i,j}}{\partial b^m}\frac{\partial^{n-k-m}v_j}{\partial b^{n-k-m}}\right)\right)d\Omega$$
$$+2\int_{\Omega}\delta\varepsilon_{ij}\left(\sum_{k=0}^{n}\binom{n}{k}\frac{\partial^k\mu}{\partial b^k}\frac{\partial^{n-k}\varepsilon_{ij}}{\partial b^{n-k}}-\frac{\partial^n p}{\partial b^n}\delta_{ij}\right)d\Omega=\int_{\Omega}\delta v_i\frac{\partial^n f_i}{\partial b^n}d\Omega+\int_{\partial\Omega_\sigma}\delta v_i\frac{\partial^n\hat{t}_i}{\partial b^n}d(\partial\Omega) \tag{2.119}$$

$$\int_{\Omega}\delta p\ \frac{\partial^n v_{i,i}}{\partial b^n}d\Omega=0, \tag{2.120}$$

$$\int_{\Omega}\delta\theta\left(\sum_{k=0}^{n}\binom{n}{k}\left(\sum_{m=0}^{k}\binom{k}{m}\frac{\partial^k\rho}{\partial b^k}\frac{\partial^{k-m}c}{\partial b^{k-m}}\right)\frac{\partial^{n-k}\dot{\theta}}{\partial b^{n-k}}\right)d\Omega+$$
$$+\int_{\Omega}\delta\theta\left(\sum_{k=0}^{n}\binom{n}{k}\left(\sum_{m=0}^{k}\binom{k}{m}\frac{\partial^k\rho}{\partial b^k}\frac{\partial^{k-m}c}{\partial b^{k-m}}\right)\left(\sum_{l=0}^{n-k}\binom{n-k}{l}\frac{\partial^l\theta_{,i}}{\partial b^l}\frac{\partial^{n-k-l}v_i}{\partial b^{n-k-l}}\right)\right)d\Omega+$$
$$+\int_{\Omega}\delta\theta_{,i}\sum_{k=0}^{n}\binom{n}{k}\frac{\partial^k k_i}{\partial b^k}\frac{\partial^{n-k}\theta_{,i}}{\partial b^{n-k}}d\Omega=\int_{\Omega}\delta\theta\frac{\partial^n g}{\partial b^n}d\Omega+\int_{\partial\Omega_q}\delta\hat{\theta}\frac{\partial^n\hat{q}_i}{\partial b^n}n_i d(\partial\Omega) \tag{2.121}$$

To transform these hierarchical equations into a form convenient for some computation, it is necessary to multiply each of them by the nth-order probabilistic moments of the input random variables or fields. Because of the great complexity of such a solution, the second-order perturbation approach is usually preferred, where computing the zeroth-order velocity, pressure, and temperature functions from Equations (2.110)–(2.112), next their first-order approximations using Equations (2.113)–(2.115), and finally the second-order terms from Equations (2.116)–(2.118), the first two probabilistic moments of these functions are derived [94].

2.2 Stochastic Finite Element Method Equations

2.2.1 Linear Potential Problems

Following the traditional FEM discretization statement [3, 180], we introduce the nodal definition of any partial order derivatives of the potential function with respect to the random input parameter as

$$\frac{\partial^m \phi}{\partial b^m} = N_\beta \frac{\partial^m \Phi_\beta}{\partial b^m}, \ \beta = 1, \ldots, R, \ m = 0, \ldots, n; \tag{2.122}$$

and quite similarly for their spatial derivatives

$$\frac{\partial^m (\phi_{,i})}{\partial b^m} = \frac{\partial N_\beta}{\partial x_i} \frac{\partial^m \Phi_\beta}{\partial b^m}, \ \beta = 1, \ldots, R, \ m = 0, \ \ldots, \ n. \tag{2.123}$$

To provide the direct differentiation method (DDM) formulation of the perturbation-based FEM, we consequently rewrite Equation (2.9) in this context as

$$\frac{\partial^m k_{\alpha\beta}^{(f)}}{\partial b^m} = \int_{\Omega_f} \left(\sum_i \frac{\partial^m k_i}{\partial b^m} \frac{\partial N_\alpha}{\partial x_i} \frac{\partial N_\beta}{\partial x_i} \right) d\Omega,$$

$$\alpha, \beta = 1, \ \ldots, R, \ m = 0, \ \ldots, \ n; \ f = 1, \ \ldots, \ F. \tag{2.124}$$

and, analogously using Equation (2.10), one can rewrite

$$\frac{\partial^m Q_\alpha^{(f)}}{\partial b^m} = \int_{\Omega_f} \left(-\frac{\partial^m q}{\partial b^m} N_\alpha \right) d\Omega + \int_{\partial\Omega_q^{(f)}} \frac{\partial^m \hat{q}_i}{\partial b^m} n_i N_\alpha \, d(\partial\Omega). \tag{2.125}$$

Finally, the hierarchical algebraic equations of up to nth order are formed and may be represented as the single recursive statement

$$\sum_{m=0}^{n} \binom{n}{m} \frac{\partial^{n-m} K_{\alpha\beta}}{\partial b^{n-m}} \frac{\partial^m \Phi_\beta}{\partial b^m} = \frac{\partial^n Q_\alpha}{\partial b^n}. \tag{2.126}$$

The zeroth-, first-, second-, and highest-order equations are extracted from this statement and, then, the first of them is solved for zeroth-order potentials, this solution is embedded into the second one on the left-hand side, and this equation is solved for first-order potentials. Consequently, this solution is inserted numerically into the next-order equations and this recursion proceeds until the highest-order equation is found. Having solved these equations for $\frac{\partial^m \Phi_\beta}{\partial b^m}$, $m = 0, \ldots, n$, these nodal solutions are used to determine the first four probabilistic moments in any degree of freedom β. Let us note that in most practical applications higher order partial derivatives vanish very fast since the system matrices and vectors depend on random variables usually by lower-order polynomials only.

Alternatively, the response function method (RFM) is proposed to implement the nth-order perturbation-based stochastic finite element method (SFEM). The nodal values of the potential function and its partial derivatives to the random input are found here in the approximate sense using polynomial response functions of this input as [82]

$$\frac{\partial^m \Phi_\beta}{\partial b^m} = \frac{\partial^m}{\partial b^m} f_\beta(b) = \frac{\partial^m}{\partial b^m} D_\beta^{(k)} b^k,\ \beta = 1, \ldots, R,\ m = 0,\ \ldots,\ n; \tag{2.127}$$

so that the random field discretization procedure of this system response starts from the following representation:

$$\frac{\partial^m \phi}{\partial b^m} = N_\beta \frac{\partial^m \Phi_\beta}{\partial b^m} = N_\beta \frac{\partial^m}{\partial b^m}\left(D_\beta^{(k)} b^k\right) = N_\beta D_\beta^{(k)} \frac{\partial^m \left(b^k\right)}{\partial b^m},\ \beta = 1, \ldots, R,\ m = 0, \ldots, n; \tag{2.128}$$

Deterministic coefficients $D_\beta^{(k)}$ were initially numerically determined from several deterministic solutions to the original matrix equation with random parameter value varying around its mean value in the interval $b = [b^0 - \Delta b, b^0 + \Delta b]$. The uniform subdivision of this interval resulting in a set of the values b_i, $i = 1, \ldots, n$ was made according to the assumed order of the perturbation method (second order analysis require three points in such a grid, etc.). The additional nodal equation systems for all degrees of freedom $\beta = 1, \ldots, R$ may be solved independently for this purpose as

$$\begin{bmatrix} b_1^{n-1} & b_1^{n-2} & \ldots & b_1^0 \\ b_2^{n-1} & b_2^{n-2} & & b_2^0 \\ \ldots & \ldots & & \ldots \\ b_n^{n-1} & b_n^{n-2} & \ldots & b_n^0 \end{bmatrix} \begin{Bmatrix} D_\beta^{(n-1)} \\ D_\beta^{(n-2)} \\ \ldots \\ D_\beta^{(0)} \end{Bmatrix} = \begin{Bmatrix} \Phi_\beta^{(1)} \\ \Phi_\beta^{(2)} \\ \ldots \\ \Phi_\beta^{(n)} \end{Bmatrix} \tag{2.129}$$

The unique solutions for these systems make it possible to calculate analytically up to n-th-order ordinary derivatives of the nodal responses $\Phi_\beta(b)$ with respect to b at the given b_0. It holds that

$$\frac{\partial^k \Phi_\beta}{\partial b^k} = \prod_{i=1}^{k} (n-i)\, D_\beta^{(n-1)} b^{n-k} + \prod_{i=2}^{k} (n-i)\, D_\beta^{(n-2)} b^{n-(k+1)} + \ldots + D_\beta^{(n-k)}. \tag{2.130}$$

2.2.2 Linear Elastostatics

Let us introduce the following approximation for the displacement field [180] and its nth-order partial derivatives with respect to the input random variable using the shape functions $\varphi_{i\alpha}(x)$:

$$u_i^0(x) = \varphi_{i\alpha}(x)\, q_\alpha^0,\; x_i \in \Omega, \tag{2.131}$$

$$\frac{\partial^k u_i(x)}{\partial b^k} = \varphi_{i\alpha}(x)\, \frac{\partial^k q_\alpha}{\partial b^k},\; x_i \in \Omega, \tag{2.132}$$

$i=1,2,3; \alpha=1,\ldots,R$ (R is the total number of degrees of freedom introduced in Ω); and $k=1,\ldots,n$. The strain tensor components are discretized analogously and it yields

$$\varepsilon_{ij}^0(x) = B_{ij\alpha}(x) q_\alpha^0, \quad x_i \in \Omega, \tag{2.133}$$

$$\frac{\partial^n \varepsilon_{ij}(x)}{\partial b^n} = B_{ij\alpha}(x)\frac{\partial^n q_\alpha}{\partial b^n}, \quad x_i \in \Omega, \tag{2.134}$$

where $B_{ij\alpha}(x_k)$ is the matrix containing the shape function derivatives

$$B_{kl\alpha}(x_i) = \frac{1}{2}\left[\varphi_{k\alpha,l}(x_i) + \varphi_{l\alpha,k}(x_i)\right], \quad x_i \in \Omega. \tag{2.135}$$

The FEM approach is finally obtained as the linear algebraic equations system [3, 180]

$$K_{\alpha\beta} q_\beta = Q_\alpha, \tag{2.136}$$

where q_β is the displacement solution vector. When some random quantities are inserted into the matrix $K_{\alpha\beta}$ and the vector Q_α, then Equation (2.136) should be rewritten and solved to determine the first consecutive orders of the random structural response.
It yields

- zeroth-order equations

$$K_{\alpha\beta}^0 q_\beta^0 = Q_\alpha^0 \tag{2.137}$$

- first-order equations

$$K_{\alpha\beta}^0 \frac{\partial q_\beta}{\partial b} = \frac{\partial Q_\alpha}{\partial b} - \frac{\partial K_{\alpha\beta}}{\partial b} q_\beta^0 \tag{2.138}$$

- second-order equations

$$K_{\alpha\beta}^0 \frac{\partial^2 q_\beta}{\partial b^2} = \frac{\partial^2 Q_\alpha}{\partial b^2} - 2\frac{\partial K_{\alpha\beta}}{\partial b}\frac{\partial q_\beta}{\partial b} - \frac{\partial^2 K_{\alpha\beta}}{\partial b^2} q_\beta^0 \tag{2.139}$$

- nth-order equations

$$\sum_{k=0}^{n} \binom{n}{k} \frac{\partial^k K_{\alpha\beta}}{\partial b^k}\frac{\partial^{n-k} q_\beta}{\partial b^{n-k}} = \frac{\partial^n Q_\alpha}{\partial b^n}. \tag{2.140}$$

Let us recall the classical definition of the stiffness matrix in the form

$$K_{\alpha\beta} = \sum_{f=1}^{F} \int_{\Omega_f} C_{ijkl}^{(f)} B_{ij\alpha} B_{kl\beta} \mathrm{d}\Omega, \tag{2.141}$$

where $C_{ijkl}^{(f)}$ denotes the elasticity tensor components for the finite element f and F is their total number within Ω. Thus, we can describe the stiffness matrix nth-order derivatives with respect to Young's modulus e as follows:

$$\frac{\partial K_{\alpha\beta}}{\partial f} = \frac{\partial C_{ijkl}}{\partial f} \sum_{f=1}^{F} \int_{\Omega_f} B_{ij\alpha} B_{kl\beta} \mathrm{d}\Omega = A_{ijkl} \sum_{f=1}^{F} \int_{\Omega_f} B_{ij\alpha} B_{kl\beta} \mathrm{d}\Omega \tag{2.142}$$

with the tensor A_{ijkl} defined in Equation (2.30) and

$$\frac{\partial^n K_{\alpha\beta}}{\partial e^n} = 0 \tag{2.143}$$

for any $n \geq 2$. Furthermore, it is seen that all partial derivatives of external load vector as independent of Young's modulus are equal to 0. Therefore, we have

- zeroth-order equations

$$K_{\alpha\beta}^{0} q_{\beta}^{0} = Q_{\alpha}^{0} \tag{2.144}$$

- first-order equations

$$K_{\alpha\beta}^{0} \frac{\partial q_{\beta}}{\partial e} = -\frac{\partial K_{\alpha\beta}}{\partial e} q_{\beta}^{0} \tag{2.145}$$

- second-order equations

$$K_{\alpha\beta}^{0} \frac{\partial^2 q_{\beta}}{\partial e^2} = -2 \frac{\partial K_{\alpha\beta}}{\partial e} \frac{\partial q_{\beta}}{\partial e} \tag{2.146}$$

- nth-order equations

$$K_{\alpha\beta}^{0} \frac{\partial^n q_{\beta}}{\partial e^n} = -(n-1) \frac{\partial K_{\alpha\beta}}{\partial e} \frac{\partial^{n-1} q_{\beta}}{\partial e^{n-1}}. \tag{2.147}$$

Finally, from the first equation of this system a zeroth-order solution is determined, which, inserted into the next equations, returns a first-order solution, and so forth, until an nth-order solution is completed. After all the solution vector components have been determined, their expected values, variances, and other probabilistic moments can be extracted. As is demonstrated further, symbolic computations packages, like Maple™, for example, having linear algebra options implemented, can be very efficient for the purpose of perturbation methodology implementations. Finally, it should be underlined that in most engineering applications, the state function and the state vector are not Gaussian variables. This makes it necessary to compute higher and especially odd probabilistic moments to proceed in a similar way (see Appendix for further equations).

Contrary to the straightforward differentiation methodology given above, let us follow the idea of the common implementation of the FEM and RFM displayed in Chapter 1. The crucial point of this approach is polynomial representation of the structural response functions – displacements, strains, and stresses in the following form, similarly to the traditional discretization:

- displacement vector discretization

$$q_\beta = D_\beta^{(p)} h^p, p = 0, \ldots, n-1; \;\; \beta = 1, \ldots, R \tag{2.148}$$

- strain tensor components discretization

$$\varepsilon_{kl} = B_{kl\beta} q_\beta = B_{kl\beta} D_\beta^{(p)} h^p, \; p = 0, \ldots, n-1; \;\; \beta = 1, \ldots, R; \;\; k,l = 1,2,3; \tag{2.149}$$

- stress tensor discretization

$$\sigma_{ij} = C_{ijkl}\varepsilon_{kl} = C_{ijkl} B_{kl\beta} q_\beta = C_{ijkl} B_{kl\beta} D_\beta^{(p)} h^p, \; p = 0, \ldots, n-1;$$
$$\alpha, \; \beta = 1, \ldots, R; \;\; i,j,k,l = 1,2,3. \tag{2.150}$$

Hence, partial derivatives of the state functions with respect to the random parameter b can be derived simply as

- first partial derivative of the displacement vector

$$\frac{\partial q_\beta}{\partial b} = \frac{\partial}{\partial b}\left(D_\beta^{(p)} b^p\right) = p D_\beta^{(p)} b^{p-1}, \; p = 0, \ldots, n-1; \;\; \beta = 1, \ldots, R \tag{2.151}$$

- first partial derivative of the strain tensor components

$$\frac{\partial \varepsilon_{kl}}{\partial b} = B_{kl\beta}\frac{\partial}{\partial b}(q_\beta) = B_{kl\beta} p D_\beta^{(p)} b^{p-1}, \; p = 0, \ldots, n-1; \;\; \beta = 1, \ldots, R; \;\; k,l = 1,2,3; \tag{2.152}$$

- first partial derivative of the stress tensor components

$$\begin{aligned} \frac{\partial \sigma_{ij}}{\partial b} &= \frac{\partial}{\partial b}\left(C_{ijkl}\varepsilon_{kl}\right) = \frac{\partial C_{ijkl}}{\partial b} B_{kl\beta} q_\beta + C_{ijkl} B_{kl\beta}\frac{\partial q_\beta}{\partial b} \\ &= \frac{\partial C_{ijkl}}{\partial b} B_{kl\beta} D_\beta^{(p)} b^p + C_{ijkl} B_{kl\beta} p D_\beta^{(p)} b^{p-1}, \end{aligned}$$
$$p = 0, \ldots, n-1; \;\; \alpha, \beta = 1, \ldots, R; \;\; i,j,k,l = 1,2,3. \tag{2.153}$$

Quite similarly, higher-order derivatives can be determined by simple differentiation according to the chain rule. All these derivatives may be embedded into the additional equations for probabilistic moments of the structural state functions following the traditional approach.

To obtain the generalized stochastic perturbation approach for the alternative stress-based formulation, we rewrite the complementary energy principle as

$$\int_{\Omega} c_{ijkl}\sigma_{kl}\delta\sigma_{ij}d\Omega = \int_{\partial\Omega_u} \hat{u}_i\delta\sigma_{ij}n_j d\,(\partial\Omega). \tag{2.154}$$

and then, by introducing the additional expansions of all random functions, one gets

- zeroth-order equation with ε^0 terms:

$$\int_{\Omega} c^0_{ijkl}\sigma^0_{kl}\delta\sigma_{ij}d\Omega = \int_{\partial\Omega_u} \hat{u}^0_i\delta\sigma_{ij}n_j d\,(\partial\Omega), \tag{2.155}$$

- first-order equation with ε^1 terms:

$$\varepsilon\int_{\Omega} c^0_{ijkl}\frac{\partial\sigma_{kl}}{\partial b}\delta\sigma_{ij}d\Omega = \varepsilon\int_{\partial\Omega_u} \frac{\partial\hat{u}_i}{\partial b}\delta\sigma_{ij}n_j d\,(\partial\Omega) - \varepsilon\int_{\Omega}\frac{\partial c_{ijkl}}{\partial b}\sigma^0_{kl}\delta\sigma_{ij}d\Omega, \tag{2.156}$$

- second-order equation with ε^2 terms:

$$\varepsilon^2\int_{\Omega} c^0_{ijkl}\frac{\partial^2\sigma_{kl}}{\partial b^2}\delta\sigma_{ij}d\Omega = \varepsilon^2\int_{\partial\Omega_u} \frac{\partial^2\hat{u}_i}{\partial b^2}\delta\sigma_{ij}n_j d\,(\partial\Omega) - \varepsilon^2\int_{\Omega}\left(2\frac{\partial c_{ijkl}}{\partial b}\frac{\partial\sigma_{kl}}{\partial b} + \frac{\partial^2 c_{ijkl}}{\partial b^2}\sigma^0_{kl}\right)\delta\sigma_{ij}d\Omega\,. \tag{2.157}$$

This procedure is repeated until the given nth-order equation is obtained with ε^n terms:

$$\varepsilon^n\int_{\Omega} c^0_{ijkl}\frac{\partial^n\sigma_{kl}}{\partial b^n}\delta\sigma_{ij}d\Omega = \varepsilon^n\int_{\partial\Omega_u} \frac{\partial^n\hat{u}_i}{\partial b^n}\delta\sigma_{ij}n_j d\,(\partial\Omega) - \varepsilon^n\int_{\Omega}\left(\sum_{p=0}^{n-1}\binom{n}{p}\frac{\partial^{n-p}c_{ijkl}}{\partial b^{n-p}}\frac{\partial^p\sigma_{kl}}{\partial b^p}\right)\delta\sigma_{ij}d\Omega. \tag{2.158}$$

Of course, the stress function variations $\delta\sigma_{ij}$ are deterministic, analogously to the displacement variations δu_i provided in Sec. 2.1.6, for instance. The increasing powers of the perturbation parameter ε are left for example here to show its influence (or its lack) on various orders' solutions to the perturbation-based equilibrium equations.

2.2.3 Non-linear Elasticity Problems

Let us start from an incremental formulation of the FEM equations [112, 138, 141] and their stochastic counterparts using additional extension of the structural polynomial response [79]. So that, instead of the displacement vector, we now need to discretize

- the structural displacement increments [79]

$$\Delta u_\zeta = a_{\zeta\beta}\Delta q_\beta = a_{\zeta\beta}D_\beta^{(p)}b^p,\ p=0,\ldots,n-1;\ \beta,\zeta=1,\ldots,R \tag{2.159}$$

- increments of the strain tensor components

$$\begin{aligned}\Delta\varepsilon_{kl} &= \Delta\hat{\varepsilon}_{kl} + \Delta\bar{\bar{\varepsilon}}_{kl} = \hat{B}_{kl\zeta}\Delta u_\zeta + \bar{\bar{B}}_{kl\xi\zeta}\Delta u_\zeta\Delta u_\xi \\ &= \hat{B}_{kl\zeta}a_{\zeta\alpha}\Delta q_\alpha + \bar{\bar{B}}_{kl\xi\zeta}a_{\zeta\alpha}\Delta q_\alpha a_{\xi\beta}\Delta q_\beta = \hat{B}_{kl\zeta}a_{\zeta\alpha}D_\alpha^{(p)}b^p + \bar{\bar{B}}_{kl\xi\zeta}a_{\zeta\alpha}D_\alpha^{(p)}a_{\xi\beta}D_\beta^{(r)}b^{p+r} \\ p,r &= 0,\ldots,n-1;\ \alpha,\beta=1,\ldots,R;\ k,l=1,2,3;\end{aligned} \tag{2.160}$$

- increments of the second Piola–Kirchhoff stress tensor components

$$\begin{aligned}\Delta\tilde{\sigma}_{ij} &= C_{ijkl}\Delta\varepsilon_{kl} = C_{ijkl}\left(\Delta\hat{\varepsilon}_{kl} + \Delta\bar{\bar{\varepsilon}}_{kl}\right) = C_{ijkl}\left(\hat{B}_{kl\zeta}\Delta u_\zeta + \bar{\bar{B}}_{kl\xi\zeta}\Delta u_\zeta\Delta u_\xi\right) \\ &= C_{ijkl}\left(\hat{B}_{kl\zeta}a_{\zeta\alpha}\Delta q_\alpha + \bar{\bar{B}}_{kl\xi\zeta}a_{\zeta\alpha}\Delta q_\alpha a_{\xi\beta}\Delta q_\beta\right) \\ &= C_{ijkl}\left(\hat{B}_{kl\zeta}a_{\zeta\alpha}D_\alpha^{(p)}b^p + \bar{\bar{B}}_{kl\xi\zeta}a_{\zeta\alpha}D_\alpha^{(p)}a_{\xi\beta}D_\beta^{(r)}b^{p+r}\right) \\ p,r &= 0,\ldots,n-1;\ \alpha,\beta=1,\ldots,R;\ i,j,k,l=1,2,3;\end{aligned} \tag{2.161}$$

Then, the partial derivatives of the state functions increments with respect to the random parameter b can be derived analytically and further, be used to determine the probabilistic moments of the structural response according to the desired order of the Taylor series expansion. The following hold:

- partial derivatives of the displacements with respect to the random input variable of the first order

$$\frac{\partial\Delta u_\zeta}{\partial b} = a_{\zeta\beta}\frac{\partial\Delta q_\beta}{\partial b} = a_{\zeta\beta}\frac{\partial\left(D_\beta^{(p)}b^p\right)}{\partial b} = pa_{\zeta\beta}D_\beta^{(p)}b^{p-1},\ p=0,\ldots,n-1;\ \beta=1,\ldots,R \tag{2.162}$$

as well as the mth-order partial derivatives

$$\frac{\partial^m\Delta u_\zeta}{\partial b^m} = a_{\zeta\beta}\frac{\partial^m\Delta q_\beta}{\partial b^m} = a_{\zeta\beta}\frac{\partial^m\left(D_\beta^{(p)}b^p\right)}{\partial b^m} = p\ldots(p-m)a_{\zeta\beta}D_\beta^{(p)}b^{p-m},$$
$$p=0,\ldots,n-1;\ \beta=1,\ldots,R \tag{2.163}$$

further, we determine the increments of the strain tensor component partial derivatives with respect to the input random variable b as

$$\frac{\partial\Delta\varepsilon_{kl}}{\partial b} = \frac{\partial\Delta\hat{\varepsilon}_{kl}}{\partial b} + \frac{\partial\Delta\bar{\bar{\varepsilon}}_{kl}}{\partial b} = \hat{B}_{kl\zeta}\frac{\partial\Delta u_\zeta}{\partial b} + \bar{\bar{B}}_{kl\xi\zeta}\frac{\partial\Delta u_\zeta}{\partial b}\Delta u_\xi + \bar{\bar{B}}_{kl\xi\zeta}\Delta u_\zeta\frac{\partial\Delta u_\xi}{\partial b}$$

$$= \hat{B}_{kl\zeta} a_{\zeta\alpha} \frac{\partial \Delta q_\alpha}{\partial b} + \bar{\bar{B}}_{kl\xi\zeta} a_{\zeta\alpha} \frac{\partial \Delta q_\alpha}{\partial b} a_{\xi\beta} \Delta q_\beta + \bar{\bar{B}}_{kl\xi\zeta} a_{\zeta\alpha} \Delta q_\alpha a_{\xi\beta} \frac{\partial \Delta q_\beta}{\partial b}$$

$$= \hat{B}_{kl\zeta} a_{\zeta\alpha} p D_\alpha^{(p)} b^{p-1} + \bar{\bar{B}}_{kl\xi\zeta} a_{\zeta\alpha} D_\alpha^{(p)} p a_{\xi\beta} D_\beta^{(r)} b^{p+r-1} + \bar{\bar{B}}_{kl\xi\zeta} a_{\zeta\alpha} D_\alpha^{(p)} a_{\xi\beta} D_\beta^{(r)} r$$

$$p,r = 0, \ldots, n-1; \quad \beta = 1, \ldots, R; \quad k,l = 1,2,3; \tag{2.164}$$

- increments of the stress tensor component partial derivatives with respect to the input random variable

$$\frac{\partial \Delta \tilde{\sigma}_{ij}}{\partial b} = \frac{\partial}{\partial b}\left(C_{ijkl} \Delta \varepsilon_{kl}\right) = \frac{\partial C_{ijkl}}{\partial b}\left(\Delta \bar{\varepsilon}_{kl} + \Delta \bar{\bar{\varepsilon}}_{kl}\right) + C_{ijkl}\left(\frac{\partial \Delta \bar{\varepsilon}_{kl}}{\partial b} + \frac{\partial \Delta \bar{\bar{\varepsilon}}_{kl}}{\partial b}\right)$$

$$= \frac{\partial C_{ijkl}}{\partial b}\left(\bar{B}_{kl\zeta} \Delta u_\zeta + \bar{\bar{B}}_{kl\xi\zeta} \Delta u_\zeta \Delta u_\xi\right.$$

$$\left. + C_{ijkl}\left(\bar{B}_{kl\zeta} \frac{\partial \Delta u_\zeta}{\partial b} + \bar{\bar{B}}_{kl\xi\zeta} \frac{\partial \Delta u_\zeta}{\partial b} \Delta u_\xi + \bar{\bar{B}}_{kl\xi\zeta} \Delta u_\zeta \frac{\partial \Delta u_\xi}{\partial b}\right)\right.$$

$$= \frac{\partial C_{ijkl}}{\partial b}\left(\bar{B}_{kl\zeta} a_{\zeta\alpha} \Delta q_\alpha + \bar{\bar{B}}_{kl\xi\zeta} a_{\zeta\alpha} \Delta q_\alpha a_{\xi\beta} \Delta q_\beta\right)$$

$$+ C_{ijkl}\left(\bar{B}_{kl\zeta} a_{\zeta\alpha} \frac{\partial \Delta q_\alpha}{\partial b} + \bar{\bar{B}}_{kl\xi\zeta} a_{\zeta\alpha} \frac{\partial \Delta q_\alpha}{\partial b} a_{\xi\beta} \Delta q_\beta + \bar{\bar{B}}_{kl\xi\zeta} a_{\zeta\alpha} \Delta q_\alpha a_{\xi\beta} \frac{\partial \Delta q_\beta}{\partial b}\right)$$

$$= \frac{\partial C_{ijkl}}{\partial b}\left(\bar{B}_{kl\zeta} a_{\zeta\alpha} D_\alpha^{(p)} b^p + \bar{\bar{B}}_{kl\xi\zeta} a_{\zeta\alpha} D_\alpha^{(p)} a_{\xi\beta} D_\beta^{(r)} b^{p+r}\right)$$

$$+ C_{ijkl}\left(\bar{B}_{kl\zeta} a_{\zeta\alpha} D_\alpha^{(p)} p b^{p-1} + \bar{\bar{B}}_{kl\xi\zeta} a_{\zeta\alpha} D_\alpha^{(p)} p a_{\xi\beta} D_\beta^{(r)} b^{p+r-1}\right.$$

$$\left. + \bar{\bar{B}}_{kl\xi\zeta} a_{\zeta\alpha} D_\alpha^{(p)} a_{\xi\beta} D_\beta^{(r)} r b^{p+r-1}\right)$$

$$= 0, \ldots, n-1; \ \alpha, \beta = 1, \ldots, R; \ i,j,k,l = 1,2,3. \tag{2.165}$$

All these derivatives may be embedded into the additional equations for probabilistic moments of the structural state functions following the traditional approach in this method; quite similarly, higher order derivatives can be determined by a simple differentiation according to the chain rule. As one may expect, these derivatives may be obtained without the RFM. To this purpose we use the initial algebraic equations in this problem and, by successful formation and solution of the increasing-order equations, the nth-order derivatives are derived. In this case, the following hold [79, 112]:

- zeroth-order equation

$$K_{\alpha\beta}^{(1)0} \Delta q_\beta^0 + K_{\alpha\beta\gamma}^{(2)0} \Delta q_\beta^0 \Delta q_\gamma^0 + K_{\alpha\beta\gamma\delta}^{(3)0} \Delta q_\beta^0 \Delta q_\gamma^0 \Delta q_\delta^0 = \Delta Q_\alpha^0 \tag{2.166}$$

- first-order equation

$$K_{\alpha\beta}^{(1)0} \frac{\partial \Delta q_\beta}{\partial b} + K_{\alpha\beta\gamma}^{(2)0} \frac{\partial \Delta q_\beta}{\partial b} \Delta q_\gamma^0 + K_{\alpha\beta\gamma}^{(2)0} \Delta q_\beta^0 \frac{\partial \Delta q_\gamma}{\partial b} + K_{\alpha\beta\gamma\delta}^{(3)0} \frac{\partial \Delta q_\beta}{\partial b} \Delta q_\gamma^0 \Delta q_\delta^0$$

$$+K^{(3)0}_{\alpha\beta\gamma\delta}\Delta q^0_\beta\frac{\partial\Delta q_\gamma}{\partial b}\Delta q^0_\delta+K^{(3)0}_{\alpha\beta\gamma\delta}\Delta q^0_\beta\Delta q^0_\gamma\frac{\partial\Delta q_\delta}{\partial b}$$
$$=\frac{\partial\Delta Q_\alpha}{\partial b}-\frac{\partial K^{(1)}_{\alpha\beta}}{\partial b}\Delta q^0_\beta-\frac{\partial K^{(2)}_{\alpha\beta\gamma}}{\partial b}\Delta q^0_\beta\Delta q^0_\gamma-\frac{\partial K^{(3)}_{\alpha\beta\gamma\delta}}{\partial b}\Delta q^0_\beta\Delta q^0_\gamma\Delta q^0_\delta \tag{2.167}$$

- a recursive nth-order formula

$$\sum_{k=0}^{n}\binom{n}{k}\frac{\partial^k K^{(1)}_{\alpha\beta}}{\partial b^k}\frac{\partial^{n-k}\Delta q_\beta}{\partial^{n-k}b}+\sum_{k=0}^{n}\binom{n}{k}\frac{\partial^k K^{(2)}_{\alpha\beta\gamma}}{\partial^k b}\sum_{l=o}^{n-k}\binom{n-k}{l}\frac{\partial^l\Delta q^0_\beta}{\partial b^l}\frac{\partial^{n-k-l}\Delta q^0_\gamma}{\partial b^{n-k-l}}$$
$$+\sum_{k=0}^{n}\binom{n}{k}\frac{\partial^k K^{(3)}_{\alpha\beta\gamma\delta}}{\partial b^k}\sum_{l=o}^{n-k}\binom{n-k}{l}\frac{\partial^l\Delta q^0_\beta}{\partial b^l}\sum_{m=o}^{n-k-l}\binom{n-k-l}{m}\frac{\partial^m\Delta q^0_\gamma}{\partial b^m}\frac{\partial^{n-k-l-m}\Delta q^0_\delta}{\partial b^{n-k-l-m}}$$
$$=\frac{\partial^n\Delta Q_\alpha}{\partial b^n}. \tag{2.168}$$

Having determined these partial derivatives, the probability density function of the input random variable is assumed to compute its moments and finally, all these quantities are embedded into the perturbation-based equations for the response probabilistic moments (cf. Equations (1.12)–(1.15)). It is also clear that the randomness modeling within the non-linear boundary value problems is essentially much more complex than for the linear elasticity. The next step toward computational analysis would be exploration of the existing constitutive theories in plasticity and viscoplasticity [15] accounting for uncertainty in especially material characteristics having large experimental evidence [128].

It is clear that the FEM formulation given above includes the specific case of the stability analysis, so that we present briefly its specific RFM version. The deformation energy of some linear elastic structure may be rewritten in matrix notation as [111, 112]

$$U^{(\alpha)}=\frac{1}{2}\mathbf{q}^T_{(\alpha)}\mathbf{K}^{(e)}_{(\alpha)}\mathbf{q}_{(\alpha)}+\frac{1}{2}\mathbf{q}^T_{(\alpha)}\mathbf{K}^{(\sigma)}_{(\alpha)}\mathbf{q}_{(\alpha)} \tag{2.169}$$

where $\mathbf{q}_{(\alpha)}$ is the generalized displacement vector. Global matrices are composed from the finite element contributions that can be defined for a two-noded 3D bar element fundamental used in further computational experiments as

$$\mathbf{k}^{(e)}_{(\alpha)}=\frac{e_{(\alpha)}A_{(\alpha)}}{l_{(\alpha)}}\begin{bmatrix}1&0&0&-1&0&0\\0&0&0&0&0&0\\0&0&0&0&0&0\\-1&0&0&1&0&0\\0&0&0&0&0&0\\0&0&0&0&0&0\end{bmatrix}, \tag{2.170}$$

being the elemental stiffness matrix and $\mathbf{k}_{(\alpha)}^{(\sigma)}$ denoting

$$\mathbf{k}_{(\alpha)}^{(\sigma)} = \frac{F_{(\alpha)}}{l_{(\alpha)}} \begin{bmatrix} 0 & 0 & 0 & 0 & 0 & 0 \\ 0 & 0 & 0 & 0 & 0 & 0 \\ 0 & 0 & 1 & 0 & 0 & -1 \\ 0 & 0 & 0 & 0 & 0 & 0 \\ 0 & 0 & 0 & 0 & 0 & 0 \\ 0 & 0 & -1 & 0 & 0 & 1 \end{bmatrix}; \tag{2.171}$$

traditionally, an index α in brackets is necessary to show the RFM iteration number. Stability analysis of the planar or spatial frames as well as plates or shells demands the relevant structural element being employed. So that, the potential energy for the entire discrete system may be expressed as

$$J_P^{(\alpha)} = \frac{1}{2}\mathbf{q}_{(\alpha)}^T \left(\mathbf{K}_{(\alpha)}^{(e)} + \mathbf{K}_{(\alpha)}^{(\sigma)}\right) \mathbf{q}_{(\alpha)} - \mathbf{Q}_{(\alpha)}^T \mathbf{q}_{(\alpha)} \tag{2.172}$$

whose minimization with respect to the generalized displacements leads to

$$\left(\mathbf{K}_{(\alpha)}^{(e)} + \mathbf{K}_{(\alpha)}^{(\sigma)}\right) \mathbf{q}_{(\alpha)} = \mathbf{Q}_{(\alpha)}. \tag{2.173}$$

Computational analysis with the FEM in linearized buckling analysis is really focused on the critical load factor $\lambda_{(\alpha)}$ such that

$$\mathbf{Q}_{(\alpha)} = \lambda_{(\alpha)} \hat{\mathbf{Q}}_{(\alpha)}, \tag{2.174}$$

where $\hat{\mathbf{Q}}_{(\alpha)}$ is the unit force series directed according to the critical forces to be determined. Then, the discrete stability equation rewritten for the entire system in the RFM version [103] becomes

$$\left(\mathbf{K}_{(\alpha)}^{(e)} + \lambda_{(\alpha)} \mathbf{K}_{(\alpha)}^{(\sigma)}\left(\hat{F}_{(\alpha)}\right)\right) \mathbf{q}_{(\alpha)} = \lambda_{(\alpha)} \hat{\mathbf{Q}}_{(\alpha)}. \tag{2.175}$$

Further, a distribution of internal forces $\hat{\mathbf{F}}_{(\alpha)}$ is equivalent to the load $\hat{\mathbf{Q}}_{(\alpha)}$ and the displacements $\mathbf{q}_{(\alpha)}$ are equivalent to the load $\lambda_{(\alpha)}\hat{\mathbf{Q}}_{(\alpha)}$. We determine the values of $\lambda_{(\alpha)}$ from the following equations:

$$\begin{cases} \left(\mathbf{K}_{(\alpha)}^{(e)} + \lambda_{(\alpha)} \mathbf{K}_{(\alpha)}^{(\sigma)}\left(\hat{F}_{(\alpha)}\right)\right) \mathbf{q}_{1(\alpha)} = \lambda_{(\alpha)} \hat{\mathbf{Q}}_{(\alpha)} \\ \left(\mathbf{K}_{(\alpha)}^{(e)} + \lambda_{(\alpha)} \mathbf{K}_{(\alpha)}^{(\sigma)}\left(\hat{F}_{(\alpha)}\right)\right) \mathbf{q}_{2(\alpha)} = \lambda_{(\alpha)} \hat{\mathbf{Q}}_{(\alpha)} \end{cases}, \text{where } \mathbf{q}_{1(\alpha)} \neq \mathbf{q}_{2(\alpha)} \tag{2.176}$$

and, by subtracting, we obtain the basic algebraic equation series representing the elastic stability

$$\left(\mathbf{K}_{(\alpha)}^{(s)} + \lambda_{(\alpha)} \mathbf{K}_{(\alpha)}^{(\sigma)}\left(\hat{F}_{(\alpha)}\right)\right) \mathbf{q}_{(\alpha)}^{*} = 0. \tag{2.177}$$

An introduction of the basic condition for the critical value $\lambda_{(\alpha)}=\lambda_{cr(\alpha)}$ leads to the following final well known condition:

$$\det\left(\mathbf{K}^{(s)}_{(\alpha)}+\lambda_{(\alpha)}\mathbf{K}^{(\sigma)}_{(\alpha)}\left(\hat{F}_{(\alpha)}\right)\right)=0. \tag{2.178}$$

The minimum positive solution to this eigenproblem allows us to determine the first critical value and, in the context of probabilistic analysis, basic characteristics of the critical force. It is known that most of the structures do not collapse when the external forces reach their first critical value (or its expectation). A definitely more interesting (and challenging) problem would be to follow a post-critical path with uncertainty, which is of paramount importance for the skeletal tower structures like this studied in further computational experiments.

2.2.4 SFEM in Elastodynamics

Let us consider the following approximation of the displacement field $u_i\left(\mathbf{x},\tau\right)$ [51]:

$$u_i(\mathbf{x},\tau)=\varphi_{i\alpha}\left(\mathbf{x}\right)\,q_\alpha(\tau), \tag{2.179}$$

where $q_\alpha(\tau)$ is the generalized coordinate vector. The strain tensor is rewritten as

$$\varepsilon_{ij}(\mathbf{x},\tau)=B_{ij\alpha}\left(\mathbf{x}\right)\,q_\alpha(\tau)\,. \tag{2.180}$$

Therefore, an equation of motion is obtained as

$$\delta\int_{t_1}^{t_2}\left[\frac{1}{2}\dot{q}_\alpha\;M_{\alpha\beta}\;\dot{q}_\beta-\frac{1}{2}q_\alpha\;K_{\alpha\beta}\;q_\beta+Q_\alpha q_\alpha\right]\;\mathrm{d}\tau=0, \tag{2.181}$$

where the matrices of mass and stiffness may be composed from the following integrals:

$$M_{\alpha\beta}=\int_\Omega\rho\varphi_{i\alpha}\varphi_{i\beta}\mathrm{d}\Omega\quad K_{\alpha\beta}=\int_\Omega B_{ij\alpha}C_{ijkl}B_{kl\beta}\mathrm{d}\Omega. \tag{2.182}$$

Integration with respect to a time variable and variation in Equation (2.181) leads to

$$\dot{q}_\beta M_{\alpha\beta}\delta q_\alpha-\int_{t_1}^{t_2}\left(\ddot{q}_\beta M_{\alpha\beta}+q_\beta K_{\alpha\beta}-Q_\alpha\right)\;\partial q_\alpha\partial\tau=0 \tag{2.183}$$

so that, in the view of

$$\delta q_\alpha\left(t_1\right)=0,\quad\delta q_\alpha\left(t_2\right)=0, \tag{2.184}$$

leads to the well-known matrix equation of motion and its stochastic RFM counterpart [89] equals to

$$M_{\alpha\beta}\ddot{q}_\beta+K_{\alpha\beta}q_\beta=Q_\alpha\quad M^{(i)}_{\alpha\beta}\ddot{q}^{(i)}_\beta+K^{(i)}_{\alpha\beta}q^{(i)}_\beta=Q^{(i)}_\alpha \tag{2.185}$$

which is frequently proposed, including the damping component as

$$M_{\alpha\beta}\ddot{q}_{\beta} + C_{\alpha\beta}\dot{q}_{\beta} + K_{\alpha\beta}q_{\beta} = Q_{\alpha} \quad M^{(i)}_{\alpha\beta}\ddot{q}^{(i)}_{\beta} + C^{(i)}_{\alpha\beta}\dot{q}^{(i)}_{\beta} + K^{(i)}_{\alpha\beta}q^{(i)}_{\beta} = Q^{(i)}_{\alpha}. \tag{2.186}$$

Upper index i in these two equations is needed to underline particular test solution in the RFM scheme. Modal analysis, being one of many techniques available now for numerical solution of the equations of motion, is based on the following transform:

$$\mathbf{q}(\tau) = \sum_{j=1}^{n} \mathbf{a}_j x_j(\tau) = \mathbf{A}\mathbf{x}(\tau), \tag{2.187}$$

where $\mathbf{A} = [\mathbf{a}_i]$ is composed of the eigenvectors of a structure, while $\mathbf{x}(\tau)$ is the vector of modal coordinates. The global equation of motion is transformed into [51]

$$\sum_{j=1}^{n} \mathbf{a}_i^T \mathbf{M}\mathbf{a}_j \ddot{x}_j(\tau) + \sum_{j=1}^{n} \mathbf{a}_i^T \mathbf{C}\mathbf{a}_j \dot{x}_j(\tau) + \sum_{j=1}^{n} \mathbf{a}_i^T \mathbf{K}\mathbf{a}_j x_j(\tau) = \mathbf{a}_i^T \mathbf{R}(\tau). \tag{2.188}$$

The orthogonality condition on the eigenvectors leads to a simplification

$$\mathbf{a}_i^T \mathbf{M}\mathbf{a}_i = \begin{cases} 0 & i \neq j \\ \tilde{m}_i & i = j \end{cases}, \quad \mathbf{a}_i^T \mathbf{K}\mathbf{a}_i = \begin{cases} 0 & i \neq j \\ \tilde{k}_i & i = j \end{cases} \tag{2.189}$$

where $\tilde{m}_i, \tilde{k}_i$ stand for the modal masses and stiffnesses. Then

$$\sum_{j=1}^{n} \mathbf{a}_i^T \mathbf{M}\mathbf{a}_j \ddot{x}_j(\tau) = \tilde{m}_i \ddot{x}_i(\tau), \quad \sum_{j=1}^{n} \mathbf{a}_i^T \mathbf{K}\mathbf{a}_j x_j(\tau) = \tilde{k}_i x_i(\tau). \tag{2.190}$$

Introducing a further modal loading vector and modal damping matrix in case of proportionality to mass and stiffness components (i.e., $\mathbf{C} = \alpha\mathbf{M} + \kappa\mathbf{K}$)

$$\tilde{p}_i(\tau) = \mathbf{a}_i^T \mathbf{R}(\tau), \quad \sum_{j=1}^{n} \mathbf{a}_i^T \mathbf{C}\mathbf{a}_j \dot{x}_j(\tau) = \tilde{c}_i \dot{x}_i(\tau), \tag{2.191}$$

the equation of motion (2.188) finally becomes

$$\tilde{m}_i \ddot{x}_i(\tau) + \tilde{c}_i \dot{x}_i(\tau) + \tilde{k}_i x_i(\tau) = \tilde{p}_i(\tau), \quad i = 1, 2, 3, \ldots, n. \tag{2.192}$$

Normalization of the eigenvectors gives $\mathbf{a}_i^T \mathbf{M}\mathbf{a}_i = 1$ and with $\tilde{m}_i = 1$, $\tilde{k}_i = \omega_i^2$, $\tilde{c}_i = 2\gamma_i\omega_i$, one obtains

$$\ddot{x}_i(\tau) + 2\gamma_i\omega_i\dot{x}_i(\tau) + \omega_i^2 x_i(\tau) = \tilde{p}_i(\tau). \tag{2.193}$$

This transformation is also used to express the boundary and initial conditions, hence

$$\mathbf{r}(0) = \sum_{j=1}^{n} \mathbf{a}_j x_j(0), \quad \dot{\mathbf{r}}(0) = \sum_{j=1}^{n} \mathbf{a}_j \dot{x}_j(0) \tag{2.194}$$

and

$$\mathbf{x}(0) = \mathbf{A}^T \mathbf{M} \mathbf{q}(0), \quad \dot{\mathbf{x}}(0) = \mathbf{A}^T \mathbf{M} \mathbf{v}(0). \tag{2.195}$$

A complete solution of Equation (2.193) is known as the Duhamel integral [3]

$$\begin{aligned} x_i(\tau) = {} & \frac{1}{\omega_i\sqrt{1-\gamma_i^2}} \int_0^t \tilde{p}_i(\tau) \exp(-\gamma_i\omega_i\tau)\, h_i(t-\tau) \sin\left(\omega_i\sqrt{1-\gamma_i^2}\,(t-\tau)\right) d\tau \\ & + \exp(-\gamma_i\omega_i t) \left\{ A_{i1} \sin\left(\omega_i\sqrt{1-\gamma_i^2} t\right) + A_{i2} \cos\left(\omega_i\sqrt{1-\gamma_i^2} t\right) \right\} \end{aligned} \tag{2.196}$$

The elastic forces vector in the initial system may be determined as

$$\mathbf{f}_s(\tau) = \sum_{i=1}^{n} \mathbf{K}\mathbf{a}_i x_i(\tau) + \sum_{i=1}^{n} \omega_i^2 \mathbf{M}\mathbf{a}_i x_i(\tau). \tag{2.197}$$

The DDM, applied in some further numerical experiment, is based on the straightforward determination of the latter exactly from the given-order state equilibrium equations. If we calculate the eigenfrequencies as a result of

$$\left(\mathbf{K} - \varpi^2 \mathbf{M}\right) \boldsymbol{\varphi} = 0, \tag{2.198}$$

then we start from the zeroth-order relation [112]

$$\left(\mathbf{K}^0 - \left(\varpi^0\right)^2 \mathbf{M}^0\right) \boldsymbol{\varphi}^0 = 0 \tag{2.199}$$

and after differentiation w.r.t. b, it holds that

$$\left(\frac{\partial \mathbf{K}}{\partial b} - \left(\frac{\partial}{\partial b}\left(\varpi^0\right)^2\right) \mathbf{M}^0 - \left(\varpi^0\right)^2 \frac{\partial \mathbf{M}}{\partial b}\right) \boldsymbol{\varphi}^0 = -\left(\mathbf{K}^0 - \left(\varpi^0\right)^2 \mathbf{M}^0\right) \frac{\partial \boldsymbol{\varphi}}{\partial b}. \tag{2.200}$$

A sequential differentiation up to nth order leads to the following recursive form [104, 105]:

$$\begin{aligned} \sum_{k=0}^{n} \binom{n}{k} \frac{\partial^{n-k}\mathbf{K}}{\partial b^{n-k}} \frac{\partial^k \boldsymbol{\varphi}}{\partial b^k} = {} & \sum_{k=0}^{n} \binom{n}{k} \sum_{l=0}^{k-1} \binom{k-1}{l} \frac{\partial^{(k-(l+1))}(2\varpi)}{\partial b^{(k-(l+1))}} \frac{\partial^{l+1}\varpi}{\partial b^{l+1}} \\ & \times \sum_{m=0}^{n-k} \binom{n-k}{m} \frac{\partial^m \mathbf{M}}{\partial b^m} \frac{\partial^{(n-k-m)}\boldsymbol{\varphi}}{\partial b^{n-k-m}} \end{aligned} \tag{2.201}$$

so that we can calculate the partial derivatives of the increasing order for the eigenvalues from Equations (2.198)–(2.201) and then, embed them directly into the particular probabilistic moment equations to recover their values. Analogously, one can rewrite the equation for the forced vibrations in the perturbation-based nth-order version as

$$\sum_{k=0}^{n}\binom{n}{k}\left\{\frac{\partial^{n-k}\mathbf{M}}{\partial b^{n-k}}\frac{\partial^{k}\ddot{\mathbf{q}}}{\partial b^{k}}+\frac{\partial^{n-k}\mathbf{C}}{\partial b^{n-k}}\frac{\partial^{k}\dot{\mathbf{q}}}{\partial b^{k}}+\frac{\partial^{n-k}\mathbf{K}}{\partial b^{n-k}}\frac{\partial^{k}\mathbf{q}}{\partial b^{k}}\right\}=\frac{\partial^{n}\mathbf{Q}}{\partial b^{n}} \tag{2.202}$$

or the modal equations system

$$\frac{\partial^{n}\ddot{x}_i(\tau)}{\partial b^{n}}+\sum_{k=0}^{n}\binom{n}{k}\left\{2\frac{\partial^{n-k}(\gamma_i\omega_i)}{\partial b^{n-k}}\frac{\partial^{k}(\dot{x}_i(\tau))}{\partial b^{k}}+\frac{\partial^{n-k}(\omega_i^2)}{\partial b^{n-k}}\frac{\partial^{k}(x_i(\tau))}{\partial b^{k}}\right\}=\frac{\partial^{n}\tilde{p}_i(\tau)}{\partial b^{n}}, \tag{2.203}$$

where usually the right-hand side vector equals 0 except for the zeroth-order equation, since the main interest is in structural randomness. The complex nature of the perturbation-based equations follows intermediately a chain differentiation of the second powers of random state functions with respect to b. Final determination of the elastic forces in the modal approach proceeds using the differential form

$$\frac{\partial^{p}\mathbf{f}_s(\tau)}{\partial b^{p}}=\sum_{i=1}^{n}\frac{\partial^{p}}{\partial b^{p}}\left\{\mathbf{Ka}_i x_i(\tau)\right\}+\sum_{i=1}^{n}\frac{\partial^{p}}{\partial b^{p}}\left\{\omega_i^2\mathbf{Ma}_i x_i(\tau)\right\}. \tag{2.204}$$

2.2.5 Transient Analysis of the Heat Transfer

Let us assume that the region Ω is discretized by use of the set of finite elements and that the scalar temperature field $\theta(x_i)$ is described by the nodal temperature vector T_α [113, 145]:

$$\theta(x_i) = \varphi_\alpha(x_i)\,T_\alpha;\;\; i=1,2,3;\;\; \alpha=1,2,\ldots,R, \tag{2.205}$$

The temperature derivatives can be written in the form

$$\theta_{,i} = \varphi_{\alpha,i}\,T_\alpha,\; i=1,2,3. \tag{2.206}$$

Moreover, let us introduce the heat capacity matrix $C_{\alpha\beta}$, the heat conductivity matrix $K_{\alpha\beta}$, and the vector P_α as follows:

$$C_{\alpha\beta} = \int_{\Omega}\rho c\;\phi_\alpha\;\phi_\beta\;d\Omega, \tag{2.207}$$

$$K_{\alpha\beta} = \int_{\Omega}k_{ij}\;\phi_{\alpha,i}\;\phi_{\beta,j}\;d\Omega, \tag{2.208}$$

and

$$P_\alpha = \int_{\Omega}g\,\varphi_\alpha d\Omega + \int_{\partial\Omega}\hat{q}_i n_i\,\varphi_\alpha d\Omega. \tag{2.209}$$

Next, let us introduce these matrices into the additional variational formulation (cf. Equation (2.70)) to obtain the following algebraic equations system:

$$C_{\alpha\beta}\dot{T}_{\beta} + K_{\alpha\beta}T_{\beta} = P_{\alpha}. \tag{2.210}$$

As it is known [3], the main issue in transient problems is the additional time discretization using some time increment Δt and the forward finite differencing scheme. Then, we can rewrite the last equation in the following manner for temperature independent system's matrices:

$$C_{\alpha\beta}\frac{T_{\beta}(t+\Delta t) - T_{\beta}(t)}{\Delta t} + K_{\alpha\beta}T_{\beta}(t) = P_{\alpha}. \tag{2.211}$$

Considering the second component in this statement we obtain the explicit method, where the nodal temperature vector in this component is taken at the beginning of this time step. However, it is possible to introduce the extra coefficient $0 \le \delta \le 1$ to include in this term the temperature vector after the time step also. It holds that

$$C_{\alpha\beta}\frac{T_{\beta}(t+\Delta t) - T_{\beta}(t)}{\Delta t} + K_{\alpha\beta}\left\{\delta T_{\beta}(t+\Delta t) + (1-\delta)T_{\beta}(t)\right\} = P_{\alpha}, \tag{2.212}$$

where $\delta = 0$ is equivalent to the explicit method, $\delta = 1/2$ serves for the Crank–Nicholson method, $\delta = 2/3$ stands for the Galerkin method, and lastly $\delta = 1$ is used in the implicit method (one can use this algorithm with δ as the extra input parameter); there are also three level schemes, where the temperatures in the moments $t+\Delta t$, t, $t-\Delta t$ are included at once. A further procedure leads to reformulation of this equation to obtain the nodal temperatures for $t+\Delta t$ on the LHS and the temperature vector in the preceding time step on the RHS, so that for the Crank–Nicholson scheme we proceed as follows:

$$C_{\alpha\beta}\left\{\frac{T_{\beta}(t+\Delta t) - T_{\beta}(t)}{\Delta t}\right\} + K_{\alpha\beta}\left\{\frac{T_{\beta}(t+\Delta t)}{2} + \frac{T_{\beta}(t)}{2}\right\} = P_{\alpha}, \tag{2.213}$$

therefore

$$\left(\frac{C_{\alpha\beta}}{\Delta t} + \frac{K_{\alpha\beta}}{2}\right)T_{\beta}(t+\Delta t) = P_{\alpha} + \left(\frac{C_{\alpha\beta}}{\Delta t} - \frac{K_{\alpha\beta}}{2}\right)T_{\beta}(t). \tag{2.214}$$

Introduction of the matrix $A_{\alpha\beta}$ instead of a combination of heat conductivity and heat capacity matrices on the LHS as well as the additional matrix $E_{\alpha\beta}$ for the RHS bracket leads to the relation

$$A_{\alpha\beta}T_{\beta}(t+\Delta t) = P_{\alpha} + E_{\alpha\beta}T_{\beta}(t). \tag{2.215}$$

Let us focus first on the DDM version of the SFEM implementation of the problem. Analogously to the previous considerations (provided also for the second-order analysis in [46, 96]), we obtain the following systems of algebraic equations describing the generalized stochastic formulation of the transient heat flow problem:

- zeroth-order equation

$$C^{0}_{\alpha\beta}\dot{T}^{0}_{\beta} + K^{0}_{\alpha\beta}T^{0}_{\beta} = P^{0}_{\alpha} \tag{2.216}$$

- first-order equation

$$C^0_{\alpha\beta}\frac{\partial \dot{T}_\beta}{\partial b} + K^0_{\alpha\beta}\frac{\partial T_\beta}{\partial b} = \frac{\partial P_\alpha}{\partial b} - \left(\frac{\partial C_{\alpha\beta}}{\partial b}\dot{T}^0_\beta + \frac{\partial K_{\alpha\beta}}{\partial b}T^0_\beta\right) \tag{2.217}$$

- second-order equation

$$C^0_{\alpha\beta}\frac{\partial^2 \dot{T}_\beta}{\partial b^2} + K^0_{\alpha\beta}\frac{\partial^2 T_\beta}{\partial b^2} = \frac{\partial^2 P_\alpha}{\partial b^2} - 2\left(\frac{\partial C_{\alpha\beta}}{\partial b}\frac{\partial \dot{T}_\beta}{\partial b} + \frac{\partial K_{\alpha\beta}}{\partial b}\frac{\partial T_\beta}{\partial b}\right)$$
$$- \left(\frac{\partial^2 C_{\alpha\beta}}{\partial b^2}\dot{T}^0_\beta + \frac{\partial^2 K_{\alpha\beta}}{\partial b^2}T^0_\beta\right) \tag{2.218}$$

and, at last, nth-order equation

$$C^0_{\alpha\beta}\frac{\partial^n \dot{T}_\beta}{\partial b^n} + K^0_{\alpha\beta}\frac{\partial^n T_\beta}{\partial b^n} = \frac{\partial^n P_\alpha}{\partial b^n} - \sum_{k=1}^{n}\binom{n}{k}\frac{\partial^k C_{\alpha\beta}}{\partial b^k}\frac{\partial^{n-k}\dot{T}_\beta}{\partial b^{n-k}} - \sum_{k=1}^{n}\binom{n}{k}\frac{\partial^k K_{\alpha\beta}}{\partial b^k}\frac{\partial^{n-k}T_\beta}{\partial b^{n-k}}. \tag{2.219}$$

We have introduced the following matrix notation in the equations posed above:

- the heat capacity matrix derivatives (after Equation (2.207))

$$\frac{\partial^n C_{\alpha\beta}}{\partial b^n} = \int_\Omega \sum_{k=0}^{n}\binom{n}{k}\frac{\partial^k \rho}{\partial b^k}\frac{\partial^{n-k} c}{\partial b^{n-k}}\varphi_\alpha\varphi_\beta d\Omega, \tag{2.220}$$

- the heat conductivity matrix derivatives (from Equation (2.208))

$$\frac{\partial^n K_{\alpha\beta}}{\partial b^n} = \int_\Omega \frac{\partial^n k_{ij}}{\partial b^n}\varphi_{\alpha,i}\varphi_{\beta,j} d\Omega, \tag{2.221}$$

- the right-hand vector derivatives (since Equation (2.209))

$$\frac{\partial^n P_\alpha}{\partial b^n} = \int_\Omega \frac{\partial^n g}{\partial b^n}\varphi_\alpha d\Omega + \int_{\partial\Omega_q}\frac{\partial^n \hat{q}_i}{\partial b^n} n_i\varphi_\alpha d\Omega, \tag{2.222}$$

where all these expressions are evaluated at the expected values of the input random variables. As is clear now, the DDM version needs the formation and solution of the increasing-order equations obtained from the initial one by a systematic differentiation with respect to the random input variable provided in a quite deterministic way. Readers interested in the second-order version of this methodology relevant to composite materials with random physical properties may find the necessary details in [97].

The main idea behind the RFM is to recover the polynomial approximation of the temperature at a given node with respect to the input random variable b of the FEM mesh in the following form:

$$T_\beta = D_\beta^{(m)} b^m, \quad m = 0, \ldots, n-1; \; \beta = 1, \ldots, R \tag{2.223}$$

so that:

$$\theta(x_i) = \varphi_\beta(x_i)\; T_\beta = \varphi_\beta(x_i) \quad D_\beta^{(m)} b^m; \; i = 1,2; \; \alpha = 1,2, \ldots, R, \\ m = 0, \ldots, n-1. \tag{2.224}$$

Therefore, the temperature gradients are similarly determined as

$$\theta_{,j} = \varphi_{\beta,j}\, T_\beta = \varphi_{\beta,j}\, D_\beta^{(m)} b^m, \quad i = 1,2, \; m = 0, \ldots, n-1. \tag{2.225}$$

The key feature of this approach is to determine numerically the coefficients $D_\beta^{(m)}$ for each node of the initial FEM mesh and each power of the polynomial representations of the nodal temperatures with respect to the random input. Starting from a random polynomial representation, it is possible to calculate up to nth-order ordinary derivatives of the nodal responses $\theta_\beta(b)$ with respect to b at the given b_0 as follows:

- first-order derivative

$$\frac{d\theta_\beta}{db} = (n-1)\, D_\beta^{(1)} b^{n-2} + (n-2)\, D_\beta^{(2)} b^{n-3} + \ldots + D_\beta^{(n-1)}, \tag{2.226}$$

- second-order derivative

$$\frac{d^2\theta_\beta}{db^2} = (n-1)(n-2)\, D_\beta^{(1)} b^{n-3} + (n-2)(n-3)\, D_\beta^{(2)} b^{n-4} + \ldots + D_\beta^{(n-2)}, \tag{2.227}$$

- kth-order derivative

$$\frac{d^k\theta_\beta}{db^k} = \prod_{i=1}^{k} (n-i)\, D_\beta^{(1)} b^{n-k} + \prod_{i=2}^{k} (n-i)\, D_\beta^{(2)} b^{n-(k+1)} + \ldots + D_\beta^{(n-k)}. \tag{2.228}$$

This differentiation has quite an analytical character and the increasing-order partial derivatives of the nodal temperatures with respect to the random variable are not so much affected by the numerical errors from the hierarchical equation solutions. It is also clear that the probabilistic transient problem needs successive polynomial responses from time increment to the next one. Therefore, for a discrete time moment τ, we have the following approximation [89]:

$$T_\beta(\tau) = D_\beta^{\tau(m)} b^m, \quad m = 0, \ldots, n-1; \; \beta = 1, \ldots, R. \tag{2.229}$$

where $D_\beta^{\tau(m)}$ denotes the coefficient relevant to the mth power of b computed in the degree of freedom β at the time moment τ.

Hence, it yields

$$\theta\left(x_i;\tau\right) = \varphi_\beta\left(x_i\right) T_\beta\left(\tau\right) = \varphi_\beta\left(x_i\right) D_\beta^{\tau(m)} b^m; \quad i = 1,2; \; \alpha = 1,2,\ldots,R, \quad m = 0,\ldots,n-1. \tag{2.230}$$

Therefore, the temperature gradients are similarly determined as

$$\theta_{,j}\left(\tau\right) = \varphi_{\beta,j} T_\beta\left(\tau\right) = \varphi_{\beta,j} D_\beta^{\tau(m)} b^m, \quad i = 1,2, \; m = 0,\ldots,n-1. \tag{2.231}$$

Finally, one realizes that the temperature-dependent physical parameters may lead to further significant numerical complications in SFEM implementations to transient problems even when using the RFM.

2.2.6 Coupled Thermo-piezoelectrostatics SFEM Equations

The SFEM discretization starts from the traditional vectorization of the displacement field $\mathbf{u}$, the electric potential $\boldsymbol{\Phi}$, and temperature $\boldsymbol{\theta}$. It is completed using the nodal values $\mathbf{q}$, Ψ, $\mathbf{T}$ and the shape functions $\boldsymbol{\varphi}_\mathbf{q}$, $\boldsymbol{\varphi}_\Psi$ as well as $\boldsymbol{\varphi}_\mathbf{T}$ as follows [95, 116]:

$$\mathbf{u} = \varphi_\mathbf{q}\mathbf{q}, \quad \boldsymbol{\Phi} = \varphi_\Psi \boldsymbol{\Psi}, \quad \boldsymbol{\theta} = \varphi_T \mathbf{T}. \tag{2.232}$$

Then, the strain tensor $\boldsymbol{\varepsilon}$, the electric field $\mathbf{E}$, and the thermal field $\mathbf{F}$ are related to the vectorized state functions as

$$\boldsymbol{\varepsilon} = \nabla\phi_\mathbf{q}\mathbf{q} = \mathbf{B}_\mathbf{q}\mathbf{q}, \tag{2.233}$$

$$\mathbf{E} = -\nabla\phi_\Psi \Psi = \mathbf{B}_\Psi \boldsymbol{\Psi}, \tag{2.234}$$

$$\mathbf{F} = -\nabla\phi_T \mathbf{T} = \mathbf{B}_\mathbf{T}\mathbf{T}. \tag{2.235}$$

The constitutive Equations (2.77)–(2.80) become, in this manner, equal to

$$\sigma = \mathbf{C}\boldsymbol{\varepsilon} - \mathbf{PE} - \boldsymbol{\kappa\theta}, \tag{2.236}$$

$$\mathbf{D} = \mathbf{P}^\mathbf{T}\boldsymbol{\varepsilon} + \mathbf{BE} + \mathbf{d}\boldsymbol{\theta}, \tag{2.237}$$

$$\mathbf{S} = \boldsymbol{\kappa}^\mathbf{T}\boldsymbol{\varepsilon} + \mathbf{d}^\mathbf{T}\mathbf{E} + \boldsymbol{\alpha}\mathbf{T}\boldsymbol{\theta}, \tag{2.238}$$

$$\mathbf{q} = \mathbf{kF}, \tag{2.239}$$

where σ, $\mathbf{D}$ are the stress tensor components and the electric displacement vector, and matrices $\mathbf{C}$, $\mathbf{B}$, $\mathbf{P}$, κ, $\mathbf{d}$, and $\mathbf{k}$ denote the elasticity, dielectric permittivity, piezoelectric constant, thermomechanical coupling, thermopiezoelectric coupling, and thermal conductivity tensors. Then, we introduce this discretization into the variational equations and, as a result, the fundamental of equilibrium equations for the thermopiezoelectric problem becomes

$$\begin{pmatrix} \mathbf{M}_{qq} & 0 & 0 \\ & 0 & 0 \\ \text{symm.} & & 0 \end{pmatrix} \begin{pmatrix} \ddot{\mathbf{q}} \\ \ddot{\boldsymbol{\Psi}} \\ \ddot{\mathbf{T}} \end{pmatrix} + \begin{pmatrix} 0 & 0 & 0 \\ 0 & 0 & 0 \\ \mathbf{C}_{Tq} & \mathbf{C}_{T\Psi} & \mathbf{C}_{TT} \end{pmatrix} \begin{pmatrix} \dot{\mathbf{q}} \\ \dot{\boldsymbol{\Psi}} \\ \dot{\mathbf{T}} \end{pmatrix} + \begin{pmatrix} \mathbf{K}_{qq} & \mathbf{K}_{q\Psi} & \mathbf{K}_{qT} \\ \mathbf{K}_{\Psi q} & \mathbf{K}_{\Psi\Psi} & \mathbf{K}_{\Psi T} \\ 0 & 0 & \mathbf{K}_{TT} \end{pmatrix} \begin{pmatrix} \mathbf{q} \\ \boldsymbol{\Psi} \\ \mathbf{T} \end{pmatrix}$$

$$= \begin{pmatrix} \mathbf{F}_q \\ \boldsymbol{\Gamma}_{\Psi} \\ \mathbf{F}_T \end{pmatrix}, \tag{2.240}$$

where $\mathbf{M}_{qq}$ is the mass matrix, $\mathbf{C}_{Tq}$, $\mathbf{C}_{T\Psi}$, and $\mathbf{C}_{TT}$ are the damping matrices resulting from the thermomechanical, thermoelectric couplings, and thermal field, respectively. The matrices $\mathbf{K}_{q\Psi}$, $\mathbf{K}_{\Psi q}$, $\mathbf{K}_{qT}$, $\mathbf{K}_{\Psi T}$, $\mathbf{K}_{qq}$, $\mathbf{K}_{\Psi\Psi}$, and $\mathbf{K}_{TT}$ are the stiffness matrices related to the piezoelectric-mechanical (first two), thermomechanical, thermoelectric couplings as well as the stiffness adjacent to the mechanical, electric, and thermal fields, correspondingly. Finally, the RHS vectors $\mathbf{F}_q$, $\mathbf{F}_{\Psi}$, as well as $\mathbf{F}_T$ represent mechanical, electric, and thermal boundary conditions. Their definitions are given as follows:

$$\mathbf{M}_{qq} = \int_{\Omega} \varphi_q^T \rho\, \varphi_q \, d\Omega, \tag{2.241}$$

$$\mathbf{C}_{Tq} = -\int_{\Omega} T_0 \varphi_T^T \mathbf{k}^T \, \mathbf{B}_q \, d\Omega \quad \mathbf{C}_{T\Psi} = -\int_{\Omega} T_0 \varphi_T^T \mathbf{d}^T \, \mathbf{B}_{\Psi} \, d\Omega \quad \mathbf{C}_{TT} = -\int_{\Omega} T_0 \varphi_T^T \alpha^T \varphi_T \, d\Omega \tag{2.242}$$

$$\mathbf{K}_{qq} = \int_{\Omega} \mathbf{B}_q^T \mathbf{C}\, \mathbf{B}_q \, d\Omega \quad \mathbf{K}_{q\Psi} = -\int_{\Omega} \mathbf{B}_q^T \mathbf{P}\, \mathbf{B}_{\Psi} \, d\Omega \quad \mathbf{K}_{qT} = -\int_{\Omega} \mathbf{B}_q^T \mathbf{k}\, \varphi_T \, d\Omega \tag{2.243}$$

$$\mathbf{K}_{\Psi q} = -\int_{\Omega} \mathbf{B}_{\Psi}^T \mathbf{P}^T \mathbf{B}_q \, d\Omega \quad \mathbf{K}_{\Psi\Psi} = \int_{\Omega} \mathbf{B}_{\Psi}^T \mathbf{B}\, \mathbf{B}_{\Psi} \, d\Omega \quad \mathbf{K}_{\Psi T} = -\int_{\Omega} \mathbf{B}_{\Psi}^T \, \mathbf{d}\, \varphi_T \, d\Omega \tag{2.244}$$

$$\mathbf{K}_{TT} = -\int_{\Omega} \mathbf{B}_T^T \kappa \, \mathbf{B}_T \, d\Omega. \tag{2.245}$$

Let us also note that

$$\mathbf{K}_{\Psi q} = \mathbf{K}_{q\Psi}^T \quad \mathbf{C}_{Tq} = T_0 \; \mathbf{K}_{qT}^T \quad \mathbf{C}_{T\Psi} = T_0 \; \mathbf{K}_{\Psi T}^T. \tag{2.246}$$

On this basis we may write the zeroth- and the nth-order stochastic equation systems corresponding to the deterministic system given as follows:

$$\begin{pmatrix} \mathbf{M}_{qq}^{(0)} & \mathbf{0} & \mathbf{0} \\ & \mathbf{0} & \mathbf{0} \\ \text{symm.} & & \mathbf{0} \end{pmatrix} \begin{pmatrix} \ddot{\mathbf{q}}^{(0)} \\ \ddot{\mathbf{\Psi}}^{(0)} \\ \ddot{\mathbf{T}}^{(0)} \end{pmatrix} + \begin{pmatrix} \mathbf{0} & \mathbf{0} & \mathbf{0} \\ \mathbf{0} & \mathbf{0} & \mathbf{0} \\ \mathbf{C}_{Tq}^{(0)} & \mathbf{C}_{T\Psi}^{(0)} & \mathbf{C}_{TT}^{(0)} \end{pmatrix} \begin{pmatrix} \dot{\mathbf{q}}^{(0)} \\ \dot{\mathbf{\Psi}}^{(0)} \\ \dot{\mathbf{T}}^{(0)} \end{pmatrix}$$
$$+ \begin{pmatrix} \mathbf{K}_{qq}^{(0)} & \mathbf{K}_{q\Psi}^{(0)} & \mathbf{K}_{qT}^{(0)} \\ \mathbf{K}_{\Psi q}^{(0)} & \mathbf{K}_{\Psi\Psi}^{(0)} & \mathbf{K}_{\Psi T}^{(0)} \\ \mathbf{0} & \mathbf{0} & \mathbf{K}_{TT}^{(0)} \end{pmatrix} \begin{pmatrix} \mathbf{q}^{(0)} \\ \mathbf{\Psi}^{(0)} \\ \mathbf{T}^{(0)} \end{pmatrix} = \begin{pmatrix} \mathbf{F}_{q}^{(0)} \\ \mathbf{F}_{\Psi}^{(0)} \\ \mathbf{F}_{T}^{(0)} \end{pmatrix} \quad (2.247)$$

and

$$\sum_{k=0}^{n} \binom{n}{k} \begin{pmatrix} \frac{\partial^k \mathbf{M}_{qq}}{\partial b^k} & \mathbf{0} & \mathbf{0} \\ & \mathbf{0} & \mathbf{0} \\ \text{symm.} & & \mathbf{0} \end{pmatrix} \begin{pmatrix} \frac{\partial^{n-k} \ddot{\mathbf{q}}}{\partial b^{n-k}} \\ \frac{\partial^{n-k} \ddot{\mathbf{\Psi}}}{\partial b^{n-k}} \\ \frac{\partial^{n-k} \ddot{\mathbf{T}}}{\partial b^{n-k}} \end{pmatrix} + \sum_{k=0}^{n} \binom{n}{k} \begin{pmatrix} \mathbf{0} & \mathbf{0} & \mathbf{0} \\ \mathbf{0} & \mathbf{0} & \mathbf{0} \\ \frac{\partial^k \mathbf{C}_{Tq}}{\partial b^k} & \frac{\partial^k \mathbf{C}_{T\Psi}}{\partial b^k} & \frac{\partial^k \mathbf{C}_{TT}}{\partial b^k} \end{pmatrix} \begin{pmatrix} \frac{\partial^{n-k} \dot{\mathbf{q}}}{\partial b^{n-k}} \\ \frac{\partial^{n-k} \dot{\mathbf{\Psi}}}{\partial b^{n-k}} \\ \frac{\partial^{n-k} \dot{\mathbf{T}}}{\partial b^{n-k}} \end{pmatrix}$$
$$+ \sum_{k=0}^{n} \binom{n}{k} \begin{pmatrix} \frac{\partial^k \mathbf{K}_{qq}}{\partial b^k} & \frac{\partial^k \mathbf{K}_{q\Psi}}{\partial b^k} & \frac{\partial^k \mathbf{K}_{qT}}{\partial b^k} \\ \frac{\partial^k \mathbf{K}_{\Psi q}}{\partial b^k} & \frac{\partial^k \mathbf{K}_{\Psi\Psi}}{\partial b^k} & \frac{\partial^k \mathbf{K}_{\Psi T}}{\partial b^k} \\ \mathbf{0} & \mathbf{0} & \frac{\partial^k \mathbf{K}_{TT}}{\partial b^k} \end{pmatrix} \begin{pmatrix} \frac{\partial^{n-k} \mathbf{q}}{\partial b^{n-k}} \\ \frac{\partial^{n-k} \mathbf{\Psi}}{\partial b^{n-k}} \\ \frac{\partial^{n-k} \mathbf{T}}{\partial b^{n-k}} \end{pmatrix} = \begin{pmatrix} \frac{\partial^n \mathbf{F}_q}{\partial b^n} \\ \frac{\partial^n \mathbf{F}_\Psi}{\partial b^n} \\ \frac{\partial^n \mathbf{F}_T}{\partial b^n} \end{pmatrix} \quad (2.248)$$

which include all the variational statements formulated before. Let us note that, contrary to the initial Equation (2.247), most of the submatrices in Equation (2.248) equal 0, especially for a single random variable, since this variable appears effectively in a few of them only, frequently as the lower-order polynomial form.

2.2.7 Navier–Stokes Perturbation-Based Equations

Let us consider a discretization of the fluid volume Ω by any type of finite elements Ω_f, such that $\Omega = \bigcup_{f=1}^{E} \Omega_f$, where N nodes are introduced. Further, we consider for simplicity only the unidirectional incompressible fluid flow including heat transfer in Cartesian coordinates along the x-axis. The following description is proposed for temperatures, velocities, and pressures in Ω:

$$\theta\,(x;t) = \varphi_\alpha\,(x)\;T_\alpha\,(t)\,; \quad (2.249)$$

$$v\,(x;t) = \varphi_\alpha\,(x)\;v_\alpha\,(t)\,; \quad (2.250)$$

$$p\,(x;t) = \tilde{\varphi}_\alpha\,(x)\;P_\alpha\,(t)\,; \quad (2.251)$$

where α runs over all degrees of freedom $\alpha = 1, \ldots, R$. It can be shown that in case of unidirectional flow, the fundamental set of corresponding variational equations can be reduced to the matrix form as follows [3, 94]:

$$\begin{bmatrix} \mathbf{M} & 0 & 0 \\ 0 & 0 & 0 \\ 0 & 0 & \mathbf{C} \end{bmatrix} \begin{bmatrix} \dot{\mathbf{V}} \\ \dot{\mathbf{P}} \\ \dot{\mathbf{T}} \end{bmatrix} + \begin{bmatrix} \mathbf{K}_{\mu V_X V_X} + \mathbf{K}_{VV_X} & \mathbf{K}_{V_X P} & 0 \\ \mathbf{K}_{V_X P}^T & 0 & 0 \\ 0 & 0 & \mathbf{K}_{VT} + \mathbf{K}_{TT} \end{bmatrix} \begin{bmatrix} \mathbf{V} \\ \mathbf{P} \\ \mathbf{T} \end{bmatrix} = \begin{bmatrix} \mathbf{R}_B + \mathbf{R}_S \\ 0 \\ \mathbf{Q}_B + \mathbf{Q}_S \end{bmatrix}. \tag{2.252}$$

The submatrices used in Equation (2.252) may be represented as

$$\mathbf{M} = \int_{\Omega} \rho \varphi^T \varphi d\Omega, \tag{2.253}$$

$$\mathbf{K}_{\mu V_X V_X} = \int_{\Omega} 2\mu \varphi_{,x}^T \varphi_{,x} d\Omega, \quad \mathbf{K}_{VV_X} = \int_{\Omega} \rho \varphi^T \varphi v_x \varphi_{,x} d\Omega, \quad \mathbf{K}_{V_x P} = -\int_{\Omega} \varphi_{,x}^T \tilde{\varphi} d\Omega \tag{2.254}$$

$$\mathbf{C} = \int_{\Omega} \rho c \varphi^T \varphi d\Omega, \quad \mathbf{K}_{TT} = \int_{\Omega} k \varphi_{,x}^T \varphi_{,x} d\Omega, \quad \mathbf{K}_{VT} = \int_{\Omega} \rho c \varphi^T \varphi v_x \varphi_{,x} d\Omega \tag{2.255}$$

as well as

$$\mathbf{R}_B = \int_{\Omega} \varphi^T \mathbf{f} d\Omega, \quad \mathbf{R}_S = \int_{\partial\Omega_\sigma} \varphi^{ST} \hat{\mathbf{t}}\, d(\partial\Omega), \tag{2.256}$$

$$\mathbf{Q}_B = \int_{\Omega} \varphi^T \mathbf{g} d\Omega, \quad \mathbf{Q}_S = \int_{\partial\Omega_q} \varphi^{ST} \hat{\mathbf{q}} d(\partial\Omega). \tag{2.257}$$

It corresponds directly to the variational formulation given by Equations (2.107)–(2.109).

When the pressure terms can be neglected, the initial equations system may be reduced to rank 2 by an elimination of the second columns and rows, respectively.

As it was demonstrated in Sec. 2.1.7, an application of stochastic perturbation methodology results in hierarchical equations which can be proposed for the second-order approach as follows:

- zeroth-order equations

$$\begin{bmatrix} \mathbf{M}^0 & 0 & 0 \\ 0 & 0 & 0 \\ 0 & 0 & \mathbf{C}^0 \end{bmatrix} \begin{bmatrix} \dot{\mathbf{v}}^0 \\ \dot{\mathbf{P}}^0 \\ \dot{\mathbf{T}}^0 \end{bmatrix} + \begin{bmatrix} \mathbf{K}_{\mu V_X V_X}^0 + \mathbf{K}_{V_X V}^0 & \mathbf{K}_{V_X P}^0 & 0 \\ \mathbf{K}_{V_X P}^{0T} & 0 & 0 \\ 0 & 0 & \mathbf{K}_{VT}^0 + \mathbf{K}_{TT}^0 \end{bmatrix} \begin{bmatrix} \mathbf{v}^0 \\ \mathbf{P}^0 \\ \mathbf{T}^0 \end{bmatrix}$$

$$= \begin{bmatrix} \mathbf{R}_B^0 + \mathbf{R}_S^0 \\ 0 \\ \mathbf{Q}_B^0 + \mathbf{Q}_S^0 \end{bmatrix} \tag{2.258}$$

- first-order statements

$$\begin{bmatrix} \mathbf{M}^0 & 0 & 0 \\ 0 & 0 & 0 \\ 0 & 0 & \mathbf{C}^0 \end{bmatrix} \begin{bmatrix} \frac{\partial \dot{\mathbf{v}}}{\partial b} \\ \frac{\partial \dot{\mathbf{P}}}{\partial b} \\ \frac{\partial \dot{\mathbf{T}}}{\partial b} \end{bmatrix} + \begin{bmatrix} \mathbf{K}^0_{\mu V_X V_X} + \mathbf{K}^0_{V_X V} & \mathbf{K}^0_{V_X P} & 0 \\ \mathbf{K}^{0T}_{V_X P} & 0 & 0 \\ 0 & 0 & \mathbf{K}^0_{VT} + \mathbf{K}^0_{TT} \end{bmatrix} \begin{bmatrix} \frac{\partial \mathbf{v}}{\partial b} \\ \frac{\partial \mathbf{P}}{\partial b} \\ \frac{\partial \mathbf{T}}{\partial b} \end{bmatrix}$$

$$= \begin{bmatrix} \frac{\partial \mathbf{R}_B}{\partial b} + \frac{\partial \mathbf{R}_S}{\partial b} \\ 0 \\ \frac{\partial \mathbf{Q}_B}{\partial b} + \frac{\partial \mathbf{Q}_S}{\partial b} \end{bmatrix} - \begin{bmatrix} \frac{\partial \mathbf{M}}{\partial b} & 0 & 0 \\ 0 & 0 & 0 \\ 0 & 0 & \frac{\partial \mathbf{C}}{\partial b} \end{bmatrix} \begin{bmatrix} \dot{\mathbf{v}}^0 \\ \dot{\mathbf{P}}^0 \\ \dot{\mathbf{T}}^0 \end{bmatrix}$$

$$- \begin{bmatrix} \frac{\partial \mathbf{K}_{\mu V_X V_X}}{\partial b} + \frac{\partial \mathbf{K}_{V_X V}}{\partial b} & \frac{\partial \mathbf{K}_{V_X P}}{\partial b} & 0 \\ \frac{\partial \mathbf{K}^T_{V_X P}}{\partial b} & 0 & 0 \\ 0 & 0 & \frac{\partial \mathbf{K}_{VT}}{\partial b} + \frac{\partial \mathbf{K}_{TT}}{\partial b} \end{bmatrix} \begin{bmatrix} \mathbf{v}^0 \\ \mathbf{P}^0 \\ \mathbf{T}^0 \end{bmatrix} \tag{2.259}$$

- second-order equations

$$\begin{bmatrix} \mathbf{M}^0 & 0 & 0 \\ 0 & 0 & 0 \\ 0 & 0 & \mathbf{C}^0 \end{bmatrix} \begin{bmatrix} \frac{\partial^2 \dot{\mathbf{v}}}{\partial b^2} \\ \frac{\partial^2 \dot{\mathbf{P}}}{\partial b^2} \\ \frac{\partial^2 \dot{\mathbf{T}}}{\partial b^2} \end{bmatrix} + \begin{bmatrix} \mathbf{K}^0_{\mu V_X V_X} + \mathbf{K}^0_{V_X V} & \mathbf{K}^0_{V_X P} & 0 \\ \mathbf{K}^{0T}_{V_X P} & 0 & 0 \\ 0 & 0 & \mathbf{K}^0_{VT} + \mathbf{K}^0_{TT} \end{bmatrix} \begin{bmatrix} \frac{\partial^2 \mathbf{v}}{\partial b^2} \\ \frac{\partial^2 \mathbf{P}}{\partial b^2} \\ \frac{\partial^2 \mathbf{T}}{\partial b^2} \end{bmatrix}$$

$$= \begin{bmatrix} \frac{\partial^2 \mathbf{R}_B}{\partial b^2} + \frac{\partial^2 \mathbf{R}_S}{\partial b^2} \\ 0 \\ \frac{\partial^2 \mathbf{Q}_B}{\partial b^2} + \frac{\partial^2 \mathbf{Q}_S}{\partial b^2} \end{bmatrix} - \begin{bmatrix} \frac{\partial^2 \mathbf{M}}{\partial b^2} & 0 & 0 \\ 0 & 0 & 0 \\ 0 & 0 & \frac{\partial^2 \mathbf{C}}{\partial b^2} \end{bmatrix} \begin{bmatrix} \dot{\mathbf{v}}^0 \\ \dot{\mathbf{P}}^0 \\ \dot{\mathbf{T}}^0 \end{bmatrix}$$

$$- \begin{bmatrix} \frac{\partial^2 \mathbf{K}_{\mu V_X V_X}}{\partial b^2} + \frac{\partial^2 \mathbf{K}_{V_X V}}{\partial b^2} & \frac{\partial^2 \mathbf{K}_{V_X P}}{\partial b^2} & 0 \\ \frac{\partial^2 \mathbf{K}^T_{V_X P}}{\partial b^2} & 0 & 0 \\ 0 & 0 & \frac{\partial^2 \mathbf{K}_{VT}}{\partial b^2} + \frac{\partial^2 \mathbf{K}_{TT}}{\partial b^2} \end{bmatrix} \begin{bmatrix} \mathbf{v}^0 \\ \mathbf{P}^0 \\ \mathbf{T}^0 \end{bmatrix}$$

$$-2\begin{bmatrix} \frac{\partial \mathbf{M}}{\partial b} & 0 & 0 \\ 0 & 0 & 0 \\ 0 & 0 & \frac{\partial \mathbf{C}}{\partial b} \end{bmatrix} \begin{bmatrix} \frac{\partial \dot{\mathbf{v}}}{\partial b} \\ \frac{\partial \dot{\mathbf{P}}}{\partial b} \\ \frac{\partial \dot{\mathbf{T}}}{\partial b} \end{bmatrix} -2 \begin{bmatrix} \frac{\partial \mathbf{K}_{\mu V_X V_X}}{\partial b} + \frac{\partial \mathbf{K}_{V_X V}}{\partial b} & \frac{\partial \mathbf{K}_{V_X P}}{\partial b} & 0 \\ \frac{\partial \mathbf{K}_{V_X P}^T}{\partial b} & 0 & 0 \\ 0 & 0 & \frac{\partial \mathbf{K}_{VT}}{\partial b} + \frac{\partial \mathbf{K}_{TT}}{\partial b} \end{bmatrix} \begin{bmatrix} \frac{\partial \mathbf{v}}{\partial b} \\ \frac{\partial \mathbf{P}}{\partial b} \\ \frac{\partial \mathbf{T}}{\partial b} \end{bmatrix}. \tag{2.260}$$

Furthermore, following Equations (2.258)–(2.260), we introduce the nth-order recursive formulas corresponding to the generalized stochastic perturbation technique:

$$\sum_{k=0}^{n}\binom{n}{k}\begin{bmatrix} \frac{\partial^k \mathbf{M}}{\partial b^k} & 0 & 0 \\ 0 & 0 & 0 \\ 0 & 0 & \frac{\partial^k \mathbf{C}}{\partial b^k} \end{bmatrix} \begin{bmatrix} \frac{\partial^{n-k} \dot{\mathbf{v}}}{\partial b^{n-k}} \\ \frac{\partial^{n-k} \dot{\mathbf{P}}}{\partial b^{n-k}} \\ \frac{\partial^{n-k} \dot{\mathbf{T}}}{\partial b^{n-k}} \end{bmatrix}$$

$$+\sum_{k=0}^{n}\binom{n}{k}\begin{bmatrix} \frac{\partial^k \mathbf{K}_{\mu V_X V_X}}{\partial b^k} + \frac{\partial^k \mathbf{K}_{V_X V}}{\partial b^k} & \frac{\partial^k \mathbf{K}_{V_X P}}{\partial b^k} & 0 \\ \frac{\partial^k \mathbf{K}_{V_X P}^T}{\partial b^k} & 0 & 0 \\ 0 & 0 & \frac{\partial^k \mathbf{K}_{VT}}{\partial b^k} + \frac{\partial^k \mathbf{K}_{TT}}{\partial b^k} \end{bmatrix} \begin{bmatrix} \frac{\partial^{n-k} \mathbf{v}}{\partial b^{n-k}} \\ \frac{\partial^{n-k} \mathbf{P}}{\partial b^{n-k}} \\ \frac{\partial^{n-k} \mathbf{T}}{\partial b^{n-k}} \end{bmatrix}$$

$$= \begin{bmatrix} \frac{\partial^n \mathbf{R}_B}{\partial b^n} + \frac{\partial^n \mathbf{R}_S}{\partial b^n} \\ 0 \\ \frac{\partial^n \mathbf{Q}_B}{\partial b^n} + \frac{\partial^n \mathbf{Q}_S}{\partial b^n} \end{bmatrix} \tag{2.261}$$

which gives the fundamental set of stochastic perturbation method equations enabling computations of 0 to n partial derivatives of the state variables and their final probabilistic moments. It is necessary to mention that the coupling matrices containing usually the linear functions of material parameters differentiated twice with respect to any of them vanish, so that the basic difference usually appears in this context between the first- and second-order equations. The second reason to highlight the second-order equations, despite the recursive formulas for the given nth order, is that very often the LSM technique returns an almost linear response function, so that, instead of the higher-order technique one may return to the classical SOSM technique. A simple numerical illustration of the DDM technique with few degrees of freedom is provided for coupled heat and fluid flow in [65].

2.3 Computational Illustrations

2.3.1 Linear Potential Problems

2.3.1.1 1D Fluid Flow with Random Viscosity

Numerical analysis deals with viscous fluid flow in the channel, where viscosity is the random Gaussian input parameter with expected value $E[\mu]=1.0$ and coefficient of variation $\alpha(\mu)=0.15$ (or varying in the interval [0.0, 0.20] in further numerical studies). From the physical point of view, we test the flow between two parallel plates sliding with relative velocity in the horizontal direction. The boundary conditions for fluid flow at the bottom and top of this channel are purely deterministic, so that theoretically any probabilistic characteristics except the expectation must equal 0 (which is documented by the computations). The SFEM discretization is provided using 11 linear stochastic two-noded finite elements and up to tenth-order perturbations are taken into account (see Figure 2.1 containing a deterministic velocity profile shown by the dashed line). The main purpose of this illustration is analysis of the probabilistic fluid flow profile (its expectation and higher-order statistics) as a function of the vertical coordinate for the specified $\alpha(\mu)$ and, separately, maximum fluid velocity probabilistic moments as functions of the input random fluctuations as well as the perturbation parameter $\varepsilon \in [0.9, 1.1]$ (traditionally substituted as 1). This is done to check which one of these parameters is more decisive for the final probabilistic characteristics.

The results of numerical analysis are presented in Figures 2.2–2.4 in the form of, in turn, expected values and coefficients of variation (Figure 2.2), third and fourth central probabilistic moments (Figure 2.3), as well as skewness and kurtosis (Figure 2.4) of the fluid velocity profiles obtained for $\alpha(b)=0.15$. All these results are shown with respect to the second-, fourth-, sixth-, eighth-, and tenth-order SFEM analyses to verify probabilistic convergence of the method with respect to

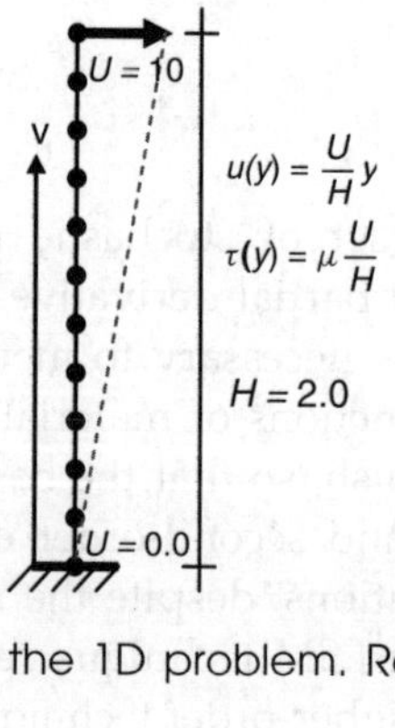

Figure 2.1 FEM discretization of the 1D problem. Reproduced with permission from Elsevier

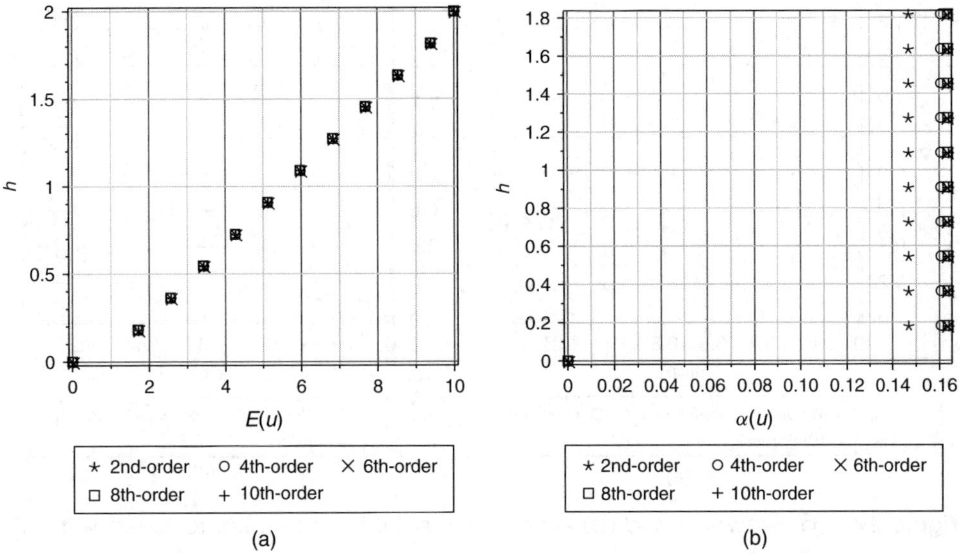

Figure 2.2 The expected (a) values and (b) coefficients of variation of the fluid velocity. Reproduced with permission from Elsevier

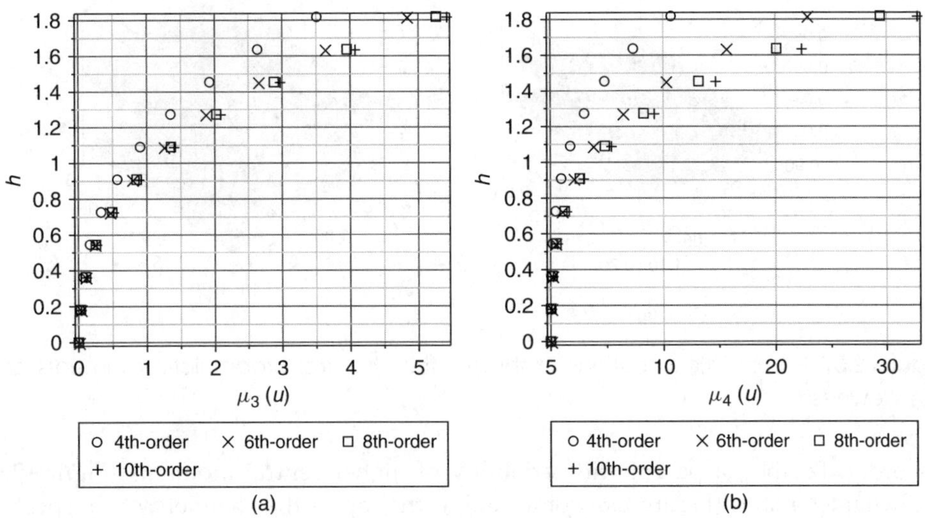

Figure 2.3 (a) Third and (b) fourth central probabilistic moments of the fluid velocity. Reproduced with permission from Elsevier

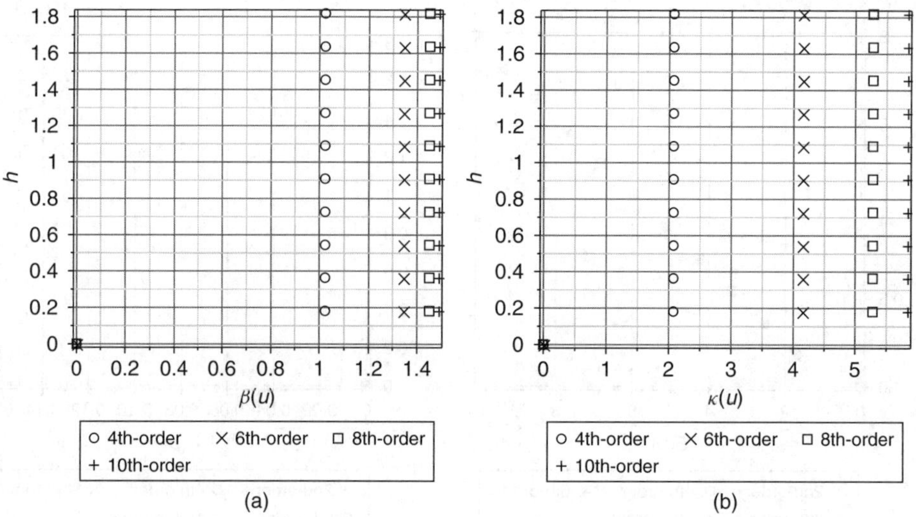

Figure 2.4 (a) Skewness and (b) kurtosis of the fluid velocity. Reproduced with permission from Elsevier

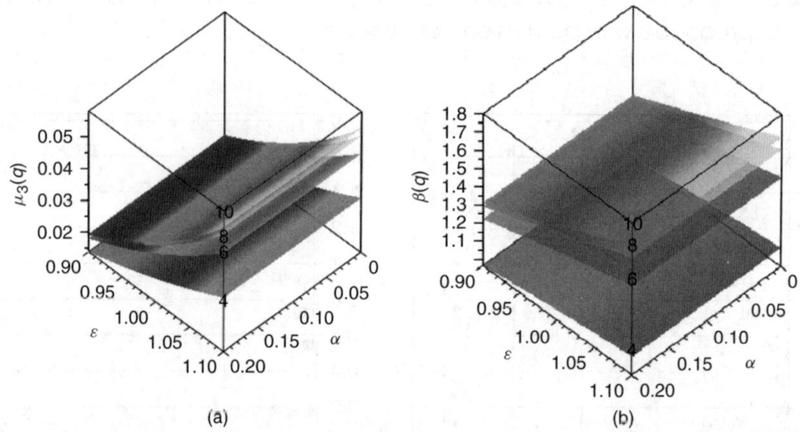

Figure 2.5 Parametric variations of the (a) third central probabilistic moments and (b) skewness

its order. Further, a parametric variability of higher central moments (Figure 2.5) and characteristics (Figure 2.6) computed at the top of the channel with respect to the perturbation parameter and the input coefficient of variation $\alpha(b) \in [0.0, 0.2]$ is given. Let us mention that the total time of computations in this case is only slightly larger than 11 times the original deterministic solution plus some small amount for probabilistic symbolic post-procedures.

As one could expect from theoretical considerations, the expected values have for all orders exactly the same linear profile as the deterministic counterpart; some small differences are noticed for the close neighborhood of the channel bottom. Now, even

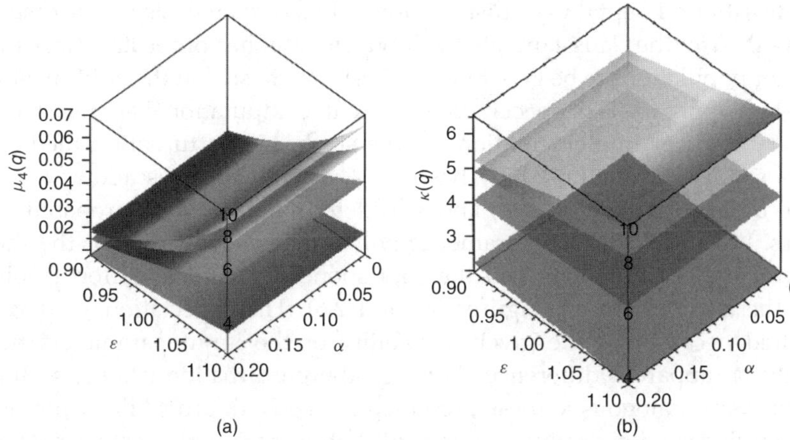

Figure 2.6 Parametric variations of the (a) fourth central probabilistic moments and (b) kurtosis

the second-order stochastic perturbation technique is correct for all values of input $\alpha(b)$ since all models with increasing perturbation orders return the same values along the channel height.

The coefficients of variation of velocity are constant along the height for all perturbation orders except the same small interval close to the bottom, where we noticed some differences between the expectations and deterministic origin. Now, we obtain essential differences for the second- and higher-order perturbation-based results, which all approach the limit almost equal to the input coefficient of variation, so that the fourth-order results are quite accurate. We may notice here also that the higher the perturbation order, the larger the final value of the studied variation coefficient. Although the variance profiles are omitted here (as partially included in the coefficient of variation), they have an almost parabolic distribution because the expectations distribution suggests a linear transform between the output final velocity and the vertical coordinate; they demonstrate a similar probabilistic convergence as this coefficient. Probabilistic convergence of higher-order moments (Figure 2.3) and statistics (Figure 2.4) is very slow – as before, the absolute values increase together with the perturbation order and the most transparent differences between the corresponding orders are detected at the top of the channel. The moment profiles are highly non-linear (the higher the central probabilistic moment, the greater the curvature of the relevant functions), while the coefficients remain constant except some fluctuations at the bottom neighborhood, exactly analogous to the variation coefficient (Figure 2.2). The third central moment is positive everywhere and for all perturbation orders, so that the final distribution of the velocity profile shows right asymmetry (a large portion of this distribution is larger than its median). Final values of the kurtosis remain clearly positive, enabling us to conclude that the velocity PDF is not Gaussian, so that higher-order statistics computations are necessary and, further, they are distributed more closely to the expectation than is typical for the normal distribution (leptokurtic distribution). Statistical analysis in this case is more expensive than for the Gaussian state functions since higher-order statistics essentially need larger populations to be generated and estimated, so that the efficient stochastic perturbation approach is of special importance in computational engineering.

Contrary to other implementations of the stochastic perturbation method [112], the proposed hybrid methodology based on symbolic analysis allows for a direct inclusion of both parameters $\alpha(b)$ and ε in determination of all probabilistic characteristics. We study the two-parametric polynomial expressions for the third and fourth probabilistic moments and the corresponding coefficients for a point very close to the channel top (cf. Figures 2.5 and 2.6). The expectations and variances are omitted because for the given variability of the input parameters there are practically no apparent differences between the perturbation orders, so that their fluctuations are analogous to these presented above. Both probabilistic moments are clearly non-linear with respect to the perturbation parameter remaining for given values almost constant with respect to the input random fluctuations; skewness and kurtosis are almost linear with respect to both input parameters. As one may expect, the larger the value of the parameter ε, the larger the final value of the probabilistic moment and the coefficient computed; they also increase together with an order of the stochastic perturbation method. From a comparison of the third- and fourth-order characteristics one can conclude that a small difference between the eighth and tenth orders for the third moment increases for the same combination of input parameters in case of the fourth-order probabilistic characteristics. From a practical point of view, the absolute values of differences for the highest perturbation orders applied here and obtained for kurtosis and the fourth moment do not qualitatively change the PDF of the fluid response. Since these differences are in the range of a few percent only for the maximum input dispersion taken into account here, we can recommend a full tenth-order approach for an efficient computation of the first four moments and characteristics. Summing up, a general conclusion that can be drawn from this example is that the higher the moment that needs to be computed, the higher the stochastic perturbation approach that is necessary.

2.3.1.2 2D Potential Problem with the Response Function Method

The second numerical illustration deals with the torsion of the isotropic, linear randomly elastic prismatic rectangular beam and is carried out using 100 three-noded triangular plane finite elements; FEM discretization is shown in Figure 2.7 [82]. The twisting moment has unit value here, and is applied at the left bottom corner of the rectangle, while the upper and right edges are free from stresses. The expected value of the Gaussian transverse modulus is taken as equal to $E[G] = 0.16080\text{E}+06$, while its standard deviation corresponds as before to 15% of this value. The entire computational procedure is carried out using the academic FEM program FENAP linked to the symbolic computing platform of Maple™, v. 11, where the nodal response functions, element fluxes, as well as the probabilistic moments and coefficients are computed and visualized together with the original FEM mesh. The resulting state variables and their functions are compared in Figures 2.8–2.11; the expected values and standard deviations (Figure 2.8), the variances and coefficients of variation (Figure 2.9), the third and fourth and fourth probabilistic moments of these fluxes (Figure 2.10), as well as their asymmetry coefficients and kurtosis within all the finite elements (Figure 2.11). A comparison of Figures 2.7 and 2.8 enables us to momentarily conclude that the expected values are practically the same as the corresponding deterministic results – the differences obtained for extremum values are negligible, while the spatial distribution of the fluxes is the same.

It is necessary to mention that the stress vectors have some physical meaning in the case of the expected values only – the remaining maps inform us about the specific properties or tendencies of the output probabilistic distributions of local stresses only. Third-order moments behave quite unsystematically and close to the positive maxima we notice the local extremum values in the opposite direction, whereas the fourth moments are always positive and the maximum stress vectors are directed accordingly.

As one could expect, the coefficient of variation of both horizontal and vertical fluxes within the entire plane region is almost constant and equal to 0.15 (exactly the input coefficient of variation). This result coincides with the results of the first test and confirms the quality of the output obtained using an application of the local version of the RFM against these obtained with the well-recognized SFEM based on the direct differentiation procedure. However, contrary to the Poiseulle flow analyzed before, now the random variable should induce planar output random field, which is confirmed by Figures 2.8–2.11. Figures 2.8 and 2.9 verify the intuitive prediction that the locations of extremum values for expectations, standard deviations, and variances must coincide (where the last two are always positive). According to the mathematical limitations only, maxima of expectations are directed accordingly to the input twisting moment and since this direction is anticlockwise then for the accompanying maxima of variances and standard deviations we obtain an inverse direction. The right bottom edge of the twisted structure only shows the consistent vertical directions and values of both moments' maximum values. As is documented further in Figure 2.10, the third and fourth central probabilistic moments exhibit almost the same tendency except for the fact that for the rest of the cross-sectional area these moments have almost zero values. Finally, Figure 2.11 illustrates the distributions of the asymmetry coefficient and the kurtosis, respectively, demonstrating a completely chaotic disorder in these characteristic distributions. It is impossible to recognize whether the final distribution has dominating positive or negative asymmetry, even with respect to some specific location within this cross-sectional area; the same notice is valid with respect to the kurtosis – most probably a finer meshing of this cross-section will not give an answer. The intervals of these parameters' variability are obtained – for the asymmetry coefficients as [−0.290, 0.205] for the horizontal stresses, and [−0.335, 0.328] for their vertical components; the additional intervals for the kurtosis are as follows: for the horizontal components [−0.390, 0.428] and [−0.398, 0.405] – in the vertical direction. So, one can conclude that some negative asymmetry prevails in both directions (larger parts of the distributions below the medians), whether the positive tendency prevails in kurtosis. Therefore, the flux distributions are more concentrated around their expected values than the input random transverse modulus.

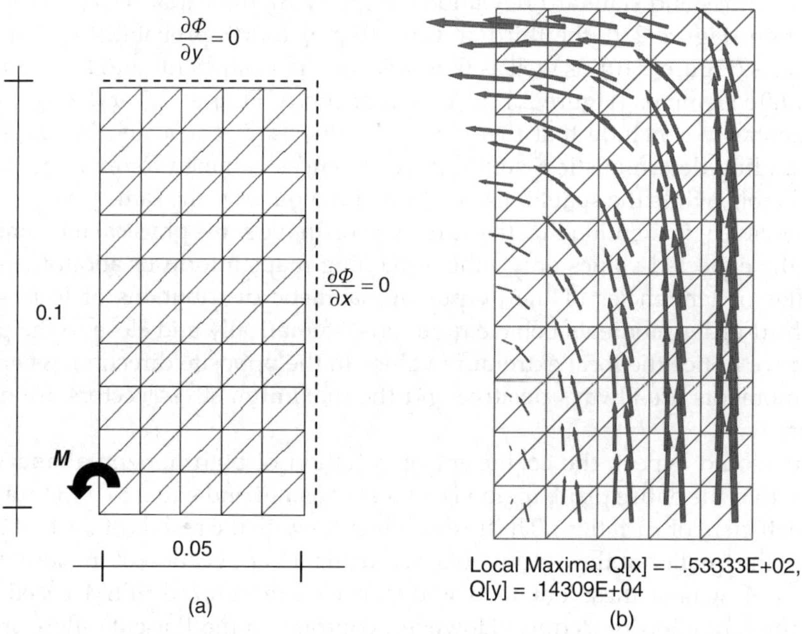

Figure 2.7 (a) FEM discretization of the 2D problem and (b) its deterministic solution. Reproduced with permission from Elsevier

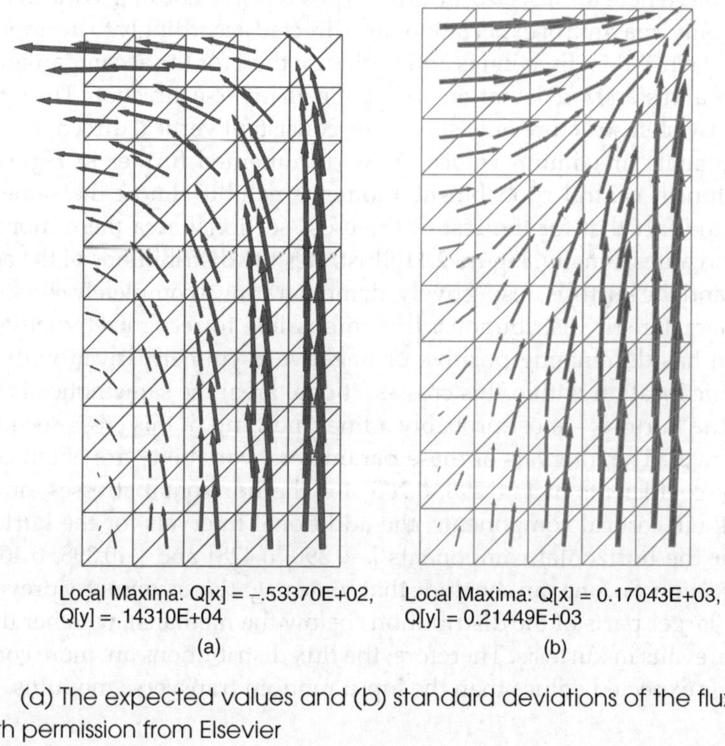

Figure 2.8 (a) The expected values and (b) standard deviations of the fluxes. Reproduced with permission from Elsevier

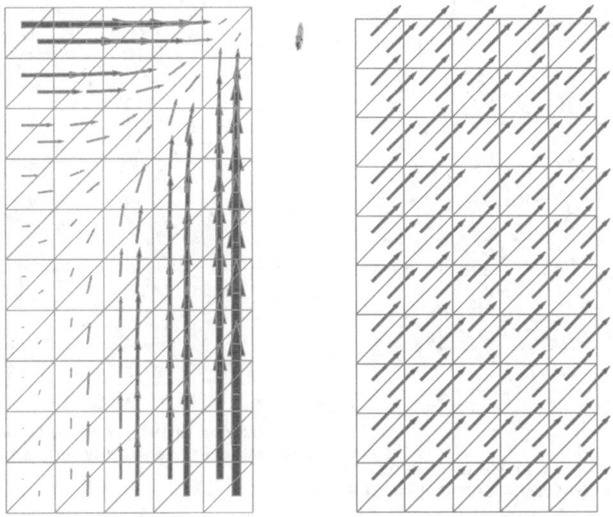

Local Maxima: Q[x] = 0.29046E+05, Q[y] = 0.46004E+05

(a)

Q[x] = 0.15523E+00, Q[y] = 0.15558E+00

(b)

Figure 2.9 (a) The variances and (b) coefficients of variation of the fluxes. Reproduced with permission from Elsevier

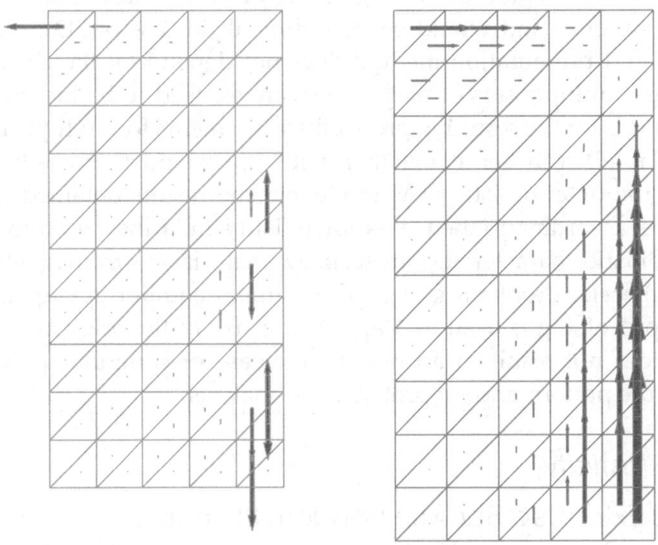

Local Maxima: Q[x] = 0.40804E+04, Q[y] = 0.13372E+06
Local Maxima: Q[x] = −0.12787E+06, Q[y] = −0.15708E+06

(a)

Local Maxima: Q[x] = 0.12285E+05, Q[y] = 0.63491+10

(b)

Figure 2.10 (a) Third and (b) fourth central probabilistic moments of the fluxes. Reproduced with permission from Elsevier

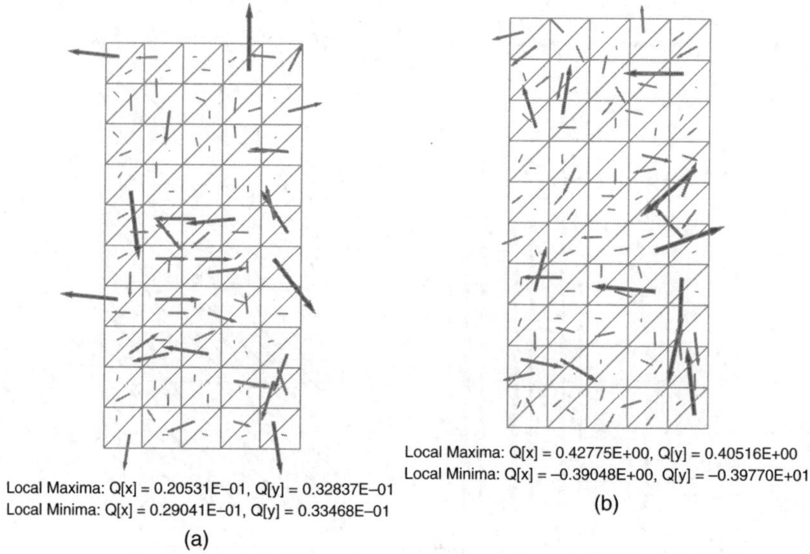

Figure 2.11 (a) The skewness and (b) kurtosis of the fluxes. Reproduced with permission from Elsevier

Finally, let us note that the visualization of higher probabilistic moments in the FEM analysis is a quite separate challenge here, the very exceptional capability in the commercial systems, and would be very difficult in case of 3D modeling. The implementation and presentation method does not depend on the choice of input random variable – multiple sets of partial derivatives need to be incorporated into analogous equations for the resulting probabilistic moments for multiple uncertainty sources, and Maple™ can serve together with its extended procedure options. Analogously to the deterministic FEM modeling, the results obtained for the first two cases may be, according to the well-known field similarities, used to predict the first four probabilistic characteristics in seepage analysis, electromagnetic analysis, and other related scalar problems governed by similar equilibrium equations. Since automatic data processing is used in Maple™, the size of the original deterministic FEM problem does not result in a dramatic increase of computer power or time consumption in the performance of probabilistic analysis.

2.3.2 Linear Elasticity

2.3.2.1 Simple Extended Bar with Random Stiffness

Computational implementation of the generalized perturbation technique in conjunction with finite element analysis is performed in this case study entirely in the Maple™ symbolic environment using the DDM. How it was documented above, it enables (i) relatively easy extension of the technique to any perturbation order, (ii) convergence analysis for particular probabilistic moments and analysis orders, but

Figure 2.12 Numerical example discretization. Reproduced with permission from Elsevier

also (iii) integration with FORTRAN codes for other FEM programs, as well as (iv) efficient visualization of the results. This code is tested on an example of 10 finite element discretization (1D linear elastic bar elements) for the simple tension of a prismatic beam with constant cross-section [74]. The following data, consistent with Figure 2.12, have been adopted: $E[e] = 200.0\text{E}9$ (Young's modulus), $A = 1.0\text{E}{-4}$ (cross-sectional area), $l = 1\text{E}{-1}$ (finite element basic dimension), $P = 10\text{E}5$ (forces applied).

The test problem is a good illustration of probabilistic convergence because of the infinite number of non-zero partial derivatives with respect to the random input parameter, since each new perturbation order introduces some new components into the probabilistic output of tensioned edge deformation. Next, it is observed that the exact analytical solution for this problem consists of inversion of the Gaussian input quantity, which means that an integral equivalent to the expected value does not converge and its value can be obtained by numerical integration only. The polynomial expression for the expected values in the tenth-order approach is returned symbolically by Maple™ for the purpose of illustration at this edge as

$$E\left[q_{11}\left(\varepsilon,\alpha(b)\right)\right] = q_{11}^{0} + 0.02392\ \varepsilon^2\alpha^2(e) + 0.07177\ \varepsilon^4\alpha^4(e) + 0.35885\ \varepsilon^6\alpha^6(e) + 2.51196\ \varepsilon^8\alpha^8(e) + 22.60766\ \varepsilon^{10}\alpha^{10}(e)$$

and all further response characteristics are studied at the same location. The computational illustration is summarized first in the form of the expectations and variances plotted in Figure 2.13, standard deviations and coefficients of variation, cf. Figure 2.14, third and fourth central probabilistic moments in Figure 2.15, as well as skewness and kurtosis, see Figure 2.16, where the perturbation parameter is taken as equal to 1. They are all computed as functions of the input coefficient of variation $\alpha(b) \in [0.00, 0.25]$ and the perturbation order of the approximation – from second up to tenth for the expected values and from fourth up to tenth for the remaining characteristics. As is documented by the first four diagrams, the second-order technique returns really satisfactory values for the first- and second-order characteristics, maybe except for the values obtained for $\alpha(b) > 0.20$, but it may follow the simplicity of the example considered. Of course, the higher the order of the SFEM analysis, the larger the values of the first two moments and all their functions are undoubtedly convex with slightly different second-order partial derivatives w.r.t. the coefficient α. It is seen that the uncertainty level is remarkably increased by this system since the ratio of output to input random dispersion is, after right-hand diagram in Figure 2.14, larger than 1.

The third and fourth central probabilistic moments behave similarly to the lower-order moments – also convex, continuous, and monotonously increasing together with α, highly non-linear but with very small absolute values close to 0. Their normalization with the additional powers of the standard deviation into the skewness

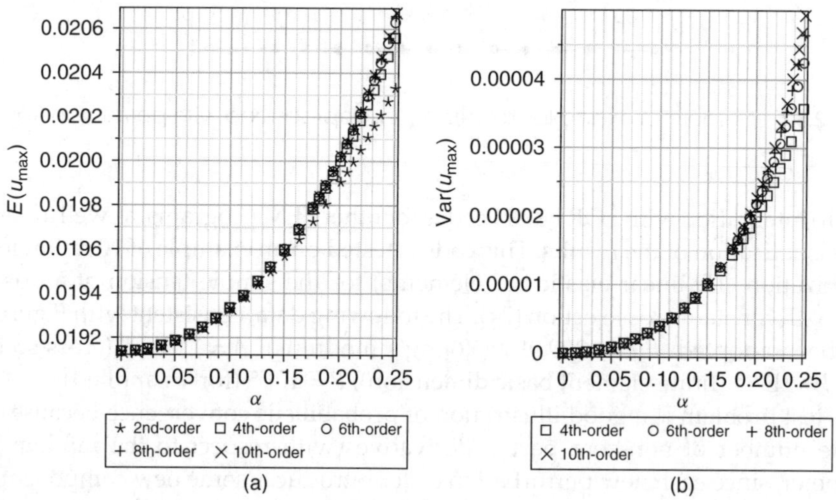

Figure 2.13 (a) Expected values and (b) variances of the maximum deformation

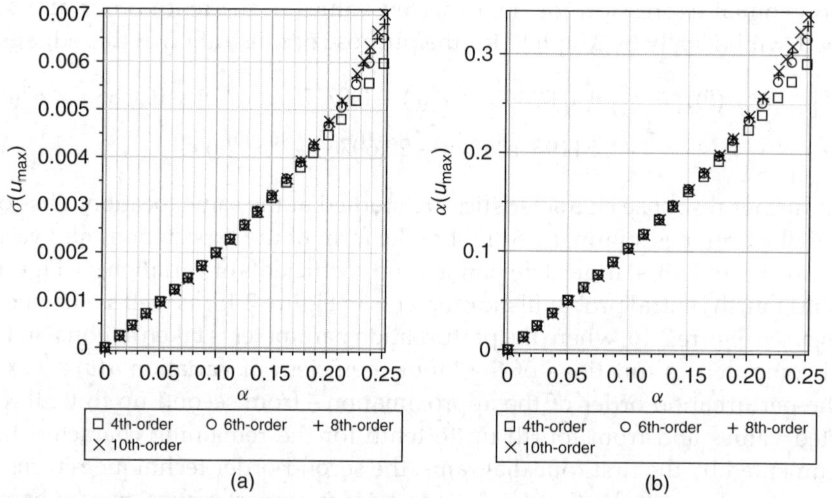

Figure 2.14 (a) Standard deviations and (b) coefficients of variation of the maximum deformation

and kurtosis confirm the initial theoretical consideration that the output random deformation is never Gaussian and the larger the input α, the larger the distance from Gaussian characteristics. It is interesting that while skewness remains always negative (left asymmetry of the PDF), the kurtosis starts from negative values for the fourth-order SFEM and through moderate values for sixth order tends to significant positive values in the tenth-order SFEM analysis. A good convergence for both

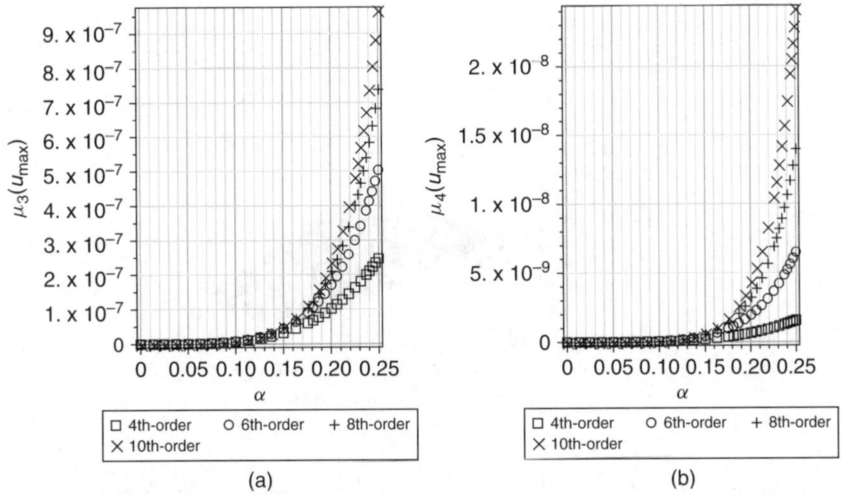

Figure 2.15 (a) Third and (b) fourth central probabilistic moment coefficients of variation of the maximum deformation

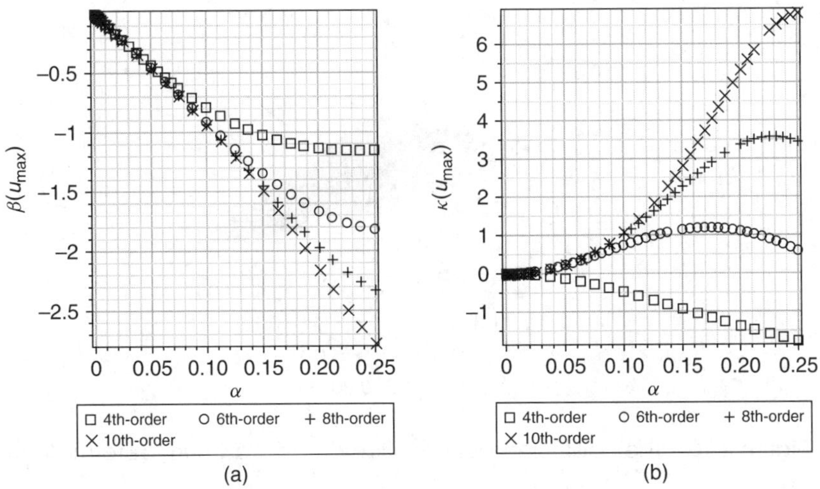

Figure 2.16 (a) Skewness and (b) kurtosis of the maximum deformation

characteristics is noticed for $\alpha < 0.15$ and we need to increase the approximation order in SFEM for larger values.

Further, we study the expectations (Figure 2.17), the coefficient of variation (Figure 2.18), skewness (Figure 2.19), as well as kurtosis (Figure 2.20) as the two-parametric functions of both perturbation parameter ε and input coefficient of variation α; the variability interval of the first variable is slightly wider than before and we have $\varepsilon \in [0.80, 1.20]$. Apparently, the first two random characteristics

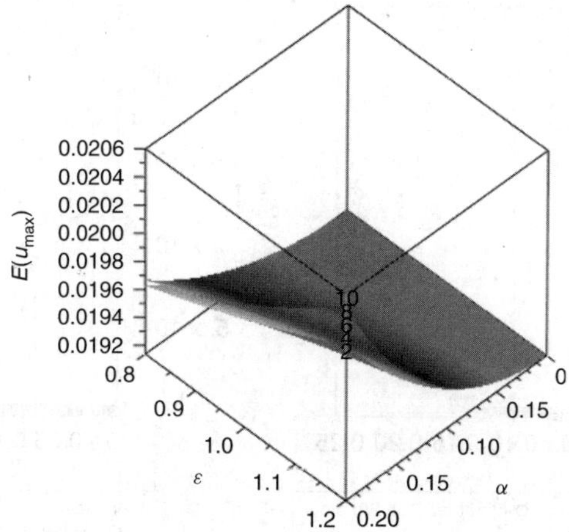

Figure 2.17 The expected values of the maximum deformation

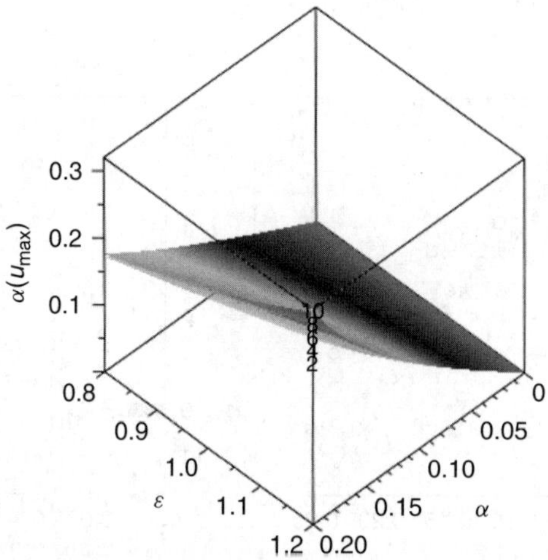

Figure 2.18 The coefficients of variation of the maximum deformation

of a deformation are non-linearly dependent upon both input parameters having positive value for the entire computational domain. It is clear that the more influential parameter is the input coefficient of variation, even for such a wide choice for the input ε.

The changes between neighboring orders of the SFEM are regular, as the figures show, in the sense that the higher the analysis order, the larger the probabilistic first- and second-order response but these differences even for the largest values of both input variables are practically negligible. The dominating role of the input coefficient of variation remains valid for the skewness and kurtosis also, but an interrelation with respect to the perturbation parameter is even closer to the linear

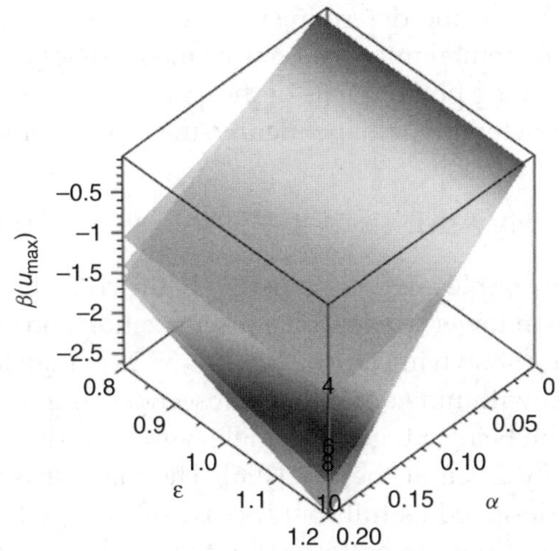

Figure 2.19 Skewness of the maximum deformation

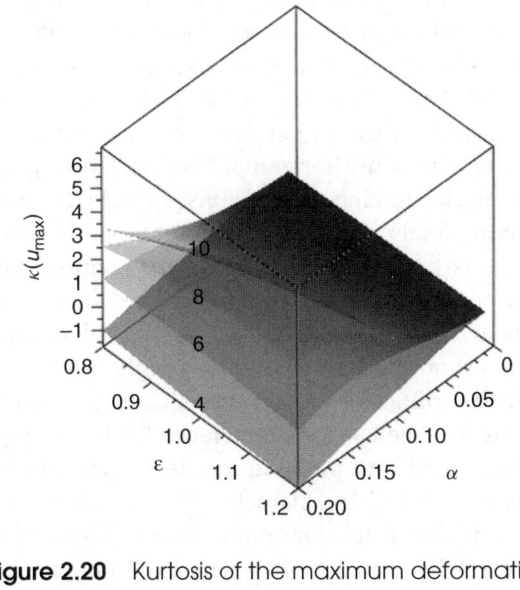

Figure 2.20 Kurtosis of the maximum deformation

function, especially for lower-order SFEM analysis. It should be mentioned that increasing the perturbation parameter value increases each time absolute value of the probabilistic response, so that for skewness we have effectively the negative extremum for a combination of maximum values of both input parameters.

The main value of the DDM implementation is a lack of response function determination for all nodes separately, so that the computational effort and time for solution are almost the same as for the deterministic origin. We also avoid any numerical error coming from polynomial interpolations or the least-squares technique, having as a benefit two-parametric polynomials of the stochastic response – with respect to the perturbation parameter and, independently, the input coefficient of variation.

2.3.2.2 Elastic Stability Analysis of the Steel Telecommunication Tower

The main aim of this example is a determination of the critical load multiplier for the steel telecommunication tower, whose FEM discretization and additional unit forces for stability analysis are shown in Figure 2.21. This 55.20 m high lightweight structure was initially designed with nine 6.0 m long subsections and surface X stiffeners, its horizontal cross-section being a regular triangle with basic dimension equal to 7.3 m at the bottom and only 2.5 m at the top [106]. The main structure is formed with semi-horizontal legs designed as full round cross-sections with diameters 110, 100, 90, and 80 mm counted from the bottom. The legs inside the particular subsections are treated as continuous elements, connected with additional small, thin steel plates (10 mm thick) via bolts. Practically, this enables very limited rotation of the higher leg with respect to the lower one, but in the computational model we treat this connection as infinitely stiff. The stiffening braces under compression/tension are performed with the use of angular cross-sections L150 × 100 × 10, L120 × 80 × 10, and L90 × 60 × 8 (the higher the location, the lower the cross-section). An additional technological platform 1.2 m high is mounted above the top of the last segment, which is perfectly attached to the upper legs and manufactured with three steel pipes Ø 82.9 mm × 4 mm with some horizontal barriers. Considering three reinforced concrete foundation blocks having dimensions 4.0 × 4.0 m and 3.00 m thick, we model this tower as completely clamped at the bottom ends of the leg. Probabilistic computational analysis is performed here using the commercial engineering FEM system ROBOT in cooperation with the system Maple™. The FEM model in ROBOT consists of 30 3D linear beam finite elements (as the tower legs) and 156 3D bar linear finite elements – connected at 298 nodal points. From a probabilistic point of view, we contrast the influence of two input Gaussian random variables – the Young's modulus of steel as well as the diameter of the lower legs (mostly exhibited by the corrosion process and any possible mechanical accidents), see Figure 2.22. We calculate the basic probabilistic characteristics of the critical load multipliers to verify how much additional telecommunication equipment we can add at the top of the tower to keep this structure in a safe exploitation regime, extending its signal transmission capabilities. Further analyses will obey some sinusoidal forced vibrations with the dynamic force located at the top also because of the future usage of the towers in global renewable energy production.

The results of computational analysis for the first and fourth critical load magnitude are given in pairs – for the randomized Young's modulus on the right and for the randomized leg diameter on the left. The response functions given in Figure 2.23 as well as the expected values plotted in Figure 2.24 are given in kN. Then, we compare the coefficients of variation (Figure 2.25), skewness (Figure 2.26), as well as kurtosis

Figure 2.21 3D FEM model of the steel tower

(Figure 2.27) – all computed for increasing stochastic perturbation technique orders, from second or fourth up to tenth. A quite analogous visualization is provided in Figures 2.28 and 2.29 (for the uncertain Young's modulus only) in the context of the fourth critical value and, similarly to most of the figures, we introduce the input coefficient of variation on the horizontal axes as the independent parameter. Its upper bound is not driven by the experimental results, more important here is the computational aspect – to enable precise evaluation of the particular moments' behavior with respect to input random fluctuation levels. The recovery of both response functions having global character is based on 13 FEM iterations of the initial deterministic problem with fluctuating random input variables according to Figure 2.23, and the least-squares approximation by the fifth-order polynomial

Figure 2.22 First critical configuration for Young's modulus (a) and for the random diameter of the structural elements (b)

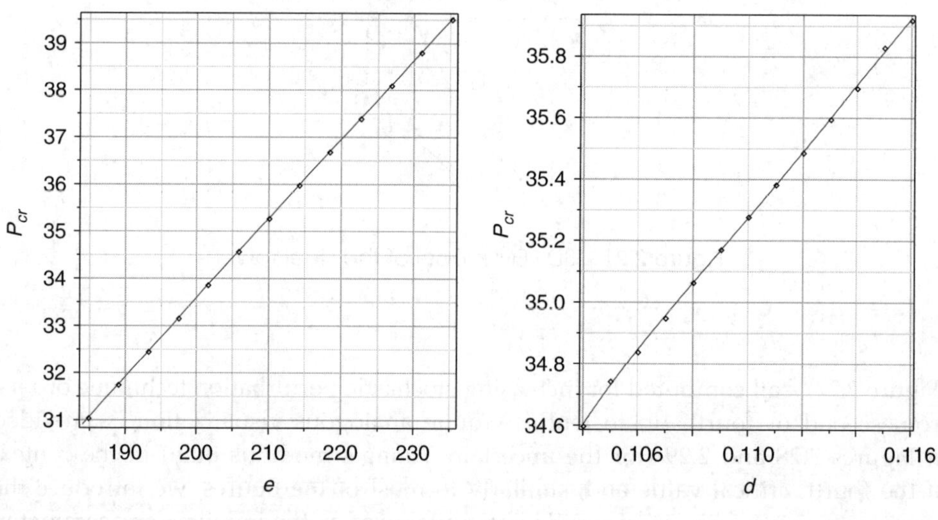

Figure 2.23 Polynomial response function for the critical load (first critical value) for the randomized Young's modulus (a) and leg diameter (b)

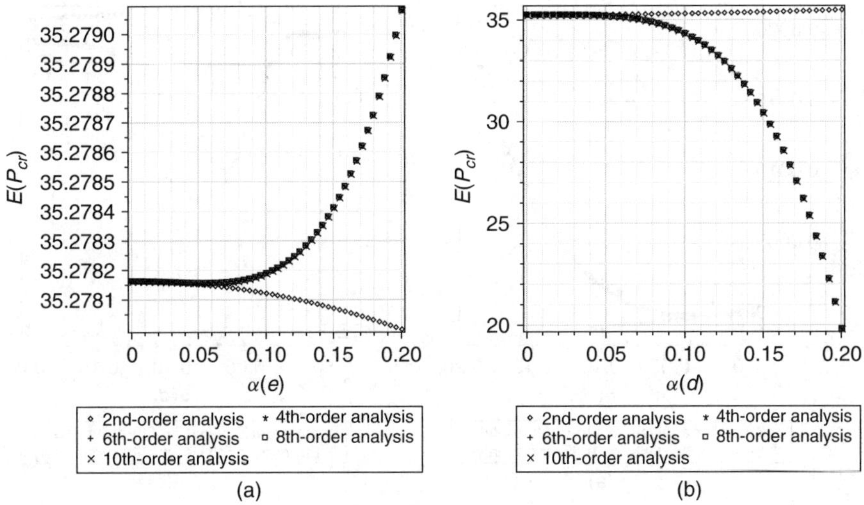

Figure 2.24 The expected values for the critical force (first critical value) for the (a) randomized Young's modulus and (b) diameter

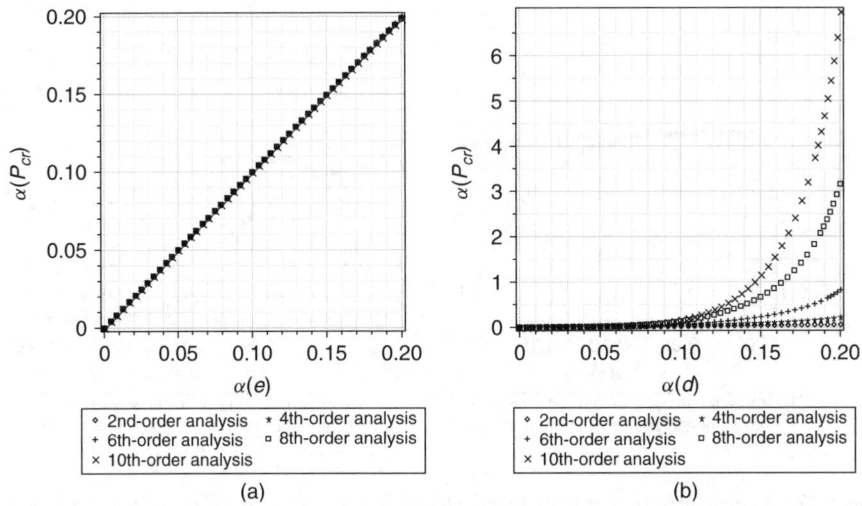

Figure 2.25 The coefficient of variation for the critical force (first critical value) for (a) the random Young's modulus and (b) random diameter

function. Let us mention that the window size variation for the polynomial response function reconstruction does not essentially change the final form of this function for the critical load, but this is not the case for all computational examples provided in this book. As one could guess, both response functions are linear (for Young's modulus)

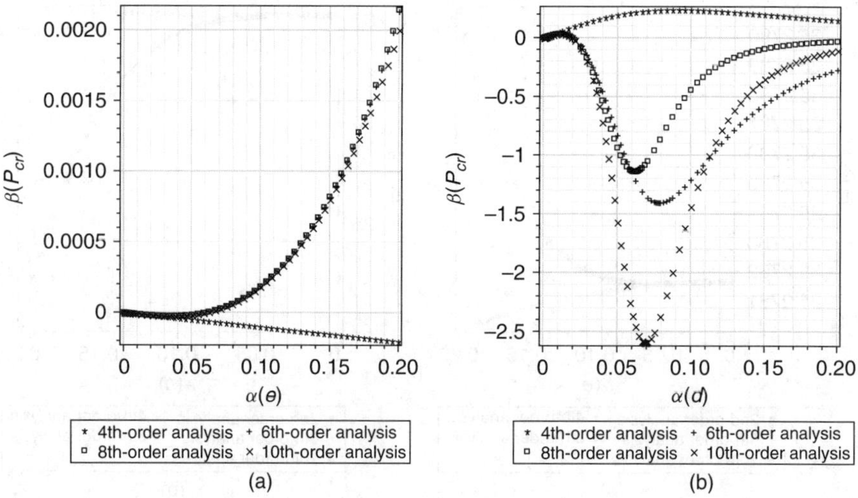

Figure 2.26 The coefficient of skewness for the critical force (first critical value) for (a) the random Young's modulus and (b) random diameter

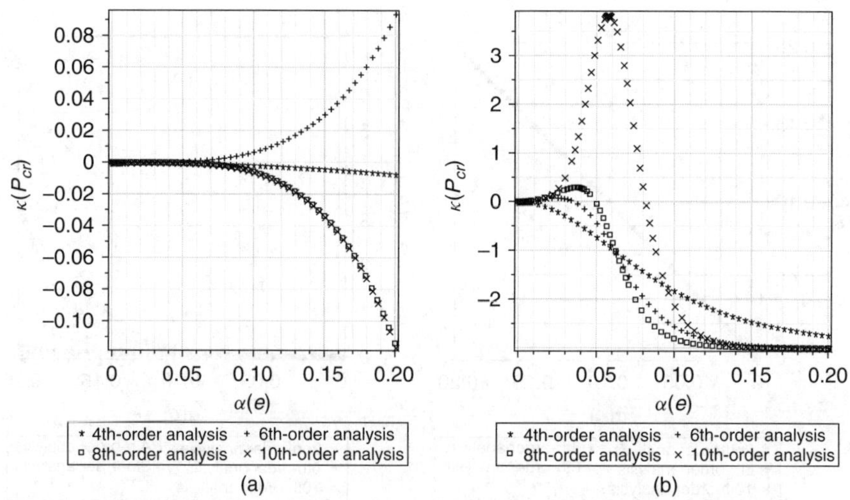

Figure 2.27 Kurtosis for the critical force (first critical value) for (a) the random Young's modulus and (b) random diameter

or almost linear (for a random diameter), so that the higher-order perturbations are not so influential on the probabilistic characteristics computed.

The most important conclusion, which partially confirms this model, is that the tower is really very sensitive to its leg diameters and a few orders less sensitive to the Young's modulus. It follows directly Figure 2.24, where expectation changes with respect to the variation coefficient are negligible for Young's modulus and extremely large in case of the random diameter. The first impression after these

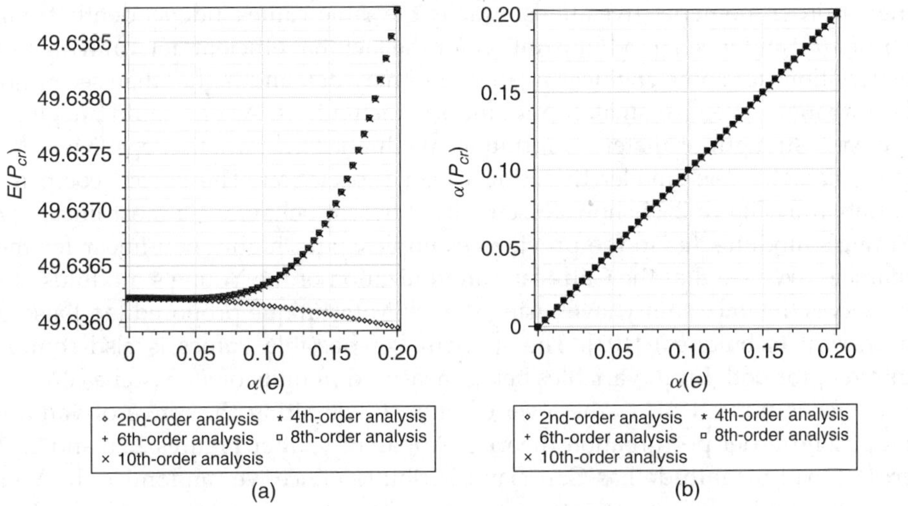

Figure 2.28 (a) The expected values and (b) coefficient of variation for the fourth critical force with randomized Young's modulus

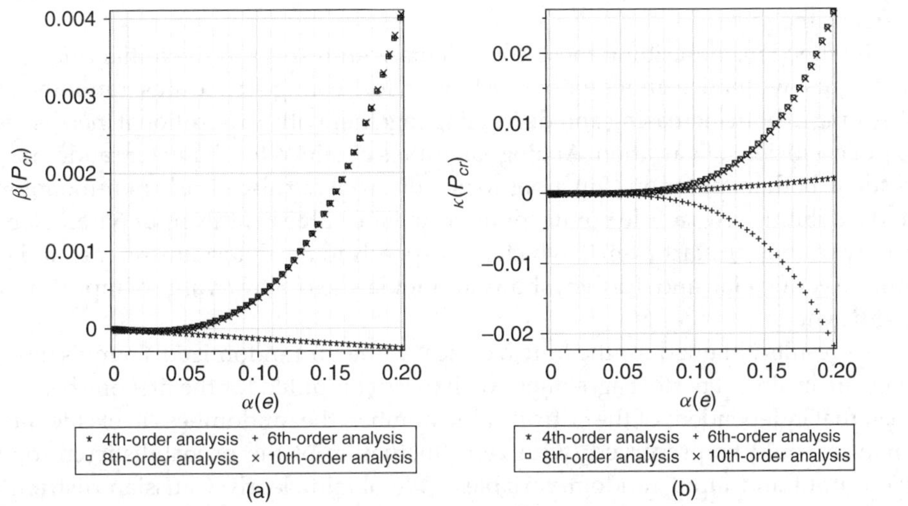

Figure 2.29 (a) The skewness and (b) kurtosis for the fourth critical force with randomized Young's modulus

figures is that the second-order analysis is extremely different from the higher-order approximations and, as such, inacceptable in both cases – where the least common value with the remaining functions is for $\alpha(b) = 0.075$. However, while the left graph shows marginal fluctuations of the critical load magnitude expectations, the right graph shows its fluctuations of paramount importance. It is interesting that all higher-order moments give almost exactly the same values independently from this input $\alpha(b)$, and it is a good illustration for the fact that efficient determination of the expectations for some engineering problems may demand higher than second-order perturbations even for small input random fluctuations. As one could imagine after the basic strength of materials formulas, the higher the variation coefficient for the leg's diameter, the smaller the critical load expectation. The output coefficient of variation in Figure 2.25 shows essentially different behavior – almost linear for the Young's modulus (as in the previous examples), and highly non-linear for the leg diameter. We see that the Gaussian randomization of the Young's modulus of steel preserves the uncertainty level using the SFEM technique proposed (and also other numerical techniques) [103]. The spectrum of possible values is also completely different for both input variables being measured in units of $d(\omega)$; some comparable behavior is noticed up to the level of $\alpha(b) < 0.1$, limiting the range of validity for the second-order perturbation theories. Apparently, after Figures 2.26 and 2.27 the critical load magnitude has Gaussian distribution once we randomize the Young's modulus, while the uncertainty in the cross-sectional area differs completely from this PDF, having a spectrum of both negative and positive values of skewness and kurtosis significantly different from Gaussian zero magnitudes. Although the larger the input dispersion the larger absolute values of the computed characteristics in this case, the final extremum values are still counted in tenths, so that the main tendency remains the same.

The functions describing these coefficients with respect to the initial $\alpha(b)$ are also different – we notice monotonous behavior for Young's modulus randomization, resulting in both increasing and decreasing together with an additional increase of the input coefficient of variation. Analogous curves obtained for diameter randomization result in both monotonous functions for smaller $\alpha(b)$ but also clearly non-monotonous and exhibiting few saddle points for intermediate values of the input parameter $\alpha(b)$. Anyway, they all start from 0, which corresponds to the deterministic case and gives simple verification and, further, show no singularities for any value of input variation coefficient.

The results obtained for the fourth critical value for randomized Young's modulus only are in almost perfect agreement with these computed for the first one before. It is seen that independent of the critical value number, the randomness in elastic stability analysis could be preserved (with zero fluctuation of the probabilistic entropy for the output and input random variables). We obtain clearly Gaussian distributions for both critical loads in all perturbation order methods and all values of the input coefficient of variation. Considering simple random transforms of the Euler formula this is a somewhat predictable result since the critical force is linearly dependent on the Young's modulus and also on the inertia moment of the bar cross-section, which means the fourth power of the leg's diameter; this fact may cause the significant differences in uncertainty results for both parameters.

2.3.3 Non-linear Elasticity Problems

A demonstration of the stochastic generalized perturbation-based FEM related to the non-linear elasticity problems is made using the plane truss structure presented in Figure 2.30, where the Young's modulus of this structure is considered as the input Gaussian random variable. Its expected value equals 30 GPa, while the coefficient of variation ranges from 0.0 (equivalent to the deterministic test) to 0.20. The entire computational procedure is performed using the classical academic FEM software FENAP with enabled elastoplastic analysis and the linear 2D truss finite elements and symbolic computational algebra system Maple™.

Using this hybrid computational procedure we examine the maximum vertical displacements at node 1 (having maximum absolute value), where the vertical concentrated force acting downwards has been applied [79]. The set of all input data is enabled for nine increments in the analysis, and each time the response function of this displacement with respect to a random input is recovered using 11-point polynomial interpolation (cf. Figure 2.31(a)). All polynomial interpolations given in this figure show that the approximants are continuous and smooth functions of the input parameter with no singularities; they exhibit monotonous behavior for all increments. As one may expect, the larger the Young's modulus (marked on the horizontal axis), the less the vertical displacements computed and they systematically increase together with the increment number.

Next, the probabilistic moments are plotted – now as functions of the input coefficient of variation given on the horizontal axis and ranging from 0.0 to 0.20. These

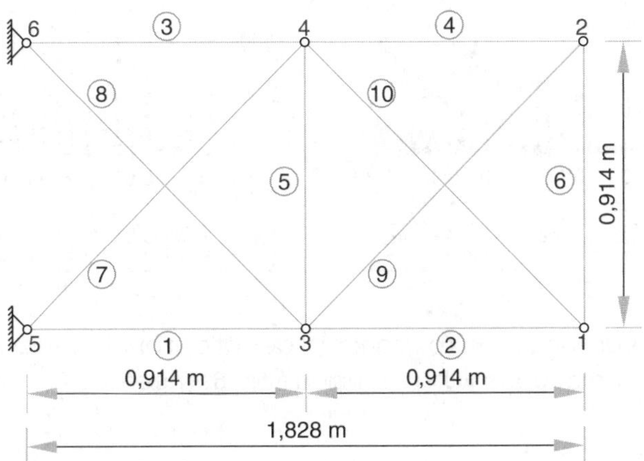

Figure 2.30 The plane truss structure. Reproduced with permission from Springer

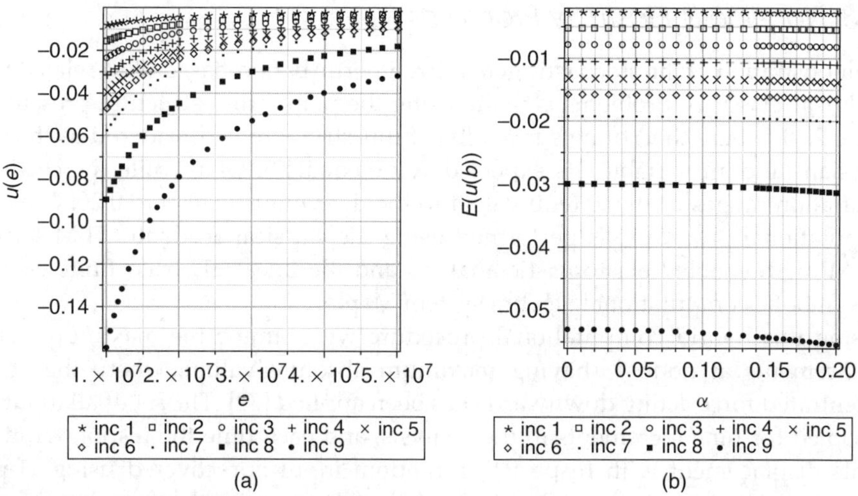

Figure 2.31 (a) Deterministic and (b) expected values for the vertical displacements at the first node. Reproduced with permission from Springer

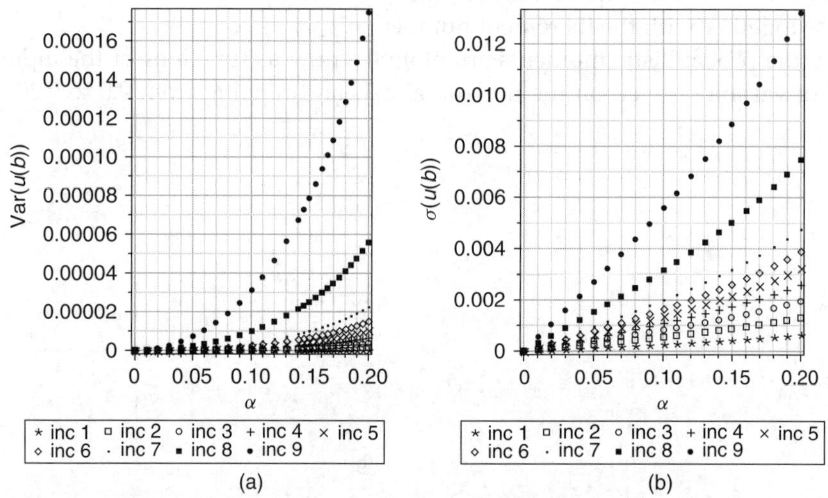

Figure 2.32 (a) Variances and (b) standard deviations of the vertical displacements at the first node. Reproduced with permission from Springer

are in turn: the expected values (Figure 2.31b), the variances and standard deviations (Figure 2.32), the coefficient of variation of the resulting displacements (Figure 2.33), the third and fourth central probabilistic moments (Figure 2.34), as well as skewness and kurtosis (cf. Figure 2.35). All probabilistic moments and characteristics are determined consecutively using the tenth-order expansion. Now it is apparent from Figure 2.31 that the higher the increment of the analysis, the larger the expectation

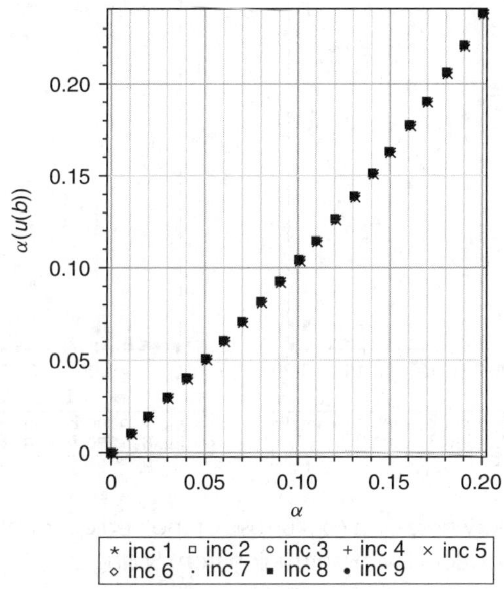

Figure 2.33 The coefficients of variation for the vertical displacements at the first node. Reproduced with permission from Springer

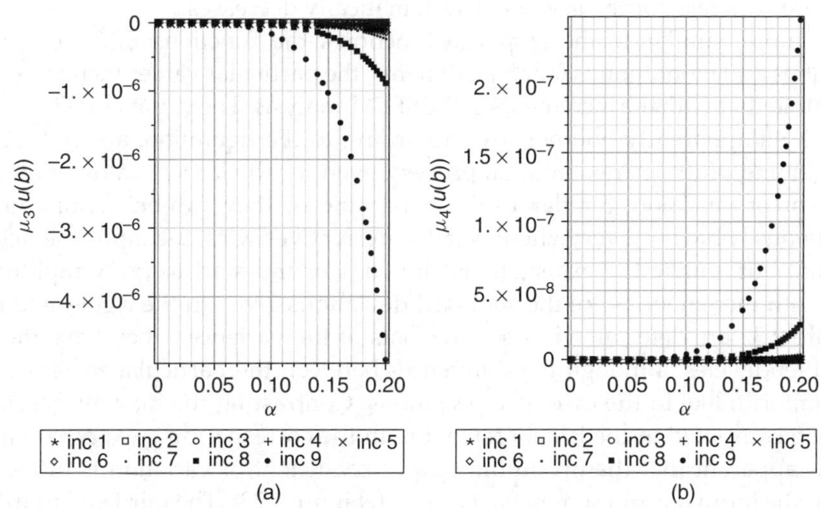

Figure 2.34 (a) The third and (b) fourth central probabilistic moments of vertical displacements at the first node. Reproduced with permission from Springer

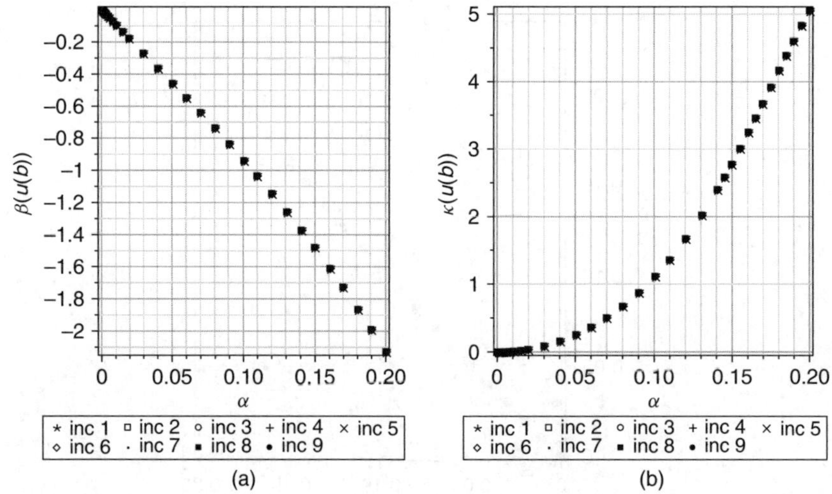

Figure 2.35 (a) The skewness and (b) kurtosis of the vertical displacements at the first node. Reproduced with permission from Springer

of the vertical displacement and, at the same time, the larger the variations of this expectation with respect to the coefficient of variation. This expectation is almost insensitive to this coefficient during the first step of the non-linear incrementing procedure, whereas for the last one it systematically decreases.

Once more, similar to the response functions, the functions $E(u) = E(u(\alpha))$ are continuous and monotonous for all increments; the particular values increase systematically together with the increments in the SFEM analysis. The variances (Figure 2.32), although all apparently non-negative according to the definition, are very close at the beginning of the incrementation process, even for the largest value of the input coefficient of variation. Similar to the expectations, they exhibit continuous and monotonous behavior (everywhere positive derivatives w.r.t. the input coefficient of variation) but, unlike the expectations, are convex and start to grow rapidly after the seventh increment. Since the standard deviations given in the right-hand graph of Figure 2.32 are determined as square roots of the variances, they have the same general tendencies, although the differences between the particular increments are more uniform, like in the case of expectations. Contrasting the maximum standard deviation computed in the last increment with the corresponding expected value, it remains apparent that the maximum output coefficient of variation is in the same range as the input one but somewhat larger (cf. Figure 2.33). The third and fourth central probabilistic moments (Figure 2.34) behave similarly to the variances – the only exceptions are the negative sign for the third moments here and the largest curvature of the particular distributions w.r.t. the input coefficient α. Their values for the entire incrementation process remain almost constant and very close to 0, while during the last two to three steps they grow significantly. All the moments presented are continuous and monotonous functions of the input Young's modulus coefficient of variation; they exhibit no local numerical oscillations as the skewness and kurtosis computed finally (see Figure 2.35). This last graph shows that the final probability distribution is non-Gaussian, has more realizations below the median, and is also leptokurtic (more concentrated around the mean than the Gaussian bell-shaped curve) and, therefore, higher-order moments and coefficients may be necessary in reliability analysis. One may compare the results obtained here thanks to polynomial interpolation with the least-squares method approximation, however, considering the expected and linear ratio between the output and input random variables, these detailed studies have been postponed and left for further numerical investigation by the reader.

2.3.4 Stochastic Vibrations of the Elastic Structures

2.3.4.1 Forced Vibrations with Random Parameters for a Simple 2 DOF System

Computational experiment for the modal superposition method is prepared using the following simple case study illustrating a steel column with partially constant cross-sectional area and subjected to certain longitudinal impulse forced vibrations. The SFEM model consists of two linear finite elements (with three nodes) and is fixed at the left end as well as loaded with the force impulse at the right-hand side (Figure 2.36). This load has zero magnitude in the intervals $t \in [0, t_0]$ and $t \in [t_1, \infty)$, while it equals P for $t \in [t_0, t_1]$. The cross-sectional area A is chosen as the truncated Gaussian input random variable with expected value equivalent to the double HEB 1000, so that $E[A] = 800\,\text{cm}^2$, $P = 1\,\text{kN}$, $E = 210\,\text{GPa}$, $l = 8.0\,\text{m}$; further, $t_0 = 5\,\text{s}$ and $t_1 = 5.1\,\text{s}$. A choice of random input coincides with the possible stochastic corrosion, which, acting on this element, may significantly decrease its effective cross-sectional area. The computational procedure based on the DDM SFEM algorithm has been entirely coded into the computer algebra system Maple™, v. 14, to calculate up to the first four probabilistic moments and characteristics of the displacements and elastic forces in this system.

The expectations of the structural response during the loading phase (Figure 2.37) appear to be quite insensitive to the input coefficient of variation, and the dynamical reactions at the top and half-height of the column have very similar character but differ by almost twice their absolute values; local extremum values appear in both locations at the same time. The coefficients of variation given in Figure 2.38 are almost time-insensitive except for larger values of input coefficient α for the very beginning of the loading phase, where the second-order coefficients start from the extremum. They take enormously large values at any time for input variation $\alpha \geq 0.15$ very close to unity rather than the traditional tenths or percents. It is characteristic that the output random dispersion at the column top, where $P(\tau)$ is applied, is almost twice

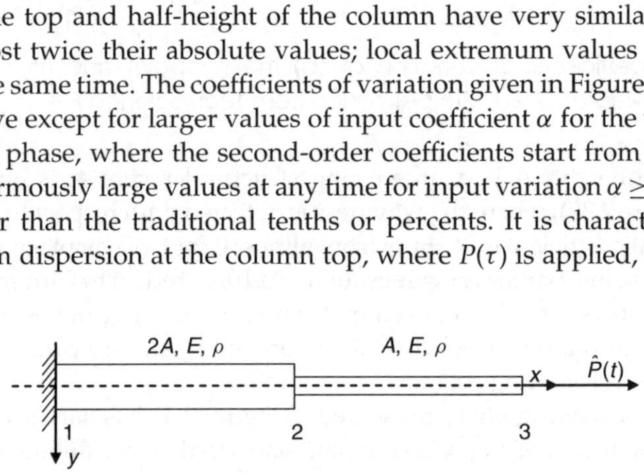

Figure 2.36 Computational model for the forced vibrations. Reproduced with permission from Tech Science Press

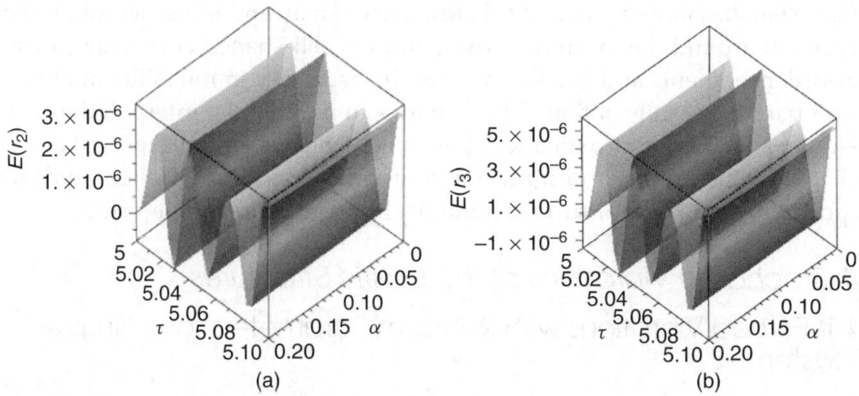

Figure 2.37 The expected values at (a) the midpoint and (b) the top during loading phase. Reproduced with permission from Tech Science Press

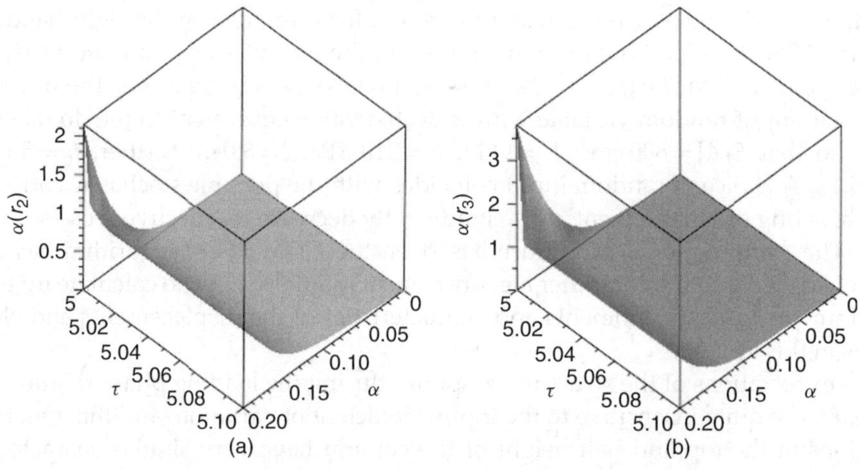

Figure 2.38 Coefficients of variation at (a) the midpoint and (b) the top during loading phase. Reproduced with permission from Tech Science Press

that for its half-height. A very interesting structural response is obtained for the skewness (Figure 2.39), where the waviness is detected with respect to both time and, especially, input coefficient α. It starts from almost 0 for both displacements r_2 and r_3, passes through some extremum values for $\alpha = 0.10$, to return to 0 for maximum $\alpha(A)$. As far as $\beta(r_2)$ shows some kind of symmetry within the range of the vertical axis (the same ranges of absolute values), $\beta(r_3)$ is dominated by negative values especially at the beginning of the loading period.

Kurtosis for the loading phase presented in Figure 2.40 has values totally different from 0, so that neither $r_2(\omega)$ nor $r_3(\omega)$ are Gaussian. Further, r_3 exhibits some enormous increase at the beginning of the loading phase and this is the only exception from the fact that this parameter is almost insensitive with respect to time within this

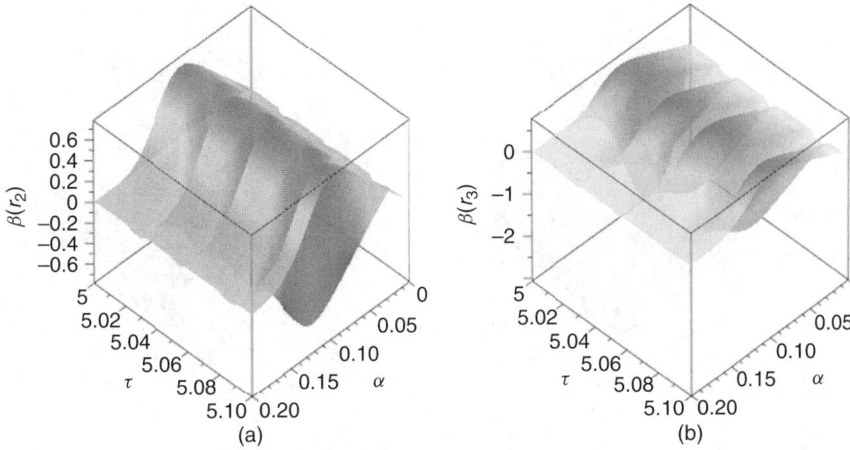

Figure 2.39 Coefficients of skewness at the (a) midpoint and (b) the top during loading phase. Reproduced with permission from Tech Science Press

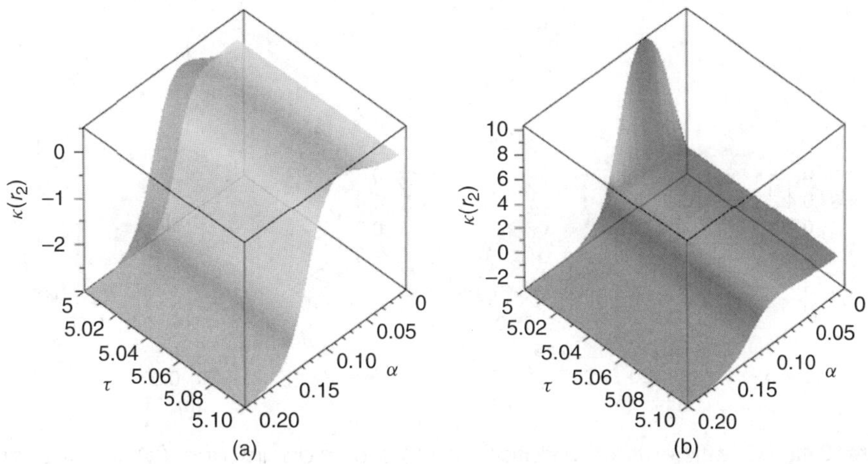

Figure 2.40 Kurtosis at (a) the midpoint and (b) the top during loading phase. Reproduced with permission from Tech Science Press

interval. The main difference in the unloading phase expectations (Figure 2.41) is observed in case of vibrations of the column top, where the amplitude significantly decreases; still an influence of the input randomness is negligibly small. Now the coefficients of variation are always highly sensitive to $\alpha(A)$, unlike with respect to τ, but have almost the same values for both locations within the entire time domain (Figure 2.42). The skewness (Figure 2.43) and kurtosis (Figure 2.44) are very similar to each other for the top and half-height of the column without any sudden jumps or local fluctuations.

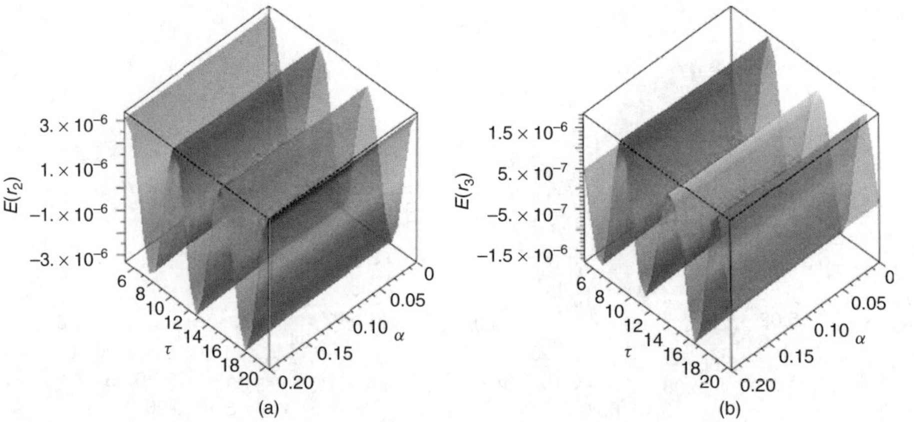

Figure 2.41 The expected values at (a) the midpoint and (b) the top during unloading phase. Reproduced with permission from Tech Science Press

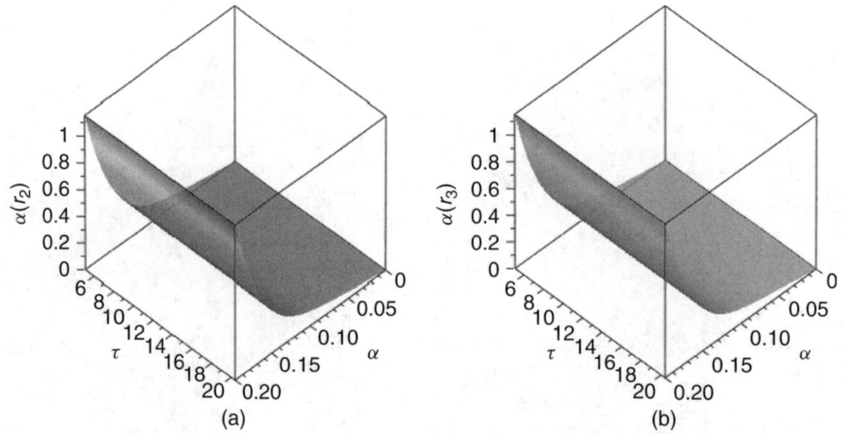

Figure 2.42 Coefficients of variation at (a) the midpoint and (b) the top during unloading phase. Reproduced with permission from Tech Science Press

Whereas skewness has the same ranges of positive and negative values, the kurtosis variability is dominated by clearly negative values. This results from the fact that higher-order partial derivatives of the structural response have negative values and start to dominate for larger values of the input coefficient of variation. So, as a final conclusion, both displacements have non-Gaussian distributions, during both loading and unloading phases. It should be mentioned that further applications of the method proposed may be validated through higher-order statistics with theoretical results available in stochastic vibration analysis [117, 153] as well as with SOSM analyses in case of the first two structural response moments [112].

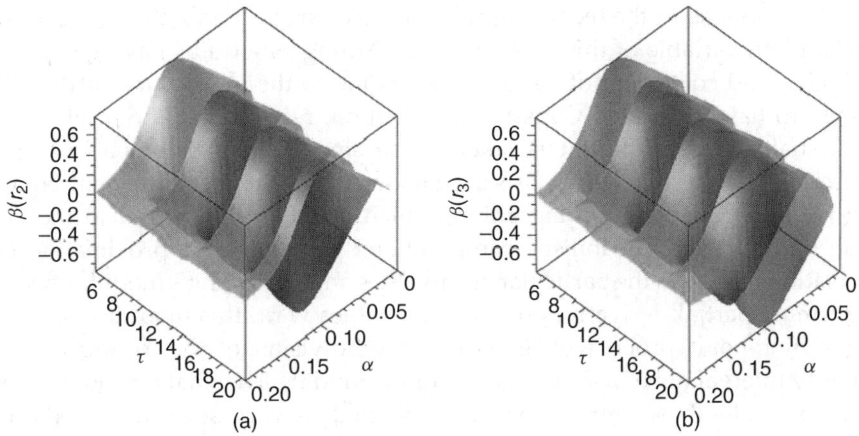

Figure 2.43 Coefficients of skewness at (a) the midpoint and (b) the top during unloading phase. Reproduced with permission from Tech Science Press

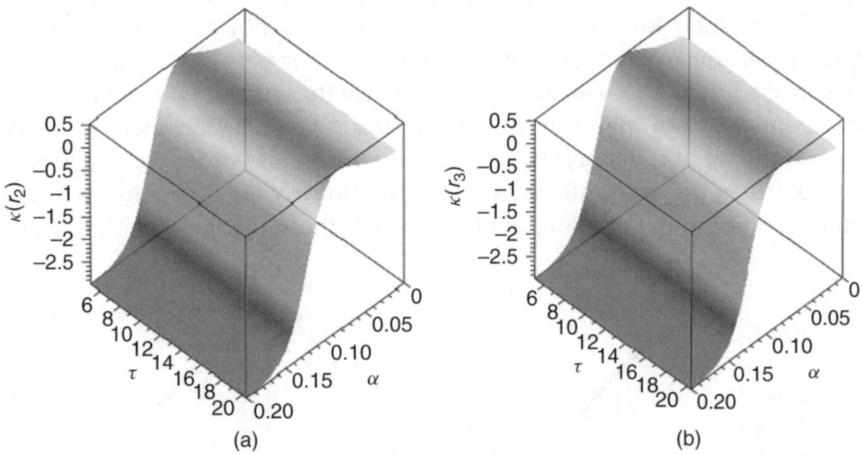

Figure 2.44 Kurtosis at (a) the midpoint and (b) the top during unloading phase. Reproduced with permission from Tech Science Press

2.3.4.2 Eigenvibrations of the Steel Telecommunication Tower with Random Stiffness

Next numerical case study is provided for a steel telecommunication tower with height equal to 52.0 m, very similar to that shown schematically in Figure 2.21. The FEM model consists of 63 two-noded linear elastic beam elements (legs, designed with full circular cross-sections) and 120 two-noded bars connected at all 67 nodes (rebars designed as angular profiles). This structure is subjected to its dead load and the additional weight of the antennas, cables, technological platforms, and their supports as well as to the technological loadings introduced by the Eurocodes. The input random variable of this problem is the Young's modulus having expectation $e = 205\,\text{GPa}$ and coefficient of variation belonging to the interval $\alpha \in [0.0, 0.2]$; it is assumed to have truncated Gaussian distribution, restricted to the positive values only. Usually, this coefficient is taken to be smaller than or equal to 0.15. The engineering FEM system Rama 3D is used to determine the first few eigenfrequencies using the subspace iteration algorithm, which are then transferred to the system Maple™ for further probabilistic computations. Polynomial approximation of the interrelations between the particular eigenvalues and the Young's modulus necessary to determine partial derivatives of the eigenvalues w.r.t. this modulus is provided using a nine-point probe with equidistant subdivision of the Young's modulus variability interval (±50% of the mean value around it). The final response functions for up to the tenth eigenfrequency are obtained as very smooth over the entire domain, without any local oscillations, so that the partial differentiation of up to the tenth order in Taylor expansions is quite straightforward and reliable. The results of computational experiments are contained in Figure 2.45 for the expected values of the first (a) and fourth (b) eigenfrequencies, in Figure 2.46 in case of the coefficients of variation for both frequencies, in Figure 2.47 for their skewness, as well as in Figure 2.48 for their kurtosis; they are all, of course, shown as functions of input random dispersion and for up to tenth-order stochastic perturbations.

The expected values as well as the variation coefficients for both eigenfrequencies are independent from the order of the stochastic perturbation method for almost the entire interval of α variability, so that here the second-order technique is totally efficient. Further, the expectations decrease non-linearly together with an increase of the input coefficient of variation, but the range of these changes is insignificant. The

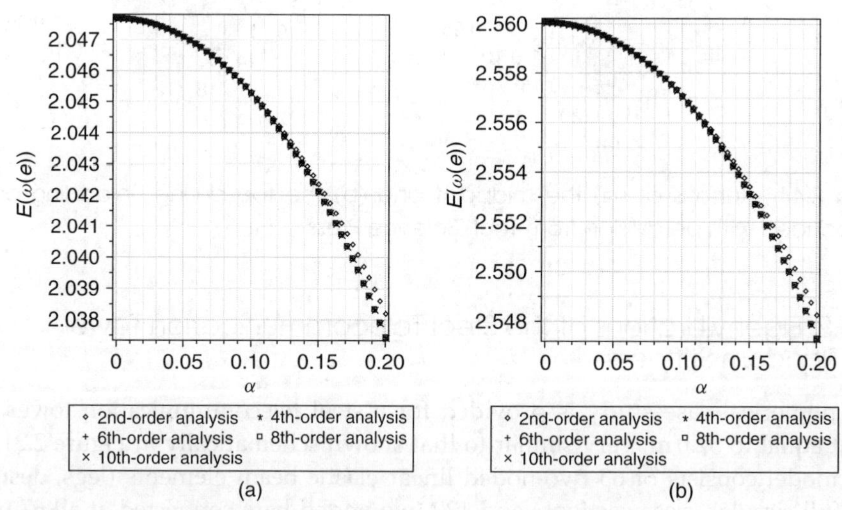

Figure 2.45 Expectations of (a) the first and (b) ninth eigenvalue. Reproduced with permission from Tech Science Press

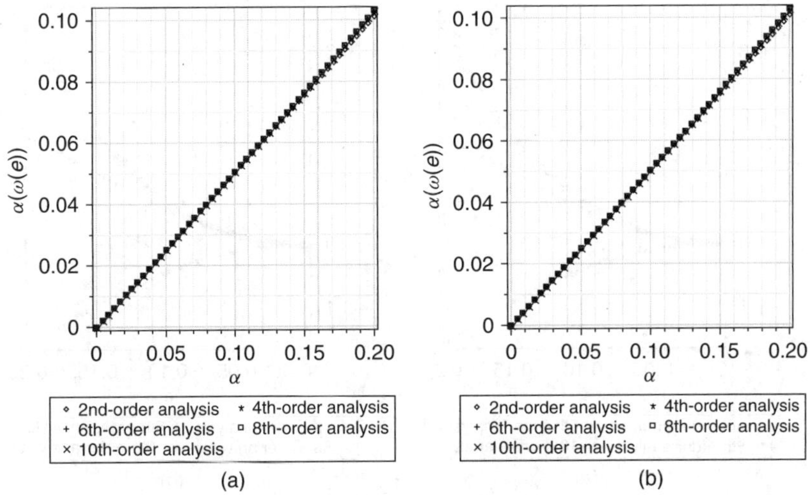

Figure 2.46 The coefficient of variation of (a) the first and (b) ninth eigenvalue. Reproduced with permission from Tech Science Press

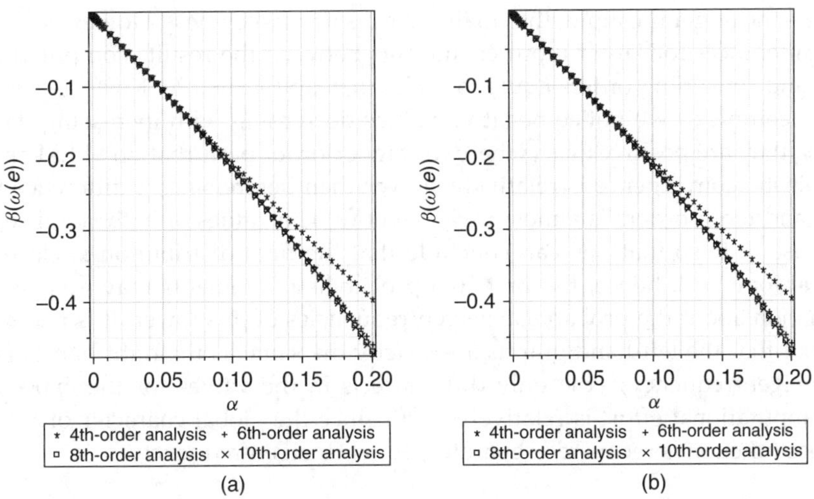

Figure 2.47 The coefficient of skewness of (a) the first and (b) ninth eigenvalue. Reproduced with permission from Tech Science Press

coefficients of variation for both eigenfrequencies increase linearly together with the input coefficient of variation, and finally are half that of the input value.

Having further computed the third and fourth central probabilistic moments, we recover numerically the skewness and kurtosis for both eigenvibrations. Skewness is somewhat antisymmetric to the variation coefficient with respect to the horizontal axis, returning negative values for the entire domain and linear dependence on the

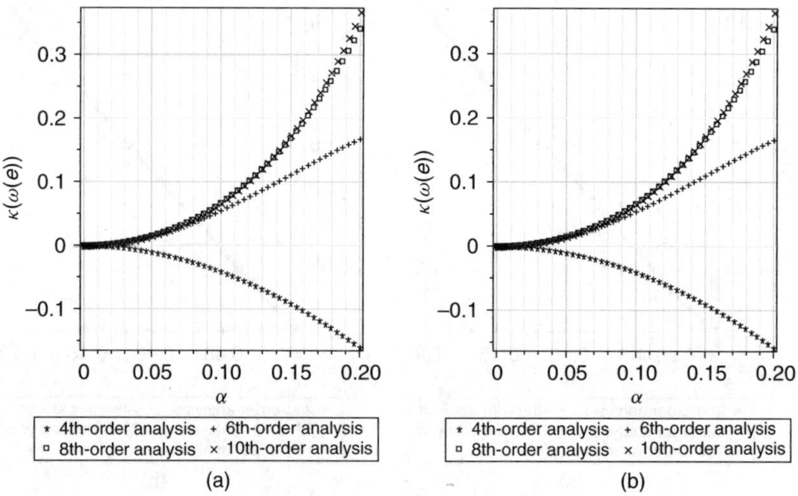

Figure 2.48 Kurtosis of (a) the first and (b) ninth eigenvalue.
Reproduced with permission from Tech Science Press

input α – the larger the input, the smaller the resulting skewness. However, we notice some differences for larger input coefficients between the results computed for the fourth and remaining orders equal to each other up to $\alpha = 0.20$. Kurtosis variability is more complex – we notice negative values decreasing with increasing values of α, while they are positive everywhere for higher-order perturbation SFEM analyses. Probabilistic convergence is relatively slower than before since a difference for the eighth and tenth orders is almost invisible now. The values of both coefficients are very close to 0, so that one can conclude that the final distribution is close to the Gaussian one and, hence, the first two probabilistic moments may give sufficient information about the probabilistic eigenfrequencies of this tower. It is necessary to point out that the left-hand graphs are exactly the same as the right-hand ones (for higher eigenfrequency) – the only difference is in the values for the expectations. The computational effort is relatively small since the global character of the single response function is recovered separately for each eigenfrequency.

2.3.5 Transient Analysis of the Heat Transfer

2.3.5.1 Heat Conduction in the Statistically Homogeneous Rod

This computational study is devoted to heating of the statistically homogeneous and isotropic rod with constant cross-sectional area and length $l = 2.0$ m, where the heat conductivity is the input Gaussian random variable with variation coefficient $\alpha(k) = 0.15$; its expected value is $E[k] = 0.10$. The temperature is fixed at the left edge as $T = 0$, the heat flux is applied at the right corner, whereas the entire structure is divided into 10 three-noded parabolic finite elements. The entire computational procedure is implemented in the symbolic platform of Maple™, v. 11, where up

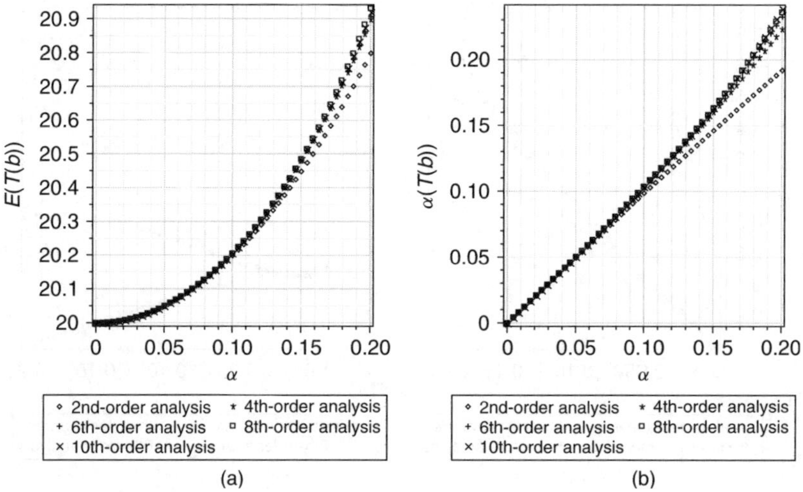

Figure 2.49 (a) Expected values and (b) coefficients of variation at the heated end

to tenth-order equations are formed and solved with analytical differentiation of all necessary system matrices; these solutions are combined into up to fourth-order probabilistic moments and coefficients. Uncertainty in a single physical parameter results here in the random field of temperatures and that is why we present probabilistic moments and coefficients for extremum values as well as probabilistic moment distributions along the rod. Therefore, we have the expectations and coefficients of variation at the heated end in Figure 2.49, together with the skewness and kurtosis at the same location in Figure 2.50, all as functions of the perturbation order and the input variation coefficient. All these functions' values increase together with the input parameter α as well as together with the stochastic analysis order (except for the fourth-order approximation for kurtosis). Undoubtedly, the final temperature distribution is non-Gaussian considering the remarkable positive values of both skewness and kurtosis, whose probabilistic convergence with analysis order is rather slow. In contrast, we notice as before faster convergence of expectations and coefficients of variation, where the fourth-order approximants are quite sufficient for the given combination of input parameters and any value of input random dispersion. Similarly to the SFEM analysis of the extended elastic bar (cf. Figures 2.13 and 2.14), the expectations exhibit convex character and the coefficients of variation are almost linear functions of the input α. The temperature distribution moments and coefficients along the heated rod are given in Figures 2.51–2.54, where Figure 2.51 contains the spatial distribution of the expected values (a) and standard deviations (b).

Now, the spatial coordinate is marked on the horizontal axes of all figures in this series, while numerical analysis is repeated for five different input coefficients of variation shown in the legends. Then we have in turn: the variances and coefficients of variation of the temperature field (Figure 2.52), the third and fourth central probabilistic moments (Figure 2.53) and, finally, the asymmetry coefficients and

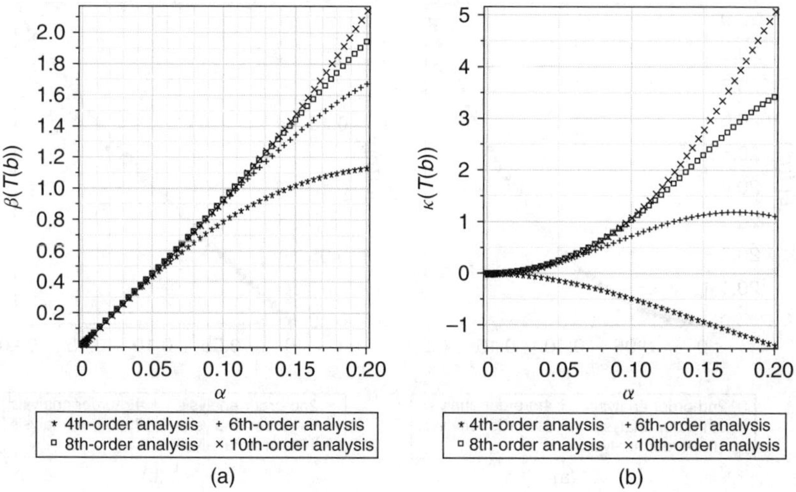

Figure 2.50 (a) Skewness and (b) kurtosis at the heated end

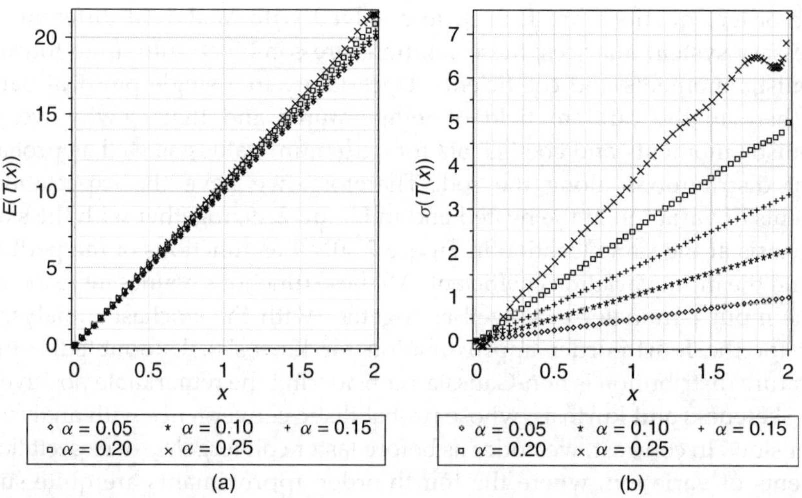

Figure 2.51 Spatial distribution of (a) the expected values and (b) standard deviations in the heated rod

kurtosis spatial variations (Figure 2.53). As one may expect, the mean values have a linear distribution along this rod – quite similarly to the deterministic case – from zero to a maximum at the right end.

The standard deviations also have an almost linear distribution, but they have clearly different values for the various input coefficients α, where of course the higher the input random fluctuation, the larger the dispersion at the output, but the increases between particular curves increase together with the value of the

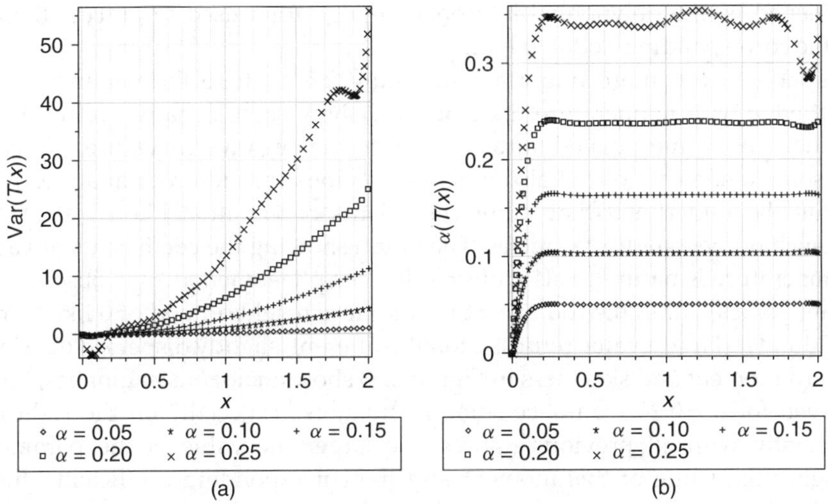

Figure 2.52 Spatial distribution of (a) the variances and (b) coefficients of variation in the heated rod

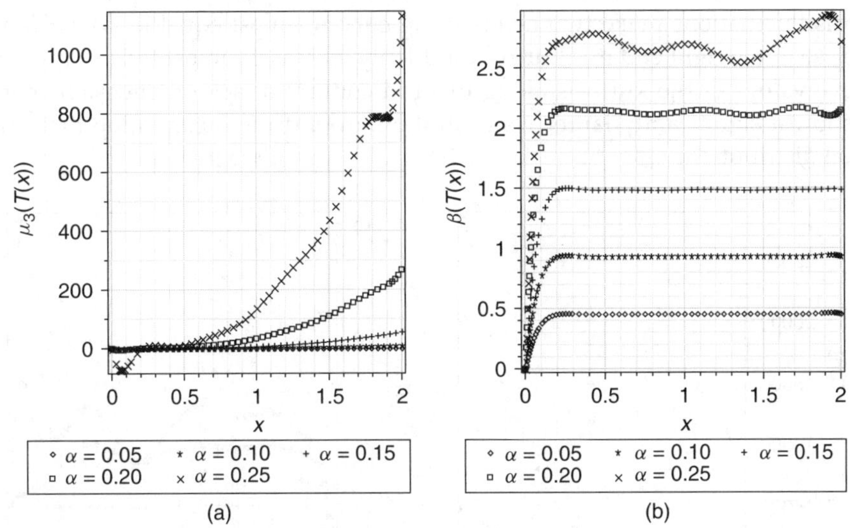

Figure 2.53 Spatial distribution of (a) the third central probabilistic moments and (b) skewness in the heated rod

coefficient α. The distribution of expectations does not show this clearly, but it is apparent from the standard deviations graph that some local fluctuations along the curve represent the given characteristic for larger values of α. This appears so especially close to the right edge, where the heat source is applied, and is partially smoothened by the polynomial interpolation algorithm used to process these data from the corresponding nodal values.

This fact is even more important in Figure 2.52, where the variances close to the left edge with temperature fixed at $T=0$ show some negative values between the nodal values interpolated automatically to the negative curvature, but anyway show some tendency noticed also in some previous second-order analyses [96, 97]. It should be underlined that input coefficients of variation like $\alpha \leq 0.20$ do not cause such numerical discrepancies. The graph showing the coefficients of variation for temperature is naturally also affected by this phenomenon – we observe almost constant values within the rod except for the close neighborhood of 0 like before for $\alpha \leq 0.20$ and its largest value returns a totally different, slightly dispersed distribution. The third moment and skewness in Figure 2.53 show analogous anomalies, however now even for $\alpha = 0.20$ we notice some fluctuations, especially on the right graph. Consistently with all previous figures, the larger the value of the parameter α, the higher the value of this moment and the corresponding coefficient – the only exception is seen for a small surrounding of the left edge (in case of the second and third moments only). Further, the skewnesses obtained for any input coefficient of variation appear to be significantly positive, so that the final distribution of the temperatures cannot be Gaussian. The fluctuating graphs for larger input variation coefficient of the heat conductivity are especially transparent in Figure 2.54, where kurtosis fluctuations make two neighboring curves almost equal in a few locations and close to the right end this kurtosis suddenly increases almost twice. Apparently, positive values of the kurtosis along this rod confirm a lack of Gaussian character and show a significantly larger concentration around the mean value than for the normal distribution.

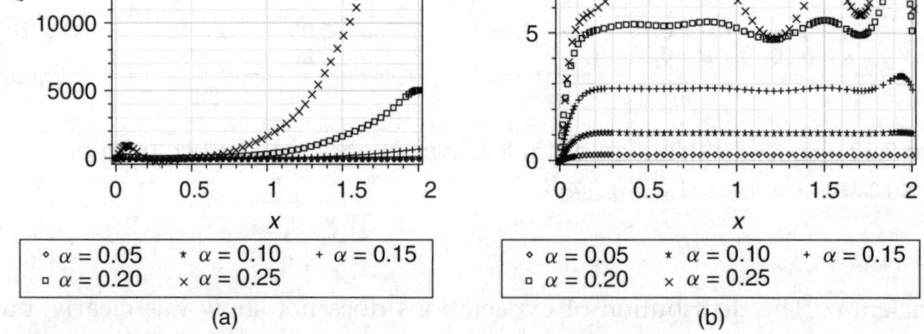

Figure 2.54 Spatial distribution of (a) the fourth central probabilistic moments and (b) kurtosis in the heated rod

The results of this test show that careful FEM discretization close to the boundary conditions location may be necessary for larger values of the input coefficient of variation (like $\alpha \geq 0.25$) even for the steady-state linear problems of heat transfer, where randomization is applied to physical parameters of the homogeneous media.

2.3.5.2 Transient Heat Transfer Analysis by the RFM

The second illustration is the transient heat transfer problem solution by SFEM for the same statistically homogeneous and isotropic rod with constant cross-sectional area and length $l = 2.0\,\text{m}$, where the time increment used was $\Delta t = 2\,\text{s}$. Now the heat conductivity and capacity are taken as truncated input Gaussian random variables with the same random dispersions, $\alpha(k) = 0.15 = \alpha(c)$; their expected values are $E[k] = 0.10$ and $E[c] = 1.0$, respectively. The temperature is fixed at the left edge as $T = 0$, the heat flux is applied at the right corner, and the entire structure is divided into 10 three-noded parabolic finite elements. The larger part of the computational experiment is conducted in the symbolic environment of Maple™, where the local response functions for the additional time increments are determined and, further, combined into the output probabilistic moments together with the additional visualization presented below; a deterministic solution to this problem has been provided externally by the academic FEM code FENAP. The 11-point discretization in the random space is used to define the numerical probing process around the expected value of the random heat conductivity and also capacity (the basic increment is equivalent to 10% of this parameter expectation), and the tenth-order stochastic perturbation method is applied consecutively. These nodal response functions are determined numerically for all time increments, which significantly increases the total computational time consumption. The results contained in Figures 2.55–2.62 contrast spatial distributions of the expected values (Figure 2.55), variances (Figure 2.56), standard deviations (Figure 2.57), coefficients of variation (Figure 2.58), third central probabilistic moments (Figure 2.59), skewness (Figure 2.60), fourth central probabilistic moments (Figure 2.61), and kurtosis (Figure 2.62) for randomized heat capacity (left-hand graphs) with uncertain heat conductivity (right-hand graphs). They are all presented along this rod for time moments corresponding to 5, 10, 20, 30, 40, 60, 80, and 100 seconds of two transient processes (the last time corresponds to the steady-state temperature distribution). Analysis of these pairs shows that the results are similar in the case of expectations only (Figure 2.55) – they exhibit really similar behavior and quite similar values during the entire heating process. They all naturally start from 0 at the left edge, undergo zero values within the first quarter of this structure, especially at the beginning of the heating process, and frequently reach a maximum at the opposite edge of this bar. They all increase monotonously along this bar to the linear function corresponding to the steady state of the temperatures – very predictable after its previous deterministic origin; these expectations are generally consistent with similar tests performed before using second-order analysis [97, 112]. Contrary to expectations, the variances shown in Figure 2.56 are quite different under uncertainty in heat capacity and conductivity. It is important that the variances for this second parameter are more than 10 times larger than for the heat capacity.

This contrast remains true for the given combination of input parameters. Secondly, the variances of temperatures all have convex distributions for random k and increase monotonously from the left edge up to the right one as well as together with the specified time increments.

The coefficients of variation (Figure 2.58) show that the randomness of the heat capacity coefficient is very important at the beginning of the heating process, when the heat starts to penetrate the medium considered; the variation coefficient has maximum value close to the left unheated end. After the initial extremum distribution these coefficients decrease monotonously at particular time moments to reach almost zero at the end. This is quite natural as the steady-state distribution, even random, does not depend at all upon the heat capacity of the material. The same coefficient of variation computed for a random parameter k has almost sinusoidal shape at the first increment of the heating simulation and rather accidentally reaches a maximum (around half that for random c) close to the left edge. Then, its distribution stabilizes along the rod and becomes very close to the results presented in the previous section, that is, almost constant everywhere with a large decrease to 0 at the left end. These results are quite consistent with these obtained before with the use of the SOSM SFEM implementation. The variance distributions obtained for the randomized c exhibit zero convexity close to the middle of the rod. Their time evolution starts with zeros corresponding to deterministically given initial temperatures, goes through maximum values for about 10–20 s of this process, to finally return to almost zero everywhere for the steady state. It naturally affects the distributions of the temperature standard deviations, cf. Figure 2.57. These computed for random k are very close, concerning shape and evolution, to the corresponding expectations, whereas uncertainty in c rather leads to similarity of the standard deviations with variances.

A spatial distribution of the third central probabilistic moments (Figure 2.59) for both random inputs is very similar to the variances except for some negative values obtained for randomized c at the heated edge for the end of the heating process. These moments for uncertain k once more decrease monotonously together with time and with the spatial coordinate x. The skewness distributions (Figure 2.60) are more regular for random c forming some kind of envelope with both positive and negative values, rather irregular for uncertain k at the beginning of heating, to become clearly positive and have shape close to the corresponding variation coefficient. Despite these apparent differences, the final temperature distribution at any moment in time of the heating process has no chance of being Gaussian after Figure 2.60 because of the large absolute values of the skewness parameter. The fourth probabilistic moments (Figure 2.61) are, as the third moments, also similar in shape and convexity to the variances. Unfortunately, the fourth moments computed for random c exhibit some numerical discrepancies resulting in negative values in the middle of the heating process. This is most probably caused by the polynomial interpolation of the response functions, where higher-order partial derivatives may have significant negative values. Kurtosis (Figure 2.62) shows, for both input variables, large variations at the beginning of the transient process taking accidentally negative values, even with random k. Kurtosis distributions, even for later moments of heating, never equal 0 and this is the second reason to treat the temperature distributions as non-Gaussian.

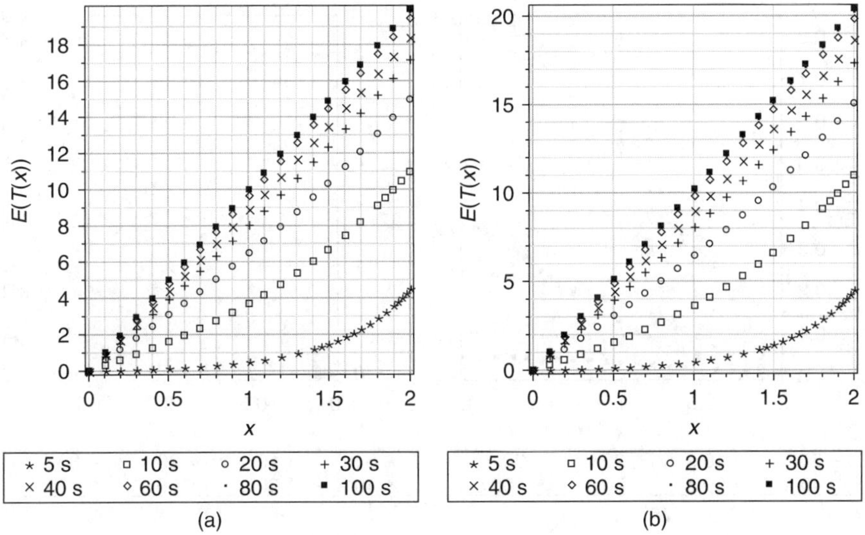

Figure 2.55 Expected values of the temperature distribution in the heated rod, (a) random c and (b) random k. Reproduced with permission from Tech Science Press

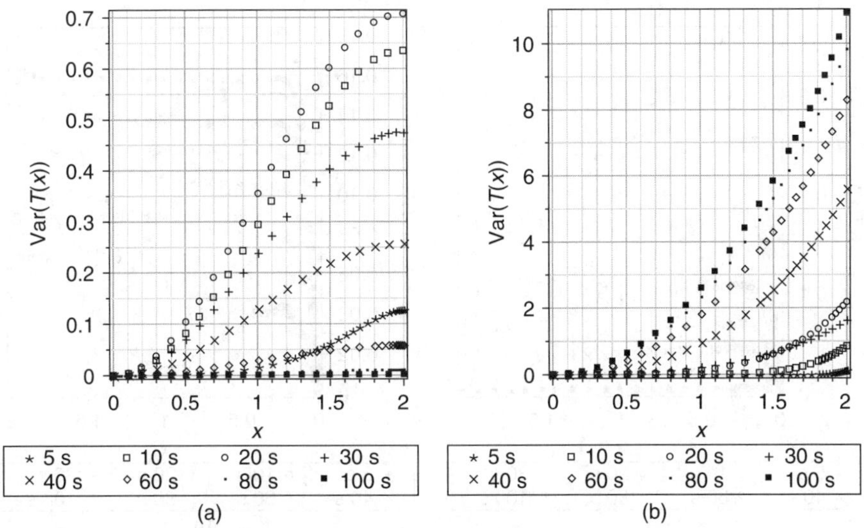

Figure 2.56 The variances of the temperature distribution in the heated rod, (a) random c and (b) random k. Reproduced with permission from Tech Science Press

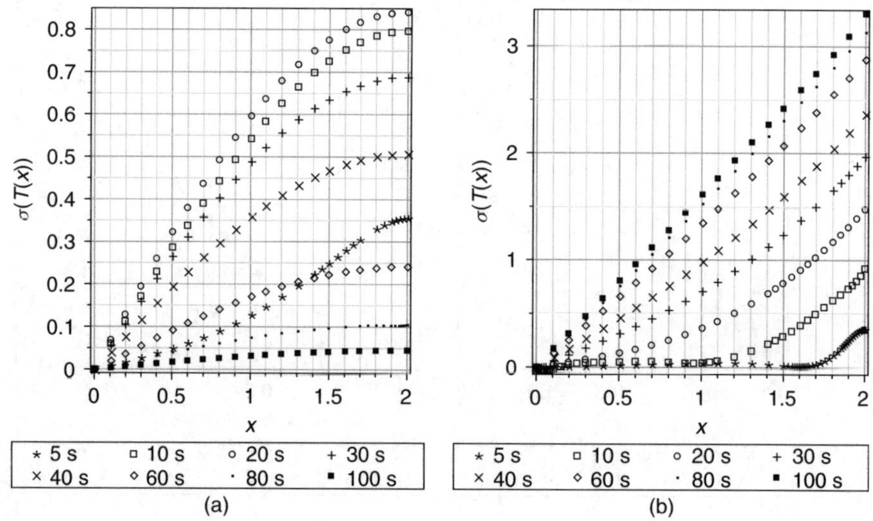

Figure 2.57 Standard deviations of the temperature distribution in the heated rod, (a) random c and (b) random k. Reproduced with permission from Tech Science Press

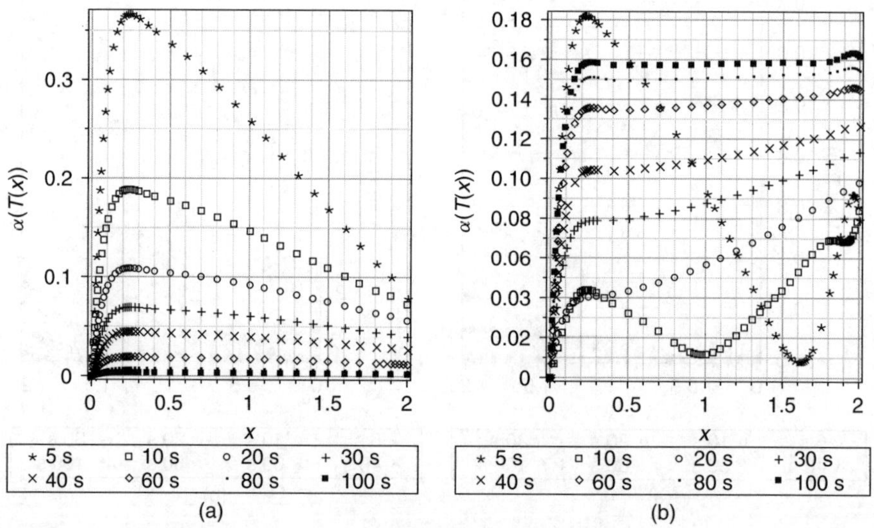

Figure 2.58 Coefficients of variation of the temperature distribution in the heated rod, (a) random c and (b) random k. Reproduced with permission from Tech Science Press

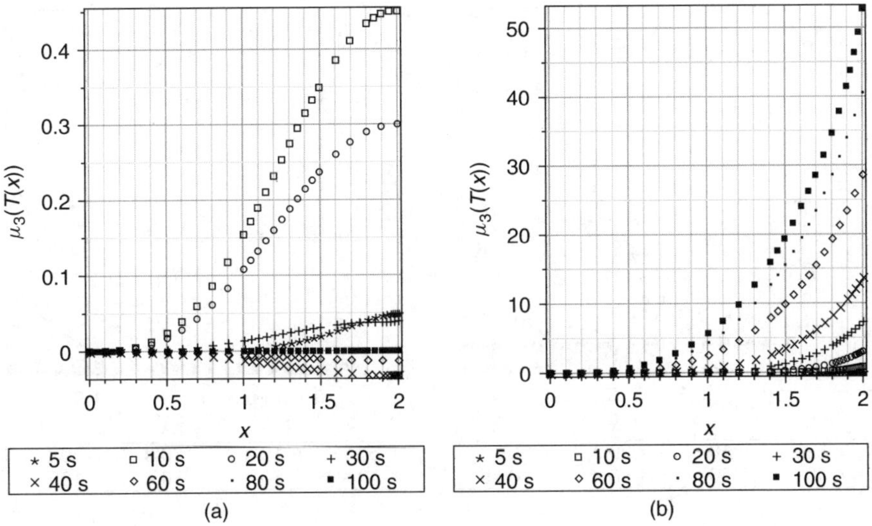

Figure 2.59 The third central probabilistic moments of the temperature distribution in the heated rod, (a) random c and (b) random k. Reproduced with permission from Tech Science Press

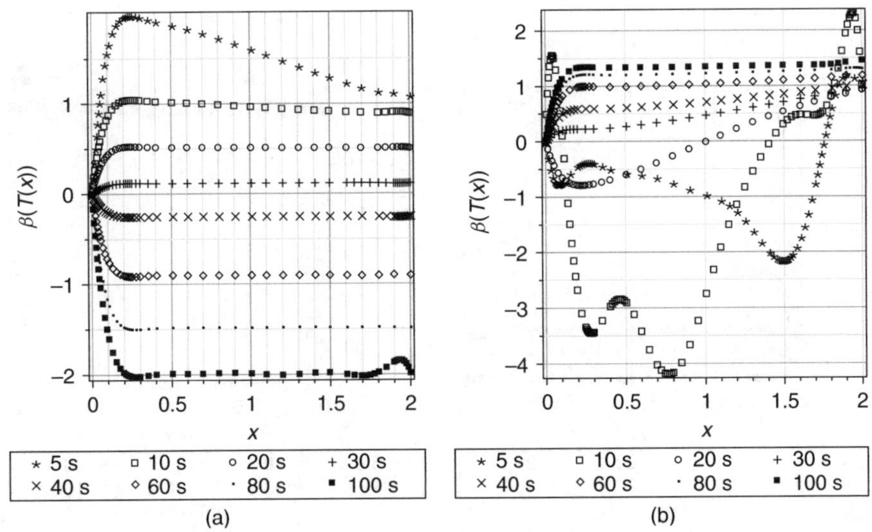

Figure 2.60 Skewness of the temperature distribution in the heated rod, (a) random c and (b) random k. Reproduced with permission from Tech Science Press

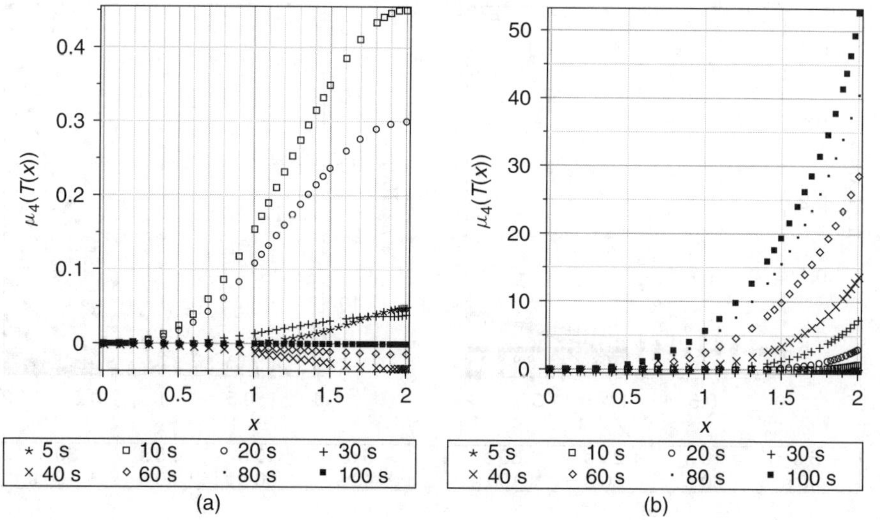

Figure 2.61 The fourth central probabilistic moments of the temperature distribution in the heated rod, (a) random c and (b) random k. Reproduced with permission from Tech Science Press

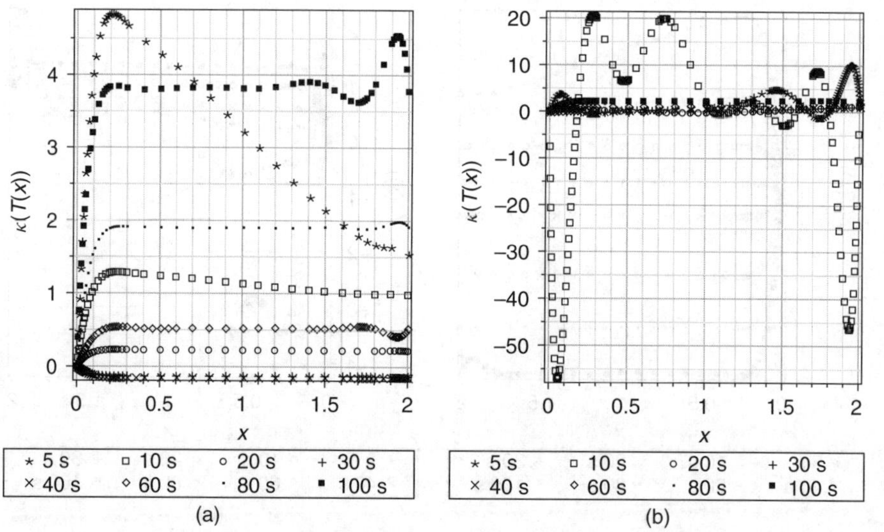

Figure 2.62 Kurtosis of the temperature distribution in the heated rod, (a) random c and (b) random k. Reproduced with permission from Tech Science Press

Finally, let us note that the overall computational effort of this analysis is relatively small in comparison to the Monte Carlo method – for 11-point discretization of the random space it equals 11 times the deterministic solution time plus a relatively small amount of post-processing time to recover the nodal (time-dependent in transient analysis) response functions and to calculate the final probabilistic moments with additional visualization. Thanks to the symbolic computations package one may also apply other probability distributions easily to include the extra components to compute the cross-correlations when heat capacity and conductivity are randomized at the same time. Further, using such an extended symbolic software like Maple™ for instance, we can make a posteriori some correction or better choice of the approximating polynomial order as well as to increase graphical resolution by appropriate window size on the input coefficient of variation.

3

Stochastic Boundary Element Method

Stochastic realizations of the Boundary Element Method (BEM) have been known for many years [11] and are available mainly in geotechnical problems, where random dispersion in civil engineering seems to be the largest area of use, for groundwater [19, 29, 108], seawater [57, 134], and porous media flows [149], in heat conduction and transfer issues [32, 124], or even for some elastodynamic [109] and wave propagation [122, 123] problems. It should be mentioned that the stochastic boundary element method (SBEM) analyses are especially efficient of course in random boundary geometry modeling, so that it is used in models with random boundary conditions or in stochastic shape design sensitivity [12]. This is the main reason to propose the new version of the generalized, perturbation-based SBEM here and test to verify its numerical efficiency. This version of SBEM offers both higher perturbation orders and a computation of higher than second probabilistic moments [61], analogously to the first version based also on the RFM [83, 87]. The basic difference is in numerical determination of this response function, which is now carried out using more efficient weighted least-squares method. It is also possible to optimize the order of the approximating polynomial during symbolic processing in the computer algebra system Maple™. Let us remember here that the order of this approximation is limited a priori and cannot be larger than the number of trial points around the mean value of the random input parameter. This order optimization undoubtedly leads in turn to minimization of the approximating error (and, effectively, speeds up probabilistic convergence of the method), since apparent linear interrelations do not need to be approximated using higher-order polynomials. Now, the optimal order of the approximating polynomials determined via the least-squares method guarantees perfect matching of the trial set of computational results with the response function. This was not the case in some previous studies, where highly non-linear interrelations between input and output parameters led to high oscillations of the response functions, somewhat against the tendency deduced from the trial points (see

The Stochastic Perturbation Method for Computational Mechanics, First Edition. Marcin Kamiński.
© 2013 John Wiley & Sons, Ltd. Published 2013 by John Wiley & Sons, Ltd.

Example 5.3, page 262 and next). Probabilistic moments of the structural response are studied using the proposed method on the examples adopted after and compared with those simulated using the Monte Carlo scheme; independently, they are derived from the analytical formulas using a stochastic perturbation technique. The very good agreement of these independent methods, as well as an application to the composite structure, make this strategy attractive and promising.

3.1 Deterministic Formulation of the Boundary Element Method

Let us consider a linear elastic homogeneous body occupying the region $\Omega \subset \Re^2$ with boundary $\Gamma \equiv \partial\Omega$ and let us assume that Γ is the Lyapunov surface. Let us also consider the body forces vector f_i with surface loadings p_i acting on this region and resulting in the displacement field $\mathbf{u}(\mathbf{x})$, strain tensor $\varepsilon_{ij}(\mathbf{x})$, and stress tensor $\sigma_{ij}(\mathbf{x})$. The linear elastostatic boundary value problem adjacent to the isotropic domain is defined using the following set of partial differential equations:

$$\operatorname{div} \boldsymbol{\sigma} + \mathbf{f} = 0, \tag{3.1}$$

$$\boldsymbol{\sigma} = \mathbf{C}\boldsymbol{\varepsilon} \tag{3.2}$$

where

$$C_{ijkl} = \delta_{ij}\delta_{kl}\ \lambda + \left(\delta_{ik}\delta_{jl} + \delta_{il}\delta_{jk}\right)\ \mu, \tag{3.3}$$

$$\varepsilon = \frac{1}{2}\left(\nabla\mathbf{u} + \nabla\mathbf{u}^{\mathrm{T}}\right) \tag{3.4}$$

with boundary conditions

$$u_i(\mathbf{x}) = \hat{u}_i(\mathbf{x}); \quad \mathbf{x} \in \Gamma_{\hat{u}}; \quad p_i(\mathbf{x}) = \hat{p}_i(\mathbf{x}); \quad \mathbf{x} \in \Gamma_{\hat{p}}. \tag{3.5}$$

Finally, it holds that

$$\Gamma_u \cup \Gamma_p = \Gamma \tag{3.6}$$

and

$$\Gamma_u \cap \Gamma_p = \emptyset. \tag{3.7}$$

The elasticity tensor given in Equation (3.4) may be composed for the two-component material with use of the following characteristic function [4]:

$$\begin{bmatrix} \lambda(\mathbf{x}) \\ \mu(\mathbf{x}) \end{bmatrix} = \chi_1(\mathbf{x}) \begin{bmatrix} \lambda_1 \\ \mu_1 \end{bmatrix} + (1 - \chi_1(\mathbf{x})) \begin{bmatrix} \lambda_2 \\ \mu_2 \end{bmatrix},$$

$$\chi_1 = \begin{cases} 1, & \mathbf{x} \in \Omega_1 \\ 0, & \mathbf{x} \in \Omega_2 \end{cases} \quad \text{where } \Omega = \Omega_1 \cup \Omega_2 \text{ and } \Omega_1 \cap \Omega_2 = \{\emptyset\}. \tag{3.8}$$

The set of Equations (3.1)–(3.7) for the homogeneous domain with regular boundary can be replaced with the following equivalent integral formulation rewritten in vector notation [10, 12, 60, 73]:

$$\mathbf{c}(\mathbf{x})\mathbf{u}(\mathbf{x}) + \int_{\Gamma} \mathbf{P}(\mathbf{x},\mathbf{y})\mathbf{u}(\mathbf{y})d\Gamma(\mathbf{y}) = \int_{\Gamma} \mathbf{U}(\mathbf{x},\mathbf{y})\mathbf{p}(\mathbf{y})d\Gamma(\mathbf{y}) + \int_{\Omega} \mathbf{U}(\mathbf{x},\mathbf{z})\mathbf{b}(\mathbf{z})d\Omega(\mathbf{z}) \tag{3.9}$$

where

$$\mathbf{P}(\mathbf{x},\mathbf{y}) = [P_{ij}(\mathbf{x},\mathbf{y})] \tag{3.10}$$

represents the stress vector components for direction j at point $\mathbf{y}$ in the unbounded elastic space resulting from a unitary force applied at the point defined by $\mathbf{x}$ in the direction defined by the index i. If the point being considered belongs to the interior of the domain Ω then $\mathbf{c} = \mathbf{I}$ but when it belongs to its boundary then

$$\mathbf{c}(\mathbf{x}) = \mathbf{I} + \lim_{\delta \to 0} \int_{\Gamma_\delta} \mathbf{P}(\mathbf{x},\mathbf{y})d\Gamma(\mathbf{y}), \tag{3.11}$$

where Γ_δ denotes a sphere with center at point $\mathbf{x}$ and radius equal to δ. It also holds that

$$\mathbf{c} = \alpha \ \mathbf{I} \ ; \alpha \in (0,1) \tag{3.12}$$

where for a boundary regular in the Lyapunov sense $\alpha = \frac{1}{2}$. Further, let us introduce a discretization of the boundary $\Gamma \in \Re^2$ using the boundary elements Γ^e such that

$$\Gamma = \sum_{e=1}^{K} \Gamma^e. \tag{3.13}$$

Hence, the displacement field $\mathbf{u}(\mathbf{x})$ can be approximated for any element Γ^e in local coordinates $\boldsymbol{\xi} = (\xi_i)$ by the nodal values $(\mathbf{u})_e^w$ as follows:

$$\mathbf{u}\left(\mathbf{x}\left(\boldsymbol{\xi}\right)\right) \approx \mathbf{N}^{\mathrm{w}}(\boldsymbol{\xi})(\mathbf{u})_{\mathrm{e}}^{\mathrm{w}} = \mathbf{I}N_w(\boldsymbol{\xi})(\mathbf{u})_{\mathrm{e}}^{\mathrm{w}} \tag{3.14}$$

where $N_w(\boldsymbol{\xi})$ are the interpolation functions analogous to these provided for the finite elements and their degrees of freedom. Introducing these statements into Equation (3.9), it is obtained that

$$\mathbf{c}(\mathbf{x})\mathbf{u}(\mathbf{x}) + \sum_{e=1}^{K}\sum_{w=1}^{W_e} (\mathbf{u})_e^w \int_{\Gamma^e} \mathbf{P}\left(\mathbf{x},\mathbf{y}\left(\boldsymbol{\xi}\right)\right)\mathbf{N}^w\left(\boldsymbol{\xi}\right)J\left(\boldsymbol{\xi}\right)d\Gamma\left(\boldsymbol{\xi}\right) = \sum_{e=1}^{K}\sum_{w=1}^{W_e} (\mathbf{p})_e^w \int_{\Gamma^e} \mathbf{U}\left(\mathbf{x},\mathbf{y}\left(\boldsymbol{\xi}\right)\right)\mathbf{N}^w\left(\boldsymbol{\xi}\right)J\left(\boldsymbol{\xi}\right)d\Gamma\left(\boldsymbol{\xi}\right) + \mathbf{B}(\mathbf{x}). \tag{3.15}$$

Variable W_e in Equation (3.15) denotes the total number of boundary elements within the applied boundary element type. The body force operator $\mathbf{B}(\mathbf{x})$ is defined as

$$\mathbf{B}(\mathbf{x}) = \int_{\Omega} \mathbf{U}(\mathbf{x},\mathbf{y})\mathbf{f}(\mathbf{y})d\Omega(\mathbf{y}) \tag{3.16}$$

in the presence of the volumetric boundary forces $\mathbf{f}(\mathbf{y})$. The Jacobian $\mathbf{J}(\boldsymbol{\xi})$ is given for two-dimensional body by

$$\mathbf{J}(\boldsymbol{\xi}) = \frac{d\boldsymbol{\Gamma}}{d\boldsymbol{\xi}} = \sqrt{\left[\left(\frac{\partial \mathbf{x}_1}{\partial \boldsymbol{\xi}}\right)^2 + \left(\frac{\partial \mathbf{x}_2}{\partial \boldsymbol{\xi}}\right)^2\right]}. \tag{3.17}$$

Further, the fundamental solutions included in Equation (3.15) adequate to the isotropic linear elastic body in the plane strain are given as

$$U_{ij}(\mathbf{x},\mathbf{y}) = -\frac{1}{8\pi(1-\nu)\mu}\left[(3-4\mu)\ln(\mathbf{r})\delta_{ij} - \mathbf{r}_{,i}\mathbf{r}_{,j}\right], \tag{3.18}$$

$$P_{ij}(\mathbf{x},\mathbf{y}) = -\frac{1}{4\pi(1-\nu)\mathbf{r}}\left\{\left[(1-2\nu)\delta_{ij} + 2\mathbf{r}_{,i}\mathbf{r}_{,j}\right]\frac{\partial \mathbf{r}}{\partial \mathbf{n}} - (1-2\nu)\left(\mathbf{r}_{,i}\mathbf{n}_j - \mathbf{r}_{,j}\mathbf{n}_i\right)\right\} \tag{3.19}$$

where

$$\mathbf{r}(\mathbf{x},\mathbf{y}) = \sqrt{\mathbf{r}_i \bullet \mathbf{r}_j}, \quad \mathbf{r}_i = \mathbf{x}_i(\mathbf{y}) - \mathbf{x}_i(\mathbf{x}) \tag{3.20}$$

and

$$\mathbf{r}_{,i} = \frac{\partial \mathbf{r}}{\partial \mathbf{x}_i(\mathbf{y})} = \frac{\mathbf{r}_i}{\mathbf{r}}. \tag{3.21}$$

We can use the parabolic shape functions $N_e^k(\xi)$ as in Example 3.4, defined as

$$\begin{cases} N_e^1(\xi) = \frac{1}{2}\xi(1-\xi), \\ N_e^2(\xi) = (1-\xi)(1+\xi), \\ N_e^3(\xi) = \frac{1}{2}\xi(1+\xi). \end{cases} \tag{3.22}$$

Equations (3.22) are valid for the boundary element defined uniquely at the ends in nodes 1, 3 and the midpoint in 2, where the local longitudinal coordinate ξ has value 0 at the midpoint and $\xi = \pm 1$ at both ends; alternatively, one may use traditional linear functions as provided in the remaining tests.

Next, let us introduce the global numbering of boundary nodal points $\mathbf{x}^\beta$, $\beta = 1, 2, \ldots, W$, where W is the total number of nodes. Then, we have [12]

$$\mathbf{r}(\mathbf{x}) = \sum_{\beta=1}^{W}\left[\mathbf{P}^\beta(\mathbf{x})\mathbf{u}_\beta - \mathbf{U}^\beta(\mathbf{x})\mathbf{p}_\beta\right] - \mathbf{B}(\mathbf{x}) = \mathbf{0} \tag{3.23}$$

with

$$\mathbf{P}^{\beta}(\mathbf{x}) = \mathbf{c}(\mathbf{x}) + \sum_{m} \int_{\Gamma^m} \mathbf{P}(\mathbf{x}, \mathbf{y}(\boldsymbol{\xi})) \mathbf{N}^{\beta}(\boldsymbol{\xi}) J(\boldsymbol{\xi}) \, d\Gamma(\boldsymbol{\xi}), \tag{3.24}$$

$$\mathbf{U}^{\beta}(\mathbf{x}) = \sum_{m} \int_{\Gamma^m} \mathbf{U}(\mathbf{x}, \mathbf{y}(\boldsymbol{\xi})) \bar{\mathbf{N}}^{\beta}(\boldsymbol{\xi}) J(\boldsymbol{\xi}) \, d\Gamma(\boldsymbol{\xi}). \tag{3.25}$$

The unknown values of displacements $\mathbf{u}(\mathbf{x})$ and forces $\mathbf{p}(\mathbf{x})$ can be found by minimization of the functional $\mathbf{r}(\mathbf{x})$ from the following condition:

$$\int_{\Gamma^e} \mathbf{T}^{\alpha}(\mathbf{x}) \mathbf{r}(\mathbf{x}) \, d\Gamma(\mathbf{x}) = 0 \tag{3.26}$$

where $\mathbf{T}^{\alpha}(\mathbf{x}) = \mathbf{I}\ T^{\alpha}(\mathbf{x}), \mathbf{x} \in \Gamma^e, e = 1, 2, \ldots, K$ is the weighted function matrix. In the case of the collocation method, this matrix has the form [4]

$$\mathbf{T}^{\alpha}(\mathbf{x}) = \mathbf{I}\ \delta\left(\mathbf{x} - \mathbf{x}^{\alpha}\right) \tag{3.27}$$

where the $\mathbf{x}^{\alpha}$ are collocation nodes. If the collocation nodal points are equivalent to the boundary nodes $\mathbf{x} = \mathbf{x}^{\alpha}, \alpha = 1, 2, \ldots, W$ then this minimization condition takes the form

$$\mathbf{r}\left(\mathbf{x}^{\alpha}\right) = 0. \tag{3.28}$$

Thus, it yields

$$c_{\alpha\beta} u_{\beta} + H_{\alpha\beta} u_{\beta} = G_{\alpha\beta} p_{\beta} + B_{\alpha}, \ \alpha, \beta = 1, \ldots, W \tag{3.29}$$

where

$$H_{\alpha\beta} = \sum_{e} \int_{\Gamma_e} \mathbf{P}\left[\mathbf{x}^{\alpha}, \mathbf{y}(\boldsymbol{\xi})\right] \mathbf{N}^{\beta}(\boldsymbol{\xi}) J(\boldsymbol{\xi}) \, d\Gamma(\boldsymbol{\xi}) \tag{3.30}$$

and

$$G_{\alpha\beta} = \sum_{e} \int_{\Gamma_e} \mathbf{U}\left[\mathbf{x}^{\alpha}, \mathbf{y}(\boldsymbol{\xi})\right] \bar{\mathbf{N}}^{\beta}(\boldsymbol{\xi}) J(\boldsymbol{\xi}) \, d\Gamma(\boldsymbol{\xi}). \tag{3.31}$$

The fundamental set of BEM algebraic equations is proposed as

$$H_{\alpha\beta} u_{\beta} = G_{\alpha\beta} p_{\beta} + B_{\alpha}; \alpha, \beta = 1, \ldots, W. \tag{3.32}$$

Further, for the needs of deterministic problem discretization, the following decomposition is usually employed:

$$\mathbf{A} \equiv \begin{bmatrix} \mathbf{H}^{11} & -\mathbf{G}^{11} \\ \mathbf{H}^{21} & -\mathbf{G}^{21} \end{bmatrix} \tag{3.33}$$

and

$$\mathbf{F} \equiv \begin{bmatrix} -\mathbf{H}^{11} & \mathbf{G}^{12} \\ -\mathbf{H}^{21} & \mathbf{G}^{22} \end{bmatrix} \begin{Bmatrix} \mathbf{u}^1 \\ \mathbf{p}^2 \end{Bmatrix} + \begin{Bmatrix} \mathbf{B}^1 \\ \mathbf{B}^2 \end{Bmatrix}, \tag{3.34}$$

where the matrices $\mathbf{H}^{ij}$, $\mathbf{G}^{ij}$, $\mathbf{u}^i$, and $\mathbf{B}^i$ for $i, j = 1, 2$ are submatrices of the matrices **H**, **G**, **u**, and **B** equivalent to the boundary segments Γ_i, $i = 1, 2$. The column matrix of body forces **B** is calculated numerically in the region Ω with use of the formula

$$B_\alpha \equiv B\left(\mathbf{x}^\alpha\right) = \sum_{w=1}^{W} f\left(\mathbf{y}_w\right) \int_{\Omega^w} U\left(\mathbf{x}^\alpha, \mathbf{y}\right) d\Omega\left(\mathbf{y}\right) \tag{3.35}$$

where the function $f(\mathbf{y})$ has a constant value inside the cell Ω^w. It may be shown that Equation (3.32) can be rewritten using the following system of linear equations with non-symmetric stiffness matrix:

$$A_{\alpha\beta} X_\beta = F_\alpha; \quad \alpha, \beta = 1, \ldots, W. \tag{3.36}$$

The solution to this system contains the displacements and stresses calculated at the boundary of a domain being modeled. Then we compute in turn, the displacements and stresses in the interior of the domain Ω as

$$\begin{aligned} \mathbf{u}(\mathbf{x}) &= \mathbf{G}(\mathbf{x})\mathbf{p} - \mathbf{H}(\mathbf{x})\mathbf{u} + \mathbf{B}(\mathbf{x}); \quad \mathbf{x} \in \Omega \\ \boldsymbol{\sigma}(\mathbf{x}) &= \mathbf{G}_D(\mathbf{x})\mathbf{p} - \mathbf{H}_S(\mathbf{x})\mathbf{u} + \mathbf{B}_D(\mathbf{x}); \quad \mathbf{x} \in \Omega \end{aligned} \tag{3.37}$$

where the matrices $\mathbf{G}(\mathbf{x})$, $\mathbf{H}(\mathbf{x})$, $\mathbf{G}_S(\mathbf{x})$, and $\mathbf{H}_D(\mathbf{x})$ generally depend on boundary integrals with nuclei $\mathbf{U}(\mathbf{x},\mathbf{y})$, $\mathbf{P}(\mathbf{x},\mathbf{y})$, $\mathbf{D}(\mathbf{x},\mathbf{y})$, and $\mathbf{S}(\mathbf{x},\mathbf{y})$, respectively, calculated for a fixed point $\mathbf{x} \in \Omega$ and for $\mathbf{y} \in \Gamma$, while the column matrices $\mathbf{B}(\mathbf{x})$ and $\mathbf{B}_D(\mathbf{x})$ depend on the integrals defined on the Ω region with nuclei $\mathbf{U}(\mathbf{x},\mathbf{y})$ and $\mathbf{D}(\mathbf{x},\mathbf{y})$, respectively. It holds that

$$\mathbf{D}\left(\mathbf{x}, \mathbf{y}\right) = \sum_{\mathbf{x}} \left[\mathbf{U}\left(\mathbf{x}, \mathbf{y}\right)\right]; \quad \mathbf{x} \in \Omega, \mathbf{y} \in \Gamma \tag{3.38}$$

$$\mathbf{S}\left(\mathbf{x}, \mathbf{y}\right) = \sum_{\mathbf{x}} \left[\mathbf{P}\left(\mathbf{x}, \mathbf{y}\right)\right]; \quad \mathbf{x} \in \Omega, \mathbf{y} \in \Gamma \tag{3.39}$$

where $\mathbf{U}(\mathbf{x},\mathbf{y})$ is the fundamental solution given by Equation (3.18) and for any additional function f

$$\sum(f) = \mu \ \left[\nabla(f) + \nabla^T(f)\right] + \lambda \ \mathbf{I}div(f). \tag{3.40}$$

3.2 Stochastic Generalized Perturbation Approach to the BEM

Traditionally, the stochastic perturbation approach to all the physical problems is entered by the respective perturbed equations of zeroth-, first-, and successively higher orders being a modification of the variational integral formulation. The following hold from Equation (3.15):

- zeroth-order equation

$$\mathbf{c}^0(\mathbf{x})\,\mathbf{u}^0(\mathbf{x}) + \sum_{e=1}^{K}\sum_{w=1}^{W_e}\left(\mathbf{u}^0\right)_e^w \int_{\Gamma^e} \mathbf{P}^0(\mathbf{x},\mathbf{y}(\boldsymbol{\xi}))\,\mathbf{N}^w(\boldsymbol{\xi})\,J(\boldsymbol{\xi})\,d\Gamma(\boldsymbol{\xi})$$
$$= \sum_{e=1}^{K}\sum_{w=1}^{W_e}\left(\mathbf{p}^0\right)_e^w \int_{\Gamma^e} \mathbf{U}^0(\mathbf{x},\mathbf{y}(\boldsymbol{\xi}))\,\mathbf{N}^w(\boldsymbol{\xi})\,J(\boldsymbol{\xi})\,d\Gamma(\boldsymbol{\xi}) + \mathbf{B}^0(x) \qquad (3.41)$$

- first-order equation

$$\frac{\partial \mathbf{c}(\mathbf{x})}{\partial b}\mathbf{u}^0(\mathbf{x}) + \mathbf{c}^0(\mathbf{x})\,\frac{\partial \mathbf{u}(\mathbf{x})}{\partial b} + \sum_{e=1}^{K}\sum_{w=1}^{W_e}\left(\frac{\partial \mathbf{u}}{\partial b}\right)_e^w \int_{\Gamma^e} \mathbf{P}^0(\mathbf{x},\mathbf{y}(\boldsymbol{\xi}))\,\mathbf{N}^w(\boldsymbol{\xi})\,J(\boldsymbol{\xi})\,d\Gamma(\boldsymbol{\xi})$$
$$+ \sum_{e=1}^{K}\sum_{w=1}^{W_e}\left(\mathbf{u}^0\right)_e^w \int_{\Gamma^e} \frac{\partial \mathbf{P}(\mathbf{x},\mathbf{y}(\boldsymbol{\xi}))}{\partial b}\mathbf{N}^w(\boldsymbol{\xi})\,J(\boldsymbol{\xi})\,d\Gamma(\boldsymbol{\xi})$$
$$= \sum_{e=1}^{K}\sum_{w=1}^{W_e}\left(\frac{\partial \mathbf{p}}{\partial b}\right)_e^w \int_{\Gamma^e} \mathbf{U}^0(\mathbf{x},\mathbf{y}(\boldsymbol{\xi}))\,\mathbf{N}^w(\boldsymbol{\xi})\,J(\boldsymbol{\xi})\,d\Gamma(\boldsymbol{\xi})$$
$$+ \sum_{e=1}^{K}\sum_{w=1}^{W_e}\left(\mathbf{p}^0\right)_e^w \int_{\Gamma^e} \frac{\partial \mathbf{U}(\mathbf{x},\mathbf{y}(\boldsymbol{\xi}))}{\partial b}\mathbf{N}^w(\boldsymbol{\xi})\,J(\boldsymbol{\xi})\,d\Gamma(\boldsymbol{\xi}) + \frac{\partial \mathbf{B}(\mathbf{x})}{\partial b} \qquad (3.42)$$

- second-order equation

$$\frac{\partial^2 \mathbf{c}(\mathbf{x})}{\partial b^2}\mathbf{u}^0(\mathbf{x}) + 2\frac{\partial \mathbf{c}(\mathbf{x})}{\partial b}\frac{\partial \mathbf{u}(\mathbf{x})}{\partial b} + \mathbf{c}^0(\mathbf{x})\,\frac{\partial^2 \mathbf{u}(\mathbf{x})}{\partial b^2}$$
$$+ \sum_{e=1}^{K}\sum_{w=1}^{W_e}\left(\frac{\partial^2 \mathbf{u}}{\partial b^2}\right)_e^w \int_{\Gamma^e} \mathbf{P}^0(\mathbf{x},\mathbf{y}(\boldsymbol{\xi}))\,\mathbf{N}^w(\boldsymbol{\xi})\,J(\boldsymbol{\xi})\,d\Gamma(\boldsymbol{\xi})$$
$$+ \sum_{e=1}^{K}\sum_{w=1}^{W_e}\left(\mathbf{u}^0\right)_e^w \int_{\Gamma^e} \frac{\partial^2 \mathbf{P}(\mathbf{x},\mathbf{y}(\boldsymbol{\xi}))}{\partial b^2}\mathbf{N}^w(\boldsymbol{\xi})\,J(\boldsymbol{\xi})\,d\Gamma(\boldsymbol{\xi})$$
$$+ 2\sum_{e=1}^{K}\sum_{w=1}^{W_e}\left(\frac{\partial \mathbf{u}}{\partial b}\right)_e^w \int_{\Gamma^e} \frac{\partial \mathbf{P}(\mathbf{x},\mathbf{y}(\boldsymbol{\xi}))}{\partial b}\mathbf{N}^w(\boldsymbol{\xi})\,J(\boldsymbol{\xi})\,d\Gamma(\boldsymbol{\xi})$$
$$= \sum_{e=1}^{K}\sum_{w=1}^{W_e}\left(\frac{\partial^2 \mathbf{p}}{\partial b^2}\right)_e^w \int_{\Gamma^e} \mathbf{U}^0(\mathbf{x},\mathbf{y}(\boldsymbol{\xi}))\,\mathbf{N}^w(\boldsymbol{\xi})\,J(\boldsymbol{\xi})\,d\Gamma(\boldsymbol{\xi})$$

$$+\sum_{e=1}^{K}\sum_{w=1}^{W_e}\left(\mathbf{p}^0\right)_e^w\int_{\Gamma^e}\frac{\partial^2\mathbf{U}\left(\mathbf{x},\mathbf{y}\left(\boldsymbol{\xi}\right)\right)}{\partial b^2}\mathbf{N}^w\left(\boldsymbol{\xi}\right)J\left(\boldsymbol{\xi}\right)\,d\Gamma\left(\boldsymbol{\xi}\right)$$

$$+2\sum_{e=1}^{K}\sum_{w=1}^{W_e}\left(\frac{\partial\mathbf{p}}{\partial b}\right)_e^w\int_{\Gamma^e}\frac{\partial\mathbf{U}\left(\mathbf{x},\mathbf{y}\left(\boldsymbol{\xi}\right)\right)}{\partial b}\mathbf{N}^w\left(\boldsymbol{\xi}\right)J\left(\boldsymbol{\xi}\right)\,d\Gamma\left(\boldsymbol{\xi}\right)+\frac{\partial\mathbf{B}\left(\mathbf{x}\right)}{\partial b}\tag{3.43}$$

- nth-order equation

$$\sum_{k=0}^{n}\binom{n}{k}\frac{\partial^{n-k}\mathbf{c}\left(\mathbf{x}\right)}{\partial b^{n-k}}\frac{\partial^k\mathbf{u}\left(\mathbf{x}\right)}{\partial b^k}+\sum_{k=o}^{n}\binom{n}{k}\sum_{e=1}^{K}\sum_{w=1}^{W_e}\left(\frac{\partial^{n-k}\mathbf{u}}{\partial b^{n-k}}\right)_e^w$$

$$\times\int_{\Gamma^e}\frac{\partial^k\mathbf{P}\left(\mathbf{x},\mathbf{y}\left(\boldsymbol{\xi}\right)\right)}{\partial b^k}\mathbf{N}^w\left(\boldsymbol{\xi}\right)J\left(\boldsymbol{\xi}\right)\,d\Gamma\left(\boldsymbol{\xi}\right)$$

$$=\sum_{k=0}^{n}\binom{n}{k}\sum_{e=1}^{K}\sum_{w=1}^{W_e}\left(\frac{\partial^{n-k}\mathbf{p}}{\partial b^{n-k}}\right)_e^w\int_{\Gamma^e}\frac{\partial^k\mathbf{U}\left(\mathbf{x},\mathbf{y}\left(\boldsymbol{\xi}\right)\right)}{\partial b^k}\mathbf{N}^w\left(\boldsymbol{\xi}\right)J\left(\boldsymbol{\xi}\right)\,d\Gamma\left(\boldsymbol{\xi}\right)+\frac{\partial^n\mathbf{B}\left(\mathbf{x}\right)}{\partial b^n}.\tag{3.44}$$

It is well known from the SFEM formulations that the system of linear algebraic equations, which is the basis of the model, can be transformed, due to the generalized stochastic perturbation-based technique, to the following systems of equations, cf. Equation (3.29):

- zeroth-order equation

$$c^0_{\alpha\beta}u^0_\beta+H^0_{\alpha\beta}u^0_\beta=G^0_{\alpha\beta}p^0_\beta+B^0_\alpha\tag{3.45}$$

- first-order equation

$$\frac{\partial c_{\alpha\beta}}{\partial b}u^0_\beta+c^0_{\alpha\beta}\frac{\partial u_\beta}{\partial b}+\frac{\partial H_{\alpha\beta}}{\partial b}u^0_\beta+H^0_{\alpha\beta}\frac{\partial u_\beta}{\partial b}=\frac{\partial G_{\alpha\beta}}{\partial b}p^0_\beta+G^0_{\alpha\beta}\frac{\partial p_\beta}{\partial b}+\frac{\partial B_\alpha}{\partial b}\tag{3.46}$$

- second-order equation

$$\frac{\partial^2 c_{\alpha\beta}}{\partial b^2}u^0_\beta+2\frac{\partial c_{\alpha\beta}}{\partial b}\frac{\partial u_\beta}{\partial b}+c^0_{\alpha\beta}\frac{\partial^2 u_\beta}{\partial b^2}+\frac{\partial^2 H_{\alpha\beta}}{\partial b^2}u^0_\beta+2\frac{\partial H_{\alpha\beta}}{\partial b}\frac{\partial u_\beta}{\partial b}+H^0_{\alpha\beta}\frac{\partial^2 u_\beta}{\partial b^2}$$

$$=\frac{\partial^2 G_{\alpha\beta}}{\partial b^2}p^0_\beta+2\frac{\partial G_{\alpha\beta}}{\partial b}\frac{\partial p_\beta}{\partial b}+G^0_{\alpha\beta}\frac{\partial^2 p_\beta}{\partial b^2}+\frac{\partial^2 B_\alpha}{\partial b^2}\tag{3.47}$$

- nth-order equation

$$\sum_{k=0}^{n}\binom{n}{k}\frac{\partial^{n-k}c_{\alpha\beta}}{\partial b^{n-k}}\frac{\partial^k u_\beta}{\partial b^k}+\sum_{k=0}^{n}\binom{n}{k}\frac{\partial^{n-k}H_{\alpha\beta}}{\partial b^{n-k}}\frac{\partial^k u_\beta}{\partial b^k}=\sum_{k=0}^{n}\binom{n}{k}\frac{\partial^{n-k}G_{\alpha\beta}}{\partial b^{n-k}}\frac{\partial^k p_\beta}{\partial b^k}+\frac{\partial^n B_\alpha}{\partial b^n}.\tag{3.48}$$

Further, we proceed with a formulation of the hierarchical equations of increasing order provided on a basis of Equation (3.36):

$$\begin{cases} A^0_{\alpha\beta} X^0_\beta = F^0_\alpha, \\ A^0_{\alpha\beta} \dfrac{\partial X_\beta}{\partial b} = \dfrac{\partial F_\alpha}{\partial b} - \dfrac{\partial A_{\alpha\beta}}{\partial b} X^0_\beta, \\ (\ldots) \\ \sum_{k=0}^{n} \binom{n}{k} \dfrac{\partial^k A_{\alpha\beta}}{\partial b^k} \dfrac{\partial^{n-k} X_\beta}{\partial b^{n-k}} = \dfrac{\partial F_\alpha}{\partial b}. \end{cases} \tag{3.49}$$

Furthermore, it is known from deterministic BEM implementation that the matrix **A** is represented by the components of the following perturbation orders:

- zeroth-order components

$$\mathbf{A}^0 \equiv \begin{bmatrix} (\mathbf{H}^{11})^0 & (-\mathbf{G}^{11})^0 \\ (\mathbf{H}^{21})^0 & (-\mathbf{G}^{21})^0 \end{bmatrix} \tag{3.50}$$

- and nth-order derivatives

$$\frac{\partial^n \mathbf{A}}{\partial b^n} = \begin{bmatrix} \dfrac{\partial^n \mathbf{H}^{11}}{\partial b^n} & -\dfrac{\partial^n \mathbf{G}^{11}}{\partial b^n} \\ \dfrac{\partial^n \mathbf{H}^{21}}{\partial b^n} & -\dfrac{\partial^n \mathbf{G}^{21}}{\partial b^n} \end{bmatrix}. \tag{3.51}$$

Quite analogously, it is possible to obtain equations describing derivatives of the RHS vector. The following hold:

- zeroth-order equations

$$\begin{cases} (\mathbf{F}^{(1)})^0 = (-\mathbf{H}^{11})^0 (\mathbf{u}^1)^0 + (\mathbf{G}^{12})^0 (\mathbf{p}^2)^0 + (\mathbf{B}^1)^0, \\ (\mathbf{F}^{(2)})^0 = (-\mathbf{H}^{21})^0 (\mathbf{u}^1)^0 + (\mathbf{G}^{22})^0 (\mathbf{p}^2)^0 + (\mathbf{B}^2)^0 \end{cases} \tag{3.52}$$

- and nth-order equations

$$\begin{cases} \dfrac{\partial^n \mathbf{F}^{(1)}}{\partial b^n} = -\sum_{k=0}^{n} \binom{n}{k} \dfrac{\partial^k \mathbf{H}^{11}}{\partial b^b} \dfrac{\partial^{n-k} \mathbf{u}^1}{\partial b^{n-k}} + \sum_{k=0}^{n} \binom{n}{k} \dfrac{\partial^k \mathbf{G}^{12}}{\partial b^k} \dfrac{\partial^{n-k} \mathbf{p}^2}{\partial b^{n-k}} + \dfrac{\partial^n \mathbf{B}^1}{\partial b^n}, \\ \dfrac{\partial^n \mathbf{F}^{(2)}}{\partial b^n} = -\sum_{k=0}^{n} \binom{n}{k} \dfrac{\partial^k \mathbf{H}^{21}}{\partial b^b} \dfrac{\partial^{n-k} \mathbf{u}^1}{\partial b^{n-k}} + \sum_{k=0}^{n} \binom{n}{k} \dfrac{\partial^k \mathbf{G}^{22}}{\partial b^k} \dfrac{\partial^{n-k} \mathbf{p}^2}{\partial b^{n-k}} + \dfrac{\partial^n \mathbf{B}^2}{\partial b^n}. \end{cases} \tag{3.53}$$

In order to calculate the expected values and higher-order probabilistic moments of displacements, strains, and stress functions, the same Taylor expansion is employed to the definitions of probabilistic moments calculated for any state random variables assuming their continuous character. It is obtained from the first of Equations (3.37) that

$$\mathbf{u}^0(\mathbf{x}) = \mathbf{G}^0(\mathbf{x})\mathbf{p}^0 - \mathbf{H}^0(\mathbf{x})\mathbf{u}^0 + \mathbf{B}^0(\mathbf{x});\ \ \mathbf{x} \in \Omega \tag{3.54}$$

as well as in the recursive form for n-th order partial derivative

$$\frac{\partial^n \mathbf{u}(\mathbf{x})}{\partial b^n} = \sum_{k=0}^{n} \binom{n}{k} \frac{\partial^k \mathbf{G}(\mathbf{x})}{\partial b^k} \frac{\partial^{n-k}\mathbf{p}}{\partial b^{n-k}} - \sum_{k=0}^{n} \binom{n}{k} \frac{\partial^k \mathbf{H}(\mathbf{x})}{\partial b^k} \frac{\partial^{n-k}\mathbf{u}}{\partial b^{n-k}} + \frac{\partial^n \mathbf{B}(\mathbf{x})}{\partial b^n};\ \ \mathbf{x} \in \Omega. \tag{3.55}$$

Similarly, the stress tensor components are calculated as, cf. Equation (3.37)

$$\boldsymbol{\sigma}^0(\mathbf{x}) = \mathbf{G}_D^0(\mathbf{x})\mathbf{p}^0 - \mathbf{H}_S^0(\mathbf{x})\mathbf{u}^0 + \mathbf{B}_D^0(\mathbf{x});\ \ \mathbf{x} \in \Omega \tag{3.56}$$

and its derivatives also as

$$\frac{\partial^n \boldsymbol{\sigma}(\mathbf{x})}{\partial b^n} = \sum_{k=0}^{n} \binom{n}{k} \frac{\partial^k \mathbf{G}_D(\mathbf{x})}{\partial b^k} \frac{\partial^{n-k}\mathbf{p}}{\partial b^{n-k}} - \sum_{k=0}^{n} \binom{n}{k} \frac{\partial^k \mathbf{H}_S(\mathbf{x})}{\partial b^k} \frac{\partial^{n-k}\mathbf{u}}{\partial b^{n-k}} + \frac{\partial^n \mathbf{B}_D(\mathbf{x})}{\partial b^n};\ \ \mathbf{x} \in \Omega. \tag{3.57}$$

These equations enable us to calculate any-order approximations for any probabilistic moments of the structural response but since the DDM may produce unknown error at higher orders the RFM is proposed and implemented below to avoid these problems.

3.3 The Response Function Method in the SBEM Equations

In this alternative approach we rewrite variational Equation (3.15) as

$$\mathbf{c}^{(m)}(\mathbf{x})\,\mathbf{u}^{(m)}(\mathbf{x}) + \sum_{e=1}^{K}\sum_{w=1}^{W_e} \left(\mathbf{u}^{(m)}\right)_e^w \int_{\Gamma^e} \mathbf{P}^{(m)}(\mathbf{x},\mathbf{y}(\boldsymbol{\xi}))\,\mathbf{N}^w(\boldsymbol{\xi})\,J(\boldsymbol{\xi})\,d\Gamma(\boldsymbol{\xi})$$

$$= \sum_{e=1}^{K}\sum_{w=1}^{W_e} \left(\mathbf{p}^{(m)}\right)_e^w \int_{\Gamma^e} \mathbf{U}^{(m)}(\mathbf{x},\mathbf{y}(\boldsymbol{\xi}))\,\mathbf{N}^w(\boldsymbol{\xi})\,J(\boldsymbol{\xi})\,d\Gamma(\boldsymbol{\xi}) + \mathbf{B}^{(m)}(\mathbf{x}) \tag{3.58}$$

where $m = 1, \ldots, M$ indexes trial solutions around the mean value of the random input. Then, the following discretization is proposed for the displacement field:

$$\mathbf{u}\left(b\left(x_j\right)\right) = \mathbf{u}\left(b_j\right) \approx N^w\left(\mathbf{u}(b_j)\right)_e^w = \mathbf{I}N_w\left(\mathbf{u}(b_j)\right)_e^w = \mathbf{I}N_w\left(a_i^{(w)}b_j^i\right)_e;\ i,j \in \mathbb{N} \tag{3.59}$$

where $N_w(\boldsymbol{\xi})$ are the classical interpolation functions used traditionally in the BEM programs and $a_i^{(w)}b_j^i$ denotes the polynomial response function. The boundary tractions are discretized in the similar manner as

$$\mathbf{p}\left(b(x_j)\right) = \mathbf{p}\left(b_j\right) \approx \mathbf{N}^w\left(\mathbf{p}(b_j)\right)_e^w = \mathbf{I}N_w\left(\mathbf{p}(b_j)\right)_e^w = \mathbf{I}N_w\left(d_i^{(w)}b_j^i\right)_e \tag{3.60}$$

where $d_i^{(w)}b_j^i$ is the nodal response function for the tractions $\mathbf{p}(\mathbf{x})$. Following the same considerations as above, the following algebraic system of equations is obtained:

$$H_{\alpha\beta j}\left(a_i^{(\beta)}b_j^i\right) = G_{\alpha\beta j}\left(d_i^{(\beta)}b_j^i\right) + B_{\alpha j};\ \alpha,\beta = 1,\ldots,W, j = 1,\ldots,n \tag{3.61}$$

the solution of which, together with the procedure of internal node displacements and tractions, enables the reliable computation of the sensitivity gradients by the modified BEM procedure. We illustrate our methodology with full tenth order expansion employed to derive first two probabilistic moments of the boundary displacements and forces. There holds for the expectations [83]

$$\begin{aligned}
E\left[u_\beta\right] &= \varepsilon^0 a_{\beta 0} + \frac{1}{2!}\varepsilon^2\left(i^2 - i\right) a_{\beta i} b^i \frac{\mu_2(b)}{b^2} + \frac{1}{4!}\varepsilon^4\left(i^4 - 6i^3 + 11i^2 - 6i\right) a_{\beta i} b^i \frac{\mu_4(b)}{b^4} \\
&+ \varepsilon^6 \frac{1}{6!}\left(i^6 - 15i^5 + 85i^4 - 225i^3 + 274i^2 - 120i\right) a_{\beta i} b^i \frac{\mu_6(b)}{b^6} \\
&+ \frac{1}{8!}\varepsilon^8\left(i^8 - 28i^7 + 332i^6 - 1960i^5 + 6769i^4 - 13132i^3 + 13068i^2 - 5040i\right) a_{\beta i} b^i \frac{\mu_8(b)}{b^8} \\
&+ \frac{1}{10!}\varepsilon^{10}\left(i^{10} - 45i^9 + 870i^8 - 9450i^7 + 63273i^6 - 269325i^5 + 723680i^4\right) a_{\beta i} b^i \frac{\mu_{10}(b)}{b^{10}} \\
&+ \frac{1}{10!}\varepsilon^{10}\left(-1172700i^3 + 1026576i^2 - 362880i\right) a_{\beta i} b^i \frac{\mu_{10}(b)}{b^{10}}.
\end{aligned} \tag{3.62}$$

The specific powers of b appearing in denominators are equivalent to the mean values of this input parameter b. Then, by inserting a definition of the coefficient of variation and putting $\varepsilon = 1$, one gets

$$\begin{aligned}
E\left[u_\beta\right] &= a_{\beta 0} + \frac{1}{2!} i(i-1) a_{\beta i} b^i \alpha^2(b) + \frac{1}{4!} i\left(i^3 - 6i^2 + 11i - 6\right) a_{\beta i} b^i \alpha^4(b) \\
&+ \frac{1}{6!} i\left(i^5 - 15i^4 + 85i^3 - 225i^2 + 274i - 120\right) a_{\beta i} b^i \alpha^6(b) \\
&+ \frac{1}{8!} i\left(i^7 - 28i^6 + 332i^5 - 1960i^4 + 6769i^3 - 13132i^2 + 13068i - 5040\right) a_{\beta i} b^i \alpha_8(b) \\
&+ \frac{1}{10!} i^4\left(i^6 - 45i^5 + 870i^4 - 9450i^3 + 63273i^2 - 269325i + 723680\right) a_{\beta i} b^i \alpha^{10}(b) \\
&+ \frac{1}{10!} i\left(-1172700i^2 + 1026576i - 362880\right) a_{\beta i} b^i \alpha^{10}(b).
\end{aligned} \tag{3.63}$$

Quite analogously, one recovers the expectations of the boundary tractions as

$$\begin{aligned}
E\left[p_\beta\right] &= d_{\beta 0} + \frac{1}{2!} i(i-1) d_{\beta i} b^i \alpha^2(b) + \frac{1}{4!} i\left(i^3 - 6i^2 + 11i - 6\right) d_{\beta i} b^i \alpha^4(b) \\
&+ \frac{1}{6!} i\left(i^5 - 15i^4 + 85i^3 - 225i^2 + 274i - 120\right) d_{\beta i} b^i \alpha^6(b) \\
&+ \frac{1}{8!} i\left(i^7 - 28i^6 + 332i^5 - 1960i^4 + 6769i^3 - 13132i^2 + 13068i - 5040\right) d_{\beta i} b^i \alpha_8(b) \\
&+ \frac{1}{10!} i^4\left(i^6 - 45i^5 + 870i^4 - 9450i^3 + 63273i^2 - 269325i + 723680\right) d_{\beta i} b^i \alpha^{10}(b) \\
&+ \frac{1}{10!} i\left(-1172700i^2 + 1026576i - 362880\right) d_{\beta i} b^i \alpha^{10}(b).
\end{aligned} \tag{3.64}$$

The formula for variance of the function $f(b)$ thanks to the tenth-order perturbation in terms of the Gaussian random input b is given as

$$
\begin{aligned}
\mathrm{Var}\left(u_{\beta}\right) &= \left(a_{\beta i} b^{i}\right)^{2} i^{2} \alpha^{2}(b)+\left\{\frac{1}{4}\left(i^{2}-i\right)^{2}+\frac{1}{3} i\left(i^{3}-3 i^{2}+2 i\right)\right\}\left(a_{\beta i} b^{i}\right)^{2} \alpha^{4}(b) \\
&+\left\{\frac{1}{36}\left(i^{3}-3 i^{2}+2 i\right)^{2}+\frac{1}{24}\left(i^{2}-i\right)\left(i^{4}-6 i^{3}+11 i^{2}-6 i\right)\right\}\left(a_{\beta i} b^{i}\right)^{2} \alpha^{6}(b) \\
&+\frac{1}{60} i\left(i^{5}-10 i^{4}+35 i^{4}-50 i^{2}+24 i\right)\left(a_{\beta i} b^{i}\right)^{2} \alpha^{6}(b) \\
&+\left\{\frac{1}{576}\left(i^{4}-6 i^{3}+11 i^{2}-6 i\right)^{2}+\frac{1}{360}\left(i^{5}-10 i^{4}+35 i^{3}-50 i^{2}+24 i\right)\right. \\
&\qquad \left.\times\left(i^{3}-3 i^{2}+2 i\right)\right\}\left(a_{\beta i} b^{i}\right)^{2} \alpha^{8}(b) \\
&+\frac{1}{2520} i\left(i^{7}-21 i^{6}+175 i^{5}-735 i^{4}+1624 i^{3}-1764 i^{2}+720 i\right)\left(a_{\beta i} b^{i}\right)^{2} \alpha^{8}(b) \\
&+\frac{1}{720}\left(i^{2}-i\right)\left(i^{6}-15 i^{5}+85 i^{4}-225 i^{3}+274 i^{2}-120 i\right)\left(a_{\beta i} b^{i}\right)^{2} \alpha^{8}(b) \\
&+\frac{1}{14400}\left(i^{5}-10 i^{4}+35 i^{3}-50 i^{2}+24 i\right)\left(a_{\beta i} b^{i}\right)^{2} \alpha^{10}(b) \\
&+\frac{1}{40320}\left(i^{2}-i\right)\left(i^{8}-28 i^{7}+322 i^{6}-1960 i^{5}+6769 i^{4}-13132 i^{3}\right. \\
&\left.+13068 i^{2}-5040 i\right)\left(a_{\beta i} b^{i}\right)^{2} \alpha^{10}(b) \\
&+\frac{1}{8640}\left(i^{4}-6 i^{3}+11 i^{2}-6 i\right)\left(i^{6}-15 i^{5}+85 i^{4}-225 i^{3}+274 i^{2}-120 i\right)\left(a_{\beta i} b^{i}\right)^{2} \alpha^{10}(b) \\
&+\frac{1}{15120}\left(i^{3}-3 i^{2}+2 i\right)\left(i^{7}-21 i^{6}+175 i^{5}-735 i^{4}+1624 i^{3}\right. \\
&\left.-1764 i^{2}+720 i\right)\left(a_{\beta i} b^{i}\right)^{2} \alpha^{10}(b) \\
&+\frac{1}{181440} i\left(i^{9}-36 i^{8}+546 i^{7}-4536 i^{6}+22449 i^{5}-67284 i^{4}\right)\left(a_{\beta i} b^{i}\right)^{2} \alpha^{10}(b) \\
&+\frac{1}{181440} i\left(118124 i^{3}-109584 i^{2}+40320 i\right)\left(a_{\beta i} b^{i}\right)^{2} \alpha^{10}(b) .
\end{aligned} \tag{3.65}
$$

The third central probabilistic moment is derived [74] as

$$
\begin{aligned}
\mu_{3}\left(u_{\beta}\right) &= \frac{3}{2} i^{3}(i-1)\left(a_{\beta i} b^{i}\right)^{3} \alpha^{4}(b) \\
&+\left\{\frac{1}{8}(i-1)^{3}+\frac{1}{2}\left(i^{2}-3 i+2\right)(i-1)+\frac{1}{8}\left(i^{3}-6 i^{2}+11 i-6\right)\right\} i^{3}\left(a_{\beta i} b^{i}\right)^{3} \alpha^{6}(b)
\end{aligned}
$$

$$+\left\{\frac{1}{24}\left(i^3-6i^2+11i-6\right)\left(i^2-3i+2\right)+\frac{1}{40}\left(i^4-10i^3+35i^2-50i+24\right)(i-1)\right\}$$
$$\times i^3\left(a_{\beta i}b^i\right)^3\alpha^8(b)$$
$$+\left\{\frac{1}{240}\left(i^5-15i^4+85i^3-225i^2+274i-120\right)+\frac{1}{32}i^3\left(i^3-6i^2+11i-6\right)(i-1)^2\right\}$$
$$\times i^3\left(a_{\beta i}b^i\right)^3\alpha^8(b)$$
$$+\frac{1}{24}i^3\left(i^2-3i+2\right)^2(i-1)\left(a_{\beta i}b^i\right)^3\alpha^8(b)$$
$$+\frac{1}{480}i^3\left(i^4-10i^3+35i^2-50i+24\right)\left(i^3-6i^2+11i-6\right)\left(a_{\beta i}b^i\right)^3\alpha^{10}(b)$$
$$+\frac{1}{1680}i^3(i-1)\left(i^6-21i^5+175i^4-735i^3+1624i^2-1764i+720\right)\left(a_{\beta i}b^i\right)^3\alpha^{10}(b)$$
$$+\frac{1}{720}i^3\left(i^5-15i^4+85i^3-225i^2+274i-120\right)\left(i^2-3i+1\right)\left(a_{\beta i}b^i\right)^3\alpha^{10}(b)$$
$$+\frac{1}{240}i^3\left(i^4-10i^3+35i^2-50i+24\right)\left(i^2-3i+2\right)(i-1)\left(a_{\beta i}b^i\right)^3\alpha^{10}(b)$$
$$+\frac{1}{13440}i^3\left(i^7-28i^6+322i^5-1960i^4+6769i^3-13132i^2+13068i-5040\right)$$
$$\times\left(a_{\beta i}b^i\right)^3\alpha^{10}(b)$$
$$+\frac{1}{960}i^3\left(i^5-15i^4+85i^3-225i^2+274i-120\right)(i-1)\left(a_{\beta i}b^i\right)^3\alpha^{10}(b)$$
$$+\left\{\frac{1}{384}i^3(i-1)\left(i^3-6i^2+11i-6\right)^2+\frac{1}{288}i^3\left(i^3-6i^2+11i-6\right)\left(i^2-3i+2\right)^2\right\}$$
$$\times\left(a_{\beta i}b^i\right)^3\alpha^{10}(b) \tag{3.66}$$

Finally, we derive the fourth central probabilistic moment in the form

$$\mu_4\left(u_\beta\right)=i^4\left(a_{\beta i}b^i\right)^4\alpha^4(b)+\left\{\frac{3}{2}i^4(i-1)^4+\frac{2}{3}i^4\left(i^2-3i+2\right)\right\}\left(a_{\beta i}b^i\right)^4\alpha^6(b)$$
$$+\left\{\frac{1}{30}i^4\left(i^4-10i^3+35i^2-50i+24\right)+\frac{1}{6}i^4\left(i^2-3i+2\right)^2+\frac{1}{16}i^4(i-1)^4\right\}$$
$$\times\left(a_{\beta i}b^i\right)^4\alpha^8(b)$$
$$+\frac{1}{1260}i^4\left(i^6-21i^5+175i^4-735i^3+1624i^2-1764i+720\right)\left(a_{\beta i}b^i\right)^4\alpha^{10}(b)$$
$$+\frac{1}{96}i^4\left(i^3-6i^2+11i-6\right)^2\left(a_{\beta i}b^i\right)^4\alpha^{10}(b)$$

$$+\frac{1}{54}i^4\left(i^2-3i+2\right)^3\left(a_{\beta i}b^i\right)^4\alpha^{10}(b)$$
$$+\frac{1}{48}i^4\left(i^3-6i^2+11i-6\right)(i-1)^3\left(a_{\beta i}b^i\right)^4\alpha^{10}(b)$$
$$+\frac{1}{24}i^4\left(i^2-3i+2\right)^2(i-1)^2\left(a_{\beta i}b^i\right)^4\alpha^{10}(b)$$
$$+\frac{1}{12}i^4\left(i^3-6i^2+11i-6\right)\left(i^2-3i+2\right)(i-1)\left(a_{\beta i}b^i\right)^4\alpha^{10}(b)$$
$$+\frac{1}{4}i^4\left(i^3-6i^2+11i-6\right)(i-1)\left(a_{\beta i}b^i\right)^4\alpha^{8}(b)$$
$$+\frac{1}{2}i^4(i-1)^2\left(i^2-3i+2\right)\left(a_{\beta i}b^i\right)^4\alpha^{8}(b)$$
$$+\frac{1}{60}i^4\left(i^4-10i^3+35i^2-50i+24\right)\left(i^2-3i+2\right)\left(a_{\beta i}b^i\right)^4\alpha^{10}(b)$$
$$+\frac{1}{120}i^4\left(i^5-15i^4+85i^3-225i^2+274i-120\right)(i-1)\left(a_{\beta i}b^i\right)^4\alpha^{10}(b)$$
$$+\frac{1}{40}i^4\left(i^4-10i^3+35i^2-50i+24\right)(i-1)^2\left(a_{\beta i}b^i\right)^4\alpha^{10}(b). \tag{3.67}$$

The very specific case of the linear response function returns the well-known relations

$$E\left[u_\beta\right]=a_{\beta 0},\ Var\left(u_\beta\right)=a_{\beta 1}^2E^2[b]\alpha^2(b),$$
$$\mu_3\left(u_\beta\right)=0,\ \ \mu_4\left(u_\beta\right)=a_{\beta 1}^4E^4[b]\alpha^4(b). \tag{3.68}$$

The readers interested in the full tenth order expansion are advised to explore the webpage accompanying this book to find the additional Maple™ sources. It needs to be mentioned that having derived these equations we can make a straightforward implementation in any programming language for further numerical realization of polynomial response functions in probabilistic computations due to all the methods dispayed in this book. Symbolic realization is necessary when one wants to replace polynomial response with exponential or harmonic functions as also easily differentiable, however then probabilistic convergence is not guaranteed.

3.4 Computational Experiments

Example 3.1: Uniformly loaded steel plate partially simply supported at lower edges

The methodology described above is tested using the example of the plane strain analysis of the structure given schematically in Figure 3.1. The BEM discretization is performed using 56 linear two-noded boundary elements −18 for the upper and lower edges and 10 for the left and right ends; supported parts of the lower edge are discretized using five boundary elements. The mean value of the Young's modulus

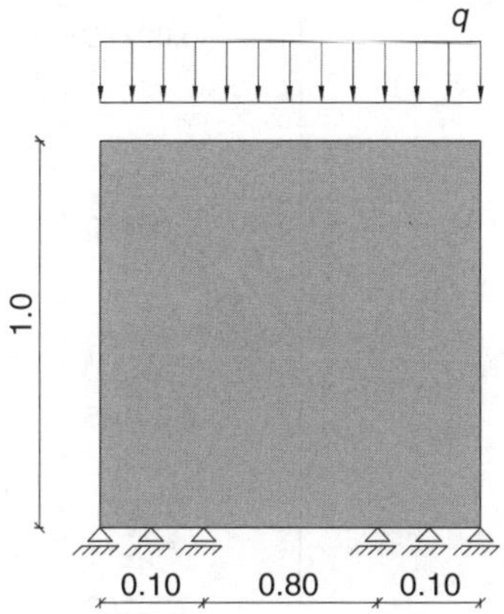

Figure 3.1 Static scheme of the computational example. Reproduced with permission from Elsevier

e is adopted as 210 GPa, the expected value of the Poisson ratio is taken as equal to 0.3, the height and length of the plate both equal to 1.0, while the mean value for the length of both side supports is adopted as 0.1. The response function for all the state parameters is approximated for $\pm 10\%$ variations from the additional expected values; all the random input quantities, that is, Young's modulus, Poisson ratio, and support depth are defined as Gaussian. The entire numerical analysis is carried out using a classical academic BEM solver [4] for the linear plane strain problem in conjunction with the symbolic environment of the system Maple™. We analyze here the probabilistic moments and coefficients of maximum displacement (for a midpoint at the lower boundary) and stress in the given plate.

The set of trial points corresponding to the structural response determination with respect to varying Young's modulus is given twice in Figure 3.2 – (a) for linear fitting of the proposed function $u(e)$ and (b) for parabolic fitting; they are both obtained thanks to the unweighted least-squares method. We study two different approximations of the smallest possible order resulting in a function close to the trial points, to verify the role of the approximating polynomial order. The case study is not trivial because an interrelation between u and e is a kind of inverse function, where parabolic fitting seems to be more accurate. This influence, together with the perturbation method order, is studied for the expected values (Figure 3.3), coefficients of variation (Figure 3.4), skewness (Figure 3.5), as well as kurtosis (Figure 3.6). These moments and coefficients are all given as functions of the input coefficient of variation $\alpha \in [0.00, 0.20]$. As one could expect, a linear response results in a constant expectation independent of both $\alpha(e)$ and the perturbation order as all higher-order partial derivatives simply vanish. The larger the input random dispersion, the higher the expectation for a parabolic function but they are all once more totally independent of the perturbation order (as higher than second-order terms do not appear). Independent of the response function order, both output coefficients of variation have very similar values, almost equal to the input one and both linearly

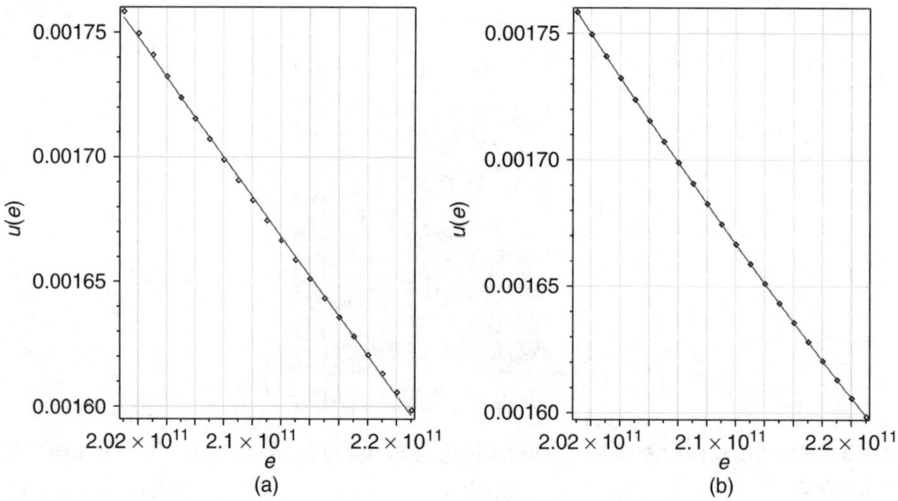

Figure 3.2 (a,b) Linear versus parabolic response functions of the maximum vertical displacement with respect to Young's modulus

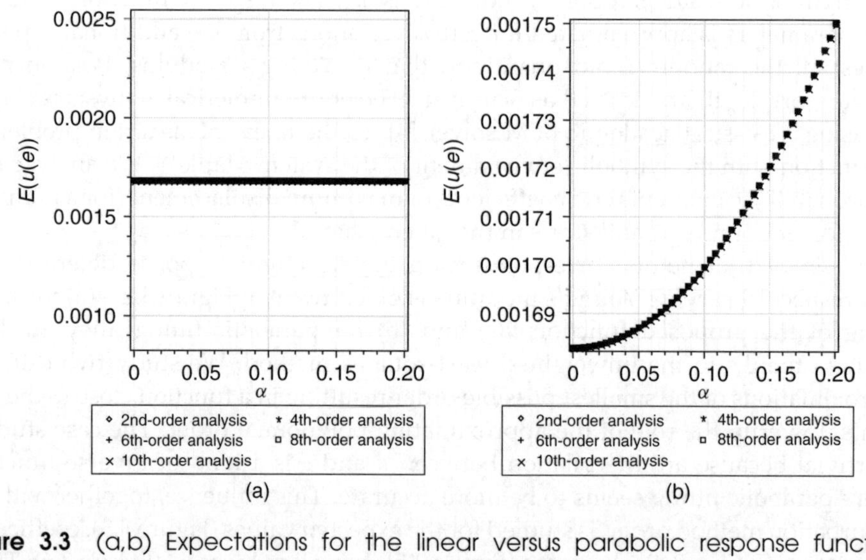

Figure 3.3 (a,b) Expectations for the linear versus parabolic response functions of the maximum vertical displacement with respect to Young's modulus

depend on the input coefficient. The difference here is observed in some marginal dependence of $\alpha(u)$ for larger values of $\alpha(e)$ on the perturbation order – practically, second-order gives slightly different results than all higher-order terms.

We study further skewness and kurtosis of this displacement (cf. Figures 3.5 and 3.6) – these coefficients automatically equal 0 for a linear approximation since higher-order partial derivatives both disappear, so they cannot depend on the

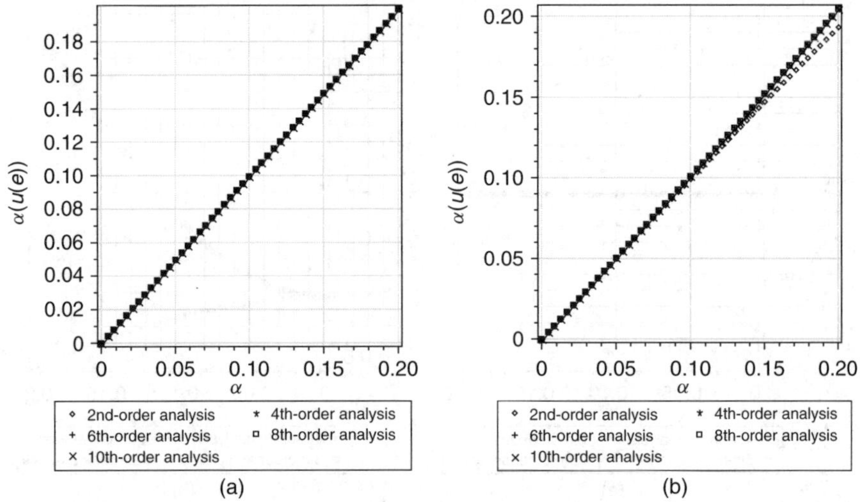

Figure 3.4 (a,b) Coefficients of variation for the linear versus parabolic response functions of the maximum vertical displacement with respect to Young's modulus

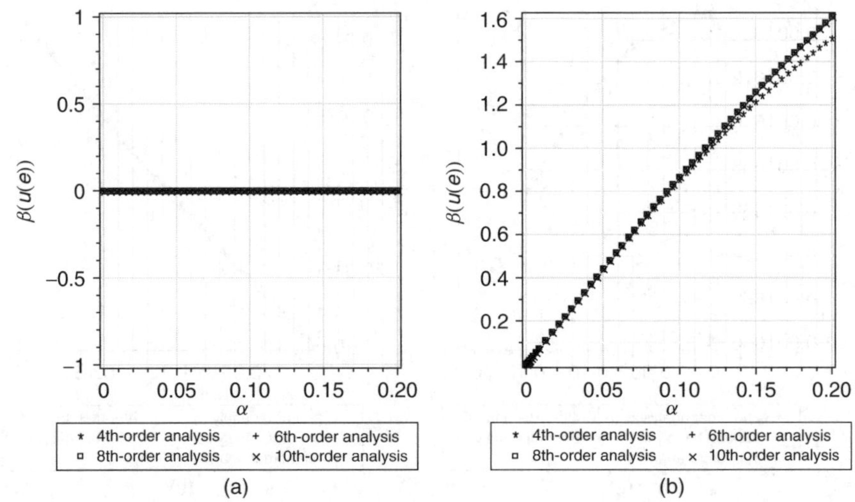

Figure 3.5 (a,b) Skewness for the linear versus parabolic response functions of the maximum vertical displacement with respect to Young's modulus

input coefficient $\alpha(e)$. Therefore, the output distribution according to this interpolation method is Gaussian and this is not the case for the parabolic response function (Figures 3.5(b) and 3.6(b)), where the higher the input $\alpha(e)$, the larger the absolute values of both coefficients. Skewness obtained from the fourth-order perturbation method is slightly different from other-order approximations (only for higher $\alpha(e)$), but probabilistic convergence of kurtosis with the perturbation order

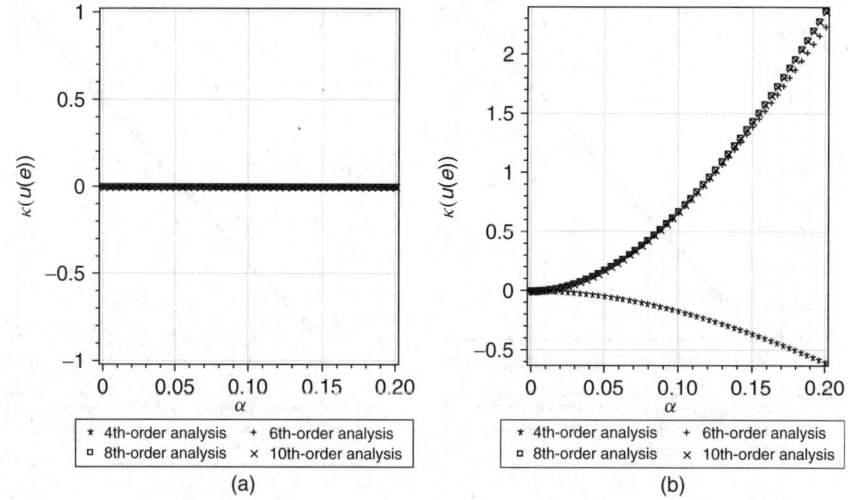

Figure 3.6 (a,b) Kurtosis for the linear versus parabolic response functions of the maximum vertical displacement with respect to Young's modulus

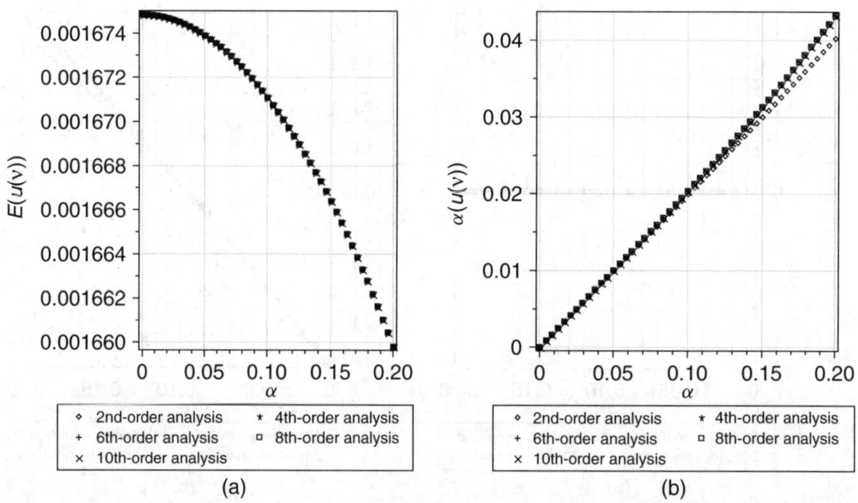

Figure 3.7 (a,b) Expectations and variation coefficients of maximum displacement to Poisson ratio

is apparently slower; both coefficients are almost always positive except the lowest order approximation for the kurtosis.

Analogous moments and coefficients computed for a single response function with respect to the input Poisson ratio behave in a different way (see Figures 3.7 and 3.8). Now the expected values (Figure 3.7(b)) decrease together with an increase of the input $\alpha(e)$, where all perturbation orders return exactly the same values. Looking for

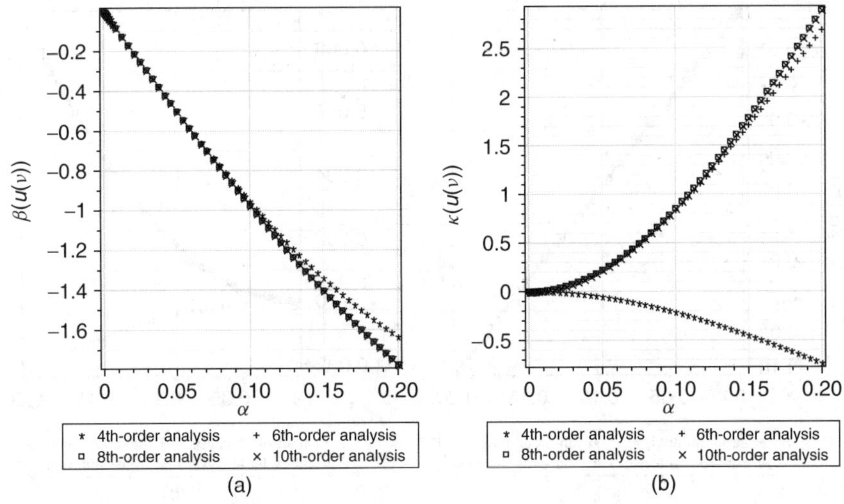

Figure 3.8 (a,b) Skewness and kurtosis of maximum displacement to Poisson ratio

the vertical scale of this figure, it is concluded that these differences may be totally disregarded. The coefficient of variation is linearly proportional to the input one as in Figure 3.4, but now the window size in the vertical axis is almost five times less than the horizontal one, so that the input random dispersion has been significantly damped by this system. Skewness and kurtosis given in Figure 3.8 have negative and positive values respectively, for the entire domain of input $\alpha(\nu)$, therefore the output

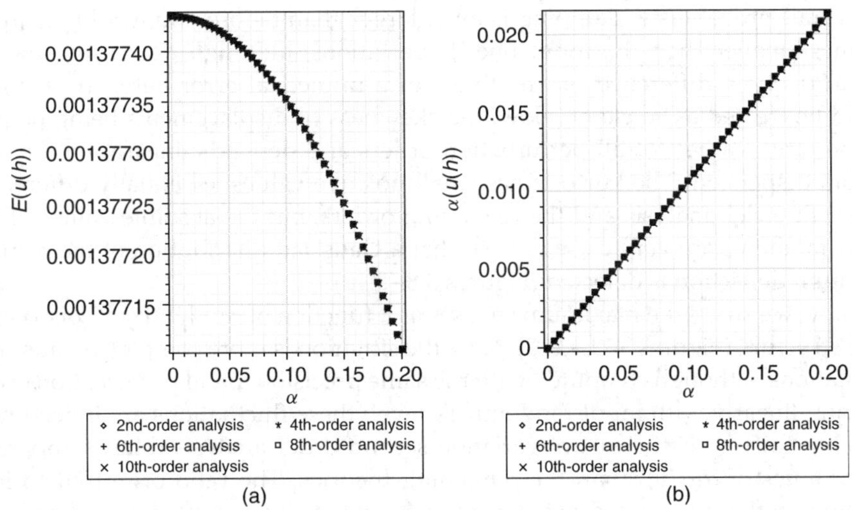

Figure 3.9 (a,b) Expectations and variation coefficients of maximum displacement with respect to Poisson ratio

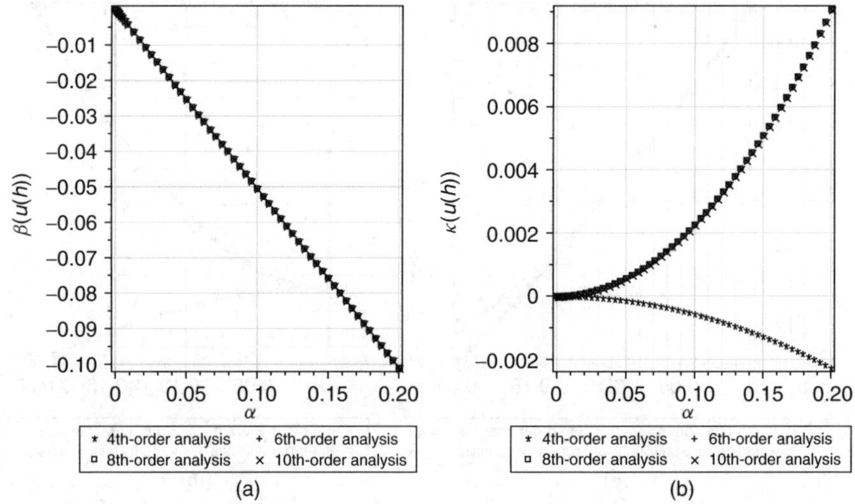

Figure 3.10 (a,b) Skewness and kurtosis of maximum displacement with respect to support length

displacement cannot be Gaussian for any combination of the input moments for the Poisson ratio. Quite similarly to other results, their absolute values increase together with an input parameter.

We study in turn the same probabilistic characteristics of the maximum displacements with respect to the length h of the supporting boundaries at the lower surface of this panel. Figure 3.9 – (a) with expectations and (b) containing coefficients of variation – shows practically no sensitivity to the perturbation order in this analysis. A ratio of output to input coefficient of variation remains constant for all $\alpha(h)$, but overall probabilistic damping is even larger than before and $\alpha(u(h))$ is around 10 times smaller than the input one. Expectations, although decreasing together with $\alpha(h)$, show differences in the limits of a numerical error here, so practically should be treated as constant. Now, the skewness of displacements being negative has the same values for all perturbation orders and depends linearly on the input random dispersion. The kurtosis has small positive values, essentially different for the fourth-order analysis and the remaining orders, but the absolute values of both higher moments are very close to 0, so that a Gaussian distribution of the resulting maximum deflection is detected (Figures 3.8–3.10).

Finally, we study the maximum stresses as a function of randomized Poisson ratio of the structure (Figures 3.11–3.12). Now the situation is more complex because even expectations, although computed with the same precision for all perturbation orders, vary non-linearly with input randomness level; these fluctuations are indeed negligible, but for the coefficients of variation second-order analysis returns apparently lower values of $\alpha(\sigma(\nu))$ than the remaining theories. The ratio of output to input variation coefficients is not constant for fourth up to tenth perturbation orders, unlike in the previous cases. Further, the output dispersion is about 20 times smaller and the coefficients of skewness and kurtosis (counted here in units) show a lack of Gaussian

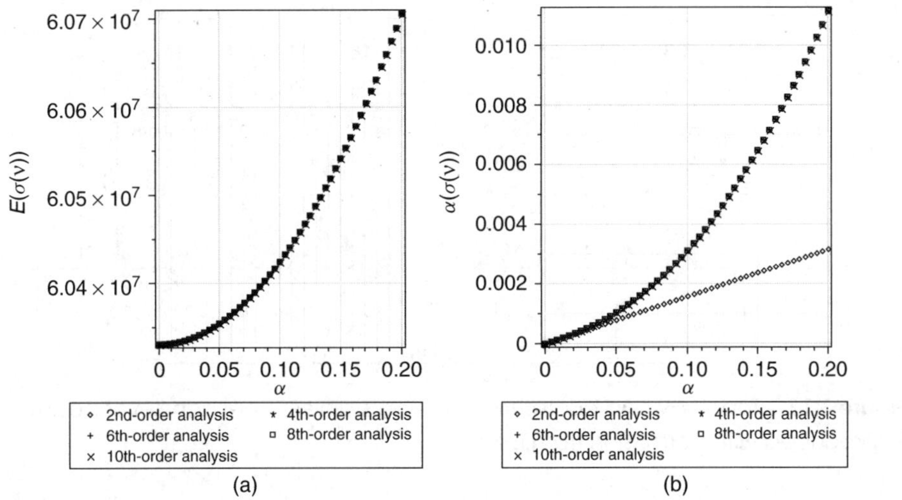

Figure 3.11 (a,b) Expectations and variation coefficients of maximum stresses with respect to support length

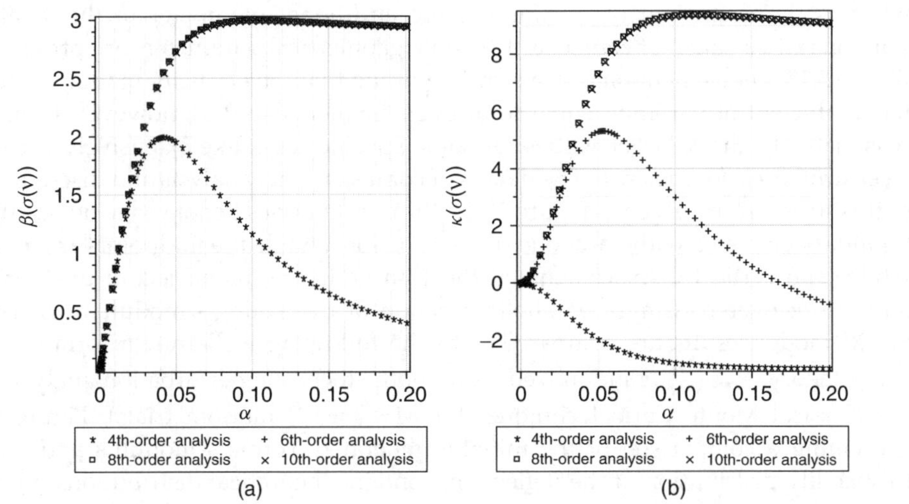

Figure 3.12 (a,b) Skewness and kurtosis of the maximum stresses with respect to Poisson ratio

character in case of stresses. Both higher-order coefficients show additionally a very slow probabilistic convergence since negligible differences are noticed for the first time between the eighth and tenth orders. Therefore, we can conclude that the reliability index computation for stresses may need much more computational effort in application of the stochastic perturbation technique than determination of the analogous index based on the plate deflections.

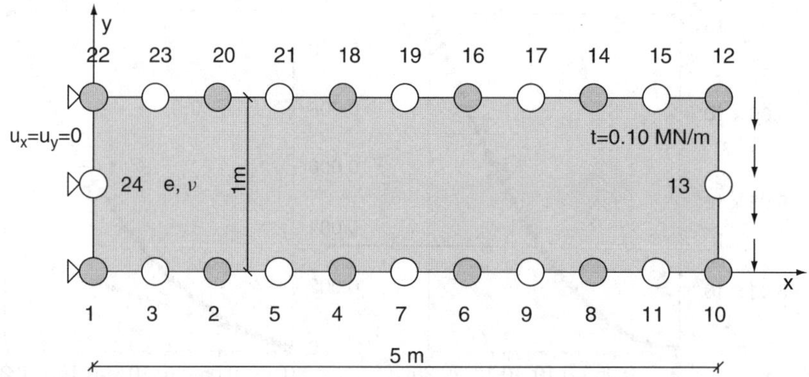

Figure 3.13 Topology and discretization of the stainless steel cantilever beam. Reproduced with permission from Elsevier

Example 3.2: Steel cantilever plate

The next numerical example deals with the static analysis of the linear elastic and isotropic stainless steel cantilever plate with Young's modulus randomized according to the Gaussian PDF (Figure 3.13). Its expected value equals 210 GPa, whereas the coefficient of variation is the additional parameter in this analysis; Poisson ratio is taken of course as 0.3, while geometrical parameters are provided in Figure 3.13. Let us mention at the very beginning that the elasticity modulus usually has coefficient of variation equal to or even smaller than 0.10, however, it may be reasonable to check larger values for some specific cases like extensively corroded steel structures, for instance. The details of computational analysis and discretization of this structure are given in Figure 3.14 – the darker nodes denote edge nodes for the boundary elements with parabolic shape functions, while the midpoints are marked white. The vertical displacements at the 13th node are the subject of this analysis and are recovered using a parabolic function of the Young's modulus, where the SBEM model results are compared with the following well-known formula from beam theory: $f = \frac{1}{3}\frac{Pl^3}{eJ}$, randomized using both stochastic perturbation analysis and the classical Monte Carlo technique (for $M = 3 \times 10^5$ random trials). Figure 3.14, presenting a comparison of the initially generated Young's modulus and output probability distribution of the deflection, contains theoretical distributions (marked as continuous lines) and simulation-based histograms. It is apparent that the final distribution is not Gaussian due to the non-zero asymmetry, and has smaller expected value than the corresponding Gaussian PDF with the same parameters. This means that determination of the first two moments is not sufficient to define uniquely the maximum deflection of this structure even for such a simple analytical formula.

The expected values presented in Figure 3.15(a) are computed according to the stochastic perturbation-based BEM, while Figure 3.15(b) shows Monte Carlo analysis and perturbation-based analytical formulas, also from the second to the tenth order. Analytical results start to diverge a little for larger values of $\alpha(e)$, while these computed using BEM are all the same for various orders. The expectations computed

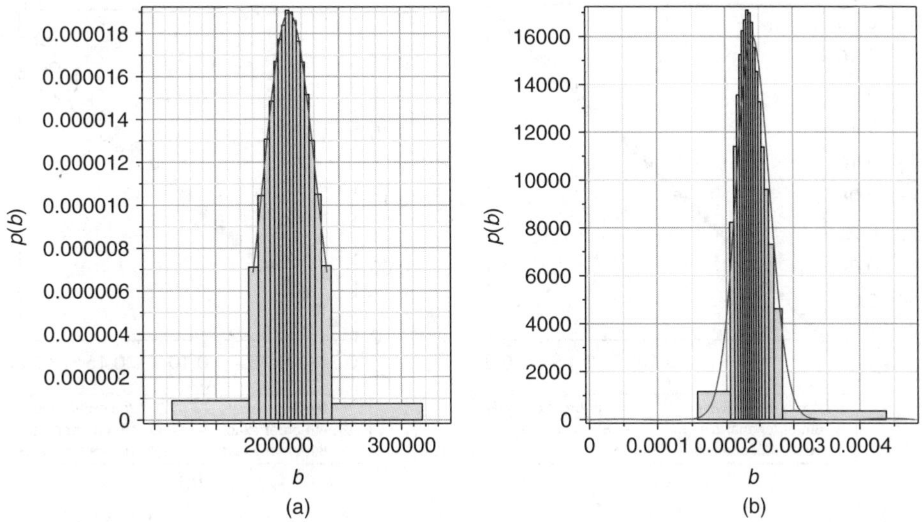

Figure 3.14 (a,b) Input and output PDFs of the stainless steel cantilever beam. Reproduced with permission from Elsevier

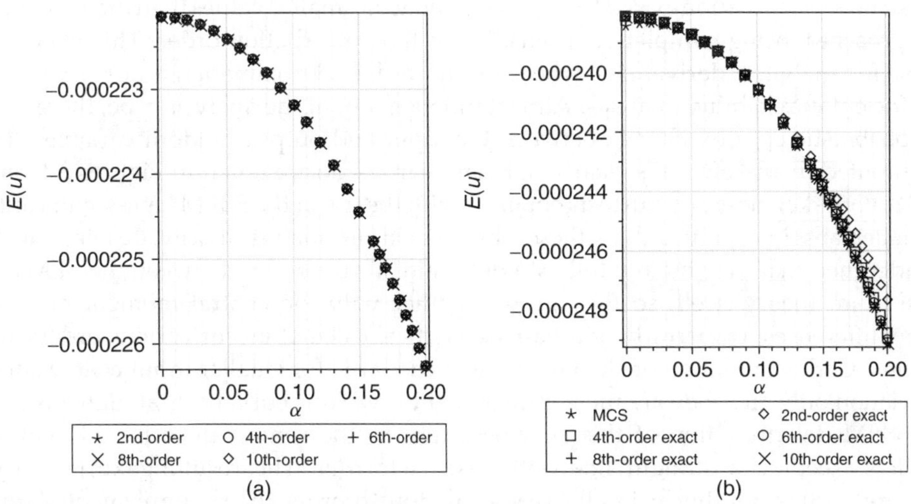

Figure 3.15 The expected values of the vertical displacement at the cantilever end via (a) MCS-SBEM and (b) perturbation-based analytical formulas

according to the SBEM second group have absolute values a little smaller than analytical computations for the entire spectrum of input coefficient variability and these simulated in Monte Carlo scheme. Of course, the larger the random input dispersion, the higher absolute value of maximum vertical deflection. Coefficients of variation divided in an analogous way into two graphs of Figure 3.16 depend

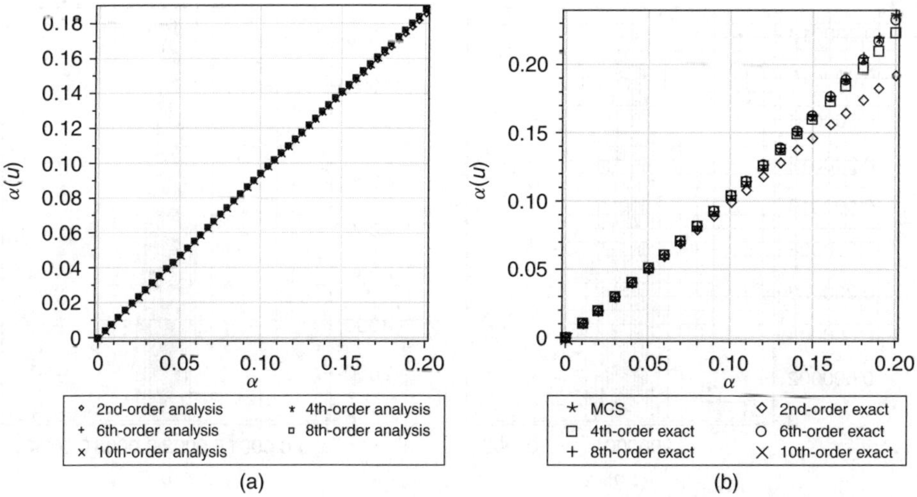

Figure 3.16 Coefficients of variation of the vertical displacement at the cantilever end via (a) MCS-SBEM and (b) perturbation-based analytical formulas

linearly on the input $\alpha(e)$ and show some gradual convexity in analytical calculus. Also now perturbation-based BEM gives somewhat smaller values than the remaining approaches, being completely insensitive to the perturbation order. This is not the case for analytical derivations, where second-order approximation is apparently not sufficient in reliability analysis. Almost independent of the analysis type, there is no probabilistic entropy change between the input and output random variables. The computed displacement is clearly not Gaussian, as documented in Figures 3.17 and 3.18. Both skewness and kurtosis obtained with the use of the SBEM have significantly smaller absolute values than these obtained in an analytical way. It needs to be underlined that they mainly follow a deterministic difference between the BEM and the beam theory itself, so they grow together with the central moment number. The differences for perturbation-based analytical calculations of various orders and Monte Carlo simulation of skewness are not so large for smaller input coefficient of variation only and they are most transparent for various curvatures at higher $\alpha(e)$.

SBEM determination of this skewness returns the same values for all analysis orders, keeping a straight line with respect to $\alpha(e)$. The resulting kurtosis (see Figure 3.18) is non-linear in all cases – now fourth-order theory implemented with the BEM gives significantly smaller results than the remaining higher-order approximations (counted in tenths). Analytical computations return kurtosis measured in units, which diverges for $\alpha(e) > 0.15$; close to the maximum value of this coefficient, Monte Carlo simulation gives enormously large values, five times larger than the perturbation analysis. Independent of the analysis type, the resulting displacement has a concentration around the mean value higher than the Gaussian distribution.

Further, we study expectations of maximum stresses (Figure 3.19(a)) and their coefficients of variation (Figure 3.19(b)), as well as the skewness (Figure 3.20(a)) and kurtosis (Figure 3.20(b)) of these stresses – all computed according to various orders

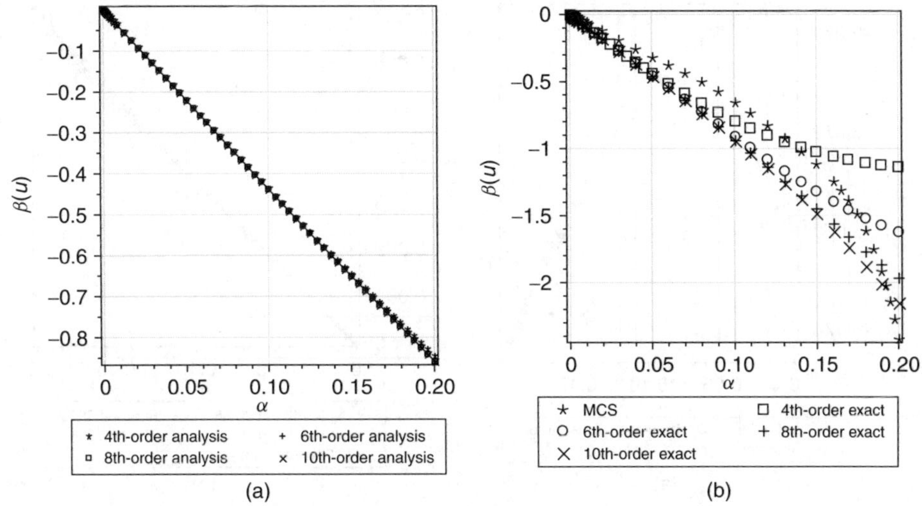

Figure 3.17 Skewness of the vertical displacement at the cantilever end via (a) MCS-SBEM and (b) perturbation-based analytical formulas

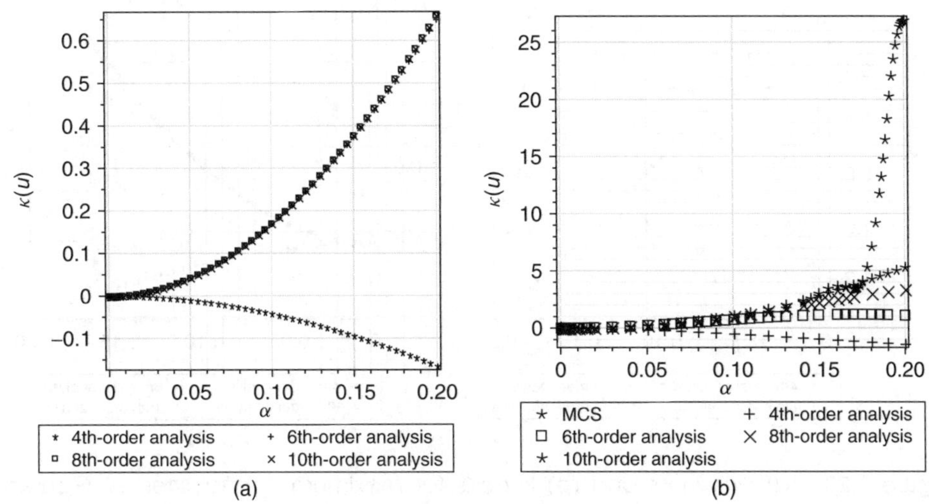

Figure 3.18 Kurtosis of the vertical displacement at the cantilever end via (a) MCS-SBEM and (b) perturbation-based analytical formulas

of the SBEM technique; it is once more based on a second-order polynomial with respect to the input random variable which is Poisson ratio ν. Absolute values of all probabilistic coefficients increase together with the input coefficient $\alpha(\nu)$. The output coefficient of variation for stresses has extremely small values increasing linearly together with $\alpha(\nu)$, analogously to the previous example. The expectation

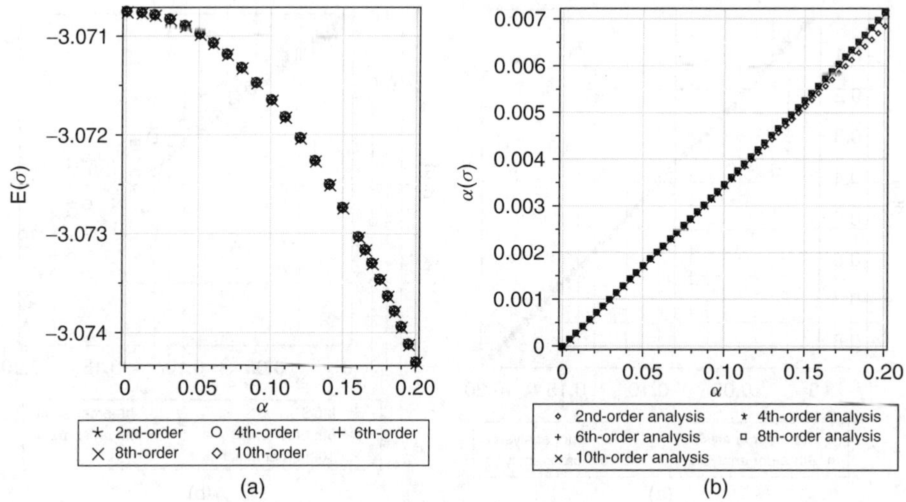

Figure 3.19 (a) The expected values and (b) coefficients of variation for maximum normal stresses, Gaussian Poisson ratio

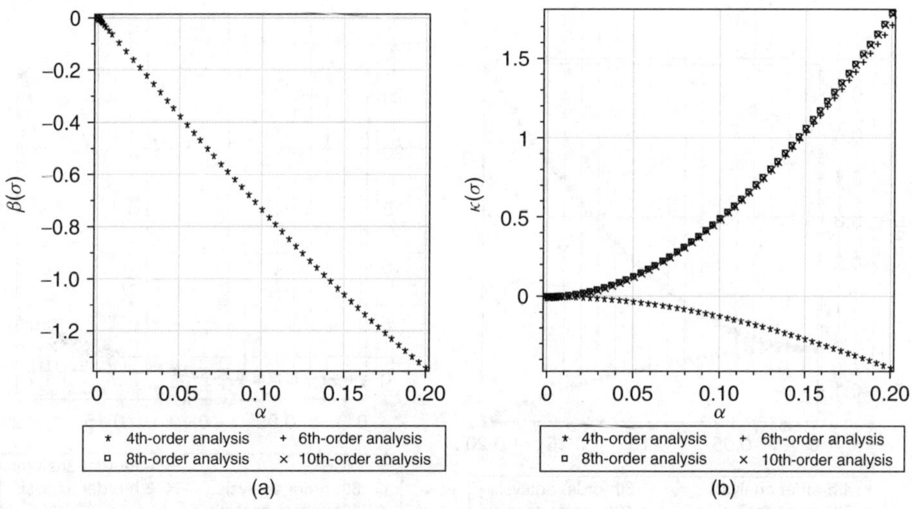

Figure 3.20 (a) Skewness and (b) kurtosis for maximum normal stresses, Gaussian Poisson ratio

and skewness are totally insensitive to the perturbation order; the coefficient of variation $\alpha(\sigma)$ shows some marginal differences for larger $\alpha(\nu)$ only, whereas the kurtosis shows full convergence starting from eighth-order SBEM analysis. The expected values decrease non-linearly together with the input dispersion, but the relative difference between the maximum and minimum values for a given window is practically negligible.

Like in many cases before, kurtosis is totally negative for the fourth-order analysis and, therefore, gives a false conclusion about smaller concentration around the mean

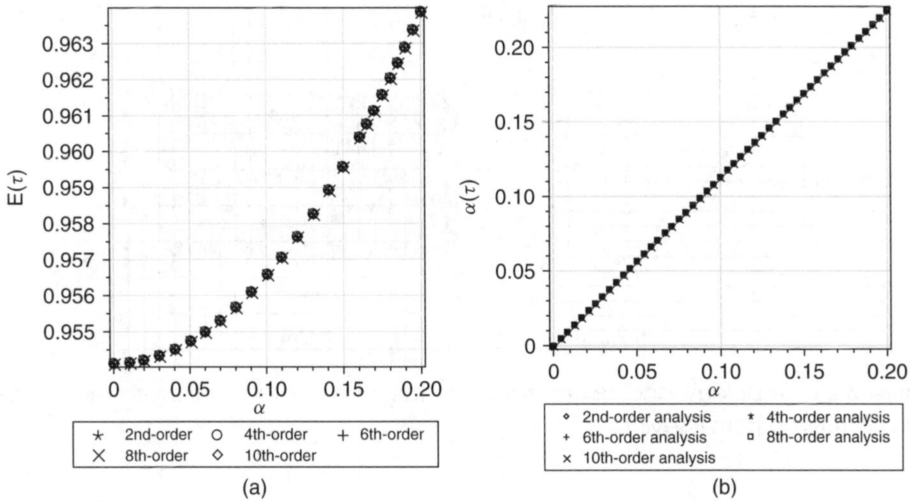

Figure 3.21 (a) The expected values and (b) coefficients of variation for maximum normal stresses, Gaussian Poisson ratio

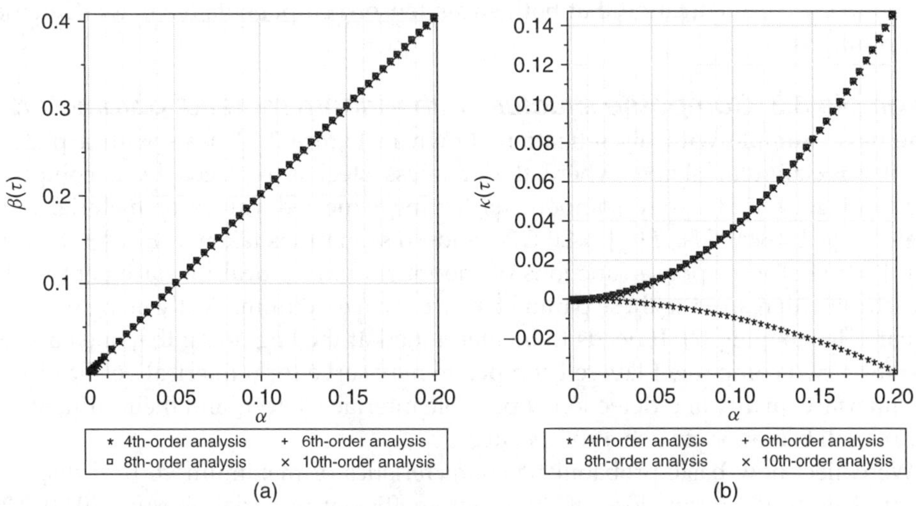

Figure 3.22 (a) Skewness and (b) kurtosis for maximum normal stresses, Gaussian Poisson ratio

value than in the case of a Gaussian distribution. For a comparison to the maximum normal stresses we contrast maximum shear stresses using expectations and coefficients of variation (Figure 3.21) as well as skewness and kurtosis (Figure 3.22). These characteristics exhibit very similar tendencies concerning probabilistic convergence with respect to the analysis order (even faster than before). The resulting coefficient of variation depends linearly upon the input one, having values slightly larger than

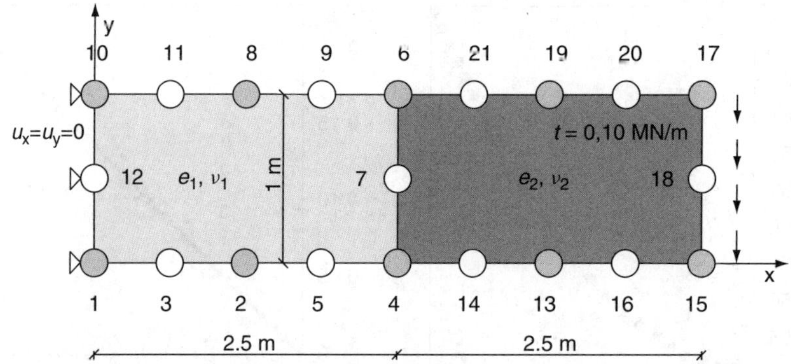

Figure 3.23 Topology and discretization of the composite cantilever. Reproduced with permission from Elsevier

$\alpha(\nu)$. The skewness $\beta(\tau)$ also has linear dependence on this coefficient but, contrary to $\beta(\sigma)$, is positive everywhere and, as before, rather distant from 0 having values two times larger than the input coefficient of variation. Kurtosis also shows some anomalies for the fourth-order stochastic perturbation and, although it is definitely smaller than $\kappa(\sigma)$, we can conclude that both stress tensor components have non-Gaussian distributions.

Example 3.3: Composite structure with random material parameters

The next example with discretization shown in Figure 3.23 deals with a plate of similar rectangular shape, where the stainless steel is replaced by a composite made of glass and epoxy components having expected values of material parameters $E[e_1]=84.0\ GPa$, $E[e_2]=4.0\ GPa$, and Poisson ratios taken as $E[\nu_1]=0.22$ and $E[\nu_2]=0.34$. The response functions of the maximum vertical displacement in this structure with respect to these parameters are recovered numerically using the least-squares technique [87]. It needs to be mentioned at the beginning that such a shear test for a two-component layered composite may serve for numerical verification of the interface quality and detection of possible interface defects and their influence on the overall behavior of such a composite.

We collect now basic probabilistic characteristics of maximum vertical displacement at node 18 as functions of the input coefficient of variation $\alpha(b) \in [0.00, 0.20]$ and for various perturbation orders using consecutively parabolic approximation between the output and input random parameters. Additionally, we analyze these expectations, coefficients of variation, skewness, and kurtosis as resulting from separate randomization of the Young's modulus of the reinforcement and matrix as well as Poisson ratios of both components. This is done to verify the most decisive input parameter in view of the reliability assessment and overall durability and integrity of a composite. The general conclusion on probabilistic convergence is that all expectations as before are practically insensitive to the given perturbation order, so that

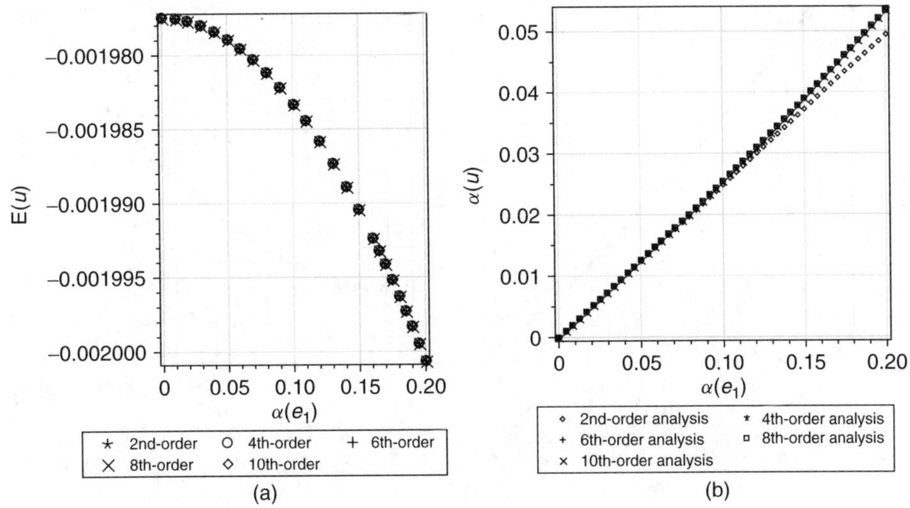

Figure 3.24 (a,b) The expectations and coefficients of variation for the cantilever end deflection for Young's modulus of the reinforcement

second-order analysis is quite sufficient (Figures 3.24a, 3.26a, 3.28a and 3.30a). This order is invalid for a precise computation of the coefficient of variation, where especially for $\alpha(b)$ close to 0.20 higher orders than the second are all the same and a little bit larger than second-order results (cf. Figures 3.24b, 3.26b, 3.28b and 3.30b). Then, for the needs of reliability analysis based on Eurocode 0 and using the Hasofer–Lind index, fourth-order analysis should be preferred. Skewness analysis used to recognize a type of output probability distribution shows unconditional convergence (independent of the input $\alpha(b)$) together with sixth-order perturbation-based SBEM. As the resulting coefficients of variation depend almost linearly on the input random dispersion, skewness as a function of $\alpha(b)$ always shows small concavity or convexity, especially for larger values of this parameter (see Figures 3.25a, 3.27a, 3.29a as well as 3.31a).

As could be expected, the lowest convergence is numerically verified for kurtosis (given in Figures 3.25b, 3.27b, 3.29b and 3.31b), where as before, second-order analysis returns initially negative values and then, depending on the input variable type is faster or slower – the fastest obtained during randomization of the Poisson ratio of reinforcement, where sixth-order analysis is precise enough. Comparison of detailed numerical values of all characteristics shows that the largest absolute values of the cantilever maximum deflections are noticed, while the Young's modulus of the matrix is random. The difference between minimum and maximum expectations is also negligible, as for the steel panel in the previous two numerical examples. Moreover, uncertainty of the Young's modulus of both components leads to an increase of this expectation absolute value together with $\alpha(e)$, whereas randomness of the Poisson ratio gives an inverse effect. A comparison of the variation coefficients shows that the most influential parameter is the Young's modulus of the matrix;

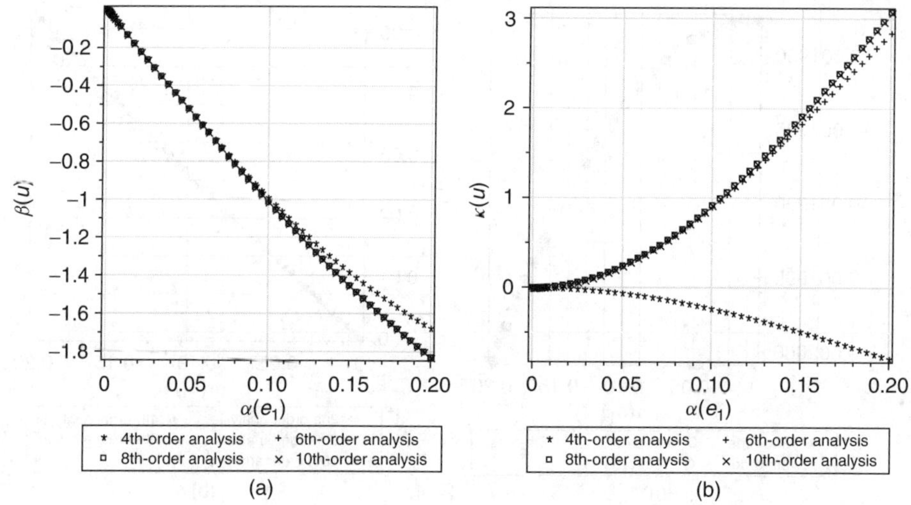

Figure 3.25 (a,b) Skewness and kurtosis for the cantilever end deflection for Young's modulus of the reinforcement

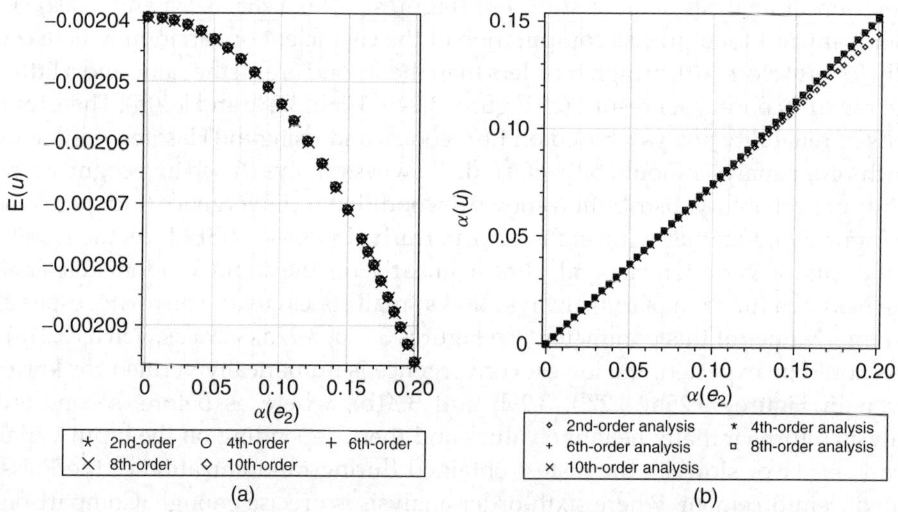

Figure 3.26 (a,b) The expectations and coefficients of variation for the cantilever end deflection for Young's modulus of the matrix

further we have Young's modulus od the reinforcement, the Poisson ratio of the reinforcement and, finally, this ratio for the matrix. The difference between the first and last parameter is dramatic, because $\alpha(u)$ for $e_2 = e_2(\omega)$ is around one-quarter smaller than the initial variation and almost 0 for $\nu_2 = \nu_2(\omega)$. Finally, we notice that none of the studied probabilistic distributions is Gaussian since the skewness and kurtosis differ significantly from 0. The skewness and kurtosis have correspondingly

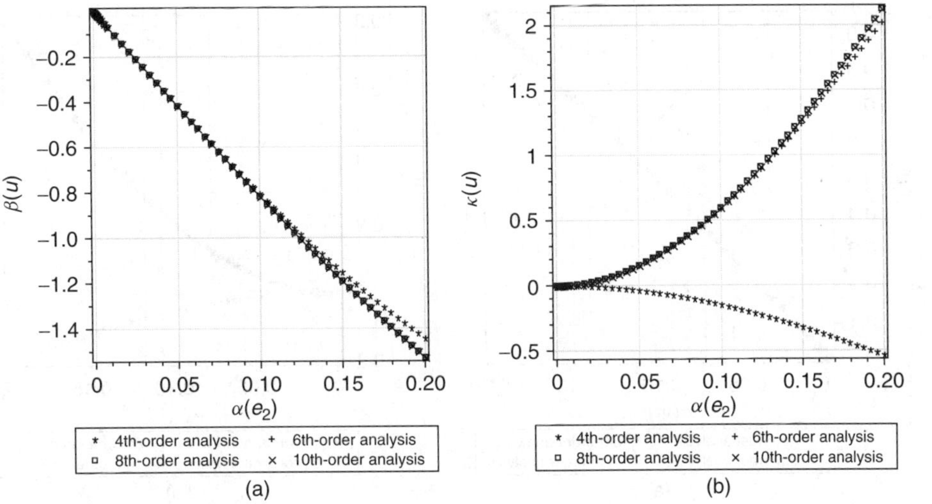

Figure 3.27 (a,b) Skewness and kurtosis for the cantilever end deflection for Young's modulus of the matrix

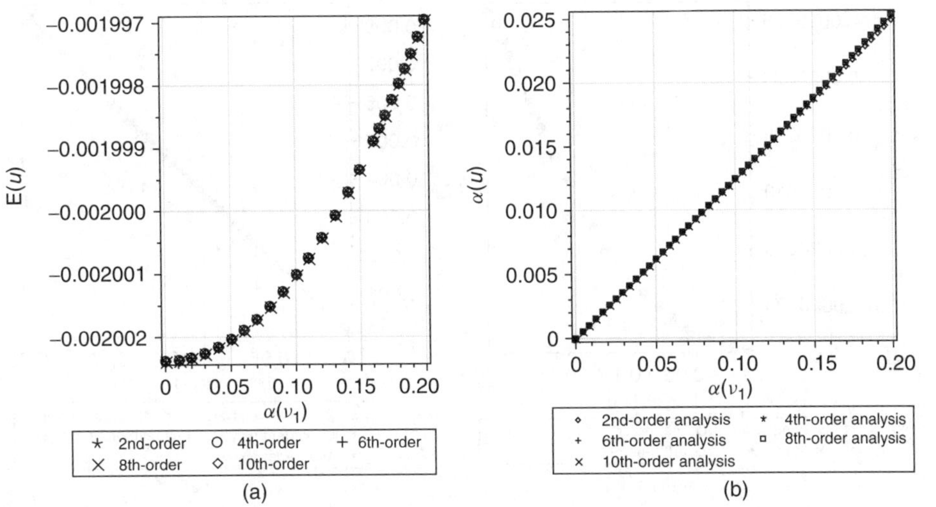

Figure 3.28 (a,b) The expectations and coefficients of variation for the cantilever end deflection for Poisson ratio of the reinforcement

very similar values, where negative third moments are obtained with randomizing both Young's moduli and positive for both Poisson ratios (Figure 3.31). The kurtosis in all four cases starts from negative values to become positive for higher than fourth-order models. Considering all the moments at once it is common that the larger the input coefficient of variation, the higher the absolute value of the given probabilistic characteristic.

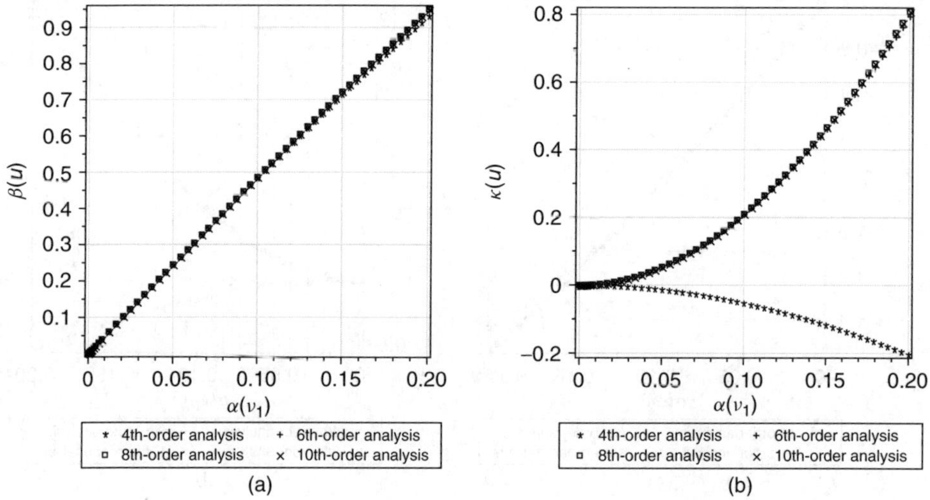

Figure 3.29 (a,b) Skewness and kurtosis for the cantilever end deflection for Poisson ratio of the reinforcement

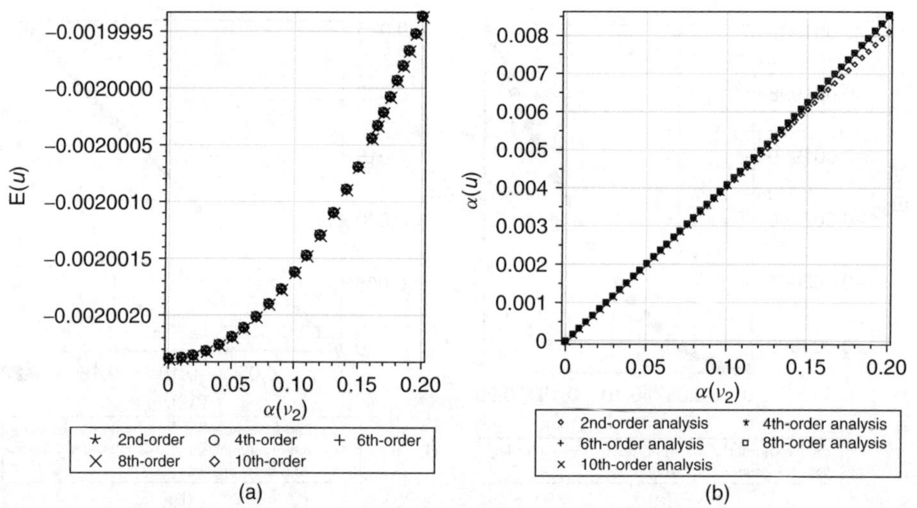

Figure 3.30 (a,b) The expectations and coefficients of variation for the cantilever end deflection for Poisson ratio of the matrix

Example 3.4: SBEM analysis of the composite with interface defects

The two-component, two-layer (for simplicity) carbon–epoxy composite given in Figure 3.32 is subjected to the SBEM analysis with the following material characteristics: $e_1 = 77.5\,\text{GPa}$, $\nu_1 = 0.24$, $e_2 = 4.1\,\text{GPa}$, and $\nu_2 = 0.34$. We use 113 nodal points linked with 64 parabolic boundary elements forming two closed regions having a common, partially continuous interface; this composite specimen is clamped at the left edge and subjected to the simple shear test realized by the tangent stresses applied

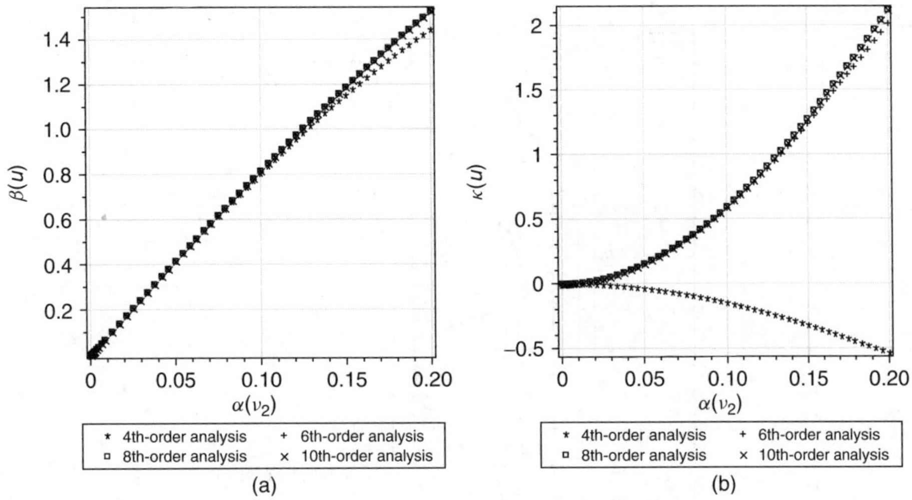

Figure 3.31 (a,b) Skewness and kurtosis for the cantilever end deflection for Poisson ratio of the matrix

at the opposite vertical edge. Four semi-circular cavities are inserted at this interface in the area of the epoxy matrix with uniform spatial distribution and the same radius ($E[r]=0.2$), each of them modeled using four curvilinear boundary elements [93]. The very detailed theoretical model of such a composite with imperfect interface is proposed in Chapter 5 of this book (see Page 245).

Using the proposed methodology implemented as the hybrid Maple™ and the multi-domain BEM solver [4], we determine first the response functions between the maximum vertical deflection in the system (Figures 3.33–3.35) and the cavities' radius. These functions are determined numerically for a uniform discretization of the given input random variable $r=[0.6, 0.7, 0.8, 0.9, 1.0, 1.1, 1.2, 1.3.1.4, 1.5, 1.6]$ and shown in Figures 3.35–3.37 (in turn, non-weighted method, triangular weights distribution around the mean value of random input parameter [1–6, 5, 4, 3, 2, 1], and Dirac distribution of the weights – [1, 1, 1, 1, 1, 6, 1, 1, 1, 1, 1]). Figures 3.33a, 3.34a and 3.35a show the variability of the different-order approximations for the entire input parameter domain and Figures 3.33b, 3.34b and 3.35b correspond to the very narrow neighborhood of the mean value. According to the opportunity of the RFM, we compare approximations of various orders – from the first until the ninth – to distinguish the differences between them and to determine for the future applications the most optimal choice for the specific application. It is apparent from the comparison of Figures 3.33–3.35 that while the non-weighted method shows some upper- and lower-end numerical discrepancies, the triangular weighted results demonstrate the same but for lower-order approximations only. The approach based on the Dirac distribution of the weights returns the most regular approximating functions without any oscillations. It is also clear that each first-order approximation is too weak to be used for reliable determination of any probabilistic moments for the structural response, whereas the differences between the second and higher orders are

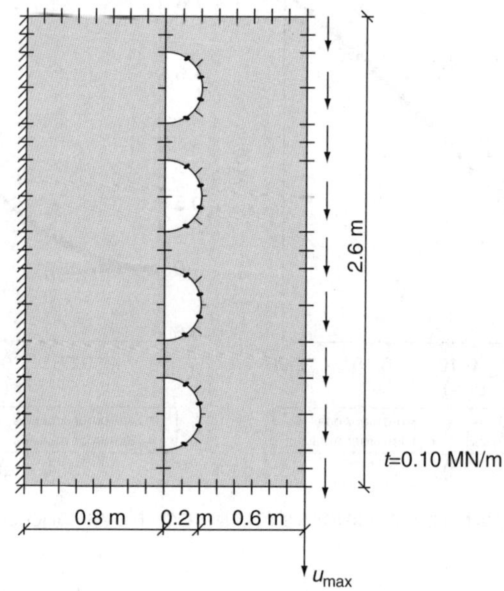

Figure 3.32 SBEM discretization model of the representative volume element (RVE) with real interface defects. Reproduced with permission from Elsevier

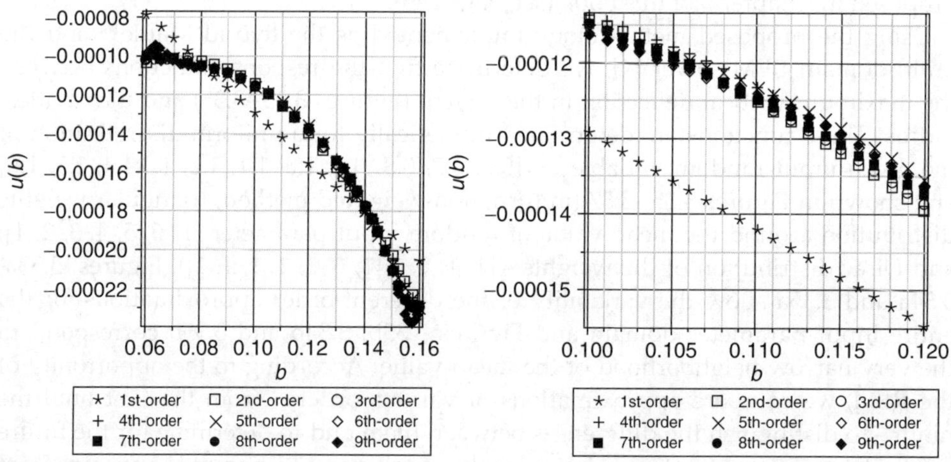

Figure 3.33 (a,b) The response function of the maximum vertical displacement versus cavities' radius, unweighted RFM–SBEM. Reproduced with permission from Elsevier

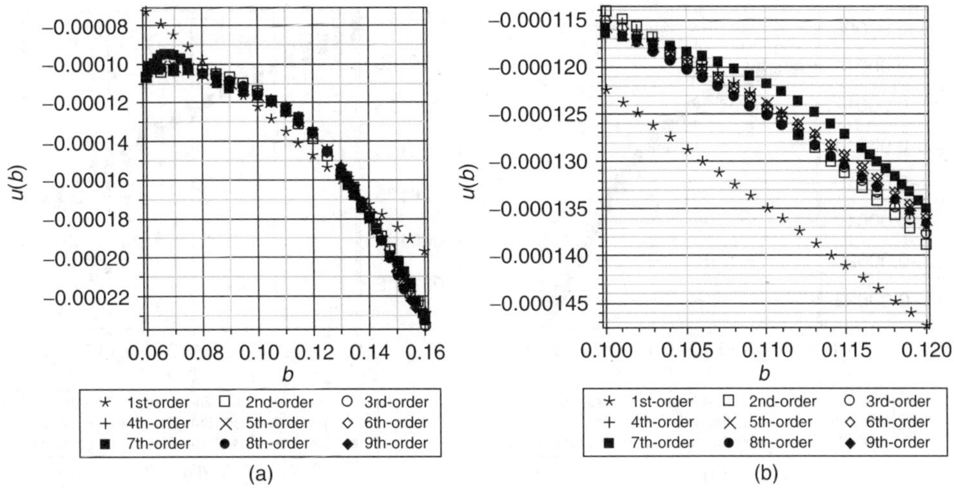

Figure 3.34 (a,b) The response function of the maximum vertical displacement versus cavities' radius, triangular weights RFM–SBEM. Reproduced with permission from Elsevier

practically negligible here. Further, we collect the expected values (Figures 3.36 and 3.37), the coefficients of variation (Figures 3.38 and 3.39), the skewnesses (Figures 3.40 and 3.41), and the kurtosis (Figures 3.42 and 3.43) of the same vertical displacements obtained in the lower right corner of this composite specimen. They are all computed and presented as functions of the input coefficient of variation of the cavities' radius (taken as $\alpha \in [0.0, 0.10]$) for perturbation parameter equal to 1 according to the non-weighted response function, the response with triangular weights as well as for the Dirac distribution of these weights. Higher-order probabilistic coefficients are computed in order to verify whether the vertical displacement at this specific point can have Gaussian distribution or not.

The expected values shown in Figures 3.36 and 3.37 exhibit very similar properties to these exposed in the response function graphs. Once more, the closest results for various-order approximations are noticed in the case of the weighted version of the SBEM according to the Dirac distribution. There are no specific orders resulting in apparently different results, like these obtained in the case of seventh order for the triangular weights distribution; of course, the first-order results are unacceptable for all three methods. It is also interesting that the absolute values of these expectations reach a maximum for the Dirac distribution, while a minimum is reached for the triangular weights and the non-weighted scheme returns intermediate values – this fact has no simple explanation. The function between the output expectation and the input coefficient of variation is entirely non-linear, where the larger the input α, the larger absolute value of the maximum vertical displacement in the shear test. Since we are interested in an optimal order of the method, then considering expectations only, the second one would be the most recommended as it is almost

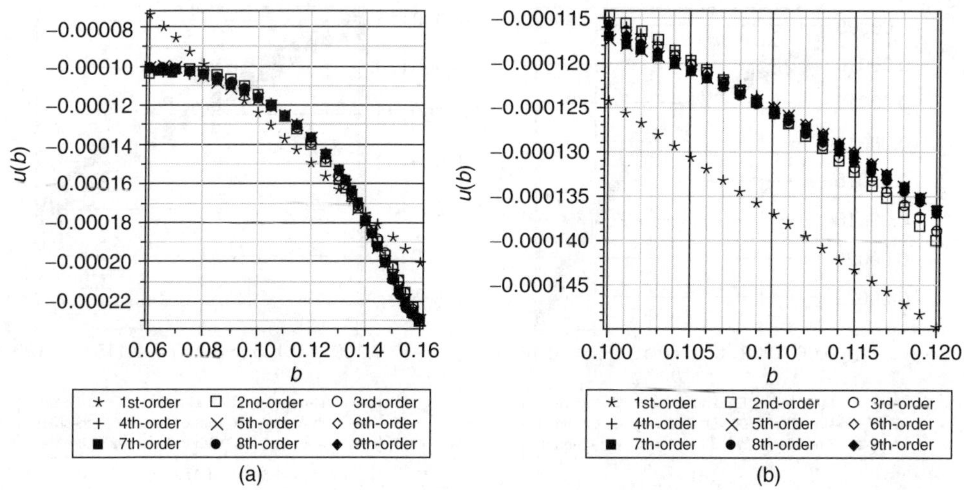

Figure 3.35 (a,b) The response function of the maximum vertical displacement versus cavities' radius, Dirac weights RFM–SBEM. Reproduced with permission from Elsevier

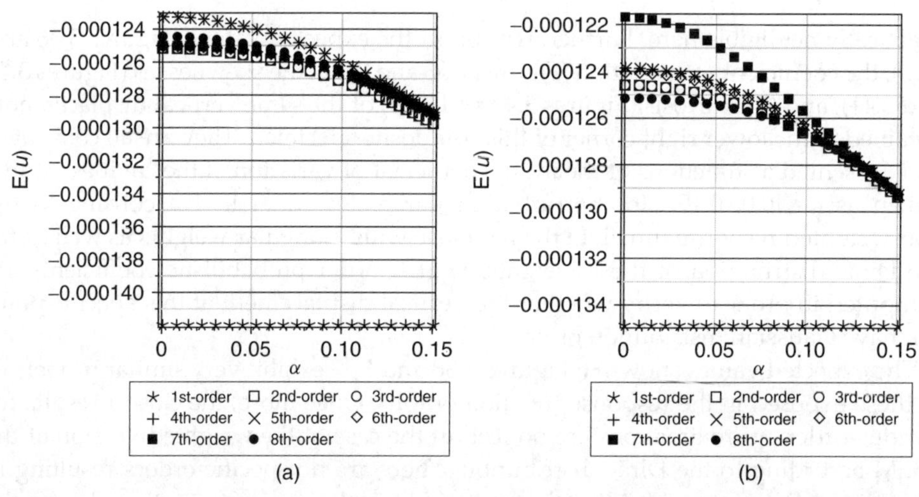

Figure 3.36 The expected values of the maximum vertical displacement, unweighted (a) and triangular weights (b) RFM–SBEM. Reproduced with permission from Elsevier

equal to the results of the highest possible order. The coefficients of variation shown in Figures 3.38 and 3.39 demonstrate that the system has neither damping nor amplifying properties in the context of uncertainty – the randomness at the input is almost the same as at the output and this interrelation is almost linear;

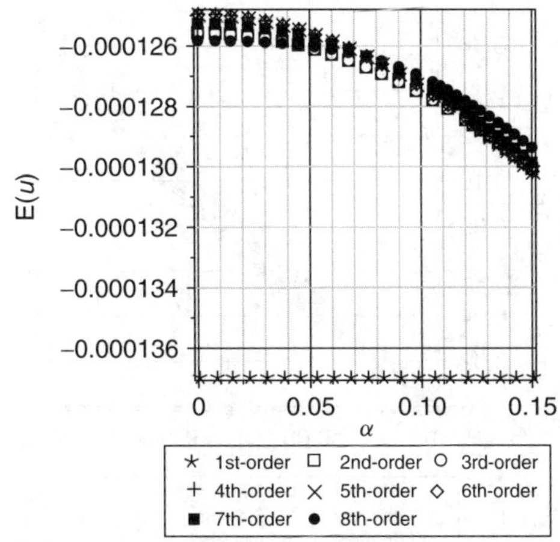

Figure 3.37 The expected values of the maximum vertical displacement, Dirac weights RFM–SBEM. Reproduced with permission from Elsevier

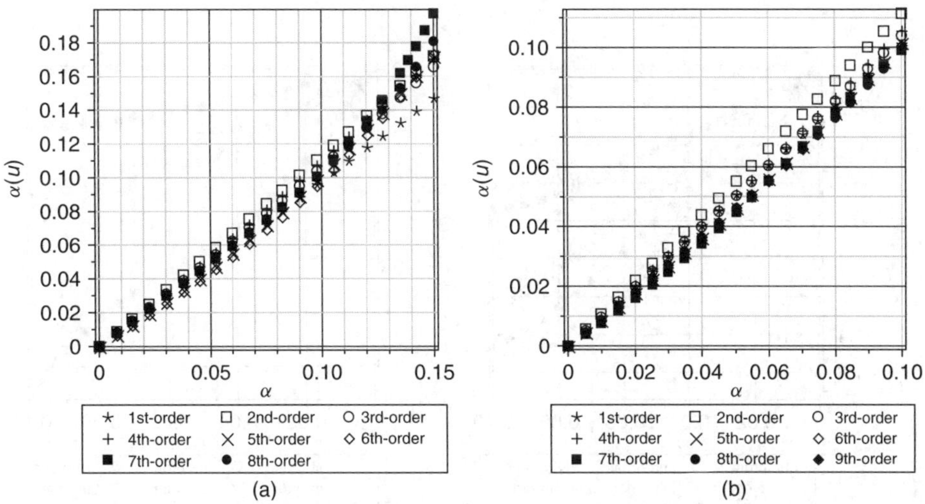

Figure 3.38 The variation coefficients of maximum vertical displacement, (a) non-weighted and (b) triangular weights RFM–SBEM. Reproduced with permission from Elsevier

nevertheless various techniques return slightly different values. As in the case of expectations, right now we have some lower-order numerical discrepancies for the non-weighted and for the triangular weights, while the Dirac weighting returns the most stable results. Now, the second order analysis is not sufficient because we notice significant differences to higher orders even in Figure 3.39. The differences for the skewness coefficients are more apparent than for the coefficients given before,

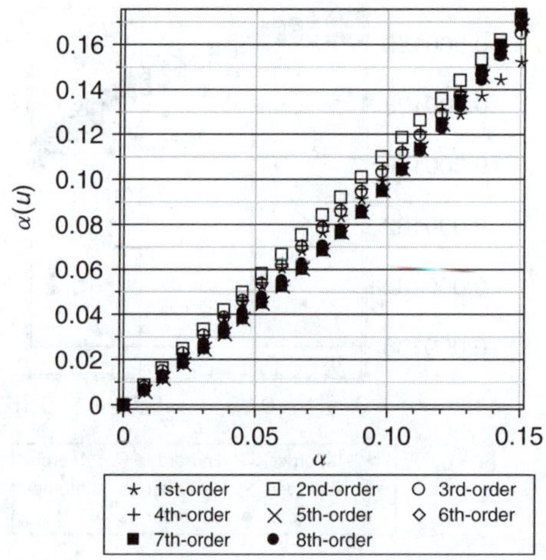

Figure 3.39 The variation coefficients of maximum vertical displacement, Dirac weights RFM–SBEM. Reproduced with permission from Elsevier

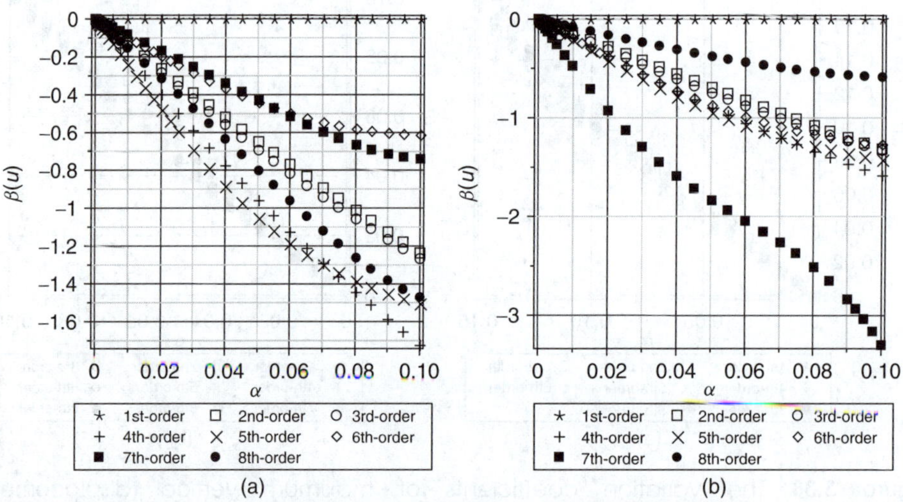

Figure 3.40 The skewness of the maximum vertical displacement, (a) non-weighted and (b) triangular weights RFM–SBEM. Reproduced with permission from Elsevier

however negative values are obtained for all cases. They all start from 0 for the deterministic test and decrease systematically together with an increase in value of the input coefficient of variation, so that we cannot treat the resulting deflection as a Gaussian variable and, unfortunately, probabilistic computational analysis and reliability index approximation cannot be limited to the rather simple formulas known from Eurocode 0. Analogously to the previous figures and to the kurtosis

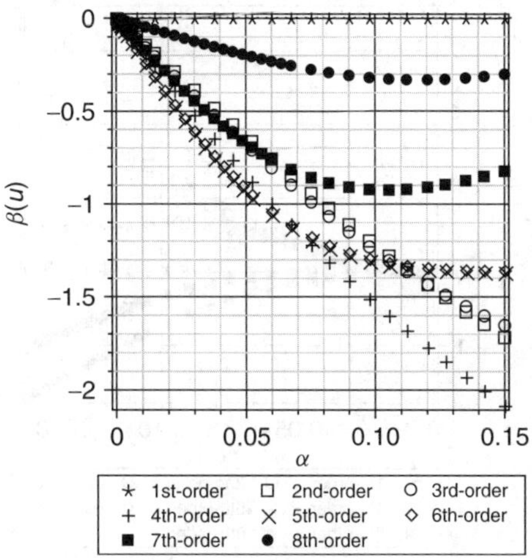

Figure 3.41 The skewness of the maximum vertical displacement, Dirac weights RFM–SBEM. Reproduced with permission from Elsevier

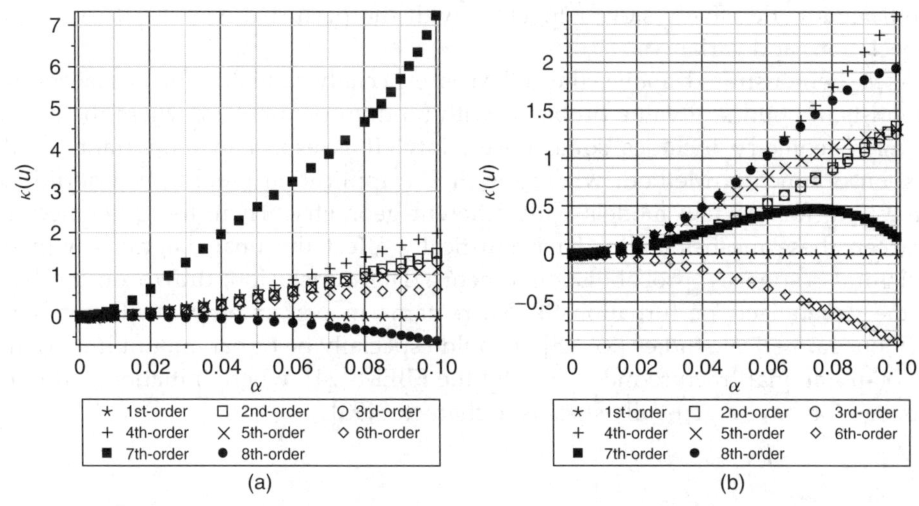

Figure 3.42 The kurtosis of the maximum vertical displacement, (a) non-weighted and (b) triangular weights RFM–SBEM. Reproduced with permission from Elsevier

given in Figures 3.42 and 3.43, the seventh-order analysis in the triangular weights scheme demonstrates some numerical instabilities and once more, the first-order results are not acceptable at all. The fourth-order coefficients also show that the final distribution is not Gaussian, but they demonstrate various properties – sometimes positive, sometimes negative – whereas their absolute values increase together with the additional increase of the input coefficient of variation. It is confirmed by Figures 3.40–3.43 that probabilistic convergence of higher order moments and

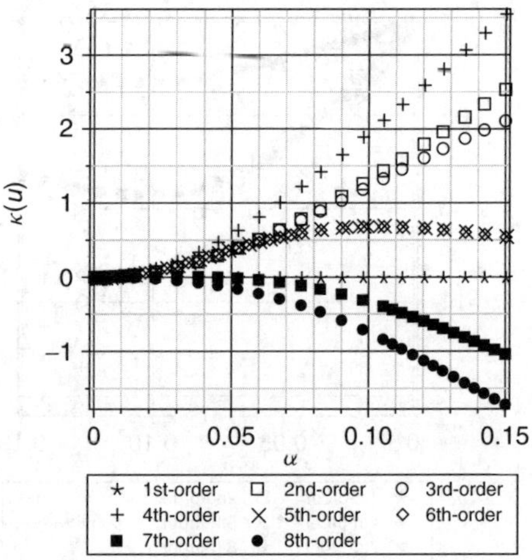

Figure 3.43 The kurtosis of the maximum vertical displacement, Dirac weights RFM–SBEM. Reproduced with permission from Elsevier

coefficients is decisively slower together with the perturbation order than for the expectations and variances.

As is demonstrated above, the SBEM is especially valuable for engineers and scientists in boundary value problems with random parameters, where the macro- or microgeometry includes some uncertainty. It is especially important for the Response Function Method, where the SFEM application would need additional remeshing for the few models with different geometry. Then, the differences in-between these meshes could deterministically affect the final diagrams with the structural responses' probabilistic moments. This conclusion, drawn on the basis of the generalized perturbation SBEM, remains in perfect agreement with some previous numerical studies [27, 48]. It could especially be of paramount importance in computational fracture mechanics by the SBEM [24], where initiation and crack propagation have a generally stochastic character [93].

4

The Stochastic Finite Difference Method

The finite difference method (FDM), being probably one of the oldest numerical techniques in engineering and scientific computations [20, 131, 167], may be still widely used in the domain of heat transfer [130], electrodynamics [163], electromagnetics [115], and hydromechanics [169], even for large-scale problems. Since engineers and scientists usually obtain an information about structural and system parameters in the form of statistical estimators or random processes, the idea of extending the FDM to modeling of random variables, fields, or even processes seems quite natural [102, 164]. Owing to such an extension, one can determine how the probabilistic parameters of the input affect the corresponding parameters of the structural response; it is known that this can be done using various mathematical and numerical method including these displayed in the preceeding chapters of this book. Alternative methods of modeling random processes, in the context of FDM stochastic extension, are via application of the Monte Carlo simulation technique or stochastic spectral finite differences (improved Neumann, Karhunen–Loeve, or polynomial chaos expansions), but their usage has a limited number of well-documented realizations and unresolved accuracy of higher probabilistic moments' determination. The generalized perturbation technique is used here analogously to its previous applications in the FEM and BEM stochastic extensions to utilize the FDM for the needs of probabilistic analysis [76, 90]. This idea follows the previous SOSM version of the stochastic FDM [66, 68, 102]. The essential difference from the FEM and BEM, as integral methods, is an assumption that all random functions and fields should be differentiable with respect to both spatial coordinates and, what is more important, at the same time to assumed random variables. We analyze the boundary value problems or transient problems with the use of unidirectional and planar regular grids including the closest and also wider neighborhood of difference operators [111]. Computational analysis in this section is entirely symbolic, because the system Maple™ with its programming language and internal linear algebra tools enables algebraic equation's

The Stochastic Perturbation Method for Computational Mechanics, First Edition. Marcin Kamiński.
© 2013 John Wiley & Sons, Ltd. Published 2013 by John Wiley & Sons, Ltd.

solutions of increasing order. The response function method (RFM) application is also quite straightforward in this context, having all difference equations coded into the Maple™ script. Further implementations and interoperability of Maple™ with FORTRAN codes give a unique opportunity to employ academic FDM software for stochastic perturbation technique applications in a symbolic environment, which is very important in view of the powerful visualization procedures.

4.1 Analysis of the Unidirectional Problems

4.1.1 Elasticity Problems

Let us consider the following ordinary fourth-order differential equation for a linear elastic isotropic and statistically homogeneous beam exposed to the transversally distributed load $q(x)$:

$$\frac{d^2}{dx^2}\left(e(x)\,J(x)\,\frac{d^2 w(x)}{dx^2}\right) = q(x), \tag{4.1}$$

fulfilling typical boundary conditions applicable in engineering theories of the elastic beams, like these idealized in Figure 4.1.

Investigation of the influence of the longitudinal and transversal forces is postponed here (but also possible), as we reduce the analysis to small deflections, which significantly simplifies the final form of the equilibrium equation subjected to the perturbation procedure. Let us divide the entire domain of length l into n equidistant subdomains having length Δx. We adopt the following notation for the ith point of this discretization:

$$e(i\Delta x) = e_i, \quad J(i\Delta x) = J_i, \quad q(i\Delta x) = q_i, \tag{4.2}$$

for the coordinate system introduced in Figure 4.1.

A series of approximations follows for derivatives of ascending order with the central finite differences, such as [23, 115]:

- first derivative

$$\left(\frac{\Delta w}{\Delta x}\right)_i = \frac{w_{i+1} - w_{i-1}}{2\Delta x} \tag{4.3}$$

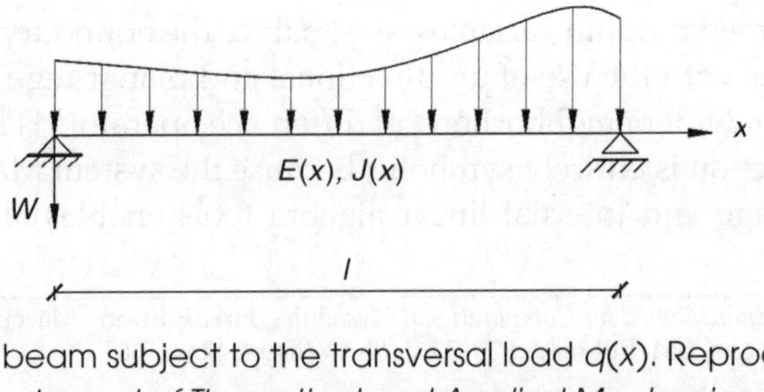

Figure 4.1 Elastic beam subject to the transversal load $q(x)$. Reproduced with permission from the *Journal of Theoretical and Applied Mechanics*

- second derivative

$$\left(\frac{\Delta^2 w}{\Delta x^2}\right)_i = \frac{w_{i+1} - 2w_i + w_{i-1}}{\Delta x^2} \tag{4.4}$$

- third derivative

$$\left(\frac{\Delta^3 w}{\Delta x^3}\right)_i = \frac{-w_{i+2} + 2w_{i+1} - 2w_{i-1} + w_{i-2}}{2\Delta x^3} \tag{4.5}$$

- fourth derivative

$$\left(\frac{\Delta^4 w}{\Delta x^4}\right)_i = \frac{w_{i+2} - 4w_{i+1} + 6w_i - 4w_{i-1} + w_{i-2}}{\Delta x^4}. \tag{4.6}$$

Equation (4.3) is an arithmetic average of the forward and backward difference equation and the next relations contain its simple consequence as a derivative of the derivative, and so forth. Introducing further for the fourth-order derivatives in Equation (4.1), and using the statement (4.2), one can obtain for the ith point of this grid the recursive difference equation as

$$\begin{aligned} e_{i-1}J_{i-1}w_{i-2} - 2\left(e_{i-1}J_{i-1} + e_iJ_i\right)w_{i-1} + \left(e_{i-1}J_{i-1} + 4e_iJ_i + e_{i+1}J_{i+1}\right)w_i \\ - 2\left(e_iJ_i + e_{i+1}J_{i+1}\right)w_{i+1} + e_{i+1}J_{i+1}w_{i+2} = q_i\Delta x^4. \end{aligned} \tag{4.7}$$

The particular case of $e(x)J(x) = \text{const.} = eJ$ enables us to transform Equation (4.7) into much simpler following formula:

$$w_{i-2} - 4w_{i-1} + 6w_i - 4w_{i+1} + w_{i+2} = \frac{q_i\Delta x^4}{eJ}. \tag{4.8}$$

The situation changes only slightly when the elastic beam rests on a single-parameter elastic foundation (known from geotechnics as the Winkler foundation). Let us assume that this foundation has a homogeneous character and is completely characterized by the compliance coefficient k. Therefore, the equilibrium equation including the deflection $w(x)$ has a single extra component on the RHS. It holds that

$$\frac{d^2}{dx^2}\left(e(x)J(x)\frac{d^2 w(x)}{dx^2}\right) = -kw(x) + q(x). \tag{4.9}$$

Inserting, as previously, the additional formulas for the derivatives, which include finite differences expressed by discrete values of the function $w(x)$ in the neighborhood of the given ith point of the grid, it is obtained that

$$\begin{aligned} e_{i-1}J_{i-1}w_{i-2} - 2\left(e_{i-1}J_{i-1} + e_iJ_i\right)w_{i-1} + \left(e_{i-1}J_{i-1} + 4e_iJ_i + e_{i+1}J_{i+1} + k\Delta x^4\right)w_i \\ - 2\left(e_iJ_i + e_{i+1}J_{i+1}\right)w_{i+1} + e_{i+1}J_{i+1}w_{i+2} = q_i\Delta x^4. \end{aligned} \tag{4.10}$$

A reduction in case of the constant stiffness $e(x)J(x) = \text{const.} = eJ$ leads to the following equation:

$$w_{i-2} - 4w_{i-1} + \left(6 + \frac{k\Delta x^4}{eJ}\right) w_i - 4w_{i+1} + w_{i+2} = \frac{q_i \Delta x^4}{eJ}. \tag{4.11}$$

Traditionally, the stochastic perturbation approach to many of the physical problems is entered by the respective perturbed equations of the zeroth, first and, successively, higher orders being a modification of the variational integral formulation. Therefore, we provide an extra differentiation of the existing differential/difference equations with respect to the input random variable of the problem and we obtain:

- zeroth-order partial differential equation

$$e^0 J^0 \frac{d^4}{dx^4}\left(w^0(x)\right) = q^0 \tag{4.12}$$

- first-order partial differential equation

$$\frac{\partial e}{\partial b} J^0 \frac{d^4}{dx^4}\left(w^0(x)\right) + e^0 \frac{\partial J}{\partial b} \frac{d^4}{dx^4}\left(w^0(x)\right) + e^0 J^0 \frac{d^4}{dx^4}\left(\frac{\partial w(x)}{\partial b}\right) = \frac{\partial q}{\partial b} \tag{4.13}$$

- second-order partial differential equation

$$\frac{\partial^2 e}{\partial b^2} J^0 \frac{d^4}{dx^4}\left(w^0(x)\right) + e^0 \frac{\partial^2 J}{\partial b^2} \frac{d^4}{dx^4}\left(w^0(x)\right) + e^0 J^0 \frac{d^4}{dx^4}\left(\frac{\partial^2 w(x)}{\partial b^2}\right)$$
$$+ 2\left\{\frac{\partial e}{\partial b}\frac{\partial J}{\partial b}\frac{d^4}{dx^4}\left(w^0(x)\right) + e^0 \frac{\partial J}{\partial b}\frac{d^4}{dx^4}\left(\frac{\partial w(x)}{\partial b}\right) + \frac{\partial e}{\partial b} J^0 \frac{d^4}{dx^4}\left(\frac{\partial w(x)}{\partial b}\right)\right\} = \frac{\partial^2 q}{\partial b^2}. \tag{4.14}$$

Quite similarly, one can derive the ascending-order partial differential equations for the elastic beam on the elastic foundation starting from relation (4.9). The following hold:

- zeroth-order partial differential equation

$$e^0 J^0 \frac{d^4}{dx^4}\left(w^0(x)\right) = -k^0 w^0(x) + q^0 \tag{4.15}$$

- first-order partial differential equation

$$\frac{\partial e}{\partial b} J^0 \frac{d^4}{dx^4}\left(w^0(x)\right) + e^0 \frac{\partial J}{\partial b} \frac{d^4}{dx^4}\left(w^0(x)\right) + e^0 J^0 \frac{d^4}{dx^4}\left(\frac{\partial w(x)}{\partial b}\right)$$
$$= \frac{\partial q}{\partial b} - \left(\frac{\partial k}{\partial b} w^0(x) + k^0 \frac{\partial w(x)}{\partial b}\right) \tag{4.16}$$

- second-order partial differential equation

$$\frac{\partial^2 e}{\partial b^2} J^0 \frac{d^4}{dx^4}\left(w^0(x)\right) + e^0 \frac{\partial^2 J}{\partial b^2} \frac{d^4}{dx^4}\left(w^0(x)\right) + e^0 J^0 \frac{d^4}{dx^4}\left(\frac{\partial^2 w(x)}{\partial b^2}\right)$$
$$+2\left\{\frac{\partial e}{\partial b}\frac{\partial J}{\partial b}\frac{d^4}{dx^4}\left(w^0(x)\right) + e^0 \frac{\partial J}{\partial b}\frac{d^4}{dx^4}\left(\frac{\partial w(x)}{\partial b}\right) + \frac{\partial e}{\partial b} J^0 \frac{d^4}{dx^4}\left(\frac{\partial w(x)}{\partial b}\right)\right\}$$
$$= \frac{\partial^2 q}{\partial b^2} - \left(\frac{\partial^2 k}{\partial b^2} w^0(x) + 2\frac{\partial k}{\partial b}\frac{\partial w(x)}{\partial b} + k^0 \frac{\partial^2 w(x)}{\partial b^2}\right). \tag{4.17}$$

The derivation of higher-order equations proceeds quite similarly – by systematic differentiation until the nth-order equation is recovered. Having solved these equations for $w^0(x)$ and higher orders, consecutively (specifically for all partial derivatives w.r.t. a random input within all discrete points of the grid), we derive the expressions for the expected values and other moments of the elastic beam deflections. The discrete equations for the stochastic FDM built upon the above equations for the perturbation-based analysis are essentially different in case of an elastic beam with and without an elastic foundation beneath it. Putting here $k=0$ returns various order relations derived on the basis of Equation (4.8) such as, for example:

- zeroth-order relation

$$w^0_{i-2} - 4w^0_{i-1} + 6w^0_i - 4w^0_{i+1} + w^0_{i+2} = \frac{q^0_i \left(\Delta x^0\right)^4}{e^0 J^0} \tag{4.18}$$

- nth-order relation

$$\frac{\partial^n w_{i-2}}{\partial b^n} - 4\frac{\partial^n w_{i-1}}{\partial b^n} + 6\frac{\partial^n w_i}{\partial b^n} - 4\frac{\partial^n w_{i+1}}{\partial b^n} + \frac{\partial^n w_{i+2}}{\partial b^n} = \frac{\partial^n}{\partial b^n}\left(\frac{q_i \Delta x^4}{eJ}\right) \tag{4.19}$$

so that the uniformly distributed randomized load gives here non-zero equations of up to first order only, random length results in up to fourth-order equations, whereas cross-sectional and/or material randomness brings an infinite number of equations. Looking for the perturbation-based specially adopted FDM (SFDM) equations for beams on the elastic foundation, one gets from Equation (4.11):

- zeroth-order equations

$$w^0_{i-2} - 4w^0_{i-1} + \left(6 + \frac{k^0 \left(\Delta x^0\right)^4}{E^0 J^0}\right) w^0_i - 4w^0_{i+1} + w^0_{i+2} = \frac{q^0_i \left(\Delta x^0\right)^4}{e^0 J^0} \tag{4.20}$$

- nth-order equations

$$\frac{\partial^n w_{i-2}}{\partial b^n} - 4\frac{\partial^n w_{i-1}}{\partial b^n} + 6\frac{\partial^n w_i}{\partial b^n} + \sum_{p=1}^{n}\binom{n}{p}\frac{\partial^p}{\partial b^p}\left(\frac{k\Delta x^4}{eJ}\right)\frac{\partial^{n-p} w_i}{\partial b^{n-p}} - 4\frac{\partial^n w_{i+1}}{\partial b^n} + \frac{\partial^n w_{i+2}}{\partial b^n}$$
$$= \frac{\partial^n}{\partial b^n}\left(\frac{q_i \Delta x^4}{eJ}\right). \tag{4.21}$$

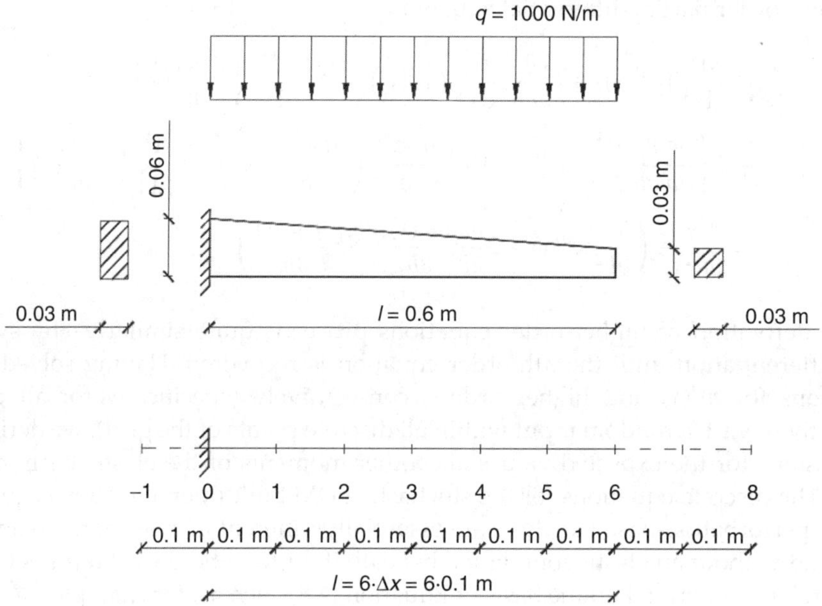

Figure 4.2 FDM discretization of the cantilever with linearly varying cross-section. Reproduced with permission from the *Journal of Theoretical and Applied Mechanics*

Let us note also that $n-1$ of the above equations may be generated automatically from the zeroth-order formula using relatively easy symbolic procedure, which is documented in the examples given in the Author's webpage accompanying this book.

We need to remember that according to the general stochastic perturbation philosophy, all partial derivatives with respect to the random input variables are calculated at its expectation. Therefore, the RFM approach needs to be focused on the response functions' determination in some close neighborhood of this value.

Example 4.1: Deflection of the linear elastic beam with linearly varying cross-sectional area

Let us determine the first four probabilistic moments for the cantilever beam with linearly varying cross-sectional area under the constant distributed load $q = 1.0\,\mathrm{kN/m}$ [146]. The Young's modulus of this beam is taken as the input Gaussian random variable, where the expected value is given as $E[e] = 206.01\,\mathrm{GPa}$; a grid for this structure consisting of six equidistant elements having length $\Delta x = 0.1\,\mathrm{m}$ is adopted (Figure 4.2). Let us note that linear variations in the beam bending stiffness need much more complex perturbation-based equations than those proposed above.

The following central finite difference equations hold true in this particular case, after Equation (4.7):

$$J_0 w_{-1} - 2(J_0 + J_1)w_0 + (J_0 + 4J_1 + J_2)w_1 - 2(J_1 + J_2)w_2 + J_2 w_3 = \frac{q\Delta x^4}{e},$$

$$J_1 w_0 - 2(J_1 + J_2)w_1 + (J_1 + 4J_2 + J_3)w_2 - 2(J_2 + J_3)w_3 + J_3 w_4 = \frac{q\Delta x^4}{e},$$

$$J_2 w_1 - 2(J_2 + J_3)w_2 + (J_2 + 4J_3 + J_4)w_3 - 2(J_3 + J_4)w_4 + J_4 w_5 = \frac{q\Delta x^4}{e},$$

$$J_3 w_2 - 2(J_3 + J_4)w_3 + (J_3 + 4J_4 + J_5)w_4 - 2(J_4 + J_5)w_5 + J_5 w_6 = \frac{q\Delta x^4}{e},$$

$$J_4 w_3 - 2(J_4 + J_5)w_4 + (J_4 + 4J_5 + J_6)w_5 - 2(J_5 + J_6)w_6 + J_6 w_7 = \frac{q\Delta x^4}{e},$$

$$J_5 w_4 - 2(J_5 + J_6)w_5 + (J_5 + 4J_6 + J_7)w_6 - 2(J_6 + J_7)w_7 + J_7 w_8 = \frac{q\Delta x^4}{e}, \tag{4.22}$$

where of course inertia moments deterministically depend on the position of a point within the grid. Introducing fictitious nodes as well as using kinematic boundary conditions at the left clamped edge, we obtain:

$$(2J_0 + 4J_1 + J_2)w_1 - 2(J_1 + J_2)w_2 + J_2 w_3 = \frac{q\Delta x^4}{e},$$

$$-2(J_1 + J_2)w_1 + (J_1 + 4J_2 + J_3)w_2 - 2(J_2 + J_3)w_3 + J_3 w_4 = \frac{q\Delta x^4}{e},$$

$$J_2 w_1 - 2(J_2 + J_3)w_2 + (J_2 + 4J_3 + J_4)w_3 - 2(J_3 + J_4)w_4 + J_4 w_5 = \frac{q\Delta x^4}{e},$$

$$J_3 w_2 - 2(J_3 + J_4)w_3 + (J_3 + 4J_4 + J_5)w_4 - 2(J_4 + J_5)w_5 + J_5 w_6 = \frac{q\Delta x^4}{e},$$

$$J_4 w_3 - 2(J_4 + J_5)w_4 + (J_4 + 4J_5)w_5 - 2J_5 w_6 = \frac{q\Delta x^4}{e},$$

$$2J_5 w_4 - 4J_5 w_5 + 2J_5 w_6 = \frac{q\Delta x^4}{e}. \tag{4.23}$$

This system of linear equations may be used for further analytical partial differentiation with respect to the random input variable and recursive formation of the higher-order equations necessary to derive probabilistic characteristics of the maximum deflection.

We compute probabilistic moments and coefficients using the Direct Differentiation Method and up to the tenth-order perturbation technique – all as a function of the input coefficient of variation $\alpha \in [0.00, 0.25]$ first. We obtain the expectation and variances given in Figure 4.3, standard deviation and coefficient of variation presented in Figure 4.4, third and fourth central probabilistic moments in Figure 4.5, as well as skewness and kurtosis shown in Figure 4.6. This is all done for the maximum deflection at the end of the cantilever analyzed, where the limit function and reliability index are to be verified. All first two moments and characteristics increase together with α – expectations and variances highly non-linearly (Figure 4.7), while standard deviations and coefficients of variation decisively closer to the linear function (Figure 4.8). Of course, they are all positive and the higher the perturbation order, the larger final values of these characteristics. Expected values converge

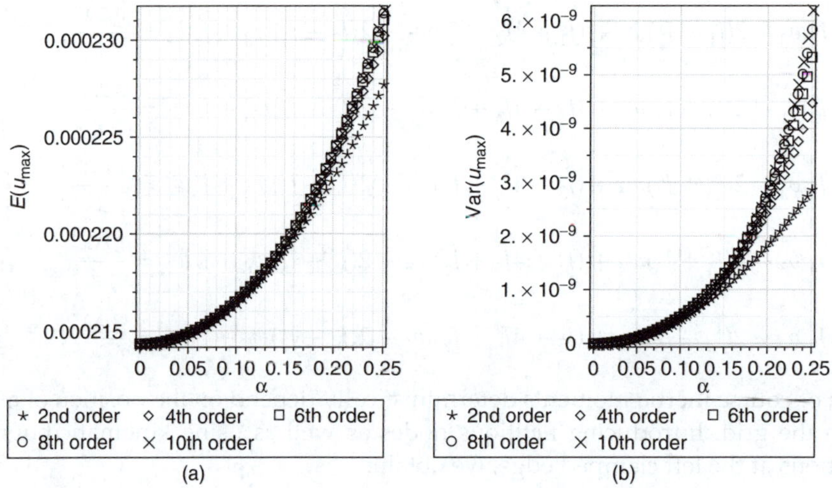

Figure 4.3 (a) Expected values and (b) variances (b) of the maximum deflection

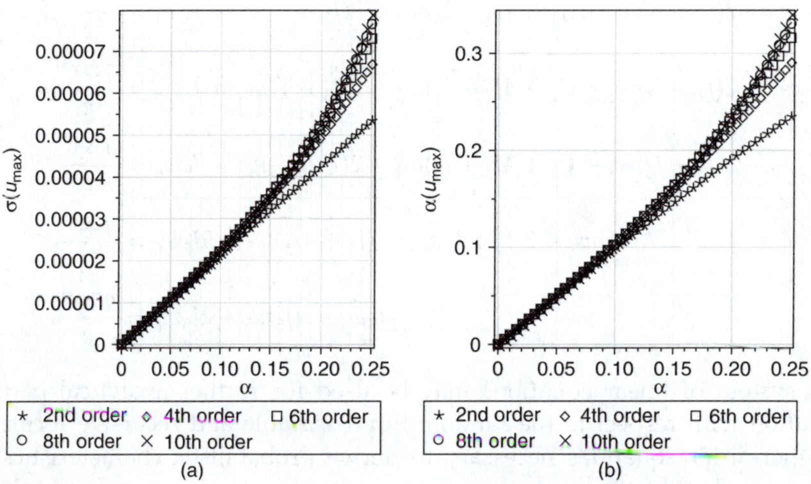

Figure 4.4 (a) Standard deviations and (b) coefficients of variation of the maximum deflection

probabilistically very fast, so that fourth-order analysis is precise enough even for the maximum value of the input coefficient of variation, while the remaining second-order characteristics need higher-order approximation and then, at least eighth order is recommended. Let us underline that the difference for the expectations computed for $\alpha = 0$ and 0.25 is almost 10% of its minimum and cannot be simply neglected. At the same time the output coefficient of variation is definitely larger than the input α, hence the system clearly amplifies input uncertainty in the Young's modulus.

As expected after Figure 4.3, the higher the central probabilistic moment considered, the smaller absolute value of this moment and also the weaker probabilistic

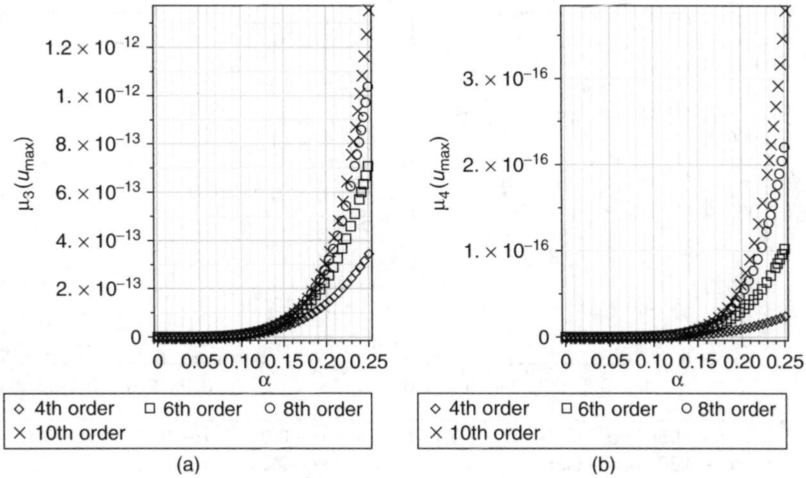

Figure 4.5 (a) Third and (b) fourth central probabilistic moments of the maximum deflection

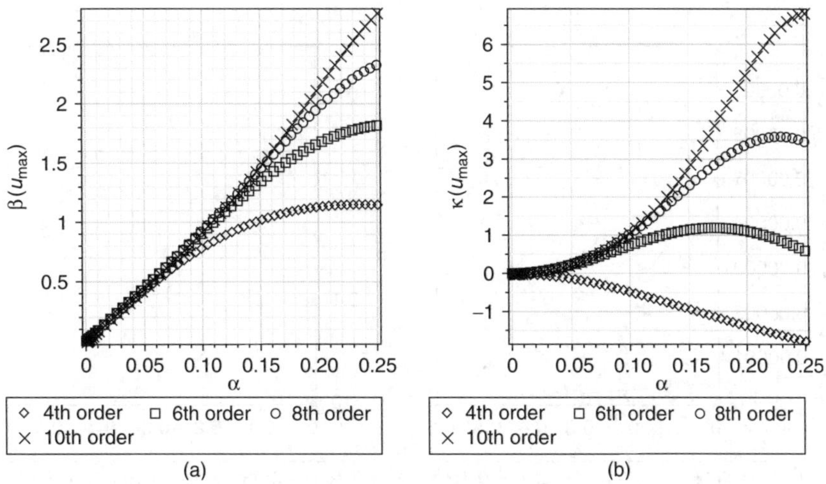

Figure 4.6 (a) Skewness and (b) kurtosis of the maximum deflection

perturbation convergence. Exceptionally, we can treat tenth-order approximation as satisfactory for the third moment, especially for smaller input α, but this is completely not the case for the fourth moment (Figure 4.9). It is also apparent that the higher the output moment of a maximum deflection, the larger the curvature of the interrelations of $\mu_k(u_{\max})$ versus $\alpha(e)$.

We need to remember that higher-order moments tending to 0 do not automatically imply that the skewness and kurtosis will behave in the same way. Now these coefficients are both dominated by the positive values (except kurtosis in the

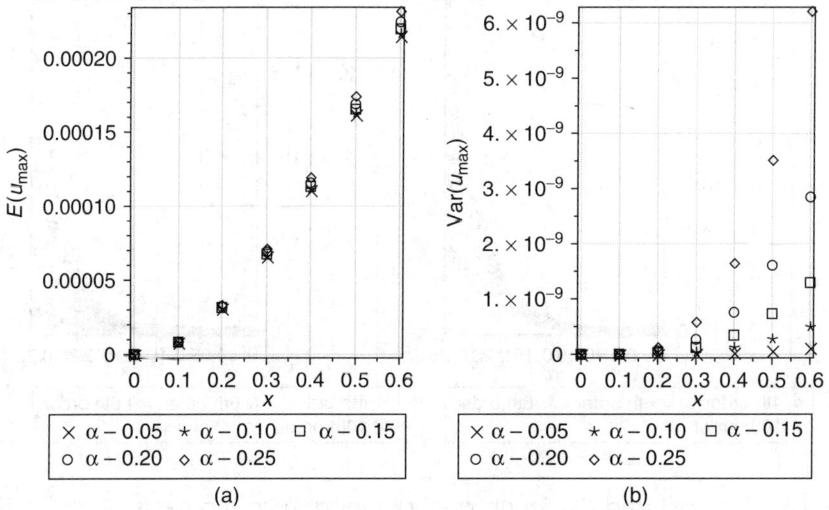

Figure 4.7 (a) Expected values and (b) variances of the maximum deflection

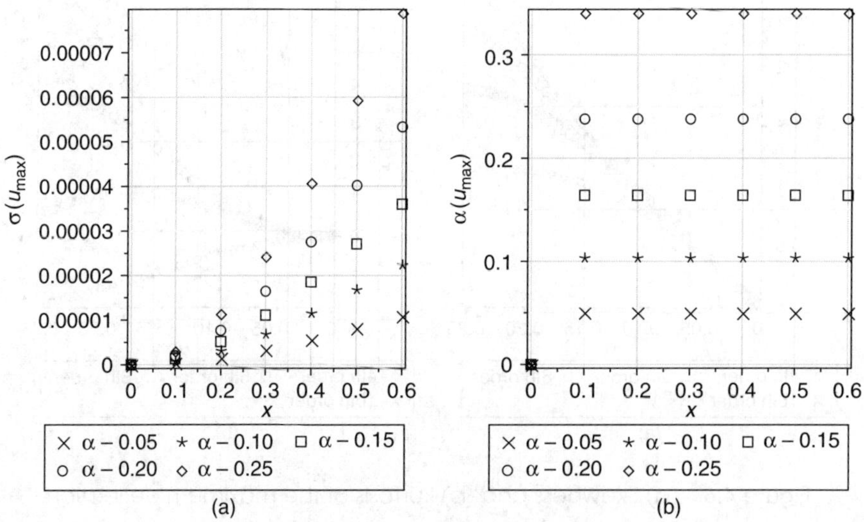

Figure 4.8 (a) Standard deviations and (b) coefficients of variation of the maximum deflection

second-order approach), which clearly states that the output maximum deflection has non-Gaussian probabilistic distribution. Skewness and kurtosis do not show any probabilistic convergence, especially for larger values of input parameter α, so that sometimes even tenth-order perturbation may be insufficient to determine higher-order statistics of the structural response precisely enough.

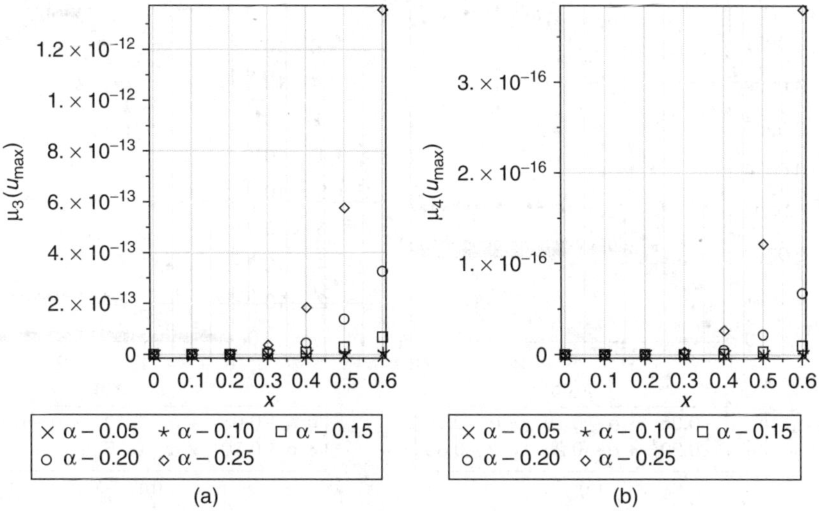

Figure 4.9 (a) Third and (b) fourth central probabilistic moments of the maximum deflection

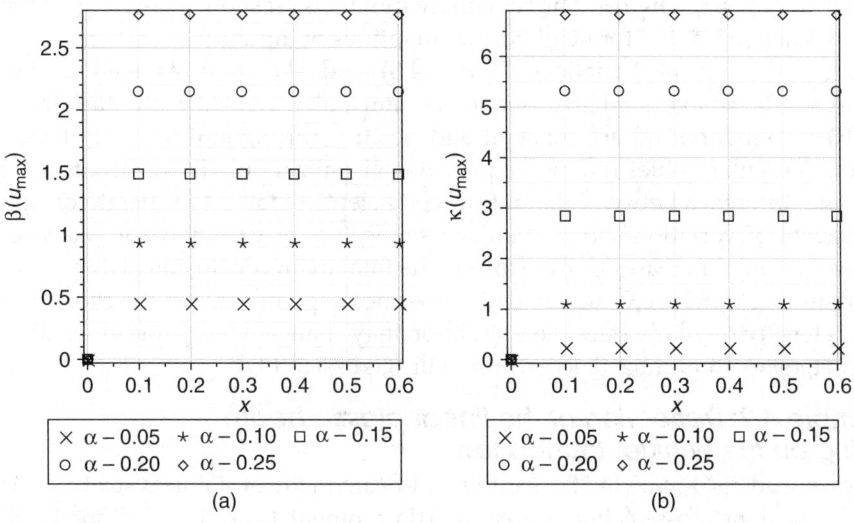

Figure 4.10 (a) Skewness and (b) kurtosis of the maximum deflection

Further, we analyze the spatial distribution of probabilistic characteristics along the cantilever beam. Somewhat similarly to the distributions of the central moments w.r.t. $\alpha(e)$, now expectations, variances, third and fourth central moments all increase non-linearly from $x = 0.0$ up to the end at $x = 3.0$. Of course, the higher the input coefficient, the larger the values of these moments and the greater the curvature of computed distributions. We notice moreover that the central moment values decrease all together with this moment order. Finally, it is noticed that only the differences

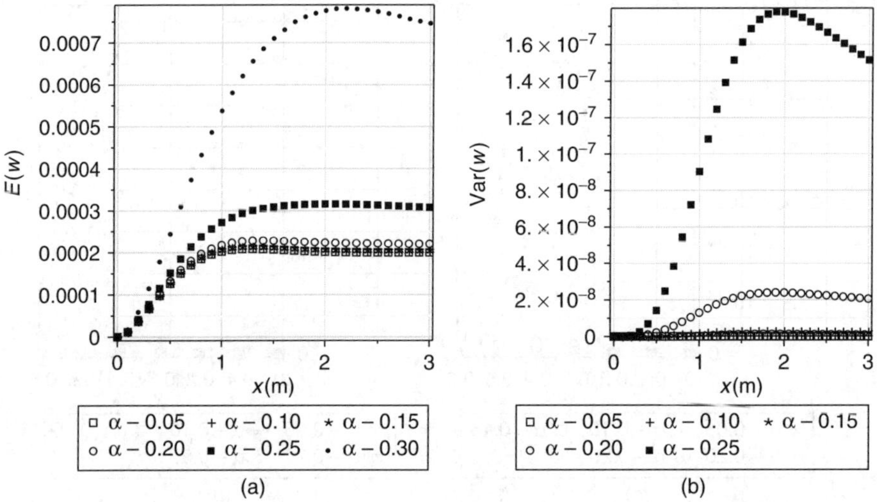

Figure 4.11 (a) Expected values and (b) variances of the beam deflections

between the expectations computed for various input coefficients of variation may be treated as almost negligible. The remaining moments all show increasing increments of the values computed for neighboring quantities of input $\alpha(e)$.

The coefficients of variation (Figure 4.8) and skewness as well as kurtosis (Figure 4.10) behave similarly but in an unusual way. They all start from 0 at the fixed boundary and are constant and positive throughout the rest of the beam length. Particular values follow the previous visualization with respect to the input $\alpha(e)$. The difference between the output coefficients obtained for neighboring input coefficients of variation shows regular (coefficients of variation and skewness) as well as irregular increases (for kurtosis). The final main conclusion is that the output distribution is nowhere Gaussian and it does not depend at all on the choice of input parameters' particular values. This situation may change when some other structural parameters are randomized according to the Gaussian PDF.

Example 4.2: Deflection of the linear elastic beam resting on the Winkler foundation

This case study is devoted to the cantilever I-beam made of stainless steel $e = 205$ GPa having constant cross-sectional area, inertia moment $J = 1.71 \times 10^{-6}$ m^4 (IPE 100), and resting on the elastic foundation characterized by expected value of compliance coefficient $k = 5 \times 10^7 \frac{\text{N}}{\text{m}^2}$. The deterministic counterpart of this example may be found with the few point of the grid only in [146]. The external distributed load is constant along this beam, $q = 10\frac{\text{kN}}{\text{m}}$ for $l = 3$ m, while its boundary conditions have been discretized as follows:

$$\left.\begin{aligned} w_0 &= 0, \\ w_{-1} &= w_1, \\ w_{n-1} &= 2w_{n-2} - w_{n-3}, \\ w_n &= 4w_{n-2} - 4w_{n-3} + w_{n-4} \end{aligned}\right\} \tag{4.24}$$

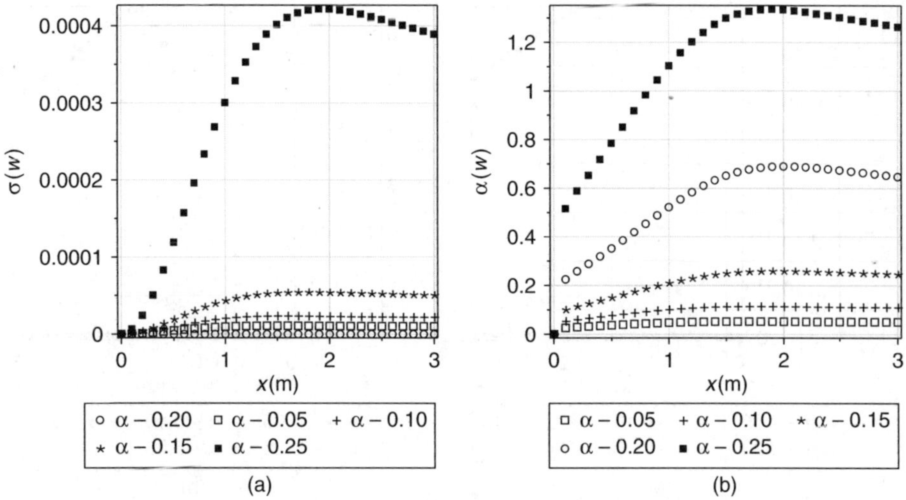

Figure 4.12 (a) Standard deviations and (b) coefficients of variation of the beam deflections

and where the nodes indexed with numbers n and $n-1$ are of course fictitious (and correspond to the free edge). The computational grid is partitioned into $n=30$ equidistant intervals and n is treated here as the parameter of this solution. The probabilistic characteristics are algebraically derived in a computational cycle taking ~4200 s for each input random dispersion on the double core *i5* processor in Maple™. The results of numerical analyses are contained in Figure 4.11 (expectations and variances of deflections), Figure 4.12 (their standard deviations and coefficients of variation), Figure 4.13 (third and fourth central probabilistic moments of the deflections), as well as Figure 4.14 (skewness and kurtosis of the overall deflection). They have all been computed for five various coefficients of variation of the input subsoil compliance $\alpha=0.05$, 0.10, 0.15, 0.20, and 0.25, correspondingly (also 0.30 for the expectations).

As one could expect, the higher the input coefficient of variation of the subsoil compliance, the higher the value of the probabilistic moments. This is not really apparent for variances, third as well as fourth central probabilistic moments for smaller values of input variation coefficient, however for $\alpha \geq 0.20$ these moments have values a few times larger than before. This result is more accurate than the previous combined-order approximations of higher-order moments, since a significant underestimation of these moments in comparison with the Monte Carlo simulations was noticed [103].

A contrast of the expected values shows that the subsoil random compliance significantly influences the final random response of this beam since a more than 50% increase is noticed for the extremum value of $\alpha(k)$, while the uncertainty in geotechnical parameters is frequently very large and may be even much larger than that resulting from randomization of mechanical and physical properties of the structures. Let us also underline that the location of the maximum deflection is slightly different for various input coefficients of variation, although the distributions of the

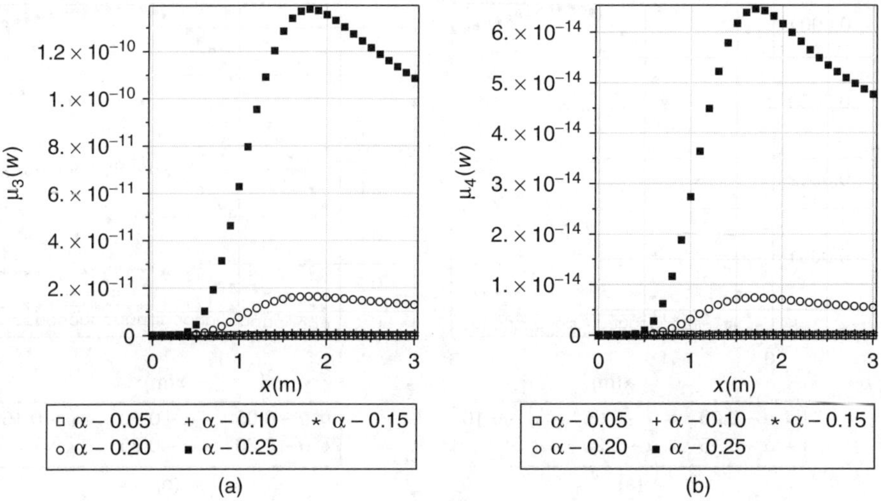

Figure 4.13 (a) Third and (b) fourth central probabilistic moments of the beam deflections

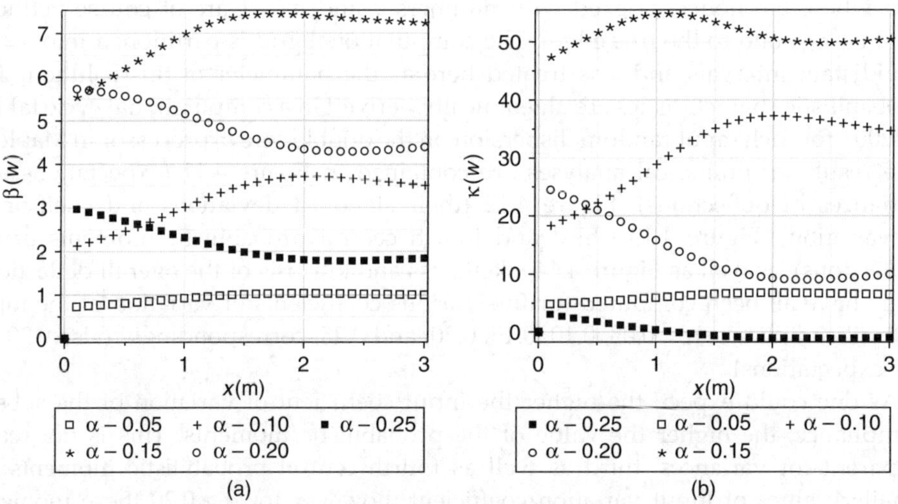

Figure 4.14 (a) Skewness and (b) kurtosis of the beam deflections

deflection functions remain very similar for all input coefficients of variation. The ratio of the variances to the expectations shows that the output coefficient of variation is significantly larger than the adjacent input value, so that the random dispersion is amplified during solution of this problem even with just a few times. Further, the third probabilistic moments are positive everywhere, so that the deflections exhibit apparent skewness in all tests and the larger part of the probability curve is above the expectation. At the same time, the fourth central moments have very large values and significantly positive kurtosis proves that the concentration around the expectation is many times higher than for the Gaussian distribution with the same expectation and standard deviation.

All the probabilistic moments start from value 0 at the clamped edge of this beam, increase further up to some extremum, and start to decrease till the very end at $l=3$ m. The higher the order of the probabilistic moment analyzed, the smaller the values of these moments, the steeper the descents of these curves, and the wider the zone with zero values at the left edge of the beam. Taking into account all these results, one can conclude that the highest values of the reliability index are undoubtedly expected at the left stiff edge of this beam, whereas the location of the smallest values (with the largest probability failure) corresponds to the expectation and variance extrema (almost the same location for $x \cong 2.0$ m in this case study). Spatial distributions of probabilistic characteristics show clearly that they are necessary for an efficient and correct computation of the reliability index. Calculation of this index in a location of extremum values obtained via deterministic analysis may be improper and can lead to some substantial modeling errors.

4.1.2 Determination of the Critical Moment for the Thin-Walled Elastic Structures

Let us consider a differential equation describing the critical moment M_{cr} under a combination of bending and twisting according to the Vlasov's theory of the thin-walled beam with constant cross-section on its length:

$$eJ_{\omega}\frac{\mathrm{d}^4\phi(x)}{\mathrm{d}x^4} - GJ_s\frac{\mathrm{d}^2\phi(x)}{\mathrm{d}x^2} - \frac{M_{cr}^2}{eJ_y}\phi(x) = 0 \tag{4.25}$$

where $\phi(x)$ denotes the angle of twist, e and G traditionally denote elastic parameters of the beam in the form of Young's and Kirchhoff moduli, while J_{ω}, J_s, J_y are basic geometrical characteristics of the cross-section. In this context M_{cr} denotes the critical value of the bending moment applied at the considered thin-walled elastic beam structure. Taking into account an application of the RFM adjacent to material and/or geometrical randomness along the structural element considered, one may propose the following extension of Equation (4.25):

$$e^{(j)}J_{\omega}^{(j)}\frac{\mathrm{d}^4\phi^{(j)}(x)}{\mathrm{d}x^4} - G^{(j)}J_s^{(j)}\frac{\mathrm{d}^2\phi^{(j)}(x)}{\mathrm{d}x^2} - \frac{\left(M_{cr}^{(j)}\right)^2}{e^{(\alpha)}J_y^{(j)}}\phi^{(j)}(x) = 0. \tag{4.26}$$

We propose the following boundary conditions from the most popular support types to solve this equation for a given case study: (i) $\phi=0$ and $\frac{\mathrm{d}\phi}{\mathrm{d}x}=0$ (support completely stiff for both twist and deplanation); (ii) $\phi=0$ and $\frac{\mathrm{d}^2\phi}{\mathrm{d}x^2}=0$ (support stiffened against twist with free deplanation); and (iii) support stiffened against the twist with a certain elastic deplanation, that is, $\phi=0$ and $\frac{\mathrm{d}\phi}{\mathrm{d}x}=\frac{e}{k}\frac{\mathrm{d}^2\phi}{\mathrm{d}x^2}$. It is clear that the same boundary conditions, where the randomness is omitted for simplicity only, concern the entire series of solutions $\phi^{(j)}$ where $j=1, \ldots, N$ indexes all deterministic solutions necessary to recover response polynomial. The FDM discretization of the initial Equation (4.25) leads to the following difference statement:

$$eJ_{\omega}\phi\left(x_{i-2}\right) - \left(4eJ_{\omega} + GJ_s\left(\Delta x\right)^2\right)\phi\left(x_{i-1}\right) + \left(6eJ_{\omega} + 2GJ_s\left(\Delta x\right)^2 - \lambda\right)\phi\left(x_i\right) - \left(4eJ_{\omega} + GJ_s\left(\Delta x\right)^2\right)\phi\left(x_{i+1}\right) + eJ_{\omega}\phi\left(x_{i+2}\right) = 0 \tag{4.27}$$

where

$$\lambda_i = \frac{M_{cr,i}^2}{eJ_y}\left(\Delta x\right)^4 \tag{4.28}$$

and λ_i denote the additional eigenvalues for Equation (4.27). Additionally, the response function determination needs a series of solutions obtained from the recursive difference formula

$$e^{(j)}J_{\omega}^{(j)}\phi^{(j)}\left(x_{i-2}\right) - \left(4e^{(j)}J_{\omega}^{(j)} + G^{(j)}J_s^{(j)}\left(\Delta x^{(j)}\right)^2\right)\phi^{(j)}\left(x_{i-1}\right) + \left(6e^{(j)}J_{\omega}^{(j)} + 2G^{(j)}J_s^{(j)}\left(\Delta x^{(j)}\right)^2 - \lambda^{(j)}\right)\phi^{(j)}\left(x_i\right) - \left(4e^{(j)}J_{\omega}^{(j)} + G^{(j)}J_s^{(j)}\left(\Delta x^{(j)}\right)^2\right)\phi^{(j)}\left(x_{i+1}\right) + e^{(j)}J_{\omega}^{(j)}\phi^{(j)}\left(x_{i+2}\right) = 0 \tag{4.29}$$

where, similarly to Equation (4.28), one shows

$$\lambda_i^{(j)} = \frac{M_{cr,i}^{(j)}}{e^{(j)}J_y^{(j)}}\left(\Delta x^{(j)}\right)^4. \tag{4.30}$$

Of course, dependence of the basic dimension of a grid Δx on the actual solution indexed here with j is necessary if and only if some global geometrical parameters are randomized – as length for 1D models; otherwise, it remains constant throughout all numerical tests.

Further, the finite difference notation for various support types leads to the following conditions:

1. $\phi=0$ and $\dfrac{d\phi}{dx} = 0$: $\phi(x_i)=0$ and $\phi(x_{i+1})=-\phi(x_{i-1})$,
2. $\phi=0$ and $\dfrac{d^2\phi}{dx^2} = 0$: $\phi(x_i)=0$ and $\phi(x_{i+1})=\phi(x_{i-1})$, and
3. $\phi=0$ and $\dfrac{d\phi}{dx} = \dfrac{e}{k}\dfrac{d^2\phi}{dx^2}$: $\phi(x_i)=0$ and $\phi\left(x_{i+1}\right) = \frac{k\Delta x+2e}{k\Delta x-2e}\phi\left(x_{i-1}\right)$.

Solving for the eigenvalues λ_i from Equation (4.27), we find consecutively the critical moments after additional inversion of Equation (4.28).

Example 4.3: Critical moments for the random thin-walled elastic beam

A linear elastic thin-walled steel straight beam is analyzed with Young's modulus expectation $E[e] = 210.0\,\mathrm{GPa}$, Poisson ratio $\nu = 0.30$, expectation of the grid distribution $E[\Delta x] = 0.20$ corresponding to its equidistant subdivision with $n = 20$ [146]. Geometrical cross-sectional parameters are all deterministic and taken as $J_\omega = 0.000265 \times 10^{-6}\mathrm{m}^6$, $J_s = 0.0172 \times 10^{-6}\mathrm{m}^4$, $J_y = 0.122 \times 10^{-6}\mathrm{m}^4$. External boundary conditions imposed on this structure show, after Figure 4.15, that both edges are fully restrained. The RFM version of the perturbation method is used here with nine trial points in case of randomized Young's modulus equal to $e = [190, 195, 200, 205, 210, 215, 220, 225, 230]$ GPa, while $\Delta x = [0.18, 0.185, 0.19, 0.195, 0.20, 0.205, 0.21, 0.215, 0.22]$ m. A quadratic approximation function is used in the non-weighted least-squares algorithm here. Figure 4.15 shows for validation a comparison of the first three eigenmodes for the Maple™ implementation of the FDM (left-hand graph) with an analytical solution – both for an initial deterministic situation and, of course, practically no difference is noticed here. Further, we compare the expected values, coefficients of variation, skewness, and kurtosis for two independently randomized systems – with respect to Young's modulus (left-hand graphs) and overall length of the structural element (right-hand graphs). This is done to validate which parameter has more influence in terms of analogous input uncertainty level. The computational effort is very similar for both input parameters as they appear explicitly in the governing difference equation; this

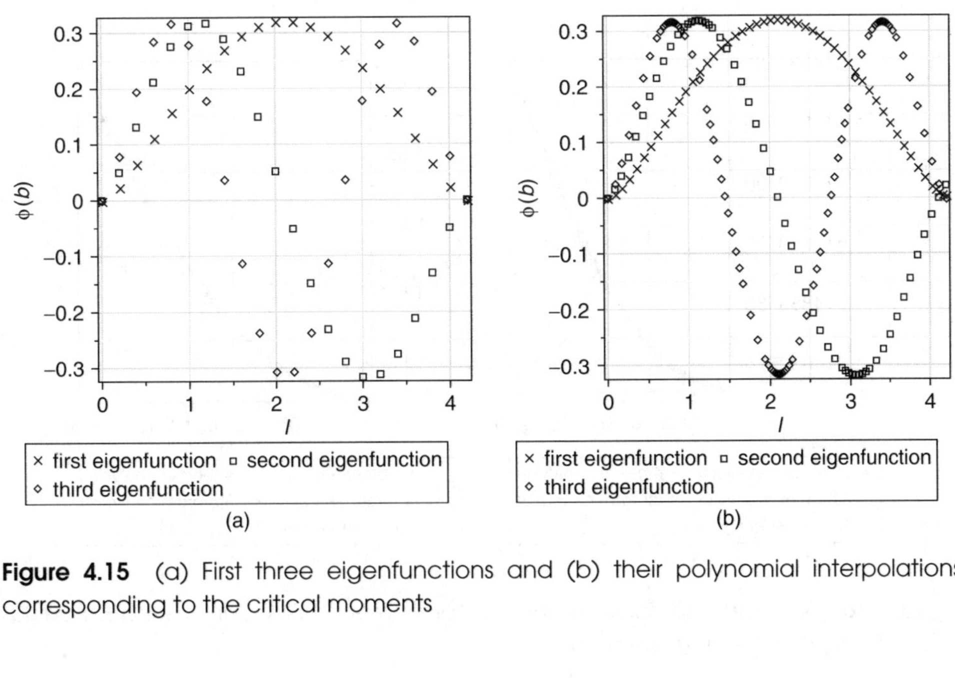

Figure 4.15 (a) First three eigenfunctions and (b) their polynomial interpolations corresponding to the critical moments

is not the case for the SFEM, where geometrical parameter randomization demands usually remesh in the RFM approach, while the Young's modulus of course does not.

The very important general result here is that most computed probabilistic coefficients are totally insensitive to the perturbation order, so that second-order analysis is quite satisfactory. The only exception is kurtosis, where sixth order guarantees the same results as higher-order analyses. Moreover, skewness and kurtosis very close to 0 for randomized elastic modulus enable us to conclude a Gaussian distribution of the resulting critical moment and this is definitely not the case for geometrical uncertainty in this problem. Generally, all probabilistic characteristics given in Figures 4.16–4.19 increase together with an increase of the input random dispersion given by the coefficient α (except for lower orders of kurtosis for the Young's modulus). It is characteristic that the expectation and kurtosis grow non-linearly according to convex curves, whereas the coefficients of variation and skewness increase linearly by only. Comparing carefully the expectations given in Figure 4.16, we see that randomization of the Young's modulus returns almost constant expectations of M_{cr}, while the mean values obtained for the random length change significantly for $\alpha \in [0.0, 0.20]$.

The output coefficients of variation show that uncertain length has much more influence on the critical moment than is obtained for Young's modulus, which coincides well with an engineering intuition. The resulting coefficients are about 50% larger than the input ones for the second random variable, while the uncertainty level is totally preserved during randomization of the elastic modulus. All the remaining characteristics are positive and have values measured in tenths for random length, which excludes a Gaussian PDF as the critical moment in this case. The very interesting result is obtained in Fig. 4.19a, where kurtosis of the critical moment differs from 0 at $\alpha(e) = 0$. It reflects a numerical error of the applied approximating method only,

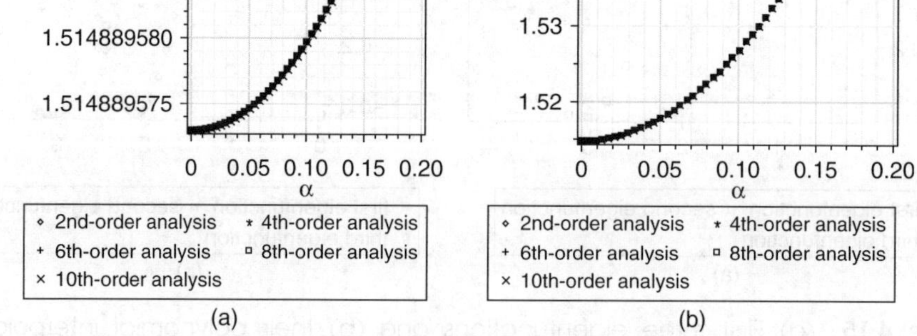

Figure 4.16 (a) Expected values of the critical moment for random Young's modulus and (b) length of the steel element

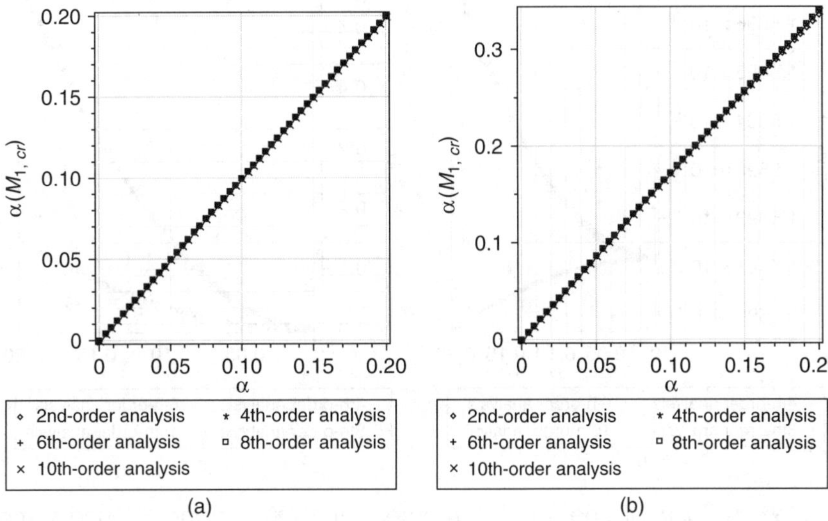

Figure 4.17 Coefficients of variation of the critical moment for (a) random Young's modulus and (b) length of the steel element

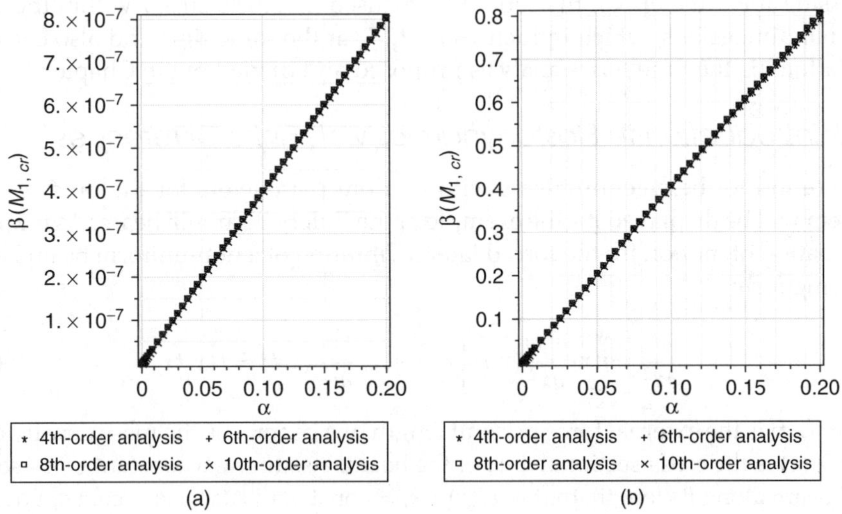

Figure 4.18 Skewness of the critical moment for (a) random Young's modulus and (b) length of the steel element

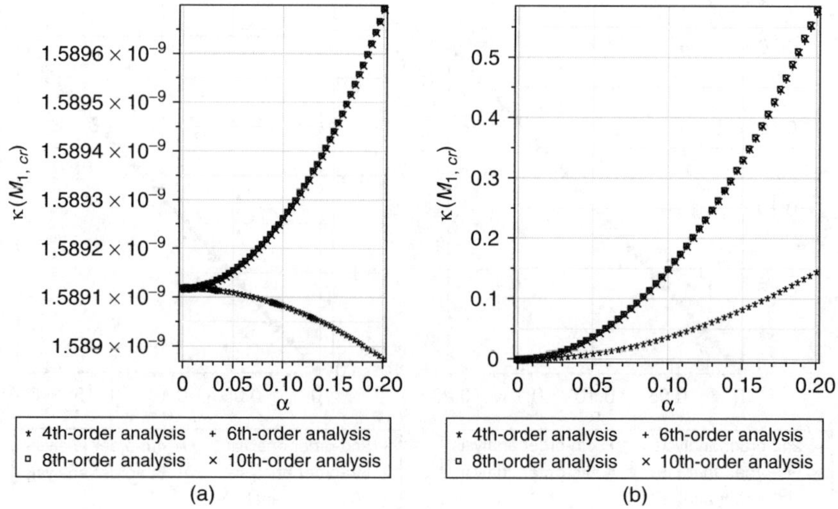

Figure 4.19 Kurtosis of the critical moment for (a) random Young's modulus and (b) length of the steel element

because perturbation based formulas return 0 in this case (cf. Equation (3.68), for instance). It needs to be mentioned also that further computational analysis in that area using the same apparatus should be focused on stochasticity within the cross-sectional dimensions, which influences J_ω, J_s, J_y at the same time and also corrosion simulation via the time series analysis proposed and carried out in Chapter 1.

4.1.3 Introduction to Elastodynamics with Finite Differences

The solution to the eigenproblems with random parameters for the unidirectional problem will be displayed for the example of the Euler–Bernoulli beam. As is known, the equation of motion for the forced lateral vibration of a non-uniform beam has the following form:

$$\frac{\partial^2}{\partial x^2}\left[eJ(x)\frac{\partial^2 w}{\partial x^2}(x,t)\right] + \rho A(x)\frac{\partial^2 w}{\partial t^2}(x,t) = f(x,t), \tag{4.31}$$

where $f(x,t)$ is the external force per unit length of a beam, ρ is the mass density of the beam, $A(x)$ is the cross-sectional area of the beam. Assuming a constant cross-section of the beam along its length, that is, $eJ(x) = eJ = \text{const.}$ and $A(x) = A = \text{const.}$, Equation (4.31) can therefore be rewritten as

$$eJ\frac{\partial^4 w}{\partial x^4}(x,t) + \rho A\frac{\partial^2 w}{\partial t^2}(x,t) = f(x,t) \tag{4.32}$$

while, for free vibration, it is obtained that

$$c^2\frac{\partial^4 w}{\partial x^4}(x,t) + \frac{\partial^2 w}{\partial t^2}(x,t) = 0 \tag{4.33}$$

where

$$c = \sqrt{\frac{eJ}{\rho A}}. \tag{4.34}$$

After traditional separation of space and time variables one gets

$$w(x,t) = W(x)T(t) \tag{4.35}$$

for the free vibrations, two equations are obtained:

$$\begin{cases} \frac{\partial^4 W(x)}{\partial x^4} - \beta^4 W(x) = 0, \\ \frac{\partial^2 T(t)}{\partial t^4} + \omega^2 T(t) = 0, \end{cases} \tag{4.36}$$

where

$$\beta^4 = \frac{\omega^2}{c^2} = \frac{\rho A \omega^2}{eJ}. \tag{4.37}$$

The characteristic function of the beam can be obtained as

$$W(x) = C_1 \exp(\beta x) + C_2 \exp(-\beta x) + C_3 \exp(i\beta x) + C_4 \exp(-i\beta x), \tag{4.38}$$

while ω being the natural frequency of vibration can be expressed as

$$\omega = \beta^2 \sqrt{\frac{EJ}{\rho A}}. \tag{4.39}$$

Applying the stochastic perturbation methodology, first of Equations (4.36) can be rewritten in the following form:

- zeroth-order equation

$$\frac{\partial^4 W^0(x)}{\partial x^4} - \left(\beta^0\right)^4 W^0(x) = 0 \tag{4.40}$$

- first-order equation

$$\frac{\partial^5 W^0(x)}{\partial b \partial x^4} - \frac{\partial\left(\beta^4\right)}{\partial b} W^0(x) - \left(\beta^0\right)^4 \frac{\partial W(x)}{\partial b} = 0 \tag{4.41}$$

- second-order equation

$$\frac{\partial^6 W^0(x)}{\partial b^2 \partial x^4} - \left\{ \frac{\partial^2\left(\beta^4\right)}{\partial b^2} W^0(x) + 2\frac{\partial\left(\beta^4\right)}{\partial b}\frac{\partial W(x)}{\partial b} + \left(\beta^0\right)^4 \frac{\partial^2 W(x)}{\partial b^2} \right\} = 0 \tag{4.42}$$

where the probabilistic derivatives of β^4 may be calculated analytically. Finally, we rewrite the recursive nth-order formula as

$$\frac{\partial^{n+4}W^0(x)}{\partial b^n \partial x^4} - \sum_{k=0}^{n} \binom{n}{k} \frac{\partial^{n-k}(\beta^4)}{\partial b^{n-k}} \frac{\partial^k W(x)}{\partial b^k} = 0. \tag{4.43}$$

In case of $\beta = \beta(b)$ we can calculate the partial derivatives of the deflection function $W(x)$ from Equation (4.38) as

$$\frac{\partial^n W(x)}{\partial b^n} = C_1 \frac{\partial^n}{\partial b^n}\{\exp(\beta x)\} + C_2 \frac{\partial^n}{\partial b^n}\{\exp(-\beta x)\} + C_3 \frac{\partial^n}{\partial b^n}\{\exp(\mathrm{i}\beta x)\} + C_4 \frac{\partial^n}{\partial b^n}\{\exp(-\mathrm{i}\beta x)\}. \tag{4.44}$$

Example 4.4: Eigenvibrations of the elastic beam with random parameters

A numerical illustration will be given here for the fixed/pinned beam, where the probabilistic moments of the free vibration frequencies are found. The analytical solution for natural frequencies and modes for the fixed/pinned beam can be obtained as

$$\omega_n^2 = (\beta_n l)^4 \left(\frac{eJ}{\rho A l^4}\right), \quad n = 1, 2, \ldots \tag{4.45}$$

If the Young's modulus of a beam is taken as a random variable, then the first two probabilistic moments of the natural frequencies square can be calculated using fundamental properties of random variables as

$$E\left[\omega_n^2\right] = \frac{J\beta_n^4}{\rho A} E[e], \quad n = 1, 2, \ldots \tag{4.46}$$

and

$$\mathrm{Var}\left(\omega_n^2\right) = \frac{J^2\beta_n^8}{\rho^2 A^2} \mathrm{Var}(e), \quad n = 1, 2, \ldots \tag{4.47}$$

Starting from these formulas, the expected values and variances of natural frequencies can be derived as

$$E[\omega] = \sqrt{E^2[\omega]} = \sqrt{E\left[\omega^2\right] - \mathrm{Var}(\omega)} \tag{4.48}$$

and

$$\mathrm{Var}\left(\omega^2\right) = 2\mathrm{Var}(\omega)\left(\mathrm{Var}(\omega) + 2E^2[\omega]\right). \tag{4.49}$$

Using a simplified stochastic second-order perturbation approach, the same probabilistic characteristics can be calculated as

$$E\left[\omega_n^2\right] = \left(\omega_n^2\right)^0 + \frac{1}{2}\frac{\partial^2\left(\omega_n^2\right)}{\partial b^2} \mathrm{Var}(e) = \frac{J\beta_n^4}{\rho A} E[e], \quad n = 1, 2, \ldots \tag{4.50}$$

as well as

$$\mathrm{Var}\left(\omega_n^2\right) = \left[\frac{\partial\left(\omega_n^2\right)}{\partial b}\right]^2 \mathrm{Var}\left(e\right) = \frac{J^2\beta_n^8}{\rho^2 A^2}\mathrm{Var}\left(e\right), \quad n = 1,2,\ldots \tag{4.51}$$

The simplicity of Equations (4.48) and (4.49) follows a linear character of the transform in-between structural output and input, but similar equations may be obtained for higher order polynomials also. These exact formulas are very important, because the Monte-Carlo simulation frequently employed in many comparative studies is only statistically convergent to the relevant moments being estimated.

As can be observed, the analytical solutions for probabilistic moments calculated from the definition and the second-order perturbation approach are exactly the same and that is why there is no need to apply the higher-order perturbation approach, which generally holds true for all outputs being linear functions of input random variables. For other type of relations between output frequencies and system random parameters, the expected values obtained from the second-order perturbation approach are higher than these derived explicitly from the definition. The simplicity of Equations (4.48) and (4.49) follows a linear character of the transform in-between structural output and the input, but similar equations may be obtained for higher order polynomials also. These exact formulas are very important, because the Monte Carlo simulation frequently employed in many comparative studies is only statistically convergent to the relevant moments being estimated.

Example 4.5: Longitudinal eigenvibrations for the elastic beam

This computational example deals with determination of the eigenfrequency squares for longitudinal vibrations of the linear elastic prismatic bar. It is carried out for the structural element having cross-sectional area $A = 113.0 \times 10^{-4}\mathrm{m}^2$ (profile HEA 340), mass density $\rho = 88.3\ \frac{kN}{m^3}$; it is partitioned into 100 equidistant intervals. This bar is simply supported at both ends and we randomized independently the Young's modulus with mean value $E[e] = 210\,\mathrm{GPa}$ and element length having expectation $E[l] = 10.0\,\mathrm{m}$ using the RFM approach with 11 trials computed for the following sets of discrete values: $e_i = [205, 206, 207, 208, 209, 210, 211, 212, 213, 214, 215]$ GPa and $l_i = [9.5, 9.6, 9.7, 9.8, 9.9, 10, 10.1, 10.2, 10.3, 10.4, 10.5]$ m; the second-order polynomial is used to recover the global response functions for the eigenfrequency squares. Probabilistic moments and coefficients are collected for the first eight eigenfrequency squares – the left-hand graphs correspond to Young's modulus uncertainty, while randomization of the element length results in the right-hand graphs of Figures 4.20–4.24 – we compare here expectations (Figure 4.20), standard deviations (Figure 4.21), coefficients of variation (Figure 4.22), skewness (Figure 4.23), as well as kurtosis (Figure 4.24).

The general observation is that, as in Example 4.3, all resulting eigenfrequency squares are distributed according to the Gaussian PDF for randomized elastic modulus, which follows negligibly small skewness and kurtosis; this is not the case for random length, where these characteristics are positive, measured in units, and significantly increasing together with an increase of the input coefficient α. Further, this increase is noticed for almost all resulting characteristics except some

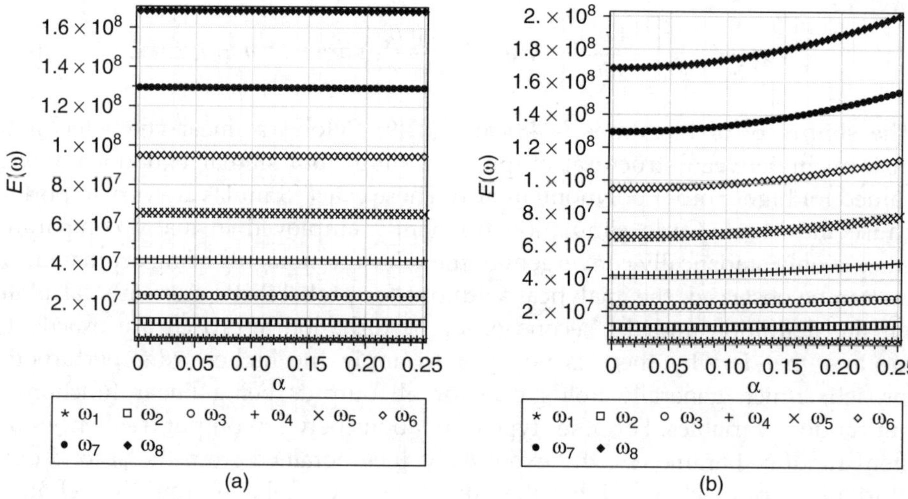

Figure 4.20 Expected values of the eigenfrequencies for (a) random Young's modulus and (b) the bar length

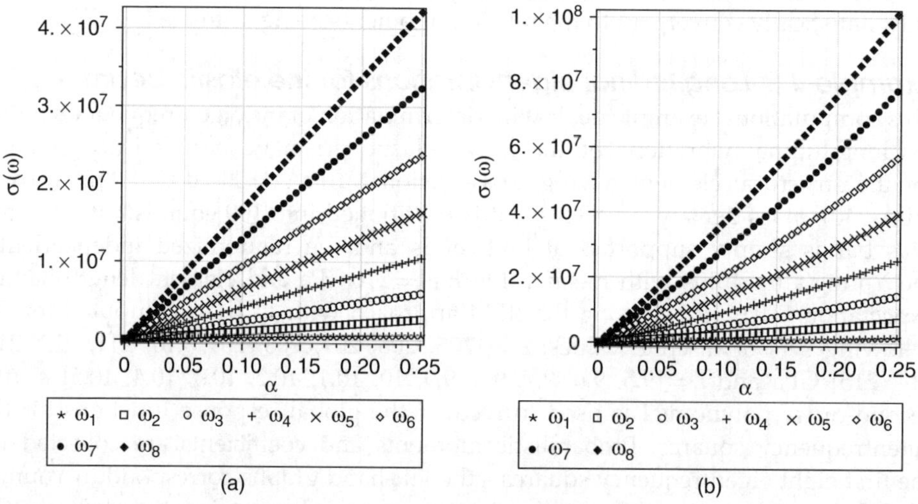

Figure 4.21 Standard deviations of the eigenfrequencies for (a) random Young's modulus and (b) the bar length

eigenfrequency skewness negative distributions (Figure 4.22(a)) as well as a complete insensitivity of kurtosis upon input coefficient of variation. Small values of these two parameters result from the RFM approximation error and may be simply postponed.

Once more the output randomness level is exactly the same as the input during randomization of the Young's modulus and amplified twice while the element length is treated as uncertain; both coefficients are linear w.r.t. the input one, however, the

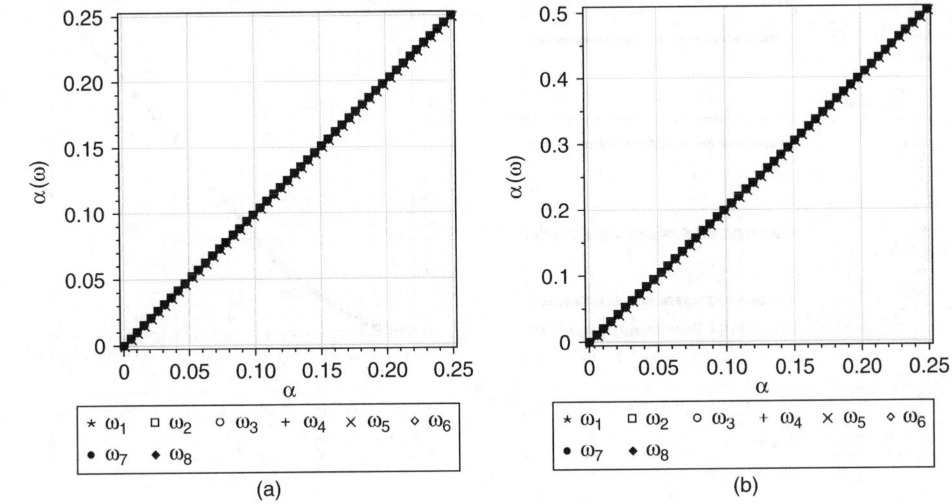

Figure 4.22 Coefficients of variation of the eigenfrequencies for (a) random Young's modulus and (b) the bar length

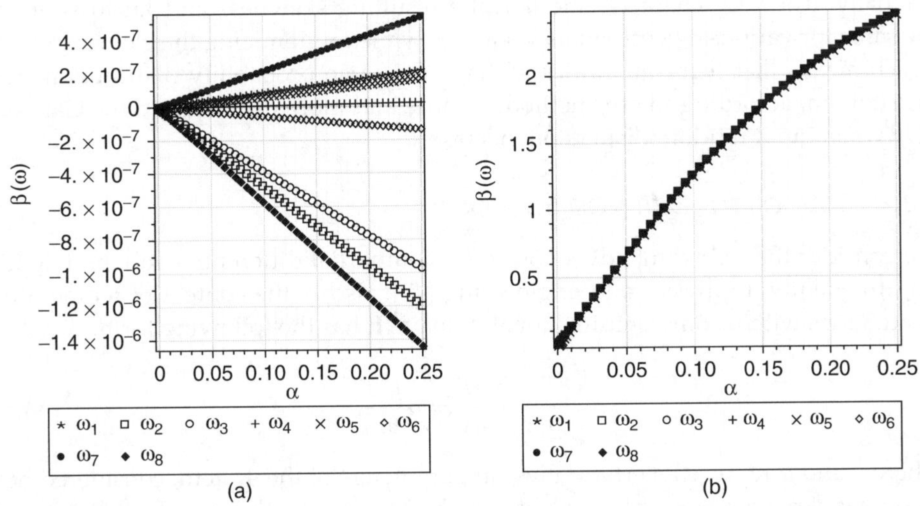

Figure 4.23 Skewness of the eigenfrequencies for (a) random Young's modulus and (b) the bar length

expectations behavior is much more complex. The mean values computed for random elastic modulus are constant, where of course the higher the eigenfrequency square, the larger the expectation. Analogous expected values from the right-hand graph increase moderately with increasing input α, where the higher the eigenvalue, the larger the increase; this interrelation is convex everywhere without any singularities and discontinuities. The last eigenfrequency square changes by about 25% from minimum $\alpha = 0$ to maximum, so that these variations cannot be neglected.

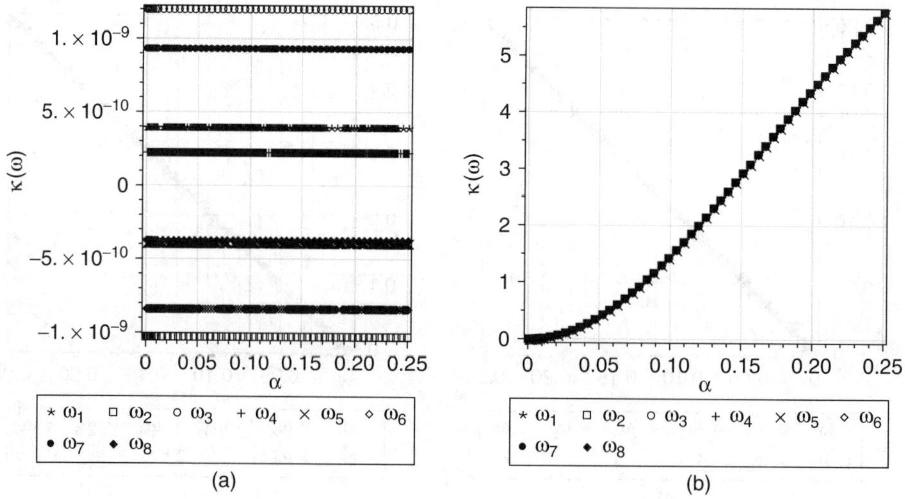

Figure 4.24 Kurtosis of the eigenfrequencies for (a) random Young's modulus and (b) the bar length

Finally, it is very characteristic that the resulting skewness and kurtosis are all the same for various eigenfrequency squares while randomizing the element length, which means that these quantities all have the same probability distribution. This distribution, according to our method, is more and more distant from the Gaussian PDF when input random dispersion increases.

4.1.4 Advection-Diffusion Equation

We consider the following advection–diffusion partial differential equation appearing frequently in physics and engineering [47], also in the context of temperature fluctuations within some unidirectional domain. It has the following form:

$$\frac{\partial T}{\partial t} + u\frac{\partial T}{\partial x} = \alpha\frac{\partial^2 T}{\partial x^2}, \tag{4.52}$$

where u and α represent certain physical parameters of the system, considered here as state independent ($\alpha = k/c\rho$, where k is the heat conductivity coefficient, c stands for the heat capacity, and ρ denotes mass density). This specific equation has been chosen as a combination of the parabolic and hyperbolic partial differential equation and the forward-time and central-space (FTCS) method is engaged to provide FDM discretization of this equation. It holds for the ith point of the grid that

$$\frac{T_{[i]}^{n+1} - T_{[i]}^{n}}{\Delta_t} = \alpha\frac{T_{[i+1]}^{n} - 2T_{[i]}^{n} + T_{[i-1]}^{n}}{\Delta_x^2} - u\frac{T_{[i+1]}^{n} - T_{[i-1]}^{n}}{2\Delta_x}, \tag{4.53}$$

and with $\alpha = 0$ we obtain the classical difference notation of the parabolic equation for unidirectional heat transfer. Upper indices denote here the given discrete time moment marked with n, while $n+1$ is the next one to be determined. Usually, it is rewritten as

$$T_{[i]}^{n+1} = T_{[i]}^{n} + \frac{\alpha \Delta_t}{\Delta_x^2}\left(T_{[i+1]}^{n} - 2T_{[i]}^{n} + T_{[i-1]}^{n}\right) - \frac{u\Delta_t}{2\Delta_x}\left(T_{[i+1]}^{n} - T_{[i-1]}^{n}\right). \tag{4.54}$$

The perturbation-based stochastic counterpart in the DDM implementation consists of an interrelation between partial derivatives computed at neighboring time moments and has the following general form convenient for the nth order:

$$\begin{aligned}\frac{\partial^m T_{[i]}^{n+1}}{\partial b^m} = \frac{\partial^m T_{[i]}^{n}}{\partial b^m} &+ \frac{\Delta_t}{\Delta_x^2}\sum_{k=0}^{m}\binom{m}{k}\frac{\partial^k \alpha}{\partial b^k}\left(\frac{\partial^{m-k} T_{[i+1]}^{n}}{\partial b^{m-k}} - 2\frac{\partial^{m-k} T_{[i]}^{n}}{\partial b^{m-k}} + \frac{\partial^{m-k} T_{[i-1]}^{n}}{\partial b^{m-k}}\right)\\ &- \frac{\Delta_t}{2\Delta_x}\sum_{k=0}^{m}\binom{m}{k}\frac{\partial^k u}{\partial b^k}\left(\frac{\partial^{m-k} T_{[i+1]}^{n}}{\partial b^{m-k}} - \frac{\partial^{m-k} T_{[i-1]}^{n}}{\partial b^{m-k}}\right).\end{aligned} \tag{4.55}$$

So, having partial derivatives for the given time moment we are able to compute them all after the next step and compose them into the final equations for the probabilistic moments. Since we apply more frequently the RFM version of the SFDM, more useful would be the following statement:

$$T_{[i]}^{n+1(k)} = T_{[i]}^{n(k)} + \frac{\alpha^{(k)}\Delta_t}{\Delta_x^2}\left(T_{[i+1]}^{n(k)} - 2T_{[i]}^{n(k)} + T_{[i-1]}^{n(k)}\right) - \frac{u^{(k)}\Delta_t}{2\Delta_x}\left(T_{[i+1]}^{n(k)} - T_{[i-1]}^{n(k)}\right), \tag{4.56}$$

where $k=1, \ldots, M$ indexes deterministic tests with varying input physical parameter to be treated as random. We never treat Δ_t as random here because time parameter in all scientific problems is always deterministic.

Example 4.6: Transient heat transfer in homogeneous rod

This computational illustration deals with the structure having length $l=1.0$, subject to the following boundary conditions: $T(0,t)=0$ and $T(l,t)=100$, while the initial conditions are adopted as $T(x,0) = \frac{100x}{L}$ for $x \in [0,l]$. The constants appearing in the differential equation are taken as $u=0.10$, $k=43$, $c=490$, and $d=7850$. The coefficient of heat conduction is chosen here as the input Gaussian random variable and numerical analysis based on the RFM 11-trials cycle is carried out twice – for $\alpha(k)=0.10$ (left-hand graphs of Figures 4.25–4.29) and for $\alpha(k)=0.30$ (right-hand graphs of Figures 4.25–4.29). We present basic probabilistic characteristics of the temperature evolution in this structure as functions of the spatial coordinate for the few time moments $\tau=[0.0, 0.5, 2.5, 5.0, 10.0, 50.0]$ s. Therefore, it is impossible to include here continuous variations with respect to $\alpha(k)$ as in the previous tests and, further, to present the few surfaces corresponding to various perturbation analysis order to maintain brevity of presentation. Time-dependent local response functions (at each point of the grid) have been recovered using second-order polynomials inserted into the unweighted least-squares algorithm. We derive expectations for temperature histories using tenth-order perturbation, the variances, third and fourth

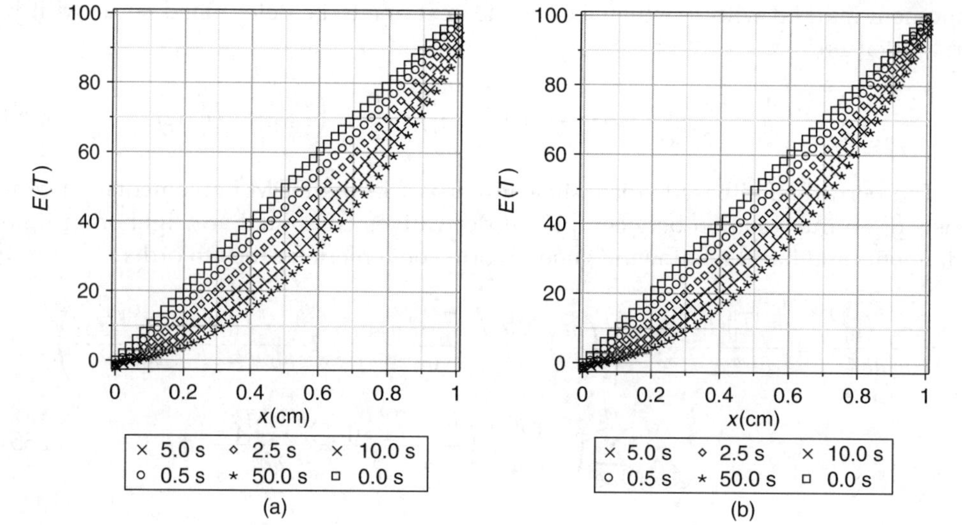

Figure 4.25 Temperature expected values for the advection-diffusion problem for (a) $\alpha = 0.1$ and (b) $\alpha = 0.30$

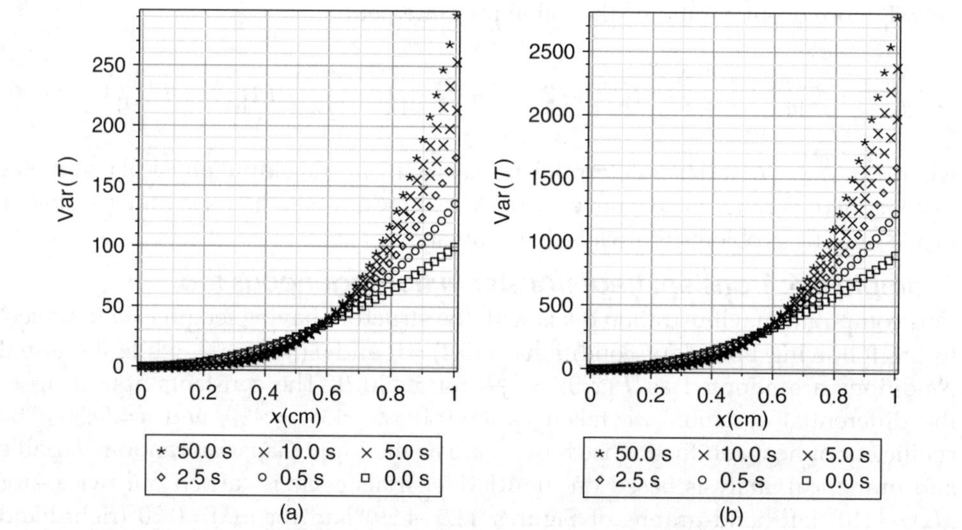

Figure 4.26 Temperature variances for the advection–diffusion problem for (a) $\alpha = 0.1$ and (b) $\alpha = 0.30$

central probabilistic moments according to the sixth-order approximations. As one may recognize from Figure 4.25, the boundary conditions applied at the left edge are matched in both cases very well by the expected value, but the temperature expectations at the right edge are slightly below the value $T(l,t) = 100$. Therefore, we need to remember that the boundary condition is imposed on $T^0(x,t)$, only, which may have different values from $E[T(x,t)]$, especially when the resulting

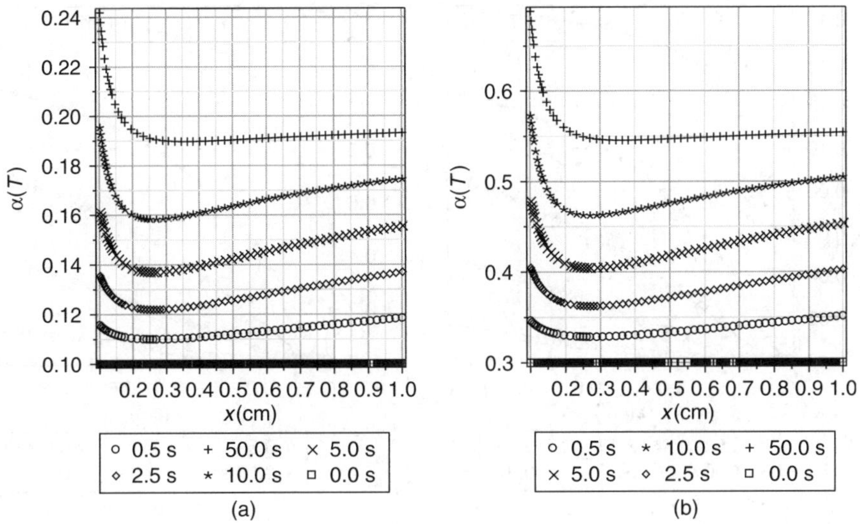

Figure 4.27 Temperature coefficients of variation for the advection–diffusion problem for (a) $\alpha = 0.1$ and (b) $\alpha = 0.30$

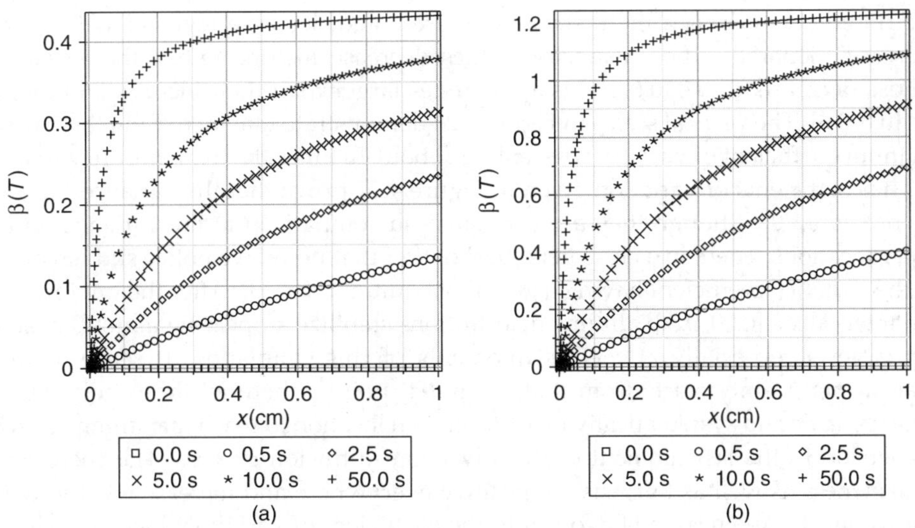

Figure 4.28 Temperature skewness for the advection–diffusion problem for (a) $\alpha = 0.1$ and (b) $\alpha = 0.30$

distribution is not Gaussian; it can be verified directly in Figures 4.28 and 4.29 for $x = l$. Comparison of both expectation spatial distributions shows moreover that for $\alpha(k) = 0.30$ the temperature mean values are apparently closer to the right edge boundary condition. In both cases we notice that the numerical method is stable as

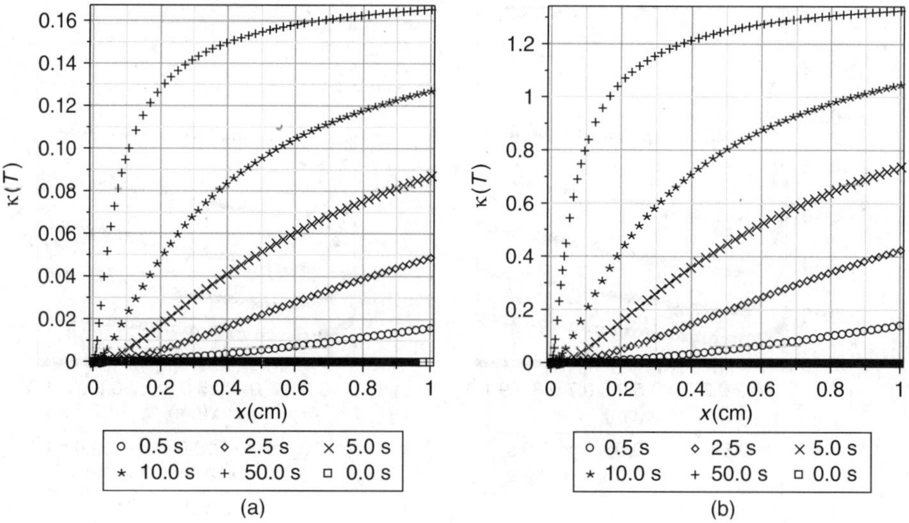

Figure 4.29 Temperature kurtosis for the advection–diffusion problem for (a) $\alpha = 0.1$ and (b) $\alpha = 0.30$

the resulting temperature expectations have no singularities in space and time and the differences between the distributions presented decrease together with a time increase. Generally, these expected values decrease in time unlike the variances, whose decrease for $x \in [0.6, 1.0]$ is a few times larger than their increase in time for $x \in [0.0, 0.6]$. The vertical scales in Figure 4.26 demonstrate that increasing three times an input coefficient of variation, we enlarge about 10 times the temperature variance.

The coefficients of variation given in Figure 4.27 prove that the variances at $x = 0$ do not equal 0, although they are very small in comparison to the variances at the other locations. Then, in a close neighborhood of that point, we notice sudden decay of the variance coefficient having a global maximum for $x = 0$. After the minimum in the interval $x \in [0.20, 0.30]$ all time distributions stabilize, appearing almost constant for larger x, especially at the last moments of this simulation. It seems natural that an uncertainty starts from 0 at the initial point, where all the conditions are deterministic and systematically increase in each location to reach maximum random dispersion in the last moment of the advection–diffusion process. The coefficients of skewness as well as kurtosis are positive everywhere and increase together with both x and t. Skewness, analogously to the coefficient of variation, increases almost three times while changing $\alpha(k)$ from 0.10 to 0.30, but the kurtosis variation is decisively larger and almost nine times higher. The final probability distributions of $T(x,t)$ after Figures 4.28 and 4.29 are not Gaussian and the larger $\alpha(k)$, the greater the distance to the combination of these parameters typical to this particular PDF type. It would be interesting indeed to extend this analysis toward the problem with temperature-dependent and random physical properties of the system as well as to use this methodology to provide stochastic reliability analysis in terms of critical temperatures for many of the engineering systems.

4.2 Analysis of Boundary Value Problems on 2D Grids

4.2.1 Poisson Equation

The Poisson equation gives an illustration of the solution of an elliptic partial differential equation with random parameters on the 2D regular grid. We consider the following problem:

$$\frac{\partial^2 u(x_1, x_2)}{\partial x_1^2} + \frac{\partial^2 u(x_1, x_2)}{\partial x_2^2} = -q(x_1, x_2), \tag{4.57}$$

which may represent a steady-state temperature distribution in the isotropic body, steady-state invisible fluid in the bounded domains (with sources), electrostatic field potential within a closed domain with charges, Newtonian gravity field potential, torsion of a prismatic elastic beam, or static deformation of a membrane under pressure and tension. The rectangular net is provided for this purpose with two different lower indices i and j numbering the given point in the directions x_1 and x_2, independently. The basic smallest cell of this grid is uniquely defined by the distances Δ_1 and Δ_2 with respect to the given coordinates' axes. Then, we discretize all necessary partial derivatives in the above equation using central finite differences and values of the function $u(x_1, x_2)$ to be determined in the closest neighborhood of the point defined uniquely by the indices $[i,j]$. It holds that

$$\frac{\partial u_{[i,j]}}{\partial x_1} = \frac{u_{[i+1,j]} - u_{[i-1,j]}}{2\Delta_1} + O\left(\Delta_1^2\right), \tag{4.58}$$

$$\frac{\partial u_{[i,j]}}{\partial x_2} = \frac{u_{[i,j+1]} - u_{[i,j-1]}}{2\Delta_2} + O\left(\Delta_2^2\right) \tag{4.59}$$

and, analogously, for the second-order partial derivatives

$$\frac{\partial^2 u_{[i,j]}}{\partial x_1^2} = \frac{u_{[i+1,j]} - 2u_{[i,j]} + u_{[i-1,j]}}{\Delta_1^2} + O\left(\Delta_1^3\right), \tag{4.60}$$

$$\frac{\partial^2 u_{[i,j]}}{\partial x_2^2} = \frac{u_{[i,j+1]} - 2u_{[i,j]} + u_{[i,j-1]}}{\Delta_2^2} + O\left(\Delta_2^3\right) \tag{4.61}$$

with mixed partial derivative of second order given as

$$\frac{\partial^2 u_{[i,j]}}{\partial x_1 \partial x_2} = \frac{u_{[i+1,j+1]} + u_{[i-1,j-1]} - u_{[i-1,j+1]} - u_{[i+1,j-1]}}{4\Delta_1\Delta_2} + O\left(\Delta_1^3\right). \tag{4.62}$$

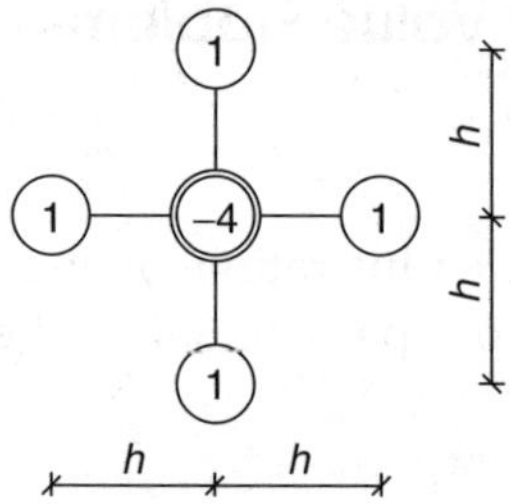

Figure 4.30 Finite difference method Laplace operator in a graphical form

An application of these formulas in the Poisson equation yields (postponing higher-order rests)

$$\frac{u_{[i+1,j]}-2u_{[i,j]}+u_{[i-1,j]}}{\Delta_1^2}+\frac{u_{[i,j+1]}-2u_{[i,j]}+u_{[i,j-1]}}{\Delta_2^2}+O\left(\Delta_1^3,\Delta_2^3\right)=-q_{[i,j]} \quad (4.63)$$

which, with net equidistant in both directions, that is, $\Delta_1=\Delta_2=a$ and by neglecting higher-order rests returns

$$4u_{[i,j]}-\left\{u_{[i+1,j]}+u_{[i,j+1]}+u_{[i-1,j]}+u_{[i,j-1]}\right\}=q_{[i,j]}a^2. \quad (4.64)$$

A graphical representation of this difference formula is given in Figure 4.30.

When the response function recovery is involved in a solution of this equation in the presence of some uncertainty sources, we solve for the deterministic series indexed with upper index m and we have

$$4u_{[i,j]}^{(m)}-\left\{u_{[i+1,j]}^{(m)}+u_{[i,j+1]}^{(m)}+u_{[i-1,j]}^{(m)}+u_{i,j-1}^{(m)}\right\}=q_{[i,j]}^{(m)}\left(a^{(m)}\right)^2. \quad (4.65)$$

Much faster is the DDM scheme, where the recursive formula for the nth-order perturbation-based solution equals

$$4\frac{\partial^n u_{[i,j]}}{\partial b^n}-\left\{\frac{\partial^n u_{[i+1,j]}}{\partial b^n}+\frac{\partial^n u_{[i,j+1]}}{\partial b^n}+\frac{\partial^n u_{[i-1,j]}}{\partial b^n}+\frac{\partial^n u_{[i,j-1]}}{\partial b^n}\right\}$$
$$=\sum_{k=0}^{n}\binom{n}{k}\frac{\partial^{n-k}q_{[i,j]}}{\partial b^{n-k}}\frac{\partial^k\left(a^2\right)}{\partial b^k}. \quad (4.66)$$

Example 4.7: Solution to the Poisson equation on a rectangular domain

This computational illustration is devoted to the solution of the Poisson equation on two different grids – 10 × 10 and 20 × 20 – having randomized basic grid dimension $a=\Delta_x=\Delta_y$. Effectively, considering fictitious nodes in the grids necessary to define given boundary conditions, these are 12×12 and 22×22 uniform quadratic nets. The global membrane dimensions equal $10\,\text{m} \times 10\,\text{m}$; it is simply supported along its external edges and the uniform unitary pressure is applied on the plate. Since deterministic values and expectations for the coarser grid appear to be larger, all probabilistic moments and coefficients in Figures 4.31–4.34 for the needs of reliability analysis are restricted to this finer grid. The RFM analysis is based on nine numerical FDM experiments in Maple™ provided that $a = [0.42, 0.44, 0.46, 0.48, 0.50, 0.52, 0.54, 0.56, 0.58]$ m with quadratic interrelation between the resulting midspan deflection and the parameter a.

All the resulting moments and coefficients presented in Figures 4.32–4.34 are given as functions of the input coefficient of variation $\alpha(a) = [0.05, 0.10, 0.15, 0.20, 0.25]$ and for the cross-section parallel to the given edges provided at the midpoint. This arrangement is proposed instead of the surface plot in Figure 4.31 to keep better resolution and for brevity of the final spatial distributions to the varying $\alpha(a)$. A distribution of the expected values is in its shape very similar to the deterministic origin-preserving zero-boundary conditions at both ends and exhibiting a maximum at the midpoint. This expectation is moderately sensitive to the input coefficient of variation, where of course the larger $\alpha(a)$, the greater the maximum deflection expectation. Spatial distributions of variances and standard deviations are similar in the context of shape and extremum locations to expectations, however now the influence of the input $\alpha(a)$ is decisive. The variances grow faster than the input coefficient of variation, while the standard deviations increase proportionally to this coefficient, which perfectly agrees with theoretical considerations. This non-proportional increase is much more apparent for higher central probabilistic

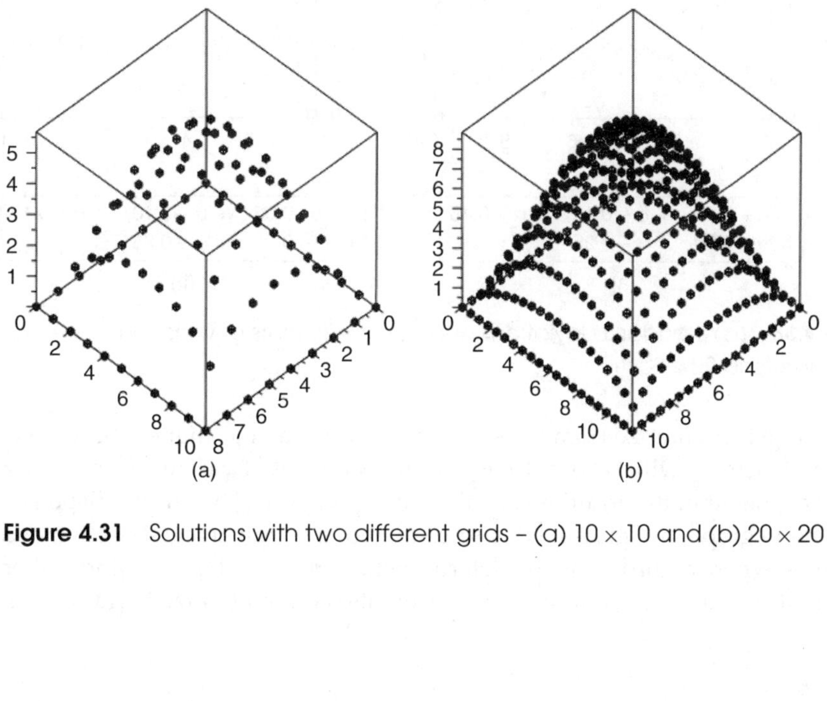

Figure 4.31 Solutions with two different grids – (a) 10×10 and (b) 20×20

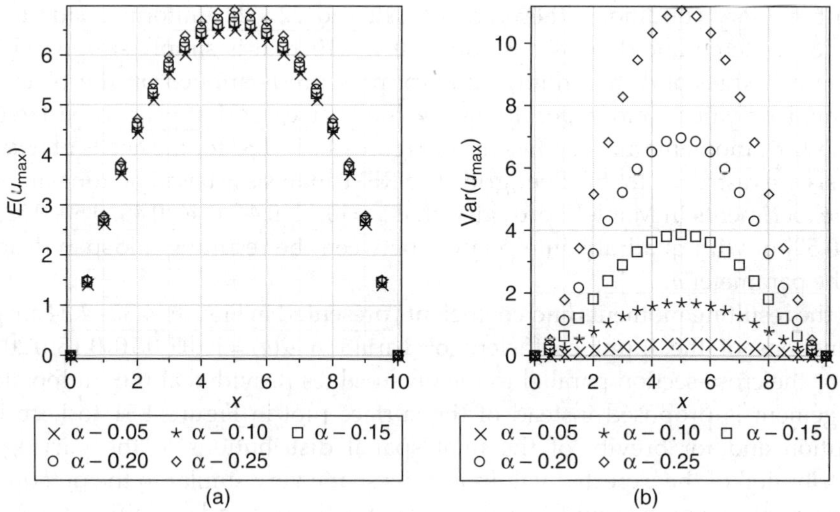

Figure 4.32 (a) Expected values and (b) variances for coarser grid 20 × 20

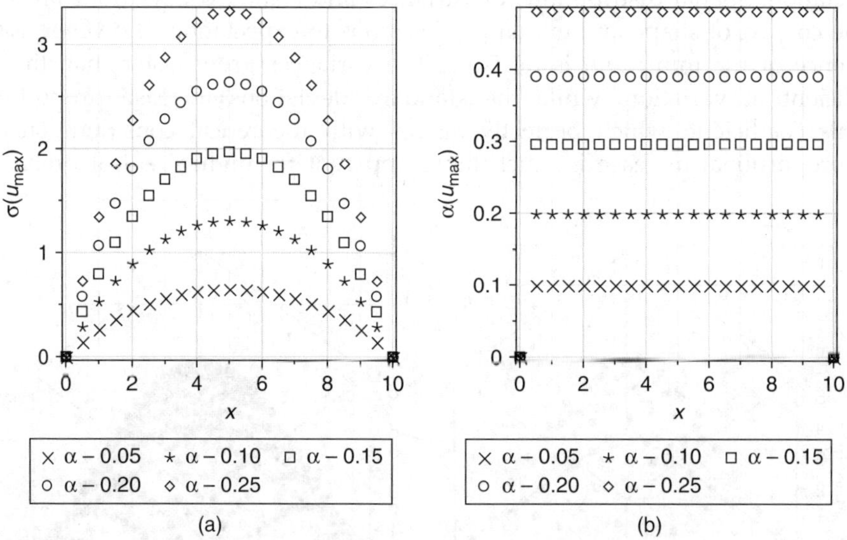

Figure 4.33 (a) Standard deviations and (b) coefficients of variation for coarser grid 20 × 20

moments (cf. Figure 4.34) and we can conclude that the higher the moment, the faster the increase with respect to the input coefficient of variation (according to the higher exponent in the additional analytical expression). The output dispersion level (right-hand graph of Figure 4.33) is constant along the given cross-section except for the edges, where according to the deterministic zero-boundary conditions there is no random fluctuation. After previous experiments with randomized grid size we notice

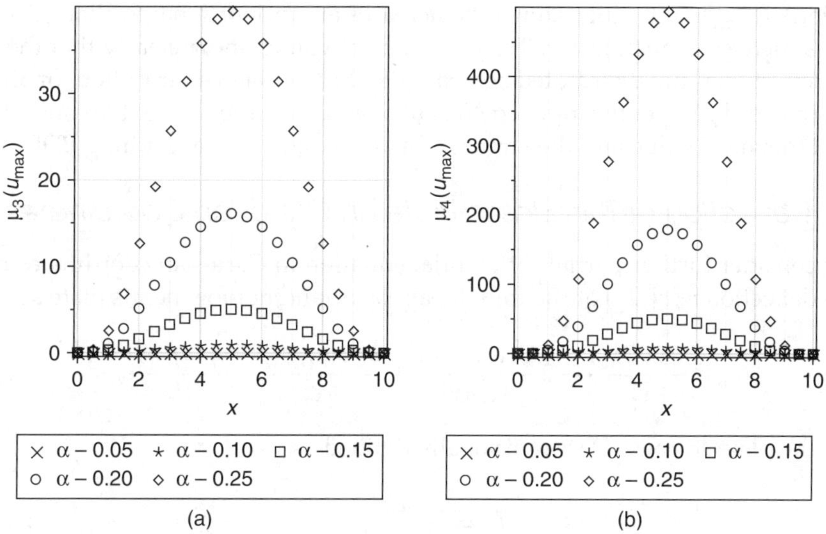

Figure 4.34 (a) Third and (b) fourth central probabilistic moments for coarser grid 20 × 20

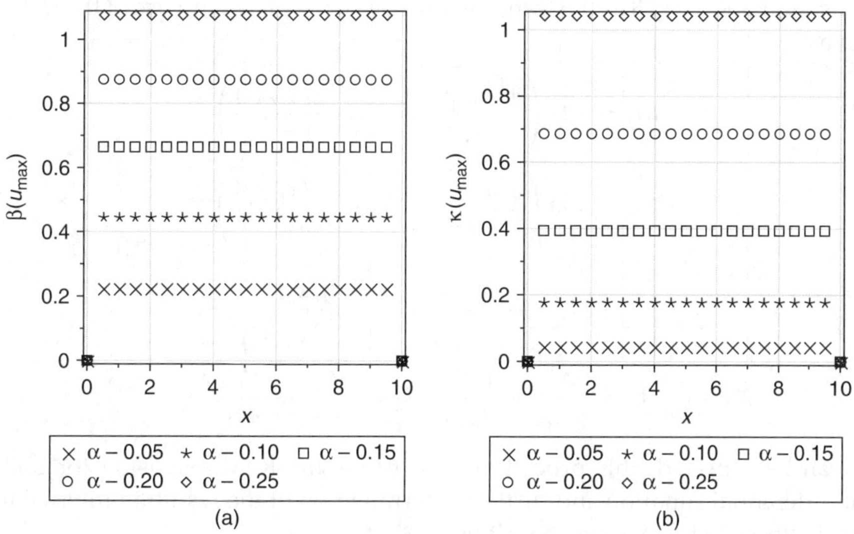

Figure 4.35 (a) Skewness and (b) kurtosis for coarser grid 20 × 20

that the ratio of output to input coefficients of variation equals almost 2; this ratio holds true for all $\alpha(a)$ and means that the perturbation-based expansions cannot be reduced to lower orders only and must be given using at least tenth-order expressions; the second-order technique in particular could be quite inefficient here. Coefficients of skewness and kurtosis have very similar spatial variations to the coefficient of variation $\alpha(u_{max})$, but while skewness increases are proportional to changes of $\alpha(a)$, kurtosis increases much faster. These coefficient values show clearly that the maximum deflection is not distributed according to the Gaussian distribution for any $\alpha(a)$. Furthermore, the larger the input coefficient of variation, the greater the concentration around the mean value and the larger right asymmetry of the resulting PDF.

4.2.2 Deflection of Thin Elastic Plates in Cartesian Coordinates

Let us consider further partial differential equation in Cartesian coordinates describing a deflection $w(x_1, x_2)$ of the thin linear elastic and homogeneous plate as

$$\frac{\partial^4 w(x_1,x_2)}{\partial x_1^4} + 2\frac{\partial^4 w(x_1,x_2)}{\partial x_1^2 \partial x_2^2} + \frac{\partial^4 w(x_1,x_2)}{\partial x_2^4} = \frac{q\,(x_1,x_2)}{D}, \tag{4.67}$$

where the plate stiffness D is traditionally defined as

$$D = \frac{eh^3}{12\,(1-\nu^2)}, \tag{4.68}$$

and e, h, and ν stand for the Young's modulus, thickness, and Poisson ratio of this structure. Having determined this deflection we derive the bending moments, twisting moment, as well as the transverse forces in the plate with use of the following formulas:

$$M_1 = -D\left(\frac{\partial^2 w(x_1,x_2)}{\partial x_1^2} + \nu\frac{\partial^2 w(x_1,x_2)}{\partial x_2^2}\right), \tag{4.69}$$

$$M_2 = -D\left(\frac{\partial^2 w(x_1,x_2)}{\partial x_2^2} + \nu\frac{\partial^2 w(x_1,x_2)}{\partial x_1^2}\right), \tag{4.70}$$

$$M_s = -\,(1-\nu)\,D\frac{\partial^2 w(x_1,x_2)}{\partial x_1 \partial x_2}, \tag{4.71}$$

$$T_1 = \frac{\partial M_1}{\partial x_1} + \frac{\partial M_s}{\partial x_2}, \quad T_2 = \frac{\partial M_2}{\partial x_2} + \frac{\partial M_s}{\partial x_1}. \tag{4.72}$$

As can be expected, this procedure also needs the RFM extension for both the initial differential equation and further determination of the resulting internal forces indexed with m in further analysis. It holds that

$$\frac{\partial^4 w^{(m)}(x_1,x_2)}{\partial x_1^4} + 2\frac{\partial^4 w^{(m)}(x_1,x_2)}{\partial x_1^2 \partial x_2^2} + \frac{\partial^4 w^{(m)}(x_1,x_2)}{\partial x_2^4} = \frac{q^{(m)}\,(x_1,x_2)}{D^{(m)}}, \tag{4.73}$$

as we also look for the series of solutions with varying design input random parameter. Hence

$$D^{(m)} = \frac{e^{(m)}\left(h^{(m)}\right)^3}{12\left(1-\left(\nu^{(m)}\right)^2\right)} \tag{4.74}$$

and, finally:

$$M_1^{(m)} = -D^{(m)} \left(\frac{\partial^2 w^{(m)}(x_1, x_2)}{\partial x_1^2} + \nu^{(m)} \frac{\partial^2 w^{(m)}(x_1, x_2)}{\partial x_2^2} \right),$$

$$M_2^{(m)} = -D^{(m)} \left(\frac{\partial^2 w^{(m)}(x_1, x_2)}{\partial x_2^2} + \nu^{(m)} \frac{\partial^2 w^{(m)}(x_1, x_2)}{\partial x_1^2} \right), \tag{4.75}$$

$$M_s^{(m)} = -\left(1 - \nu^{(\alpha)}\right) D^{(m)} \frac{\partial^2 w^{(m)}(x_1, x_2)}{\partial x_1 \partial x_2}, \tag{4.76}$$

$$T_1^{(m)} = \frac{\partial M_1^{(m)}}{\partial x_1} + \frac{\partial M_s^{(m)}}{\partial x_2}, \quad T_2^{(m)} = \frac{\partial M_2^{(m)}}{\partial x_2} + \frac{\partial M_s^{(m)}}{\partial x_1}. \tag{4.77}$$

We consider three types of boundary conditions: (i) completely free boundary, (ii) simply supported edge, and (iii) fully restrained boundary. So, considering the straight boundary parallel to the given x_2 axis, one may introduce a zero internal forces condition in case (i) as

$$M_1^{(m)} = -D^{(m)} \left(\frac{\partial^2 w^{(m)}(x_1, x_2)}{\partial x_1^2} + \nu^{(m)} \frac{\partial^2 w^{(m)}(x_1, x_2)}{\partial x_2^2} \right) = 0, \tag{4.78}$$

$$T_1^{(m)} + \frac{\partial M_s^{(m)}}{\partial x_1} = \frac{\partial M_1^{(m)}}{\partial x_1} + \frac{\partial M_s^{(m)}}{\partial x_2} + \frac{\partial M_s^{(m)}}{\partial x_1} = 0. \tag{4.79}$$

The boundary conditions relevant to case (ii) are given as

$$w^{(m)} = 0, \quad M_1^{(m)} = -D^{(m)} \left(\frac{\partial^2 w^{(m)}(x_1, x_2)}{\partial x_1^2} + \nu^{(m)} \frac{\partial^2 w^{(m)}(x_1, x_2)}{\partial x_2^2} \right) = 0, \tag{4.80}$$

whereas case (iii) uses

$$w^{(m)} = 0, \quad \frac{\partial w^{(m)}}{\partial x_1} \equiv \varphi_1 = 0. \tag{4.81}$$

For the needs of the FDM we employ a rectangular grid with basic axial distances Δx_1, Δx_2 looking for the set of discrete deflections denoted for a point indexed with $[i,j]$ as $w_{[i,j]}$. So, the following holds for the distributed load concentrated in the point $[i,j]$ denoted as $q_{[i,j]}$:

$$\frac{w_{[i-2,j]} - 4w_{[i-1,j]} + 6w_{[i,j]} - 4w_{[i+1,j]} + w_{[i+2,j]}}{\Delta x_1^4}$$
$$- \frac{4w_{[i,j-1]} + 4w_{[i-1,j]} - 8w_{[i,j]} + 4w_{[i+1,j]} + 4w_{[i,j+1]}}{\Delta x_1^2 \Delta x_2^2}$$

$$
+\frac{2w_{[i-1,j-1]}+2w_{[i-1,j+1]}+2w_{[i+1,j+1]}+2w_{[i+1,j-1]}}{\Delta x_1^2\Delta x_2^2}
+\frac{w_{[i-2,j]}-4w_{[i-1,j]}+6w_{[i,j]}-4w_{[i+1,j]}+w_{[i+2,j]}}{\Delta x_2^4}=\frac{q_{[i,j]}}{D}. \tag{4.82}
$$

When the grid is equidistant in both directions, that is, $\Delta x_1=\Delta x_2=a$, one obtains

$$
\begin{aligned}
&20w_{[i,j]}-8\left(w_{[i-1,j]}+w_{[i,j-1]}+w_{[i+1,j]}+w_{[i,j+1]}\right)\\
&\quad+2\left(w_{[i-1,j-1]}+w_{[i-1,j+1]}+w_{[i+1,j+1]}+w_{[i+1,j-1]}\right)\\
&\quad+w_{[i-2,j]}+w_{[i+2,j]}+w_{[i,j-2]}+w_{[i,j+2]}=\frac{q_{[i,j]}}{D}a^4.
\end{aligned} \tag{4.83}
$$

Analogously to the Poisson equation, one can propose a graphical form of the difference operator multipliers as in Figure 4.36.

Finally, we introduce a modification corresponding to the response function recovery by indexing all necessary values once more with upper index m. This yields

$$
\begin{aligned}
&20w^{(m)}_{[i,j]}-8\left(w^{(m)}_{[i-1,j]}+w^{(m)}_{[i,j-1]}+w^{(m)}_{[i+1,j]}+w^{(m)}_{[i,j+1]}\right)\\
&\quad+2\left(w^{(m)}_{[i-1,j-1]}+w^{(m)}_{[i-1,j+1]}+w^{(m)}_{[i+1,j+1]}+w^{(m)}_{[i+1,j-1]}\right)\\
&\quad+w^{(m)}_{[i-2,j]}+w^{(m)}_{[i+2,j]}+w^{(m)}_{[i,j-2]}+w^{(m)}_{[i,j+2]}=\frac{q^{(m)}_{[i,j]}}{D^{(m)}}\left(a^{(m)}\right)^4.
\end{aligned} \tag{4.84}
$$

Alternatively, one may prefer the DDM and then the stochastic perturbation-based version of this finite difference scheme yields

$$
\sum_{k=0}^{n}\binom{n}{k}\frac{\partial^k D}{\partial b^k}\left\{20\frac{\partial^{n-k}w_{[i,j]}}{\partial b^{n-k}}-8\left(\frac{\partial^{n-k}w_{[i-1,j]}}{\partial b^{n-k}}+\frac{\partial^{n-k}w_{[i,j-1]}}{\partial b^{n-k}}+\frac{\partial^{n-k}w_{[i+1,j]}}{\partial b^{n-k}}+\frac{\partial^{n-k}w_{[i,j+1]}}{\partial b^{n-k}}\right)\right\}
$$

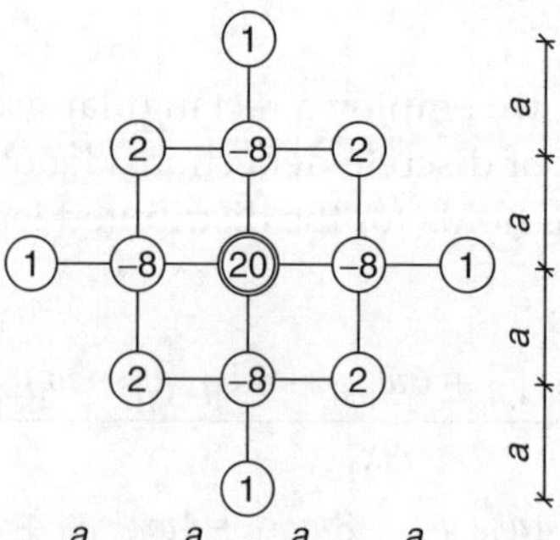

Figure 4.36 Finite difference method biharmonic operator in a graphical form

$$+2\sum_{k=0}^{n}\binom{n}{k}\frac{\partial^k D}{\partial b^k}\left(\frac{\partial^{n-k}w_{[i-1,j-1]}}{\partial b^{n-k}}+\frac{\partial^{n-k}w_{[i-1,j+1]}}{\partial b^{n-k}}+\frac{\partial^{n-k}w_{[i+1,j+1]}}{\partial b^{n-k}}+\frac{\partial^{n-k}w_{[i+1,j-1]}}{\partial b^{n-k}}\right)$$

$$+\sum_{k=0}^{n}\binom{n}{k}\frac{\partial^k D}{\partial b^k}\left\{\frac{\partial^{n-k}w_{[i-2,j]}}{\partial b^{n-k}}+\frac{\partial^{n-k}w_{[i+2,j]}}{\partial b^{n-k}}+\frac{\partial^{n-k}w_{[i,j-2]}}{\partial b^{n-k}}+\frac{\partial^{n-k}w_{[i,j+2]}}{\partial b^{n-k}}\right\}$$

$$=\sum_{m=0}^{n}\binom{n}{m}\frac{\partial^m q_{[i,j]}}{\partial b^m}\frac{\partial^{n-m}\left(a^4\right)}{\partial b^{n-m}}. \tag{4.85}$$

The boundary conditions expressed in finite difference notation for the completely free boundary (i) parallel to the given x_2 axis become

$$w_{[i+1,j]}^{(m)}=-w_{[i-1,j]}^{(m)}+2\left(1+\nu^{(m)}\right)w_{[i,j]}^{(m)}-\nu^{(m)}\left(w_{[i,j-1]}^{(m)}+w_{[i,j+1]}^{(m)}\right), \tag{4.86}$$

$$w_{[i+2,j]}^{(m)}=-w_{[i-2,j]}^{(m)}-4\left(3-\nu^{(m)}\right)w_{[i-1,j]}^{(m)}+6\left(2+2\nu^{(m)}-\left(\nu^{(m)}\right)^2\right)w_{[i,j]}^{(m)}$$
$$+2\left(2-\nu^{(m)}\right)\left(w_{[i-1,j-1]}^{(m)}+w_{[i-1,j+1]}^{(m)}\right)-4\left(1+2\nu^{(m)}-\left(\nu^{(m)}\right)^2\right)\left(w_{[i,j-1]}^{(m)}+w_{[i,j+1]}^{(m)}\right)$$
$$+\nu^{(m)}\left(2-\nu^{(m)}\right)\left(w_{[i,j-2]}^{(m)}+w_{[i,j+2]}^{(m)}\right). \tag{4.87}$$

Case (ii) uses

$$w_{[i,j]}^{(m)}=0,\quad w_{[i+1,j]}^{(m)}=-w_{[i-1,j]}^{(m)}, \tag{4.88}$$

while case (iii) needs

$$w_{[i,j]}^{(m)}=0,\quad w_{[i+1,j]}^{(m)}=w_{[i-1,j]}^{(m)}. \tag{4.89}$$

Example 4.8: Bending of the elastic thin rectangular plate with random parameters

The stainless steel plate ($e=210\,\text{GPa}$, $\nu=0.30$) with dimensions 5.0 m × 5.0 m, thickness $h=0.025$ m subject to uniformly distributed load $q=15.0\,\text{kN/m}^2$ is discretized with a regular grid of 26 × 26 points. The plate is fully clamped at all external edges and we study separately an uncertainty in the structure thickness and Poisson ratio. The RFM version of the SFDM is employed in the computations, so we use an 11-point cycle for randomized thickness taking the discrete values $h=$ [0.020, 0.021, 0.022, 0.023, 0.024, 0.025, 0.026, 0.027, 0.028, 0.029, 0.030] m and the best least-squares fit with smallest order is obtained for the fourth-order polynomial. The RFM discretization for uncertainty in Poisson ratio has the same number of realizations equal to $\nu=$ [0.25, 0.26, 0.27, 0.28, 0.29, 0.30, 0.31, 0.32, 0.33, 0.34, 0.35] and has the same polynomial order. A comparison of basic probabilistic characteristics for both random inputs, for varying input coefficients $\alpha=$ [0.05, 0.10, 0.15, 0.20, 0.25] and for the midpoint cross-section is given in Figures 4.38–4.45 (all probabilistic moments and coefficients of up to fourth order); the deterministic basic solution is presented in Figure 4.37.

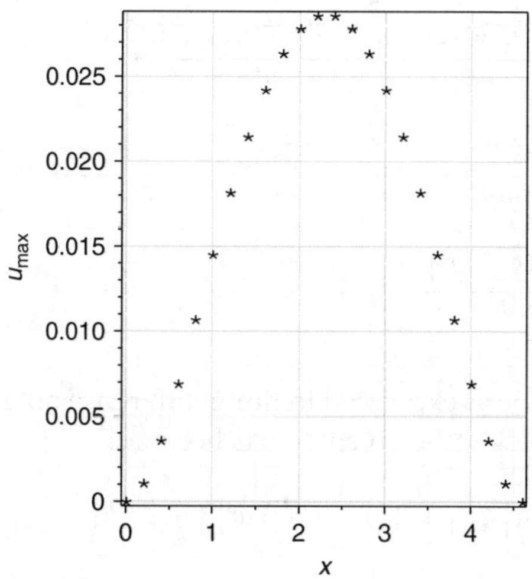

Figure 4.37 FDM solution for the grid 22 × 22

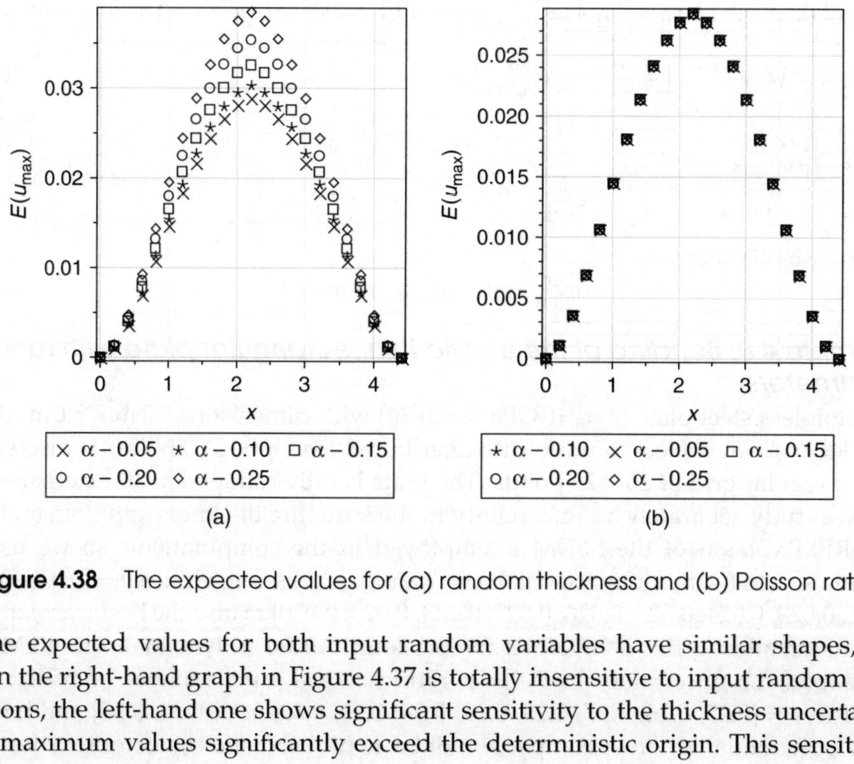

Figure 4.38 The expected values for (a) random thickness and (b) Poisson ratio

The expected values for both input random variables have similar shapes, but when the right-hand graph in Figure 4.37 is totally insensitive to input random fluctuations, the left-hand one shows significant sensitivity to the thickness uncertainty and maximum values significantly exceed the deterministic origin. This sensitivity is of course more apparent close to the midpoint and less clear close to the clamped edges of a structure.

The influence of the input parameter α affects all second-order characteristics – once more the shapes of variances and standard deviations are really very similar but

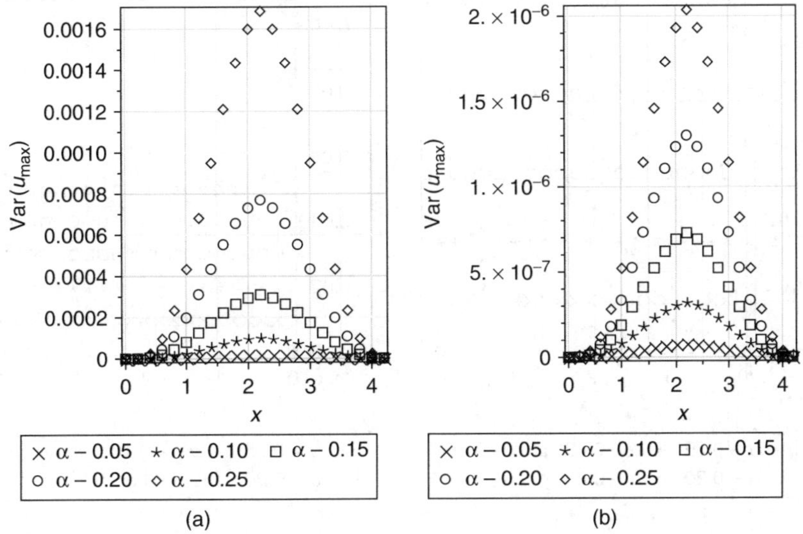

Figure 4.39 The variances for (a) random thickness and (b) Poisson ratio

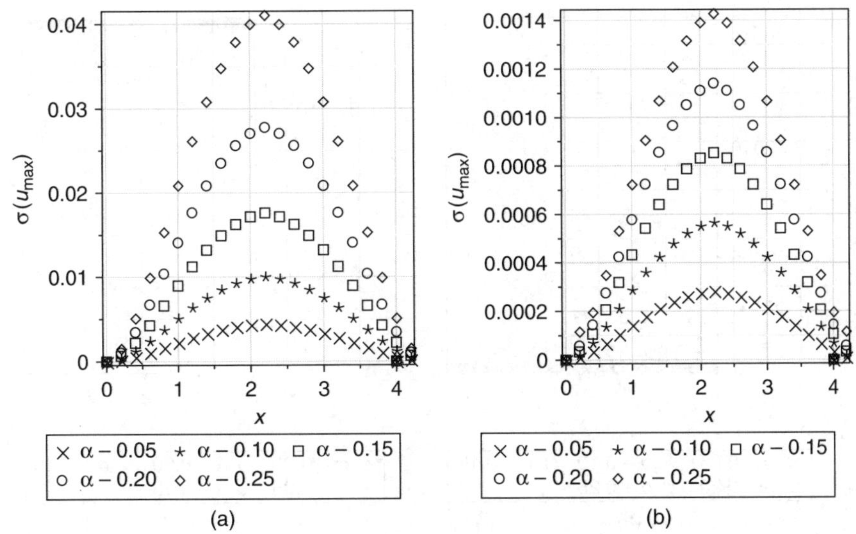

Figure 4.40 The standard deviations for (a) random thickness and (b) Poisson ratio

the values obtained with randomized plate thickness are many times larger than for uncertainty in its Poisson ratio. The largest fluctuations area in the input random dispersion reduces for variances at the midpoints' neighborhood much closer than for the expectations, while the region close to the edges with negligible values becomes wider. This tendency is especially clear after analysis of Figures 4.42 and 4.43 with third and fourth central probabilistic moments. Analogously to the expectations, the

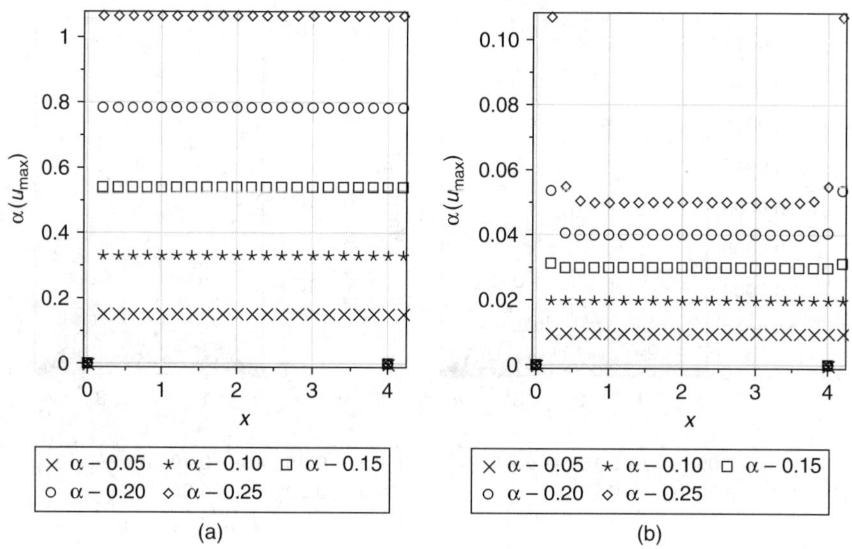

Figure 4.41 The coefficients of variation for (a) random thickness and (b) Poisson ratio

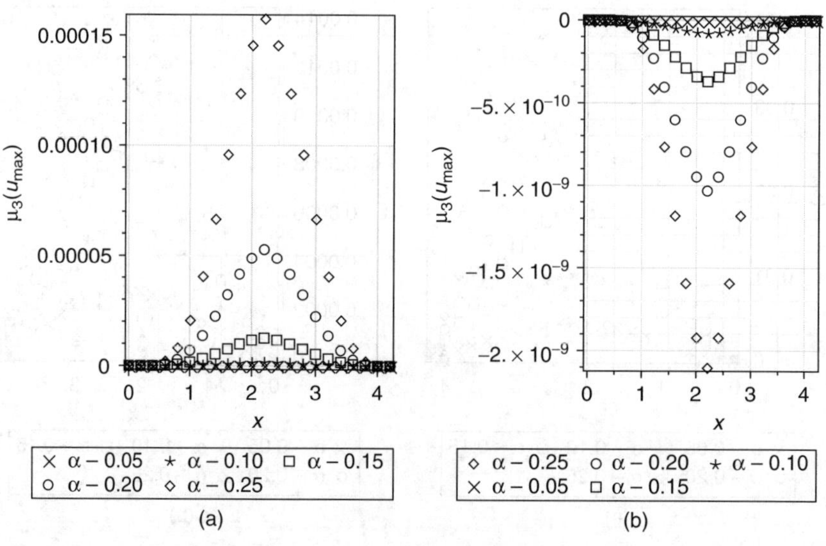

Figure 4.42 The third moments for (a) random thickness and (b) Poisson ratio

variances and standard deviations for random thickness increase much faster than the input coefficient of variation; this is not the case for the same characteristics with random Poisson ratio, which are more proportional to the input coefficients of variation. The coefficients of variation (Figure 4.41) of the maximum deflections profile are totally incomparable for both random input parameters. The first case study results in constant $\alpha(u_{max})$ throughout the entire profile except at both edges,

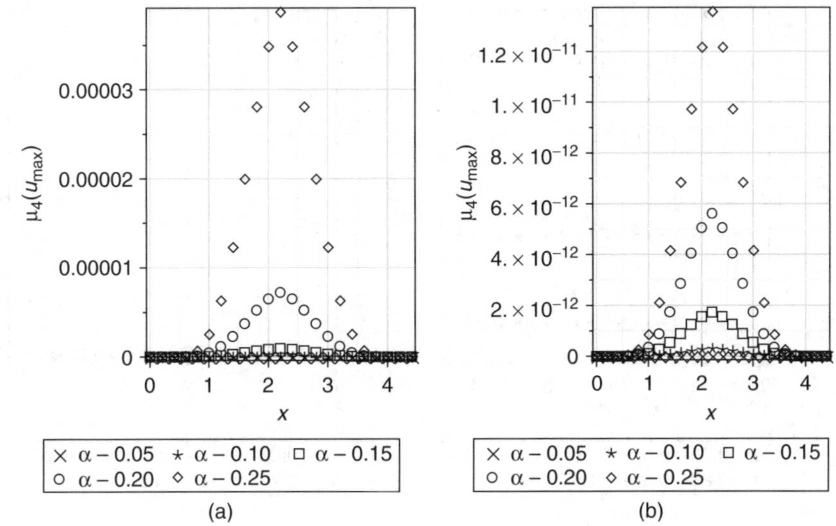

Figure 4.43 The fourth moments for (a) random thickness and (b) Poisson ratio

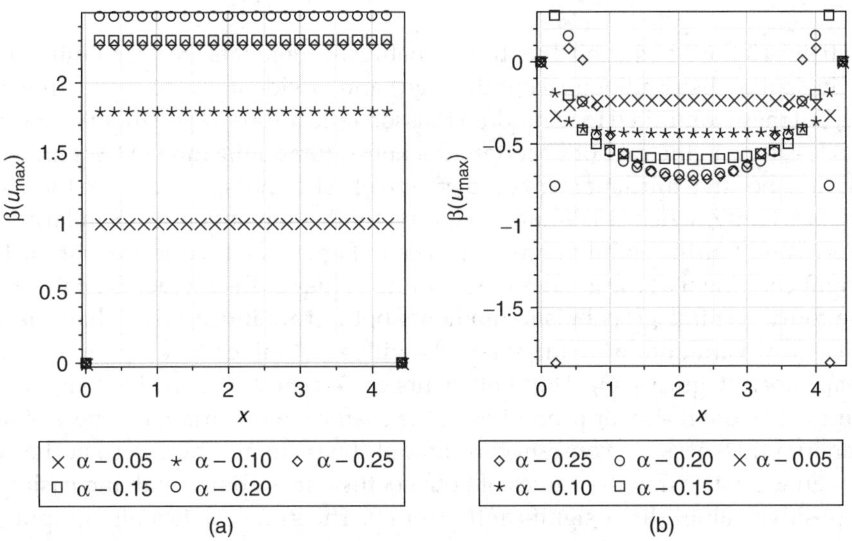

Figure 4.44 The skewness for (a) random thickness and (b) Poisson ratio

where we observe no randomness at all. These profiles for randomized Poisson ratios are mostly also constant, but close to the fixed edges they show some sudden large deviations (especially for the largest proposed input $\alpha(\nu)$); thereafter, they decrease to 0 at both edges. All the resulting coefficients of course increase together with an additional increase of the input coefficient α, but for randomized thickness this is almost four times larger than the input 1 and 10 times larger than the maxima reached by $\alpha(u_{max})$ in the second case study.

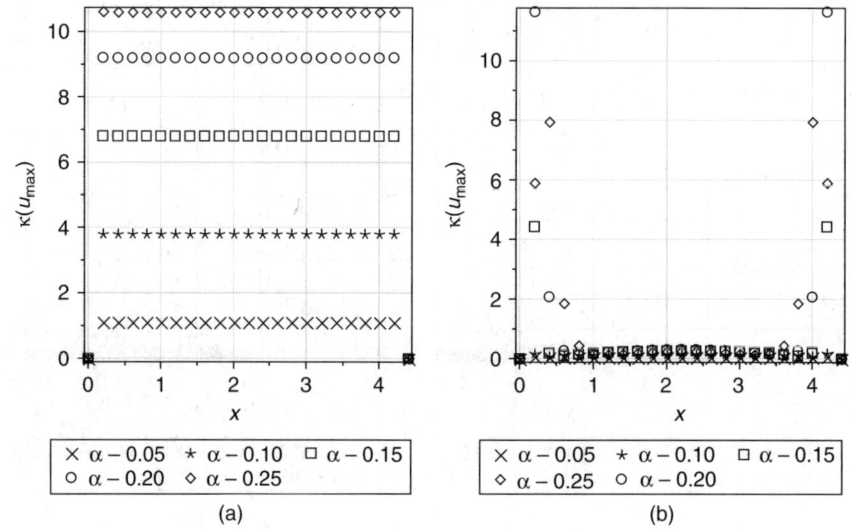

Figure 4.45 The kurtosis for (a) random thickness and (b) Poisson ratio

Spatial variations of the third central probabilistic moments are very similar to each other in spatial distribution (except the sign) and, accidentally, to the Gaussian bell curve, but these computed for a random Poisson ratio are almost negligible – being six orders less than in the case of uncertain thickness. Randomization of the parameter h results in a positive third moment (and effectively skewness), contrary to the Poisson ratio uncertainty, when it is totally negative (with skewness also). Similarly to the variances and fourth central moments (given in Figure 4.43) we notice that the larger the input coefficient α, the greater the absolute values of the moments analyzed.

The fourth central probabilistic moments both grow much faster than the input coefficient of variation; now the intervals with zero values take up almost half of the plate span (Figure 4.43). The coefficients of skewness (Figure 4.44) and kurtosis (Figure 4.45) exhibit similar properties to the coefficient of variation – these obtained for random thickness are constant throughout the plate, reaching zero at the edges. Although we notice that the values of both coefficients stabilize with increasing $\alpha(h)$, their positive values differ significantly from 0. The same coefficients computed for randomized Poisson ratio have some enormously large local variations close to both edges, where their values are almost equal to these computed in case of $h(\omega)$. For the remaining part of the plate cross-section, skewness is slightly negative and stabilizes with an increase of $\alpha(\nu)$, whereas kurtosis keeps close to 0. Therefore, considering the reliability condition on a maximum deflection for the plate midspan, one can assume with the relatively small error that this deflection has Gaussian distribution.

4.2.3 Vibration Analysis of Elastic Plates

Starting from the Love's equations for a plate equilibrium and applying zero curvatures, the resulting expression can be rewritten as [159]

$$D\nabla^4 w + \rho h\ddot{w} = q, \tag{4.90}$$

where ρ denotes the mass density of the plate. The relation (4.90) is usually rewritten in rectangular coordinates in the following manner:

$$D\left(\frac{\partial^4 w}{\partial x^4} + 2\frac{\partial^4 w}{\partial x^2 \partial y^2} + \frac{\partial^4 w}{\partial y^4}\right) + \rho h\ddot{w} = q. \tag{4.91}$$

Looking for the natural frequencies, the following transformation is applied:

$$w = W \exp\left(j\omega t\right), \tag{4.92}$$

which results in

$$D\left(\frac{\partial^4 W}{\partial x^4} + 2\frac{\partial^4 W}{\partial x^2 \partial y^2} + \frac{\partial^4 W}{\partial y^4}\right) + \rho h\omega^2 W = 0. \tag{4.93}$$

Further solution can be carried out if only detailed boundary conditions are specified. If, for instance, we have a plate with dimensions $a \times b$ corresponding to the axes x and y and it is simply supported along the axes parallel to direction y, then the solution can be expressed as

$$W\left(x, y\right) = Y(y) \sin\frac{m\pi x}{a}, \tag{4.94}$$

which gives an equation of motion in the following form:

$$\frac{\partial^4 Y}{\partial y^4} - 2\left(\frac{m\pi}{a}\right)^2 \frac{\partial^2 Y}{\partial y^2} + \left[\left(\frac{m\pi}{a}\right)^4 - \frac{\rho h}{D}\omega^2\right] Y = 0 \tag{4.95}$$

where

$$Y(y) = \sum_{i=1}^{4} C_i \exp\left(\lambda_i (y/b)\right). \tag{4.96}$$

Further, it is obtained that

$$\lambda_i = \pm\frac{b}{a} m\pi \sqrt{1 \pm K} \tag{4.97}$$

and

$$K = \frac{\omega}{\left(\frac{m^2\pi^2}{a^2}\right)\sqrt{\frac{D}{\rho h}}}. \tag{4.98}$$

Application of the generalized stochastic perturbation technique in Equation (4.91) results in:

- zeroth-order relation

$$D^0 \left(\frac{\partial^4 W^0}{\partial x^4} + 2\frac{\partial^4 W^0}{\partial x^2 \partial y^2} + \frac{\partial^4 W^0}{\partial y^4} \right) + \rho^0 h^0 \left(\omega^0\right)^2 W^0 = 0 \tag{4.99}$$

- first-order relation

$$D^0 \left(\frac{\partial^5 W}{\partial b \partial x^4} + 2\frac{\partial^5 W}{\partial b \partial x^2 \partial y^2} + \frac{\partial^5 W}{\partial b \partial y^4} \right) + \frac{\partial \left(\rho h \omega^2\right)}{\partial b} W^0 + \rho^0 h^0 \left(\omega^0\right)^2 \frac{\partial W}{\partial b}$$
$$= -\frac{\partial D}{\partial b} \left(\frac{\partial^4 W^0}{\partial x^4} + 2\frac{\partial^4 W^0}{\partial x^2 \partial y^2} + \frac{\partial^4 W^0}{\partial y^4} \right) \tag{4.100}$$

- nth-order relation

$$\sum_{k=0}^{n} \binom{n}{k} \frac{\partial^k D}{\partial b^k} \left(\frac{\partial^{n-k+4} W}{\partial b^{n-k} \partial x^4} + 2\frac{\partial^{n-k+4} W}{\partial b^{n-k} \partial x^2 \partial y^2} + \frac{\partial^{n-k+4} W}{\partial b^{n-k} \partial y^4} \right)$$
$$+ \sum_{k=0}^{n} \binom{n}{k} \frac{\partial^k}{\partial b^k} \left(\rho h \omega^2\right) \frac{\partial^{n-k} W}{\partial b^{n-k}} = 0. \tag{4.101}$$

Calculating zeroth-order terms from the first equation and next, sequentially, first- and second-order terms, and so on, all the probabilistic moments of the dynamic structural response may be obtained. It is also seen that prior to determination of probabilistic characteristics for the final solution W, we need to determine partial derivatives of the plate eigenfrequencies as they appear explicitly in Equation (4.101).

5

Homogenization Problem

The variability of elastic characteristics and geometrical dimensions of homogeneous as well as composite elements is a frequent problem in designing new and inspecting existing structures. This variability in the design process is included during the optimization phase, where, thanks to non-gradient or gradient techniques, the most optimal distribution (in functional graded material – FGM applications, for instance [7]) or the best choice and contrast (for composites with two or more constituents) is to be found. Such variability in design parameters is taken into account a priori and almost always has a clearly deterministic character. In contrast, experimental testing and a posteriori inspection of engineering structures return statistical or sometimes even stochastic information about spatial or spatio-temporal [178] random distributions of their material and/or geometrical parameters. Therefore, one can notice that sensitivity analysis [136] and probabilistic modeling can be very similar from a quantitative point of view [77]. This similarity is even more transparent in numerical analysis when the discussed perturbation method is applied to determine sensitivity coefficients or gradients with respect to some design parameters as well as some basic probabilistic moments of structural response; these design parameters are simply treated as input random variables or fields. The relation between sensitivity and probabilistic analysis appears to be even closer when we realize that using higher than second-order perturbation methodology [99] it is possible to calculate both higher-order sensitivities and, after minor changes of algebraic formulas in the relevant definitions, higher-order probabilistic moments.

The problem raised above is of special importance in the area of composite materials [21], where not only single particular parameters but multiple characteristics of the same type and, furthermore, their composition may be the subject of such analysis. Even if phenomena like delamination or soft matrix penetration by rigid fibers, typical for specific composites only, are neglected, the sensitivity or probabilistic analysis still has a number of parameters. As is known, the homogenization method has been discovered and extended to reduce significantly the number of composite design parameters by the introduction of effective characteristics. Although this technique in

The Stochastic Perturbation Method for Computational Mechanics, First Edition. Marcin Kamiński.
© 2013 John Wiley & Sons, Ltd. Published 2013 by John Wiley & Sons, Ltd.

its modern view is almost 40 years old [129], there are still some new ideas and applications in the food industry [107], some composites made of wood, metallic components [100], superconductors [101], fully and partially saturated heterogeneous solids [148], and so forth. After fundamental discoveries concerning elastic, thermal [120], and electric effective properties [114], now thermodynamic wave propagation, various multiscale problems [179], even for time-dependent cases with an "equation-free" approach as well as a variety of materially non-linear multi-component composites [14, 37, 54, 121] are homogenized. Following numerous engineering applications the strength of composites by a homogenization method can be estimated [36]; atomistic and nano levels [22, 160] appear to be the smallest resolutions, some new issues are explored [150] as well as well-known methods recently revisited [119, 157, 170]. Much attention is obviously paid to random composites because of the uncertainty in reinforcement location/shape and/or pore spatial distribution in matrices, and randomness in the components' physical and mechanical characteristics [71, 176].

The homogenization method is sometimes connected with sensitivity analysis and optimization (see the problem of a homogeneous plate with holes), used in conjunction with probabilistic analysis using Monte Carlo simulation [59, 98] or some expansion methods. Let us note that even if the mathematical apparatus has strictly deterministic character, some issues like the basic correlation dimension or representativeness of the basic cell subject to the homogenization process are nevertheless discussed. The observations collected above make it negligible to formulate a single universal and general approach to calculate the effective characteristics, where sensitivity gradients and probabilistic moments can be extracted from the same equations using similar algebra. Let us note here that an analogous homogenization approach can be provided starting from the complementary energy principle.

The perturbation technique based on the nth-order Taylor expansion displayed in Chapter 1 and particularly for the FEM throughout Chapter 2 is applied here in conjunction with the effective modules method to determine practically any-order partial derivatives of the homogenized constitutive tensor. The composite subjected to this procedure is linearly elastic and transversely isotropic, where elastic characteristics of the components are assumed as truncated Gaussian random variables and they are defined uniquely by the first two probabilistic moments. The composite remains periodic in all cases in the sense that long round fibers with constant radius have periodic distribution in the plane transversal to the fiber directions; all the fibers are perfectly parallel. The material characteristics are the same in each cell for the sensitivity computations and they have the same first two moments in each cell for random analysis to maintain this perfect periodicity in a composite. The key feature here is the response function reconstruction, where using multiple solutions of the deterministic homogenization problem each effective tensor component is represented as a polynomial function of the input random variables. As we know, this is in significant contradiction to previous applications of the perturbation technique, where zeroth- and higher-order equilibrium equations were derived and solved numerically. Now, an nth-order polynomial function is proposed and implemented without any mixed terms, which reflects the case of a single random input variable or the series of uncorrelated random parameters. The symbolic computations package Maple™ is used here to efficiently solve for the coefficients of this polynomial expression as well as to process normalization of the sensitivity gradients and/or determination of probabilistic moments of the homogenized tensor. Thanks to the application of symbolic computations, it is possible to insert the perturbation parameter ε into the polynomial expansion of the random structural response (homogenized tensor here) and, further, to calculate higher-order partial derivatives analytically. As can be seen at this point, the approach enables us to eliminate the limitations of the second-order perturbation technique, to join sensitivity analysis and probabilistic modeling, to shorten the entire computational process (related to the Monte Carlo simulation procedure), as well as to provide the computational process with a given a priori accuracy. Since the FEM-based plane strain numerical analysis is the core for a solution of the homogenization problem, all error analysis procedures including h, p, or hp adaptivity may also be further adopted here [126].

5.1 Composite Material Model

Let us consider a periodic random composite in plane strain in the unstressed and undeformed state and assume that the section of this structure is constant along the x_3 axis, and finally consider the section $Y \subset R^2$ in the plane $x_3 = 0$. Let us assume that the representative volume element (RVE) [40, 71] Ω of Y has a rectangular external boundary. Further, let Ω contain n ($n < \infty$) coherent regions satisfying the following conditions:

$$\Omega = \bigcup_{a=1}^{n} \Omega_a \cup \bigcup_{a=2}^{n} \Gamma_{(a-1,a)}, \text{ for } a = 1,2, \ldots, n \tag{5.1}$$

$$\Omega_a \cap \Omega_b = \emptyset, \text{ for } a, b = 1,2, \ldots, n \tag{5.2}$$

where $\Gamma_{(a-1,a)}$ is the boundary (interface) between the regions Ω_{a-1} and Ω_a. Let us consider such a class of Ω that for every a the interior of the contour $\Gamma_{(a-2,a-1)}$ is contained in $\Gamma_{(a-1,a)}$ and these contours are disjoint (Figure 5.1). Considering further aspects of this model it is assumed that, although $\Omega_a = \Omega_a(\omega; t)$ as well as $\Gamma_{(a-1,a)} = \Gamma_{(a-1,a)}(\omega; t)$ for any $a = 1, \ldots, n$ are random processes, these two relations hold true for any time $t \in [0, \infty)$ and the external geometrical dimensions of Ω remain constant.

Let every component of this composite contain linear elastic and transversely isotropic material and their elastic parameters (Young's modulus and Poisson ratio) are given using the following processes:

$$e_a(\omega; t) = e_{0a}(\omega) - \dot{e}_a(\omega)t, \tag{5.3}$$

$$\nu_a(\omega; t) = \nu_{0a}(\omega) - \dot{\nu}_a(\omega)t. \tag{5.4}$$

The quantities $e_{0a}(\omega)$ and $\nu_{0a}(\omega)$ are the Gaussian random variables given by their first two probabilistic moments and truncated according to their physical limitations.

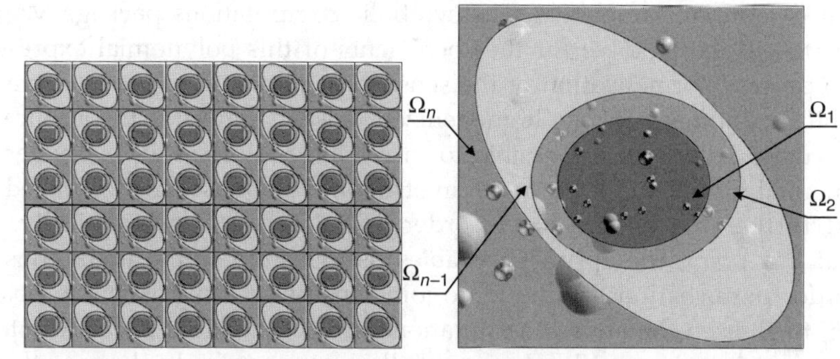

Figure 5.1 The macroscale and microscale of the periodic composite with defects. Reproduced with permission from Elsevier

The two additional variables $\dot{e}_a(\omega)$ and $\dot{\nu}_a(\omega)$ denote for any composite component $a = 1, \ldots, n$ the aging rates and are non-negative Gaussian variables defined by their expectations and standard deviations. Therefore, the elasticity tensor $C_{ijkl}(\mathbf{x};\omega;t)$ is defined as

$$\begin{aligned} C_{ijkl}(\mathbf{x};\,\omega;\,t) &= \chi_a\,(\mathbf{x})\,C^{(a)}_{ijkl}\,(\omega;\,t) \\ &= \chi_a\,(\mathbf{x})\,e_a(\omega;\,t)\left\{\delta_{ij}\delta_{kl}\frac{\nu_a(\omega;\,t)}{(1+\nu_a(\omega;\,t))\,(1-2\nu_a(\omega;\,t))}\right. \\ &\quad \left.+\left(\delta_{ik}\delta_{jl}+\delta_{il}\delta_{jk}\right)\frac{1}{2\,(1+\nu_a(\omega;\,t))}\right\} \\ &\quad \text{for } i,j,k,l = 1,2 \end{aligned} \tag{5.5}$$

where $\chi_a(\mathbf{x})$ is the characteristic function for the composite component indexed by a and equals

$$\chi_a\,(\mathbf{x}) = \begin{cases} 1, & \mathbf{x} \in \Omega_a \\ 0, & \text{elsewhere.} \end{cases} \tag{5.6}$$

Further, let us consider some material microdefects [133] having random shape and location distributed along the interfaces inside the specimen Ω [69, 71]. According to the engineering evidence such defects in the reinforcement are decisively smaller, so that they are neglected in this approach.

Considering the matrix defects (microcavities) idealized as bubbles (cf. Figure 5.2 versus Figure 5.3), it is assumed that (i) the stochastic nucleation rule for the defects along the interface $\Gamma_{(a-1,a)}$ is defined as

$$n_{\Gamma(a-1,a)}(\omega;\,t) = n^0_{\Gamma(a-1,a)}(\omega) + \dot{n}_{\Gamma(a-1,a)}(\omega)t, \tag{5.7}$$

where both Gaussian variables $n^0_{\Gamma(a-1,a)}(\omega)$ (initial defects number) and $\dot{n}_{\Gamma(a-1,a)}(\omega)$ (aging rate) are truncated to positive values only. The second of these parameters

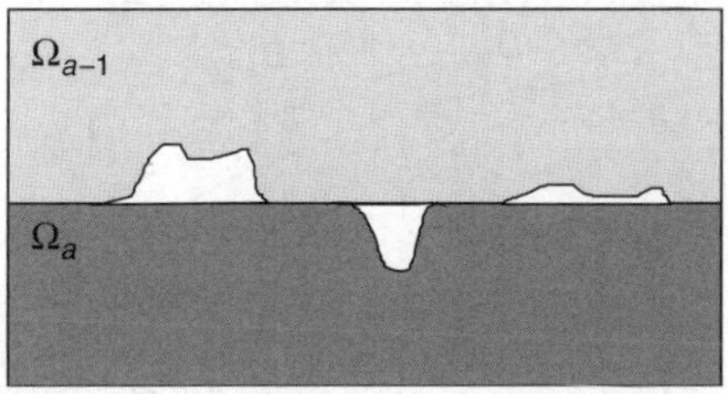

Figure 5.2 The real defects at the composite interface. Reproduced with permission from Springer

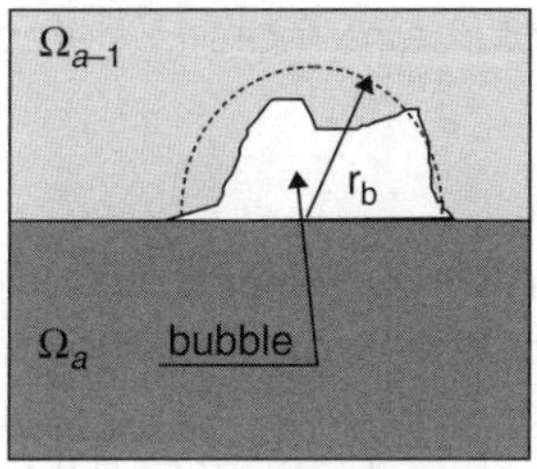

Figure 5.3 Defects approximation by the bubble. Reproduced with permission from Springer

is relevant to random nucleation of the new defects inside the existing interphase; (ii) these defects are geometrically approximated by the semicircles (bubbles) lying with their diameters on the relevant interfaces; the stochastic growth of the bubbles is described by the following similar law:

$$r_{\Gamma(a-1,a)}(\omega;t) = r^0_{\Gamma(a-1,a)}(\omega) + \dot{r}_{\Gamma(a-1,a)}(\omega)t. \tag{5.8}$$

It is possible to consider these defects in a more general form, that is, as semi-ellipses lying with their main axis on the interface (major or minor [72]). We denote here $r^0_{\Gamma(a-1,a)}(\omega)$ as the initial average radius of the bubbles at the interface $\Gamma_{(a-1,a)}$ in the weaker material $\Omega_a(t)$, while $\dot{r}_{\Gamma(a-1,a)}(\omega)$ is its aging rate (they are both Gaussian variables with additional truncation). Finally we assume that (iii) the geometric dimensions of every defect belonging to any $\Omega_a(t)$ are small in comparison with the minimal distance between $\Gamma_{(a-2,a-1)}$ and $\Gamma_{(a-1,a)}$ boundaries for $a=3, \ldots, n$ and also with Ω_1 geometric dimensions; (iv) all elastic characteristics equal to 0 if $\mathbf{x} \in D_a(t)$, for $a=1, 2, \ldots, n$. The signs of the aging rates are defined depending on the dominating external loading character and other physical phenomena – we may assume non-negative values of the aging velocity if the composite is subject to tension and aggressive corrosion, for instance. It should be underlined that the model introduced approximates the real defects rather precisely; however, the cavities should be appropriately averaged over the corresponding materials, which they belong to, to build up a reliable numerical procedure.

Let us consider in particular stochastic material non-homogeneities located in any $\Omega_a(t) \subset \Omega$ for $t \in [0,\infty)$. The set $D_a(t)$ of all the defects considered for any

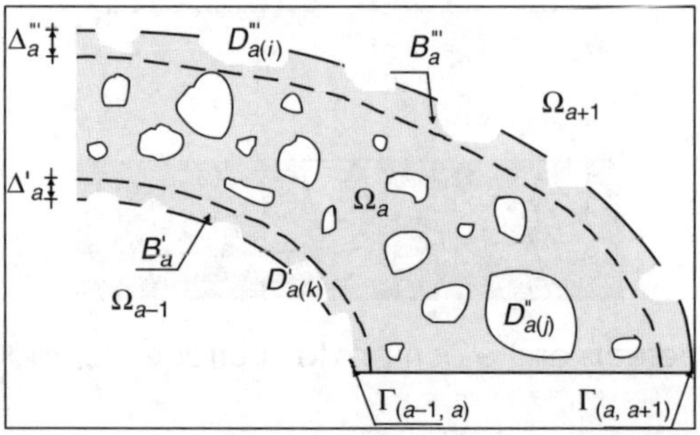

Figure 5.4 The interface and volumetric material defects. Reproduced with permission from Springer

component can be divided into three disjoint subsets $D'_a(t)$, $D''_a(t)$, and $D'''_a(t)$, where $D'_a(t)$ contains all the defect areas having non-zero intersection with the boundary $\Gamma_{(a-1,a)}$, $D''_a(t)$ having zero intersection with $\Gamma_{(a-1,a)}$ and $\Gamma_{(a,a+1)}$, and $D'''_a(t)$ having non-zero intersection with $\Gamma_{(a,a+1)}$ only. Further, all the defects belonging to subsets $D'_a(t)$ and $D'''_a(t)$ are called the stochastic interface defects, and these belonging to $D''_a(t)$ – the volumetric stochastic defects, respectively. Let us consider such $\Omega'_a(t)$, $\Omega''_a(t)$, and $\Omega'''_a(t)$, where $\Omega_a(t) = \Omega'_a(t) \cup \Omega''_a(t) \cup \Omega'''_a(t)$, such that the relations $D'_a(t) \subset \Omega'_a(t)$, $D''_a(t) \subset \Omega''_a(t)$, and $D'''_a(t) \subset \Omega'''_a(t)$ (cf. Figure 5.4) hold true with probability equal to 1.

The subsets $\Omega'_a(t)$, $\Omega''_a(t)$, and $\Omega'''_a(t)$ can be geometrically constructed using probabilistic moments of the defect parameters (their geometric dimensions). Let us introduce for this purpose the random processes $\Delta'_a(\omega; t)$ and $\Delta'''_a(\omega; t)$ as upper bounds on norms of normal vectors defined on the boundaries $\Gamma_{(a-1,a)}$ and $\Gamma_{(a,a+1)}$ and the boundaries of the defects belonging to $D'_a(t)$ and $D'''_a(t)$, respectively. Next, let us consider the upper bounds of probabilistic distributions of $\Delta'_a(\omega; t)$ and $\Delta'''_a(\omega; t)$, respectively, given as follows:

$$\Delta'_a(t) = E\left[\Delta'_a(\omega; t)\right] + 3\sigma\left(\Delta'_a(\omega; t)\right), \tag{5.9}$$

$$\Delta'''_a(t) = E\left[\Delta'''_a(\omega; t)\right] + 3\sigma\left(\Delta'''_a(\omega; t)\right). \tag{5.10}$$

Thus, $\Omega'_a(t)$ and $\Omega'''_a(t)$ can be expressed in the following form:

$$\Omega'_a(t) = \left\{P(x_i) \in \Omega_a(t) : d(P, \Gamma_{(a-1,a)}) \leq \Delta'_a(t)\right\}, \tag{5.11}$$

$$\Omega'''_a(t) = \left\{P(x_i) \in \Omega_a(t) : d(P, \Gamma_{(a,a+1)}) \leq \Delta'''_a(t)\right\}, \tag{5.12}$$

where $i=1$, 2 and $d(P,\Gamma)$ denotes the distance from the given point P to the appropriate interface. Let us note that $\Omega''_a(t)$ can be obtained as

$$\Omega''_a(t) = \Omega_a(t) - \left\{\Omega'_a(t) \cup \Omega'''_a(t)\right\}. \tag{5.13}$$

The deterministic spatial averaging of properties $Y_a(\omega;t)$ on the continuous and disjoint subsets $\Omega_a(t)$ of Ω is employed to formulate the stochastic averaging method. The effective process $Y^{(eff)}(\omega;t)$ characterizing Ω is given by the following equation:

$$Y^{(eff)}(\omega;t) = \frac{\sum_{a=1}^{n} Y_a(\omega;t)\ |\Omega_a(t)|}{|\Omega|};\mathbf{x} \in \Omega, \tag{5.14}$$

where $|\Omega|$ is the two-dimensional Lebesgue measure of Ω (its area or length). Using the equations given above, the mechanical and physical properties are stochastically averaged within the regions $\Omega_a(t)$ and their subsets $\Omega'_a(t)$, $\Omega''_a(t)$, $\Omega'''_a(t)$ for $a=1, \ldots, n$. Finally, the primary stochastic geometry of the composite considered is replaced by the deterministic one – in this way the n-component composite with stochastic interface defects on both sides of every phase boundary and with volume non-homogeneities is replaced with an artificial $3n - 2$-component structure. Let us consider spatial averaging of the elastic characteristics for $\Omega_a(t)$ containing the bubbles defined by the parameters $n_a(\omega;t)$ and $r_a(\omega;t)$. Let us denote by $Y_{m(a)}(\omega;t)$ the elastic parameters of the solid in $\Omega_a(t)$ and by $Y_{d(a)}(\omega;t)$ – within the defects. Then, we can calculate

$$\begin{aligned} Y^{(eff)a}(\omega;t) &= \frac{\pi}{2\Omega_a} n_a(\omega;t) r_a^2(\omega;t) Y_{d(a)}(\omega;t) + \left(1 - \frac{n_a(\omega;t)\pi r_a^2(\omega;t)}{2\Omega_a(t)}\right) Y_{m(a)}(\omega;t) \\ &= \left(1 - \frac{n_a(\omega;t)\pi r_a^2(\omega;t)}{2\Omega_a(t)}\right) Y_{m(a)}(\omega;t) \end{aligned} \tag{5.15}$$

according to the remark (iv) above. So, the effective Young's modulus in the ath component subjected to the stochastic aging equals

$$\begin{aligned} e_a^{(eff)}(\omega;t) = &\left(1 - \frac{\left(n_a^0(\omega) + \dot{n}_a(\omega)t\right)\pi\left(r_a^0(\omega) + \dot{r}_a(\omega)t\right)^2}{2\Omega_a(t)}\right) \\ &\left(e_a^0(\omega) + \dot{e}_a(\omega)t\right). \end{aligned} \tag{5.16}$$

Next, we can rewrite the effective elasticity tensor (EET) components in this material as [62]

$$\begin{aligned} C_{ijkl}^{a(eff)}(\omega;t) = e_a^{(eff)}(\omega;t)\,\delta_{ij}\delta_{kl} &\left\{ \frac{\nu_a^{(eff)}(\omega;t)}{\left(1+\nu_a^{(eff)}(\omega;t)\right)\left(1-2\nu_a^{(eff)}(\omega;t)\right)} \right. \\ &\left. + \left(\delta_{ik}\delta_{jl} + \delta_{il}\delta_{jk}\right) \frac{1}{2\left(1+\nu_a^{(eff)}(\omega;t)\right)} \right\} \end{aligned} \tag{5.17}$$

which equals (time and random space dependence is omitted for clarity of presentation only)

$$C_{ijkl}^{(eff)a} = \left(1 - \frac{n_a \pi r_a^2}{2\Omega_a}\right) e_a^{(eff)} \left\{ \delta_{ij}\delta_{kl} \frac{\left(1 - \frac{n_a \pi r_a^2}{2\Omega_a}\right) \nu_a^{(eff)}}{\left(1 + \left(1 - \frac{n_a \pi r_a^2}{2\Omega_a}\right) \nu_a^{(eff)}\right)\left(1 - 2\left(1 - \frac{n_a \pi r_a^2}{2\Omega_a}\right) \nu_a^{(eff)}\right)} + \frac{\left(\delta_{ik}\delta_{jl} + \delta_{il}\delta_{jk}\right)}{2\left(1 + \left(1 - \frac{n_a \pi r_a^2}{2\Omega_a}\right) \nu_a^{(eff)}\right)} \right\}. \tag{5.18}$$

Moreover, we can observe that for any $a = 2, \ldots, m$ the total area of $\Omega_a(t)$ can be described as

$$\Omega_a(t) = \pi \left[\left(R_a + \Delta_a(t)\right)^2 - R_a^2 \right] \tag{5.19}$$

where R_a denotes the internal radius identifying $\Omega_a(t)$, while $\Delta_a(t)$ denotes the thickness of the given interphase. Further, the elasticity tensor of Ω_a may be described using the geometry model parameters n_a, r_a, R_a, Δ_a, and input elastic parameters as [62]

$$C_{ijkl}^{(eff)a} = \left(1 - \frac{n_a r_a^2}{2\left[\left(R_a + \Delta_a\right)^2 - R_a^2\right]}\right) e_a^{(eff)} \times \left\{ \delta_{ij}\delta_{kl} \frac{\left(1 - \frac{n_a \pi r_a^2}{2\Omega_a}\right) \nu_a^{(eff)}}{\left(1 + \left(1 - \frac{n_a r_a^2}{2\left[(R_a+\Delta_a)^2 - R_a^2\right]}\right) \nu_a^{(eff)}\right)\left(1 - 2\left(1 - \frac{n_a r_a^2}{2\left[(R_a+\Delta_a)^2 - R_a^2\right]}\right) \nu_a^{(eff)}\right)} + \frac{\left(\delta_{ik}\delta_{jl} + \delta_{il}\delta_{jk}\right)}{2\left(1 + \left(1 - \frac{n_a r_a^2}{2\left[(R_a+\Delta_a)^2 - R_a^2\right]}\right) \nu_a^{(eff)}\right)} \right\}. \tag{5.20}$$

Let us note that the denominator does not include multiplier 2 when we average the stochastic volumetric defects for all composite components, it appears only for the interphases. Unfortunately, it seems to be very complicated or even impossible to describe analytically the first two probabilistic moments of the elasticity tensor components as functions of expected values and variances of Poisson ratios, while it is quite straightforward for the Young's modulus using fundamental properties introduced in Chapter 1. Then, quite similarly to Equation (5.20) one can derive the random process of $C_{ijkl}(\omega;t)$ having defined time series of geometrical and material parameters as proposed in Equation (5.8). Some alternative way of stochastic material defects FEM analysis for linear elasticity and viscoplasticity may be found in [43].

5.2 Statement of the Problem and Basic Equations

Let us introduce a geometrical scaling parameter $\zeta > 0$ between the micro- and macroscale of the composite (see Figure 5.5) and introduce two coordinate systems $\mathbf{y}=(y_1, y_2, y_3)$ on the microscale of the composite and $\mathbf{x}=(x_1, x_2, x_3)$ on the macroscale.

Next, let us express any state function G defined on composite region Y as

$$G^{\zeta}(\mathbf{x}) = G\left(\frac{\mathbf{x}}{\zeta}\right) = G(\mathbf{y}). \tag{5.21}$$

The linear elasticity problem for the periodic composite structure is given as follows:

$$\begin{cases} \dfrac{\partial \sigma_{ij}^{\zeta}}{\partial x_j} + f_i = 0 \\ \sigma_{ij}^{\zeta}\, n_j = \bar{t}_i; & \mathbf{x} \in \partial Y_{\sigma} \\ u_i^{\zeta} = 0; & \mathbf{x} \in \partial Y_u \qquad i,j,k,l = 1,2. \\ \sigma_{ij}^{\zeta} = C_{ijkl}^{\zeta}(\mathbf{x})\ \varepsilon_{kl}^{\zeta} \\ \varepsilon_{kl}^{\zeta} = \frac{1}{2}\left(u_{k,l}^{\zeta} + u_{l,k}^{\zeta}\right) \end{cases} \tag{5.22}$$

Assuming perfect interfaces between matrix and the fibers as well as no cracks and other defects (after stochastic averaging) in these constituents we solve the problem

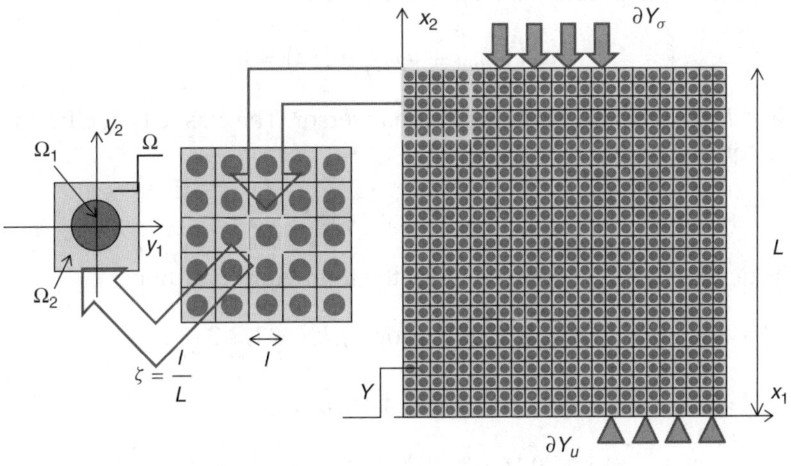

Figure 5.5 Periodic composite structure Y. Reproduced with permission from IFTR PAS

(Equation (5.22)) using a bilinear form

$$a^{\zeta}(\mathbf{u},\mathbf{v}) = \int_{\Omega} C_{ijkl}\left(\frac{\mathbf{x}}{\zeta}\right) \varepsilon_{ij}(\mathbf{u})\varepsilon_{kl}(\mathbf{v})d\Omega \tag{5.23}$$

and the linear one

$$L(\mathbf{v}) = \int_{\Omega} f_i v_i d\Omega + \int_{\partial\Omega_\sigma} \bar{t}_i v_i d(\partial\Omega) \tag{5.24}$$

in the following Hilbert space of admissible displacements defined on Y:

$$V = \left\{\mathbf{v}|\ \mathbf{v} \in \left(H^1(Y)\right)^2,\ \mathbf{v}|_{\partial\Omega_u} = 0\right\}, \tag{5.25}$$

$$\|\mathbf{v}\|^2 = \int_{\Omega} \varepsilon_{ij}(\mathbf{v})\varepsilon_{ij}(\mathbf{v})d\Omega. \tag{5.26}$$

Then, the variational statement equivalent to the equilibrium problem (Equation (5.22)) is to find $\mathbf{u}^{\zeta} \in V$, being a solution of the following equation:

$$a^{\zeta}\left(\mathbf{u}^{\zeta},\mathbf{v}\right) = L(\mathbf{v}), \quad \mathbf{v} \in V. \tag{5.27}$$

Let us define the additional space of the admissible displacement functions $P(\Omega) = \left\{\mathbf{v},\ \mathbf{v} \in \left(H^1(\Omega)\right)^2\right\}$ periodic on the composite cell Ω. So, we introduce the new bilinear form for any $\mathbf{u},\mathbf{v} \in P(\Omega)$:

$$a_y(\mathbf{u},\mathbf{v}) = \int_{\Omega} C_{ijkl}(\mathbf{y})\varepsilon_{ij}(\mathbf{u})\varepsilon_{kl}(\mathbf{v})d\Omega \tag{5.28}$$

and the homogenization function $\chi_{(ij)k} \in P(\Omega)$ (also of displacement type) as a solution for the so-called local problem on a periodicity cell

$$a_y\left(\left(\chi_{(ij)k} + y_j\delta_{ki}\right)\mathbf{n}_k, \mathbf{w}\right) = 0 \tag{5.29}$$

for any $\mathbf{w} \in P(\Omega)$, where $\mathbf{n}_k$ is the additional versor. The elasticity coefficients can be defined according to Equation (5.21) as

$$C^{\zeta}_{ijkl}(\mathbf{x}) = C_{ijkl}(\mathbf{y}). \tag{5.30}$$

The elasticity tensor components fulfill the following conditions:

$$C_{ijkl} \in L^{\infty}(\Re^2), \text{ for } i,j,k,l = 1,2,3 \tag{5.31}$$

$$C_{ijkl} = C_{klij} = C_{jikl}, \tag{5.32}$$

$$\exists C_0 > 0;\ C_{ijkl}\xi_{ij}\xi_{kl} \geq C_0\xi_{ij}\xi_{ij}\ \forall_{i,j}\ \xi_{ij} = \xi_{ji}. \tag{5.33}$$

Moreover, for any of the composite constituents this tensor is defined after Equation (4.24) as

$$C_{ijkl}(\mathbf{x}) = e(\mathbf{x}) \left\{ \delta_{ij}\delta_{kl} \frac{\nu(\mathbf{x})}{(1+\nu(\mathbf{x}))(1-2\nu(\mathbf{x}))} + \left(\delta_{ik}\delta_{jl} + \delta_{il}\delta_{jk}\right) \frac{1}{2(1+\nu(\mathbf{x}))} \right\}. \quad (5.34)$$

If, analogously to Equation (2.30), it is introduced that

$$A_{ijkl}(\mathbf{x}) = \delta_{ij}\delta_{kl} \frac{\nu(\mathbf{x})}{(1+\nu(\mathbf{x}))(1-2\nu(\mathbf{x}))} + (\delta_{ik}\delta_{jl} + \delta_{il}\delta_{jk}) \frac{1}{2(1+\nu(\mathbf{x}))}, \quad (5.35)$$

then the partial derivatives of the elasticity tensor with respect to the Young's modulus in homogeneous material is equal to

$$\frac{\partial C_{ijkl}}{\partial e} = A_{ijkl}, \quad (5.36)$$

and any higher-order partial derivatives equal 0:

$$\frac{\partial^n C_{ijkl}}{\partial e^n} = 0. \quad (5.37)$$

The case of differentiation w.r.t. Poisson ratio is essentially more complex and returns

$$\frac{\partial C_{ijkl}}{\partial \nu} = e \left\{ \delta_{ij}\delta_{kl} \frac{1}{(1+\nu)(1-2\nu)} \left(1 - \frac{\nu}{(1+\nu)} + 2\frac{\nu}{(1-2\nu)}\right) - \left(\delta_{ik}\delta_{jl} + \delta_{il}\delta_{jk}\right) \frac{1}{2(1+\nu)^2} \right\}, \quad (5.38)$$

and further successive differentiation enlarges its algebraic expression complexity. Next, we find the EET components from the following theorem.

The solution $\mathbf{u}^\zeta$ of Equation (5.22) converges weakly in space V:

$$\mathbf{u}^\zeta \to \mathbf{u} \text{ as } \zeta \to 0 \quad (5.39)$$

for the Ω-periodic tensor $C^\zeta_{ijkl}(\mathbf{x})$, where the solution $\mathbf{u}$ is the unique one for the problem:

$$\mathbf{u} \in V, S(\mathbf{u}, \mathbf{v}) = L(\mathbf{v}) \quad (5.40)$$

for any $\mathbf{v} \in V$ and

$$S(\mathbf{u}, \mathbf{v}) = \int_Y S_{ijkl}\varepsilon_{ij}(\mathbf{u})\varepsilon_{kl}(\mathbf{v})dY, \quad (5.41)$$

where

$$S_{ijkl} = \frac{1}{|\Omega|} a_y \left(\left(\chi_{(ij)p} + y_i\delta_{pj}\right) \mathbf{n}_p, \left(\chi_{(kl)r} + y_l\delta_{rk}\right) \mathbf{n}_r \right). \quad (5.42)$$

Now we consider the boundary value problem in its differential form

$$\frac{\partial}{\partial x_j}\left(C_{ijkl}\left(\frac{\mathbf{x}}{\zeta}\right)\varepsilon_{kl}\left(\mathbf{u}^{\zeta}\right)\right)+f_i=0;\ u_i^{\zeta}=0 \text{ for } \mathbf{y}\in\partial Y \tag{5.43}$$

and the following representation for the displacement using the parameter ζ:

$$u^{\zeta}(\mathbf{x})=\sum_{m=0}^{\infty}\zeta^m u^{(m)}(\mathbf{x},\mathbf{y}), \tag{5.44}$$

where any $\mathbf{u}^{(m)}(\mathbf{x},\mathbf{y})$ is periodic in $\mathbf{y}$ on Ω and $m=0,\ 1,\ 2,\ \ldots$ Let us note that contrary to the stochastic perturbation methods, this expansion is provided with a small positive-scale perturbation parameter but the general idea remains the same. A separation of the variables $\mathbf{x}$ and $\mathbf{y}$ leads to the statement

$$\varepsilon_{ij}(\mathbf{v})=\varepsilon_{ij}^x(\mathbf{v})+\frac{1}{\zeta}\varepsilon_{ij}^y(\mathbf{v}), \tag{5.45}$$

where, for instance.

$$\varepsilon_{ij}^x(\mathbf{v})=\frac{1}{2}\left(\frac{\partial v_i}{\partial x_j}+\frac{\partial v_j}{\partial x_i}\right). \tag{5.46}$$

Thus, relation (5.43) is expanded as

$$\left(\zeta^{-2}L_1+\zeta^{-1}L_2+L_3\right)\times\sum_{m=1}^{\infty}\zeta^m u^{(m)}(\mathbf{x},\mathbf{y})+\mathbf{f}=0, \tag{5.47}$$

where

$$L_1\mathbf{u}=\frac{\partial}{\partial y_i}\left(C_{ijkl}(\mathbf{y})\right)\ \varepsilon_{kl}^y(\mathbf{u}), \tag{5.48}$$

$$L_2\mathbf{u}=C_{ijkl}(\mathbf{y})\frac{\partial}{\partial x_j}\left(\varepsilon_{kl}^y(\mathbf{u})\right)+\frac{\partial}{\partial y_i}\left(C_{ijkl}(\mathbf{y})\varepsilon_{kl}^x(\mathbf{u})\right), \tag{5.49}$$

$$L_3\mathbf{u}=C_{ijkl}(\mathbf{y})\frac{\partial}{\partial x_j}\ \left(\varepsilon_{kl}^x(\mathbf{u})\right). \tag{5.50}$$

Next, equating the terms with the same order of ζ to 0, the following equations of zeroth, first, and second order are obtained:

$$L_1\mathbf{u}^{(0)}=0, \tag{5.51}$$

$$L_1\mathbf{u}^{(1)}+L_2\mathbf{u}^{(0)}=0, \tag{5.52}$$

$$L_1\mathbf{u}^{(2)}+L_2\mathbf{u}^{(1)}+L_3\mathbf{u}^{(0)}+\mathbf{f}=0 \tag{5.53}$$

to determine the solutions $\mathbf{u}^{(0)}$, $\mathbf{u}^{(1)}$, and $\mathbf{u}^{(2)}$. Let us also note that the equation

$$L_1\mathbf{u} + \mathbf{P} = 0 \tag{5.54}$$

with $\mathbf{u}$ a Ω-periodic function has a unique solution for

$$\langle \mathbf{P} \rangle = \frac{1}{|\Omega|}\int_{\Omega} \mathbf{P}\ \mathrm{d}\mathbf{y} = 0. \tag{5.55}$$

Starting from the above equations yields

$$\mathbf{u}^{(0)}(\mathbf{x},\mathbf{y}) = \mathbf{u}(\mathbf{x}), \tag{5.56}$$

while Equation (5.52) takes the following form:

$$L_1\mathbf{u}^{(1)}(\mathbf{x},\mathbf{y}) + \frac{\partial}{\partial y_i}\left(C_{ijkl}(\mathbf{y})\right)\ \varepsilon^{x}_{kl}\left(\mathbf{u}^{(0)}(\mathbf{x})\right) = 0. \tag{5.57}$$

The separation process of the variables $\mathbf{x}$ and $\mathbf{y}$ leads to

$$u_i^{(1)}(\mathbf{x},\mathbf{y}) = \chi_{(kl)i}(\mathbf{y})\varepsilon^{x}_{kl}\left(\mathbf{u}^{(0)}\right) + u_i\left(\mathbf{x}\right). \tag{5.58}$$

The last two equations give the formulation for the Ω-periodic functions $\chi_{(kl)i}(\mathbf{y})$ as

$$\frac{\partial}{\partial y_i}\left(C_{ijkl}(\mathbf{y})\frac{\partial \chi_{(kl)m}(\mathbf{y})}{\partial y_m}\right) + \frac{\partial}{\partial y_i}C_{ijkl}(\mathbf{y}) = 0. \tag{5.59}$$

So, the local problem of the homogenization function $\chi_{(kl)i}(\mathbf{y})$ reduces to the equations posed above and the following conditions hold true:

$$\chi_{(kl)i}\ \text{periodic on}\ \Omega \tag{5.60}$$

and

$$\sigma_{ij}\left(\chi_{(pq)}\right)n_j = \left[C_{pqij}\right]|_{\Gamma_{12}}n_j = F_{(pq)i}|_{\Gamma_{12}} = C^{(2)}_{pqij} - C^{(1)}_{pqij}\ \text{on}\ \Gamma_{12}. \tag{5.61}$$

Hence, the variational formulation necessary for the displacement version of the FEM analysis of the local problem can be written for two-component composite as

$$C^{(1)}_{ijkl}\int_{\Omega_1}\varepsilon_{kl}\left(\boldsymbol{\chi}_{(pq)}\right)\varepsilon_{ij}\left(\mathbf{v}\right)\mathrm{d}\Omega + C^{(2)}_{ijkl}\int_{\Omega_2}\varepsilon_{kl}\left(\boldsymbol{\chi}_{(pq)}\right)\varepsilon_{ij}\left(\mathbf{v}\right)\mathrm{d}\Omega = -\int_{\Gamma_{12}}\sigma_{ij}\left(\boldsymbol{\chi}_{(pq)}\right)n_j v_i \mathrm{d}\Gamma. \tag{5.62}$$

After solving for $\chi_{(pq)i}$ from Equation (5.62) one computes the EET as

$$C^{(eff)}_{ijpq} = \frac{1}{|\Omega|}\int_{\Omega}\left(C_{ijpq} + C_{ijkl}\varepsilon_{pq}\left(\boldsymbol{\chi}_{(kl)}\right)\right)\mathrm{d}\Omega. \tag{5.63}$$

Further, it is clear that if the second component of the RHS integrand function is omitted, well-known upper bounds for the EET for the composite are returned (as simple spatial averages of the elasticity tensor). The experimental and computational analyses prove that these bounds are easy to calculate even for the case of random spaces of composite constituent elastic characteristics, but their values significantly overestimates the real effective properties. Let us finally note that since our probabilistic method based on the RFM uses several deterministic solutions to complete the entire homogenization of a random composite, this mathematical apparatus does not need further improvement and the homogenization theorem remains valid (the same as for the Monte Carlo simulation).

We consider the special case of the multi-component unidirectional composite beam with prismatic cross-section constant along the beam length, which has perfectly periodic structure and material properties varying along the spatial macro-coordinate x_3 and constant with respect to the remaining axes. Now, the RVE consists of the few layers with deterministically defined thicknesses corresponding to the composite constituents and having perfect interfaces perpendicular to x_3. The following system of partial differential equations is employed to calculate the effective properties [146]:

$$\left(C_{ijkl}\left(\frac{x_3}{\zeta}\right)u^{\zeta}_{k,l}\right)_{,j}=f_i(\mathbf{x}),\quad \mathbf{u}^{\zeta}(\mathbf{x})=\mathbf{u}^0(\mathbf{x}),\quad \mathbf{x}\in\partial\Omega. \tag{5.64}$$

Similarly to the procedure displayed above (cf. Equation (5.59)), the periodic homogenization functions $\boldsymbol{\chi}_{(mn)}(\mathbf{y})$ are employed and determined as the solution to the local problem on the RVE:

$$\frac{\partial}{\partial y_j}\left(C_{ijkl}(y_3)\frac{\partial}{\partial y_l}(\chi_{(mn)k})+C_{ijmn}(y_3)\right)=0. \tag{5.65}$$

Obviously, a solution is expected in the form $\boldsymbol{\chi}_{(mn)}(\mathbf{y})=\boldsymbol{\chi}_{(mn)}(y_3)$. It yields

$$\frac{\partial}{\partial y_3}\left(C_{i3k3}(y_3)\frac{\partial}{\partial y_3}(\chi_{(mn)k})+C_{i3mn}(y_3)\right)=0. \tag{5.66}$$

for any periodic $\boldsymbol{\chi}_{(mn)}(y_3)$. Therefore, Equation (5.66) is transformed into the form

$$C_{i3k3}(y_3)\,\chi_{(mn)k,3}+C_{i3mn}(y_3)=A_{imn} \tag{5.67}$$

and may be solved explicitly, so that

$$\chi_{(mn)k,3}=-\left\{C_{k3j3}\right\}^{-1}C_{j3mn}+\left\{C_{k3j3}\right\}^{-1}A_{jmn}. \tag{5.68}$$

One obtains from the periodicity condition $\left\langle\boldsymbol{\chi}_{(mn),3}\right\rangle_\Omega=0$:

$$-\left\langle\left\{C_{k3j3}\right\}^{-1}C_{j3mn}\right\rangle_\Omega+\left\langle\left\{C_{k3j3}\right\}^{-1}\right\rangle_\Omega A_{jmn}=0. \tag{5.69}$$

Therefore

$$A_{imn} = \left\langle \left\{ C_{i3k3} \right\}^{-1} \right\rangle_{\Omega}^{-1} \left\langle \left\{ C_{k3j3} \right\}^{-1} C_{j3mn} \right\rangle_{\Omega} \tag{5.70}$$

and it holds that

$$\chi_{(mn)k,3} = -\left\{ C_{k3j3} \right\}^{-1} C_{j3mn} + \left\{ C_{k3j3} \right\}^{-1} \left\langle \left\{ C_{j3q3} \right\}^{-1} \right\rangle_{\Omega}^{-1} \left\langle \left\{ C_{q3p3} \right\}^{-1} C_{p3mn} \right\rangle_{\Omega}. \tag{5.71}$$

Taking into account that the state functions depend only on the y_3 axis, the following holds:

$$C_{ijkl}^{(eff)} = \left\langle C_{ijkl} + C_{ijm3} \chi_{(kl)m,3} \right\rangle_{\Omega}. \tag{5.72}$$

Finally, the homogenized elasticity tensor components are given by

$$C_{ijkl}^{(eff)} = \left\langle C_{ijkl} \right\rangle_{\Omega} - \left\langle C_{ijm3} \left\{ C_{m3p3} \right\}^{-1} C_{p3kl} \right\rangle_{\Omega} + \left\langle C_{ijm3} \left\{ C_{m3p3} \right\}^{-1} \right\rangle_{\Omega} \left\langle \left\{ C_{p3n3} \right\}^{-1} \right\rangle_{\Omega}^{-1} \left\langle \left\{ C_{n3q3} \right\}^{-1} C_{q3kl} \right\rangle_{\Omega}. \tag{5.73}$$

In case of isotropic and linear elastic constituent materials of this composite, it is obtained after some algebraic manipulation that [58]

$$C_{1111}^{(eff)} = C_{2222}^{(eff)} = \left\langle \frac{(1-\nu)e}{(1+\nu)(1-2\nu)} \right\rangle_{\Omega} - \left\langle \frac{e(1-2\nu)}{1-\nu^2} \right\rangle_{\Omega} + \frac{\left\langle \frac{1-2\nu}{1-\nu} \right\rangle_{\Omega}^{2}}{\left\langle \frac{(1+\nu)(1-2\nu)}{(1-\nu)e} \right\rangle_{\Omega}}, \tag{5.74}$$

$$C_{3333}^{(eff)} = \left\langle \frac{(1+\nu)(1-2\nu)}{(1-\nu)e} \right\rangle_{\Omega}^{-1}, \tag{5.75}$$

$$C_{1133}^{(eff)} = C_{3311}^{(eff)} = C_{2233}^{(eff)} = C_{3322}^{(eff)} = \frac{\left\langle \frac{1-2\nu}{1-\nu} \right\rangle_{\Omega}}{\left\langle \frac{(1+\nu)(1-2\nu)}{(1-\nu)e} \right\rangle_{\Omega}}, \tag{5.76}$$

$$C_{1122}^{(eff)} = C_{2211}^{(eff)} = \left\langle \frac{e}{1-\nu} \right\rangle_{\Omega} - \left\langle \frac{e(1-2\nu)}{1-\nu^2} \right\rangle_{\Omega} + \frac{\left\langle \frac{1-2\nu}{1-\nu} \right\rangle_{\Omega}}{\left\langle \frac{(1+\nu)(1-2\nu)}{(1-\nu)e} \right\rangle_{\Omega}}, \tag{5.77}$$

$$C_{1212}^{(eff)} = C_{2121}^{(eff)} = \left\langle \frac{e}{1+\nu} \right\rangle_{\Omega}, \tag{5.78}$$

$$C_{1313}^{(eff)} = C_{3131}^{(eff)} = C_{2323}^{(eff)} = C_{3232}^{(eff)} = \left\langle \frac{1}{\frac{1+\nu}{e}} \right\rangle_{\Omega}, \tag{5.79}$$

with the remaining components of the EET equal to 0. The equations given above enable both analytical derivations of the probabilistic moments for the given random variable in the system, Monte Carlo simulation of the homogenized tensor and generalized stochasticperturbation-based computational analysis; all these techniques are contrasted in computational experiments provided later in this chapter. Another model of effective in-plane characteristics of the laminates to be considered in the presence of some uncertainty may be found in [162].

5.3 Computational Implementation

Let us introduce the following approximation of homogenization functions $\chi_{(rs)i}$ at any point of the considered continuum Ω in terms of a finite number of generalized coordinates $q_{(rs)\alpha}$ and shape functions $\varphi_{i\alpha}$ [71, 92]:

$$\chi_{(rs)i} = \varphi_{i\alpha} q_{(rs)\alpha}, \quad i, r, s = 1,2, \ \alpha = 1, \ldots, N, \tag{5.80}$$

and the strain $\varepsilon_{ij}(\chi_{(rs)})$ as well as stress $\sigma_{ij}(\chi_{(rs)})$ tensors

$$\varepsilon_{ij}(\chi_{(rs)}) = B_{ij\alpha} q_{(rs)\alpha}, \tag{5.81}$$

$$\sigma_{ij(rs)} = \sigma_{ij}(\chi_{(rs)}) = C_{ijkl}\varepsilon_{kl}(\chi_{(rs)}) = C_{ijkl} B_{kl\alpha} q_{(rs)\alpha}, \tag{5.82}$$

where $B_{kl\alpha}$ is the conventional FEM strain–nodal displacement operator provided in Equation (2.135). Therefore, a variational statement for the homogenization problem is proposed as

$$\int_{\Omega} \delta\chi_{(rs)i,j} C_{ijkl} \chi_{(rs)k,l} \, d\Omega = \sum_{a=2}^{n} \int_{\Gamma_{(a-1,a)}} \delta\chi_{(rs)i} \left[F_{(rs)i} \right] |_{\Gamma_{(a-1,a)}} d\Gamma \ \text{(no sum on } r, s). \tag{5.83}$$

Next, let us define the global stiffness matrix as

$$K_{\alpha\beta} = \sum_{f=1}^{E} K_{\alpha\beta}^{(f)} = \sum_{f=1}^{E} \int_{\Omega_f} C_{ijkl}^{(f)} B_{ij\alpha} B_{kl\beta} d\Omega. \tag{5.84}$$

Introducing this matrix into Equation (5.83) together with the discretization made in Equation (5.80) and minimizing it with respect to the generalized coordinates we arrive at

$$K_{\alpha\beta} q_{(rs)\alpha} = Q_{(rs)\alpha}, \tag{5.85}$$

where $Q_{(rs)\alpha}$ is the external load vector which includes the stress boundary conditions applied along the interface (if only the neighboring components are perfectly bonded)

Table 5.1 The components of the forces $F_{(pq)i}$

	$\chi_{(11)}$	$\chi_{(12)}$	$\chi_{(22)}$
$F_{(pq)1}$	$C^{(2)}_{1111} - C^{(1)}_{1111}$	$C^{(2)}_{1212} - C^{(1)}_{1212}$	$C^{(2)}_{1122} - C^{(1)}_{1122}$
$F_{(pq)2}$	$C^{(2)}_{2211} - C^{(1)}_{2211}$	$C^{(2)}_{1212} - C^{(1)}_{1212}$	$C^{(2)}_{2222} - C^{(1)}_{2222}$

in the following form:

$$\sigma_{ij}\left(\chi_{(pq)}\right) n_j = \left[C_{ijpq}\right] n_j = F_{(pq)i}; \quad \mathbf{x} \in \partial\Omega_{12}, \tag{5.86}$$

where n_j is the component of the unit vector normal to the given interface and directed to the RVE interior, while $[f]$ denotes the difference of the function f values:

$$[f] = f^{(2)} - f^{(1)}. \tag{5.87}$$

The stress boundary conditions corresponding to different homogenization problems are specified in Table 5.1.

It should be underlined that taking into account the interface phenomena in engineering composites, the fiber and matrix boundaries may be partially different contours (lack of contact between the components caused by a delamination), which may be the result of composite processing thermal stresses or extensive fatigue processes. Finally let us note that to assure the symmetry conditions on periodicity cell edges, the displacements perpendicular to the external RVE boundary are fixed for every nodal point.

As could be noticed during derivation of the equations for the generalized perturbation-based approach, one of the most complicated issues is numerical determination of up to nth-order partial derivatives of the structural response function with respect to the randomized parameter. As we demonstrated in the preceeding chapters, it is possible to determine this function by a multiple solution of the boundary value problem around the expectation of the random parameter. The response function for each component of the homogenized tensor is built up from uniform symmetric discretization of this expectation in its close neighborhood with equidistant intervals. A set of ordinary deterministic computations of the homogenized tensor components leads to the final formation of the response function for all $C^{(eff)}_{ijkl}$. That is why we consider further the problem of unknown response function approximation with the following polynomial of $(n-1)$th order [80]:

$$C^{(eff)}_{ijkl} = D^{(n-1)}_{ijkl} b^{n-1} + D^{(n-2)}_{ijkl} b^{n-2} + \cdots + D^{(0)}_{ijkl} \tag{5.88}$$

it is solved using the least-squares method in both weighted and non-weighted schemes displayed in Chapter 1. Having determined the coefficients in Equation (5.88) it is possible to calculate up to nth-order ordinary derivatives of the homogenized elasticity tensor with respect to the random input parameter b at the given mean b_0 as follows:

- first-order derivative

$$\frac{\partial C_{ijkl}^{(eff)}}{\partial b} = (n-1)\, D_{ijkl}^{(n-1)} b^{n-2} + (n-2)\, D_{ijkl}^{(n-2)} b^{n-3} + \cdots + D_{ijkl}^{(1)} \tag{5.89}$$

- second-order derivative

$$\frac{\partial^2 C_{ijkl}^{(eff)}}{\partial b^2} = (n-1)\,(n-2)\, D_{ijkl}^{(n-1)} b^{n-3} + (n-2)\,(n-3)\, D_{ijkl}^{(n-2)} b^{n-4} + \cdots + D_{ijkl}^{(2)} \tag{5.90}$$

- kth-order derivative

$$\frac{\partial^k C_{ijkl}^{(eff)}}{\partial b^k} = \prod_{i=1}^{k} (n-i)\, D_{ijkl}^{(n-i)} b^{n-k} + \prod_{i=2}^{k} (n-i)\, D_{ijkl}^{(n-2)} b^{n-(k+1)} + \cdots + D_{ijkl}^{(n-k)}. \tag{5.91}$$

Providing that the response function of the EET has a single independent argument being the input random variable of the problem, it is possible to employ the stochastic perturbation technique based on the Taylor representation to compute up to the mth central probabilistic moments $\mu_m(C_{ijkl}^{(eff)})$.

5.4 Numerical Experiments

Example 5.1: Aging analysis of the interphase parameters

The main aim of the first example is to study the effective Young's modulus of the interphase using some combination of the input probabilistic moments using Equation (5.15) [78]. The output probabilistic moments from this approximation may be used in further homogenization of the entire periodicity cell – the only modification in relation to the traditional RVE meshing would be an additional discretization of the interphase between the fiber and the matrix. The following combination of the interface defects input data is used to check the time fluctuations of the first two probabilistic moments for the homogenized Young's modulus: (i) for the expected values $E\left[r_a^0(\omega)\right] = 0.004$, $E\left[\dot{r}_a(\omega)\right] = 4.0E-5/\text{year}$, $E\left[n_a^0(\omega)\right] = 100$ and (ii) for all the coefficients of variation for the defect parameters taken equal to 0.2, so that $\text{Var}\left(n_a^0(\omega)\right) = \left(0.2\ E\left[n_a^0(\omega)\right]\right)^2$, $\text{Var}\left(\dot{n}_a(\omega)\right) = \left(0.2\ E\left[\dot{n}_a(\omega)\right]\right)^2$, $\text{Var}\left(r_a^0(\omega)\right) = \left(0.2\ E\left[r_a^0(\omega)\right]\right)^2$, $\text{Var}\left(\dot{r}_a(\omega)\right) = \left(0.2\ E\left[\dot{r}_a(\omega)\right]\right)^2$. The internal radius of the boundary, where these defects are located, is adopted as $R = 0.4$, and material data for the original matrix are taken as $E\left[e_a^0(\omega)\right] = 200 \times 10^{10}$, $E[\dot{e}_a(\omega)] = 0.015 \times 10^{10}/\text{year}$, whereas the variances are equal to $\text{Var}\left(e_a^0(\omega)\right) = \left(0.1 \times E\left[e_a^0(\omega)\right]\right)^2$ and $\text{Var}(\dot{e}_a(\omega)) = (0.2 \times E[\dot{e}_a(\omega)])^2$; we assume finally a lack of any cross-correlations between all these variables.

The results of numerical analysis performed in the symbolic system Maple™ are presented in Figure 5.6 (expected values) and Figure 5.7 (coefficient of variation) – they are determined as functions of time (given in years) and expected value of the cavities nucleating at the considered interface. A computer script

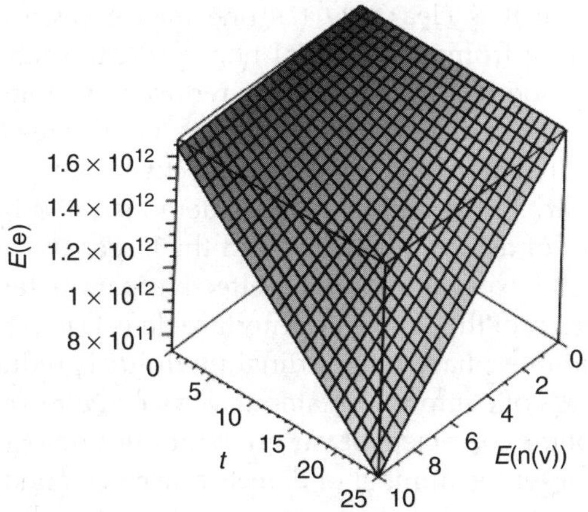

Figure 5.6 The expected values for time-dependent effective Young's modulus. Reproduced with permission from Begell House

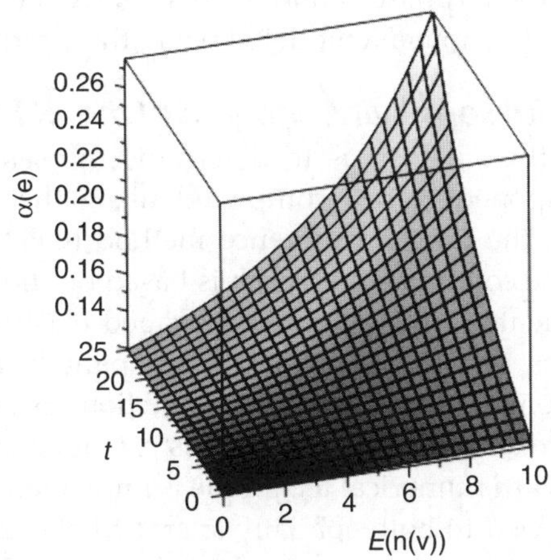

Figure 5.7 The coefficients of variation for time-dependent effective Young's modulus. Reproduced with permission from Begell House

written in the Maple™ internal language enables analytic determination of the first two moments of the homogenized parameter random process given on the LHS of Equation (5.15) and may be expanded further toward determination of higher moments and characteristics also. These analytic derivations are possible mainly due to the fact that the interphase area appearing in the RHS denominator is given as deterministically dependent on time, so that the first two central moments are obtained as algebraic combinations of the first two moments for the defect variations as well as the Young's modulus of the virgin material (the truncation effect of these processes is not included in this study).

Analyzing Figure 5.6 it is clear that (as one may expect) even for expectation equal to 0 and resulting from some initial non-zero expectations of these defects, the primary Young's modulus is significantly reduced even at the beginning of the aging process – for $t=0$. An increase of time for the same value $E\left[n_a^0(\omega)\right]=0$ leads to some further small reduction of the interphase effective Young's modulus, where this change is almost linear. The general tendency demonstrated by this surface is that the larger the number of defects nucleated and the larger the time of the composite history, the smaller the expectation of the effective parameter being studied. It is seen that the total impact of the stochastic interface defects is very significant because the expectation of the interphase longitudinal modulus is reduced here three times within the first 10 years of composite material service. Quite an inverse observation can be made in case of the coefficient of variation for this homogenized parameter. As one can predict, the larger the number of defects nucleated and the longer the time of the composite history, the higher the coefficient of variation for $\alpha\left(e_a^{(eff)}(\omega;t)\right)$. As is easy to detect, the output coefficient of variation is not proportional and rather larger than the input coefficient of variation, so that considering a decrease in expected value for the same time moment of this process means the random dispersion of the Young's modulus in the interphase cannot be disregarded in further computational analysis of the interface phenomena modeled using the interphase concept.

Example 5.2: Material sensitivity analysis of the EET

The main purpose of this analysis is to determine the sensitivity coefficients for the in-plane EET components. It is completed using the RFM displayed above and, for a comparison, the central difference method (CFD) implemented for this problem before [77]. The combined approach is based on the Maple™ computations of the sensitivity gradients of the spatially averaged elasticity tensor components with respect to various design parameters. The spatially averaged stress tensor components coming from the homogenization function are approximated using the response function approach and differentiation is provided symbolically in Maple™ also. A full straightforward numerical technique is implemented in this system using the RFM technique applied to both spatially averaged elasticity and homogenizing stress tensor components. The separate solver for the RFM computations based on the least-squares method is implemented in the Maple™ symbolic environment together with normalization procedures for all sensitivity coefficients computed. It is necessary to notice that the core of the homogenization process is carried out with use of the FEM-based program MCCEFF, where plane strain elements are employed.

We consider a composite with quarter of the periodicity cell (Figure 5.8) – the fiber has a round cross-section and the entire cell is square; the reinforcement ratio is equal to 50% of the RVE total area. Material characteristics for the computational analysis are as follows: $e_1=84.0\,\mathrm{GPa}$, $\nu_1=0.22$ as well as $e_2=4.0\,\mathrm{GPa}$, $\nu_2=0.34$; the FEM discretization using 62 four-noded elements with 76 nodal points for the plane strain analysis implemented in the system MCCEFF is presented below.

The normalized sensitivity coefficients for the EET are collected in Table 5.2, and for the spatially averaged over the RVE elasticity tensor components (AET) in Table 5.3; the design variables are taken separately as Young's moduli of the fiber and the

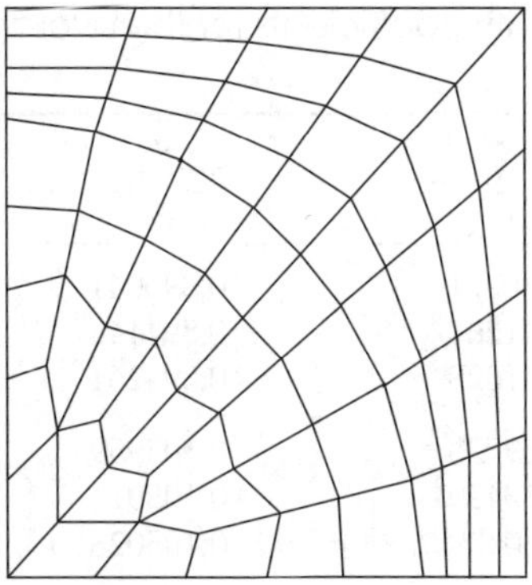

Figure 5.8 FEM discretization of the RVE quarter. Reproduced with permission from Elsevier

Table 5.2 Sensitivity coefficients for the effective elasticity tensor

h	$\frac{\partial C_{1111}^{(eff)}}{\partial h}$	$\frac{\partial C_{1122}^{(eff)}}{\partial h}$	$\frac{\partial C_{1212}^{(eff)}}{\partial h}$
e_1	0.110404	0.025965	0.954875
	0.110171	0.029391	0.954875
ν_1	0.011347	0.141546	−0.178031
	0.012469	0.138449	−0.178211
e_2	0.880843	0.932087	0.039481
	0.882215	0.939607	0.040033
	(0.867)	(0.926)	(0.044)
ν_2	1.081097	2.505468	−0.008692
	1.101308	2.556652	−0.009940
	(1.205)	(2.814)	(−0.011)

matrix together with the additional Poisson ratios. The first table is so arranged that the first value in both tables corresponds to the RFM method, the second to the CFD method (for the quarter of the cell), whereas the CFD results for the entire RVE are included in the brackets below. The next table collects the analytical results for the AET components, the RFM implemented and, finally, the CFD computational results. Let us note that all the results computed using the central finite difference scheme are obtained for the increment of the perturbed parameter equal to 1%, which follows the conclusions from other numerical models.

The first and most important conclusion which can be drawn from these results is almost perfect agreement of various numerical approaches for the homogenized characteristics' response function approach, the finite difference technique in the first table as well as both of them with pure analytical differentiation implemented in the system Maple™ (provided in Table 5.3). In practical terms it means that for the needs

Table 5.3 Sensitivity coefficients for the averaged elasticity tensor

h	$\frac{\partial C_{1111}^{(eff)}}{\partial h}$	$\frac{\partial C_{1122}^{(eff)}}{\partial h}$	$\frac{\partial C_{1212}^{(eff)}}{\partial h}$
e_1	0.940271	0.896041	0.958866
	0.941264	0.894144	0.957908
	0.941028	0.894261	0.957907
ν_1	0.059728	0.103959	0.041134
	0.060429	0.105011	0.041660
	0.060439	0.105028	0.041667
e_2	0.304023	1.438492	−0.172910
	0.304494	1.436554	−0.173195
	0.303889	1.434796	−0.173158
ν_2	0.080999	0.298495	−0.010437
	0.080488	0.296135	−0.010260
	0.082141	0.302260	−0.010470

of the homogenization method, computational implementation of the sensitivity analysis with all these methods is accurate and can be used equivalently depending on the engineering software employed for a simulation. It should be clearly exposed here that the usage of the RFM is independent of any further numerical parameters like the increments in the CFD computations, no closed formulas are necessary like in the analytical approach, and no technical interventions are really needed in any of the source codes for FEM homogenization-oriented programs. Therefore, the apparent efficiency of this technique compared with the remaining methodologies gives a new modeling tool for sensitivity analysis as well as for further random modeling, as will be seen in the next subsection.

In further numerical simulation it would be interesting to check the influence of the number of fibers of the RVE discretized on the values of these gradients, also in case of random distribution of these fibers in the computational domain. According to previous studies in this area, it can be confirmed on the RVE quarter that the most important role in such a composite is played by the elastic characteristics of the matrix, with a smaller influence of the fiber Young's modulus, whereas its Poisson ratio can be practically neglected (during the optimization process). Some of these parameters, namely the Poisson ratio, can result in negative sensitivity coefficients, so that its increase will decrease some homogenized elasticity tensor components.

Example 5.3: Computations of expected values of the EET using polynomial interpolation

The probabilistic RFM-related technique is implemented here in two quite separate ways. The first approach can be classified as a combined analytical–numerical methodology, where the zeroth-order spatially averaged elasticity tensor together with its higher-order derivatives with respect to the input random variable are determined all using the system Maple™. The second part, consisting of the spatially averaged homogenizing stresses over the RVE, is partially computed in the finite element-based system MCCEFF and then included in the Maple™ system to approximate the response functions of the spatially averaged homogenizing stress tensor components w.r.t. random input quantities. The procedure has been programmed with use of the homogenization-oriented computer program MCCEFF used previously for computations of the EET component probabilistic moments via the Monte Carlo simulation technique.

Let us consider as an illustration a composite with quarter of the periodicity cell – the fiber has a round cross-section and the entire cell is square; the reinforcement ratio is equal to 50% of the total area of the RVE as before. Elastic properties of the glass fiber and epoxy matrix are adopted as follows: the Young's moduli expected values $E[e_1]=84.0\,\text{GPa}$, $E[e_2]=4.0\,\text{GPa}$, while the deterministic Poisson ratios are taken as equal to $E[\nu_1]=0.22$ for the fiber and $E[\nu_2]=0.34$ for the matrix (each parameter is randomized separately and then the expectations of the remaining properties become simply their deterministic values) [77].

The preliminary results of the computational analysis are presented in Figure 5.9 as the response functions of all the homogenized elastic tensor components, where the Poisson ratio of the matrix is taken as the input random variable. This primary choice was justified by the fact that all previous computational studies show that this particular composite is the most sensitive to this ratio's variations. As is clear from all the graphs, the very smooth function is obtained at the expectation of this parameter but at both edges of the computational domain the resulting polynomial representation of the tenth order returns some fluctuations of the response function. It results here from the fact that simple polynomial interpolation is used to recover all necessary response functions.

Next, Figures 5.10–5.13 show the expected values of the first component for the EET, $E\left[C^{(eff)}_{1111}\right]$, as functions of the perturbation order of the method – from the second until the tenth – as well as of the input coefficient of variation of the random input

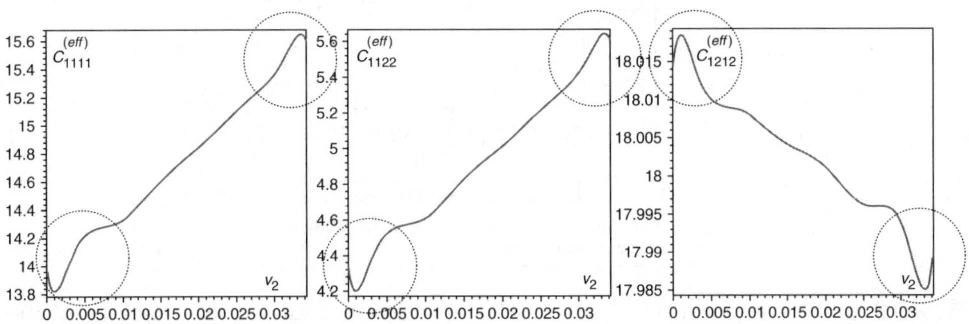

Figure 5.9 The probabilistic response functions around the expectations for the homogenized tensor components. Reproduced with permission from Elsevier

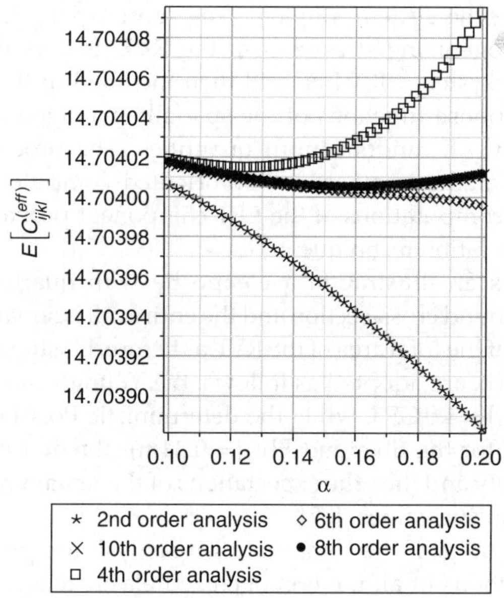

Figure 5.10 The expected values of $C_{1111}^{(eff)}$ for randomized Young's modulus of the fiber. Reproduced with permission from Elsevier

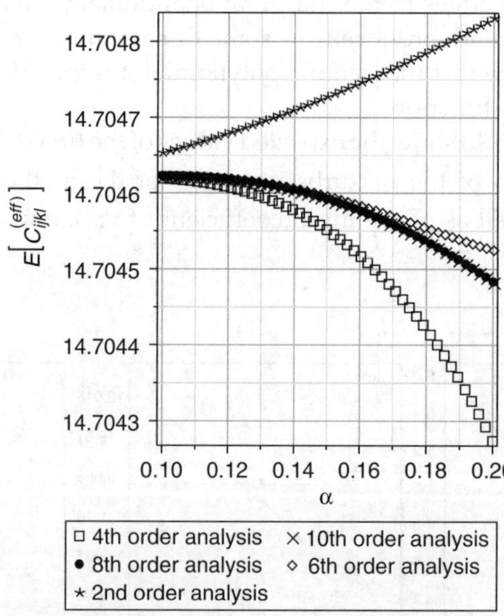

Figure 5.11 The expected values of $C_{1111}^{(eff)}$ for randomized Young's modulus of the matrix. Reproduced with permission from Elsevier

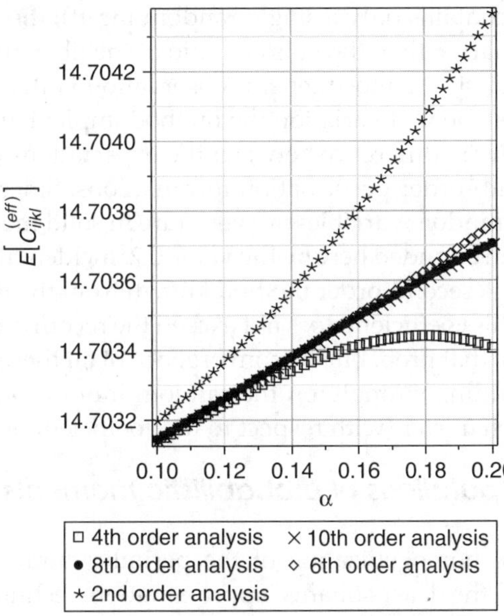

Figure 5.12 The expected values of $C_{1111}^{(eff)}$ for randomized Poisson ratio of the fiber. Reproduced with permission from Elsevier

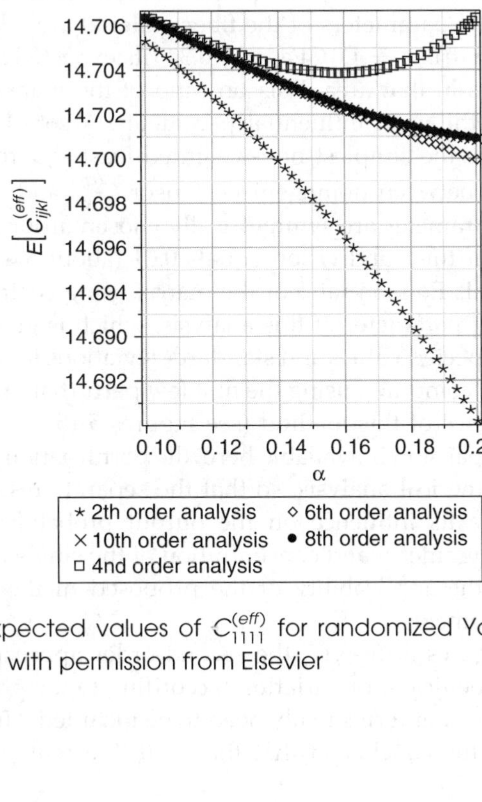

Figure 5.13 The expected values of $C_{1111}^{(eff)}$ for randomized Young's modulus of the matrix. Reproduced with permission from Elsevier

parameter (each test contains only a single random input); the Young's modulus of the fiber, next of the matrix; then the Poisson ratio of the fiber and finally the Poisson ratio for the matrix. First, the most general observation is that even for the largest value of the input coefficient of variation the method implemented is convergent, so that there is practically no difference between the expectations computed according to the eighth- and tenth-order perturbation formulations. Since large coefficients of variation of the input random variables are very rare in solid mechanics applications, this coefficient has been bounded here by the value 0.2. Incidentally, one can notice the differences between the second-order method known from the literature and higher-order results, even if this coefficient does not exceed the recommended 0.1 value [112]. It is very characteristic that probabilistic convergence of all these expected values has a different type depending strongly on the random input type but generally has a definitely non-linear character (with respect to the coefficient of variance).

Example 5.4: Computations of probabilistic moments of the EET using LSM

Numerical analysis of homogenization of the periodic random fiber composite is provided now using the least-squares approximation technique and performed using the program MCCEFF and the symbolic computing environment of system Maple™. The internal automatic generator of this program for the square RVE with centrally located round fiber occupying 34% of the RVE is used to prepare the mesh consisting of 144 four-noded rectangular plane strain finite elements and 153 nodes (see Figure 5.14). Elastic parameters of the fiber material are taken as $e_1 = 84.0\,\text{GPa}$, $\nu_1 = 0.22$ and for the matrix $e_2 = 4.0\,\text{GPa}$, $\nu_2 = 0.34$ (its expected value). According to many previous studies in that area, Poisson ratio of the matrix has been detected for this composite as the most influential parameter. A set of 11 trial equidistant points is used to make the simplest non-weighted least-squares approximation of the response function between homogenized tensor $C^{(eff)}_{ijkl}$ and ν_2 [84]. The discrete values of this input parameter are symmetrically chosen around its expected value and the basic length of this subdivision equals 0.01 (about 3% of the basic value). Now we randomize this Poisson ratio of the matrix using coefficient of variation α as the additional input parameter of this analysis, which is given each time on the horizontal axis. The expected values and standard deviations for all the homogenized tensor components are computed using the first few perturbation orders to verify the probabilistic convergence of this method (see Figures 5.15–5.20, correspondingly). Since full analytical expansion is available here, the perturbation parameter may also be included in the numerical analysis, so that the separate results (cf. Figures 5.21 and 5.22) demonstrate its influence on the output probabilistic moments. They show coefficients of asymmetry and concentration of the single homogenized tensor component to prove the applicability of the proposed method for computing of higher probabilistic moments.

As is clear from Figures 5.15–5.17, the second-order approach is acceptable for the very small input coefficient of variation (according to the previous predictions), but for $\alpha > 0.10$ higher-order terms really need to be included. Higher-order analysis leads immediately to the conclusion that for $\alpha < 0.15$ a tenth-order approach has

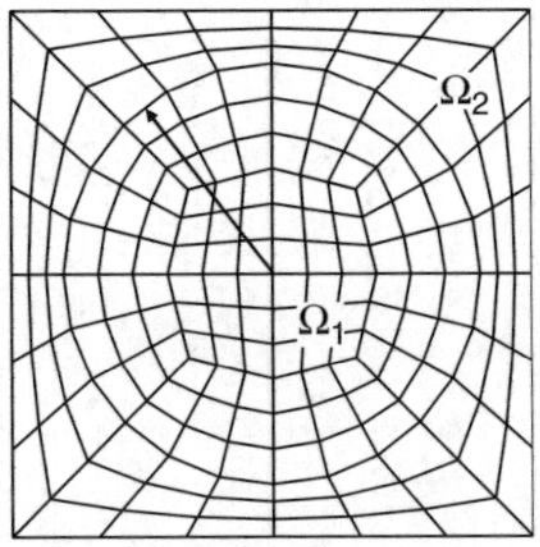

Figure 5.14 Discretization of the RVE with a single fiber

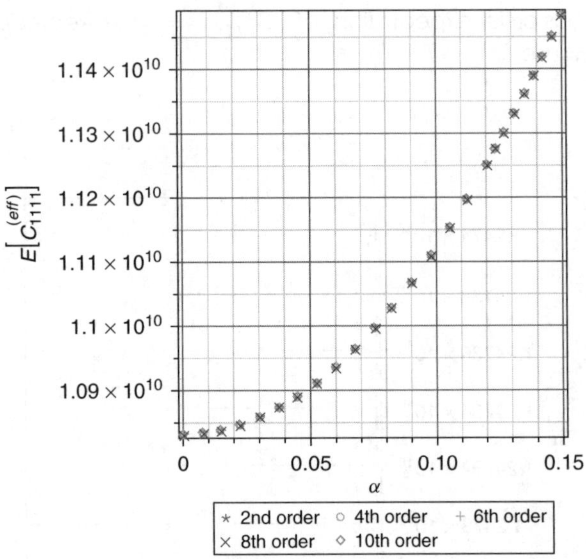

Figure 5.15 Various order expectations of $C_{1111}^{(eff)}$, $\nu_2 = \nu_2(\omega)$. Reproduced with permission from Begell House

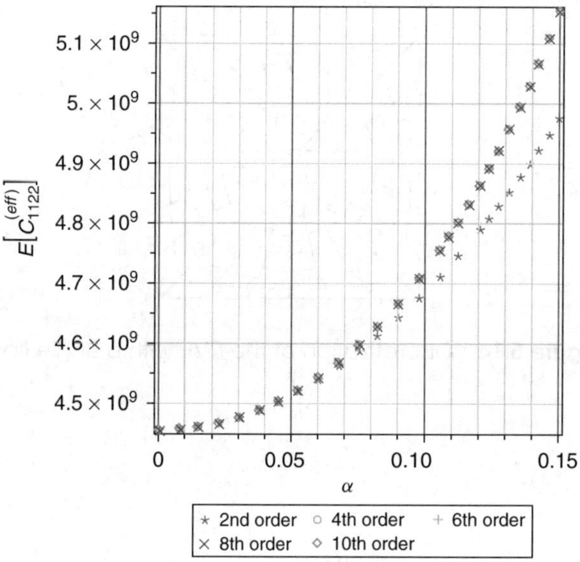

Figure 5.16 Various order expectations of $C_{1122}^{(eff)}$, $\nu_2 = \nu_2(\omega)$. Reproduced with permission from Begell House

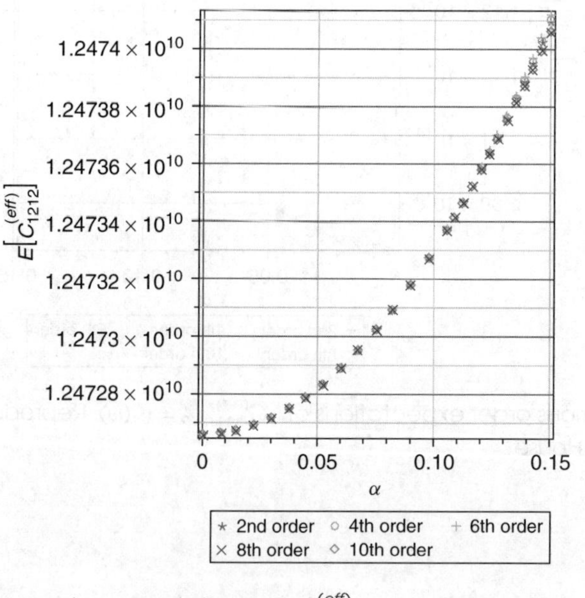

Figure 5.17 Various order expectations of $C_{1212}^{(eff)}$, $\nu_2 = \nu_2(\omega)$. Reproduced with permission from Begell House

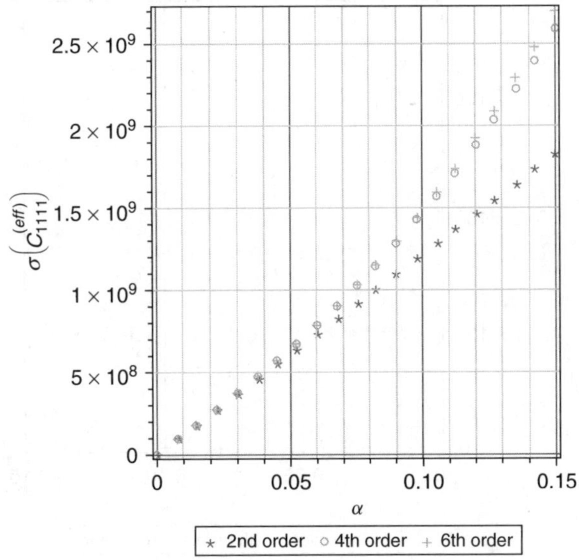

Figure 5.18 Various order standard deviations of $C_{1111}^{(eff)}$, $\nu_2 = \nu_2(\omega)$. Reproduced with permission from Begell House

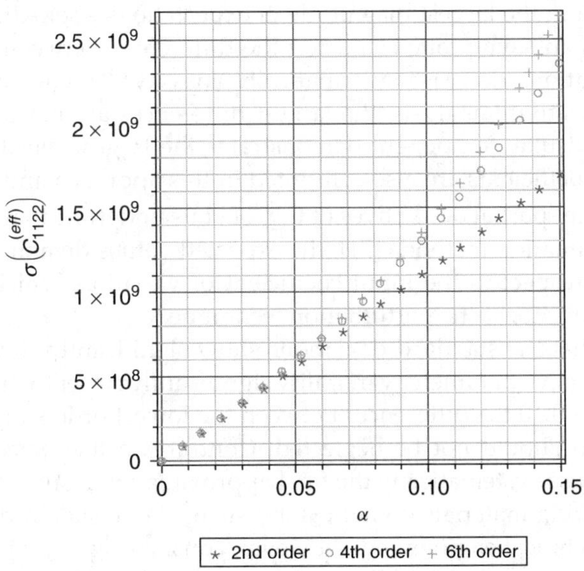

Figure 5.19 Various order standard deviations of $C_{1122}^{(eff)}$, $\nu_2 = \nu_2(\omega)$. Reproduced with permission from Begell House

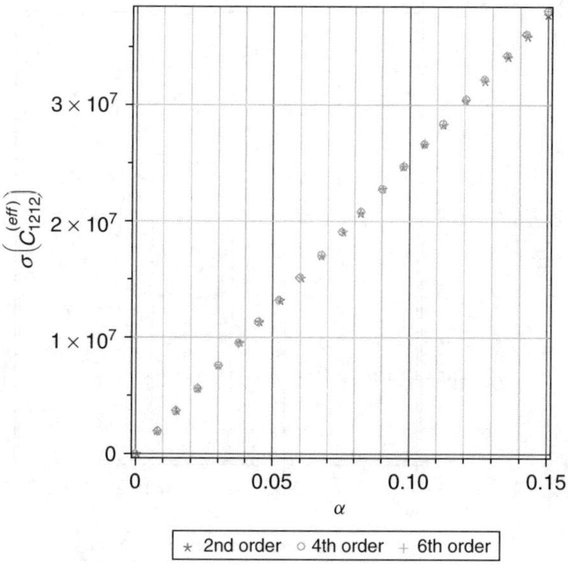

Figure 5.20 Various order standard deviations of $C_{1212}^{(eff)}$, $\nu_2 = \nu_2(\omega)$. Reproduced with permission from Begell House

sufficient accuracy for the expected values of all components of the homogenized tensor. Contrasting Figures 5.16 and 5.17 shows that tenth-order analysis not always really does result in the largest magnitude of expectations – sometimes probabilistic convergence has an asymptotic character; the differences between the neighboring order approximations decrease systematically anyway. As one could expect after deterministic sensitivity analysis, the largest differences are noticed in Figure 5.16 because this particular component demonstrates the largest sensitivity coefficients w.r.t. ν_2. These coefficients are also computed in this approach and can be extracted from the first-order partial derivatives of $C_{ijkl}^{(eff)}$. Let us underline that here, contrary to the statistical estimation methods [71], the expected values demonstrate some small variability with respect to the input coefficient of variation, which is the inherent aspect of the entire stochastic perturbation technique.

It is obvious that the standard deviations show significantly slower probabilistic convergence and have parameter variability significantly closer to the linear function. Now even for $\alpha > 0.10$ the differences between the lowest orders of these deviations are apparent and should not be neglected. Contrary to the expectations, now all new orders increase systematically the final approximation results. Let us remember that real engineering materials do not exhibit such a large standard deviation as 0.2 (except for geotechnical engineering), however, considering other input parameters, it shows possible range of the homogenized tensor random fluctuations.

Next, we examine in Figures 5.21 and 5.22 the coefficients of asymmetry and concentration, parametrized also with the perturbation parameter $\varepsilon = 0.9, \ldots, 1.1$. It confirms that the fourth-order characteristic is almost entirely influenced by this

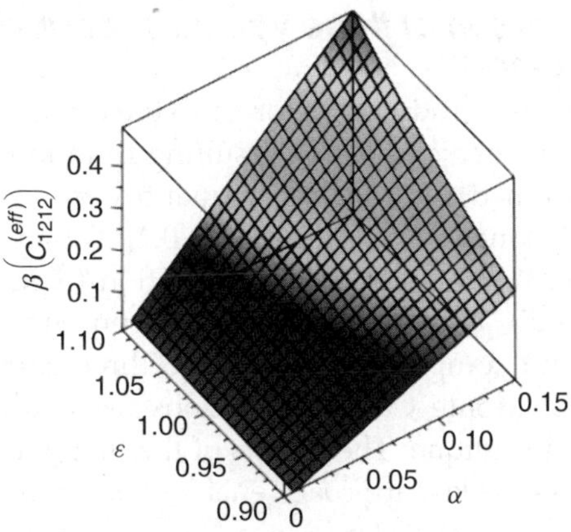

Figure 5.21 Coefficients of asymmetry for $C_{1212}^{(eff)}$, $\nu_2 = \nu_2(\omega)$. Reproduced with permission from Begell House

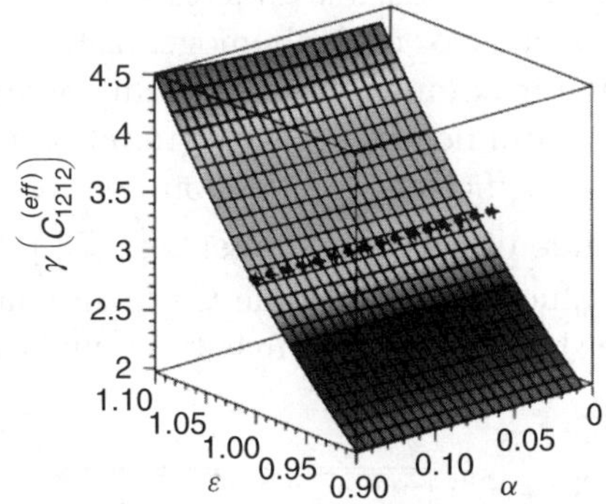

Figure 5.22 Coefficients of concentration for $C_{1212}^{(eff)}$, $\nu_2 = \nu_2(\omega)$. Reproduced with permission from Begell House

parameter choice and completely insensitive to the input coefficient of variation. Computer analysis returns here $\gamma = 3$ for $\varepsilon = 1$, which is typical for the Gaussian distribution. The third-order coefficient is dominated by α and less influenced by ε, however for all combinations it appears to be positive and nowhere exactly equal to 0 as for the Gaussian variables. The results obtained for the generalized stochastic perturbation technique are contrasted in Figure 5.22 with the Monte Carlo simulation results obtained for 10^4 random trials marked here with the asterisks. All the simulations have been provided for $\alpha = 0.0, 0.01, \ldots, 0.15$ and $\varepsilon = 1$ and, as is apparent, the simulation results are each time slightly larger than the stochastic perturbation technique results. So, the output PDFs are recognized thanks to both methods as very close to the Gaussian distributions, which remains in perfect agreement with the previous Monte Carlo simulations.

Example 5.5: Comparison of three various probabilistic homogenization methods

We analyze the same case study as before to study the basic four probabilistic characteristics of the homogenized tensor resulting from randomization of the fiber Young's modulus e_1 assumed as the Gaussian quantity with given expected value; its coefficient of variation belongs to the interval [0.0, 0.2]. They are collected in turn – the expectations (Figure 5.23), the coefficient of variation (Figure 5.24), the coefficient of skewness (Figure 5.25), and the coefficient of concentration (Figure 5.26) – all as functions of $\alpha(e_1)$, and computed according to three various methods: (i) semi-analytical approach, (ii) Monte Carlo simulations ($M = 10^4$), and (iii) generalized stochastic perturbation technique. The results of the first two methods are shown in a discrete mode, whereas the last approach enables continuous approximation of the basic random characteristics of $C_{ijkl}^{(eff)}$. This is due to the fact that the computer algebra system uses automatic differentiation to process the response functions, which of course preserves that continuity during the computations. The expected values of $C_{1111}^{(eff)}$ decrease together with an increase of $\alpha(e_1)$ in the Monte Carlo simulation and remain the same in the other methods; this underestimation is nevertheless negligible as comparable to a numerical error of the homogenization method itself. However, the difference between the maximum and minimum values of this expectation for the simulation method is around one or two promiles, and so can be neglected in practice. The resulting coefficients of variation (Figure 5.24) represent different trend – the higher the value of $\alpha(e_1)$, the larger $\alpha\left(C_{1111}^{(eff)}\right)$. Semi-analytical method returns here the same values as both Monte Carlo simulation and the stochastic perturbation method. The homogenization may be treated as linear transform on the

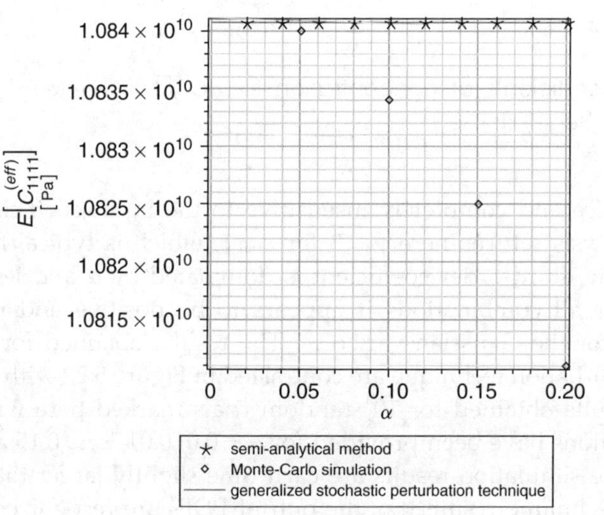

Figure 5.23 The expected values of the component $C_{1111}^{(eff)}$ with randomized e_1. Reproduced with permission from John Wiley & Sons Ltd.

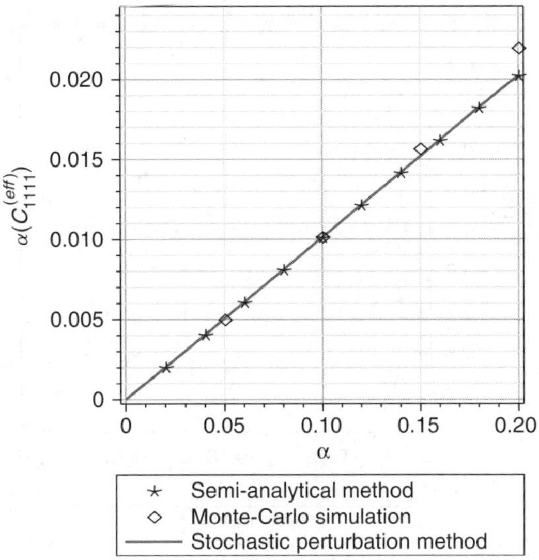

Figure 5.24 The coefficients of variation of the component $C_{1111}^{(eff)}$ with randomized e_1. Reproduced with permission from John Wiley & Sons Ltd.

basis of these results which perfectly agrees with other results obtained using Monte Carlo simulation for different volumetric ratios of the reinforcement.

The coefficient of skewness (Figure 5.25) is essentially different now for the simulation-based method, where we obtain clearly negative values decreasing for an additional increase of $\alpha(e_1)$, while for the other methods $\beta\left(C_{1111}^{(eff)}\right) = 0$. It is apparent that this decrease is also somewhat non-linear, but it does not influence further computations devoted to the entropy variations, since $\beta\left(C_{1111}^{(eff)}\right)$ never appears in the additional equations. Even more apparent differences between statistical and non-statistical models are shown in Figure 5.26 – in case of the concentration coefficient. The non-statistical methods clearly return here $\gamma\left(C_{1111}^{(eff)}\right) = 3$, which is typical for the Gaussian distribution, whereas Monte Carlo simulation gives a convex non-linear function of this coefficient with respect to $\alpha(e_1)$. Effectively, the simulation method almost doubles the coefficient γ for the input α taken from the interval [0.0, 0.2]; once more, however, these differences do not influence the final entropy computations. One may conclude at this point that non-statistical methods need to be improved, because the Monte Carlo simulation is theoretically a precise approximation of the probabilistic moments. We need to recall the asymptotic convergence of this technique, which may be efficient in determination of the first two moments for $M = 10^4$, but for higher moments essentially a larger number of random trials is required (in the crude version of this simulation). It is clear that the very long expansion is not really necessary because of the perfect agreement with the semi-analytical method here. We need to emphasize that the dominating part of the numerical error in this technique is in deterministic least-squares approximation of

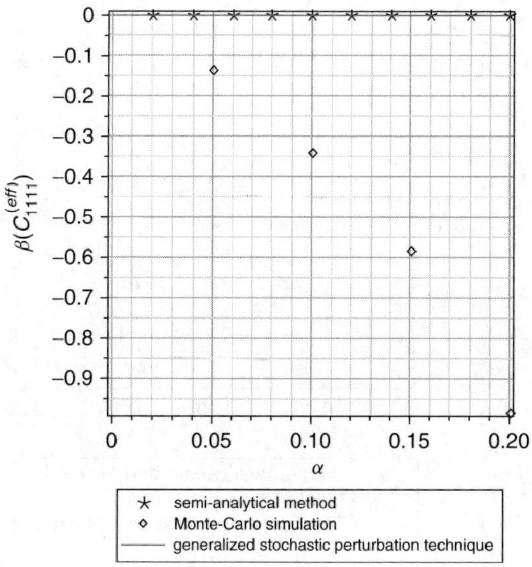

Figure 5.25 The coefficients of skewness of the component $C_{1111}^{(eff)}$ with randomized e_1. Reproduced with permission from John Wiley & Sons Ltd.

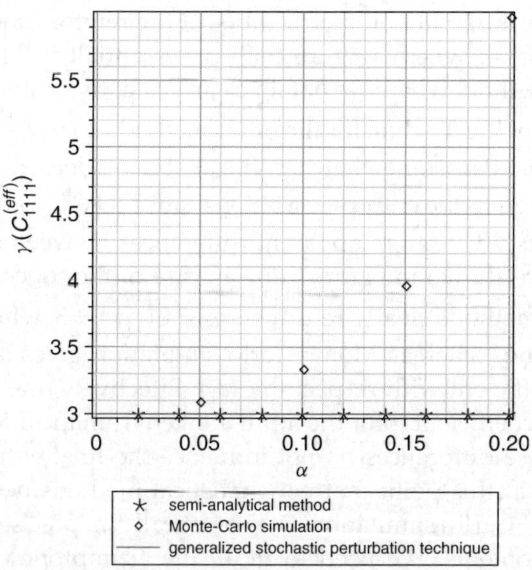

Figure 5.26 The coefficients of concentration of the component $C_{1111}^{(eff)}$ with randomized e_1. Reproduced with permission from John Wiley & Sons Ltd.

the analytical function between random input and output; the probabilistic part is provided straightforwardly from the definitions.

Further, we discuss the entropy fluctuations (difference between output and input probabilistic entropies) of the components $C_{1111}^{(eff)}$ (Figure 5.27), $C_{1122}^{(eff)}$ (Figure 5.28), and $C_{1212}^{(eff)}$ (Figure 5.29) with respect to $\alpha(e_1)$ and, separately, with respect to $\alpha(e_2)$ (Figures 5.30–5.32). They are calculated all from Equation (1.9) having a priori confirmed that the output PDF is Gaussian. As expected after previous probabilistic models, the entropy variations with respect to e_1 have larger absolute values (but are all negative) than the entropies computed for the variable e_2, because e_1, as the input random variable, is more influential than e_2 in the homogenization problem and, at the same time, e_1 itself is associated with significantly larger uncertainty than $C_{ijkl}^{(eff)}$. The largest absolute values within Figures 5.27–5.29 are noticed for $C_{1122}^{(eff)}$, then for $C_{1111}^{(eff)}$, while the smallest are in case of $C_{1212}^{(eff)}$. The detailed comparison of three numerical methods shows that the semi-analytical method always returns constant entropy loss and that the generalized stochastic perturbation method gives once more almost identical values. The exception from this rule is in the case of $C_{1212}^{(eff)}$, where the perturbation method results increase slowly non-linearly together with an increase of $\alpha(e_1)$. The Monte Carlo simulation results in the largest absolute values each time, always increasing together with $\alpha(e_1)$ (not really apparently for $C_{1212}^{(eff)}$), but entropy variations together with $\alpha(e_1)$ are really in the range of the few percents, so that they are not qualitatively different from the other methods. We need to remember that the Young's modulus for most of the structural components exhibits a randomness equivalent to $\alpha(e) \leq 0.10$, so that practically we obtain each time really almost the same results; the larger window for $\alpha(e)$ is taken to verify numerical efficiency here only.

Finally, we study the entropy fluctuations with respect to the variable $e_2 = e_2(\omega)$ and we have three different situations – all positive for $C_{1111}^{(eff)}$, almost equal to 0 (or tending to 0) in the case of $C_{1122}^{(eff)}$, and apparently negative – as shown in Figure 5.32 for $C_{1212}^{(eff)}$. The perfect agreement between the semi-analytical method and the generalized stochastic perturbation technique remains the same and, further, they both return constant values, totally independent of $\alpha(e_2)$. Contrary to the previous case of $e_1 = e_1(\omega)$, the Monte Carlo simulations start each time from essentially different values (for $\alpha(e_2)$ close to 0) than these returned by the first two methods. All they meet each other for about $\alpha(e_2) = 0.15$ and keep very close for larger values of the initial random dispersion of e_2. Looking for the vertical window sizes of Figures 5.30–5.32, one notices that now the differences are given in tenths (contrary to Figures 5.27–5.29) and cannot be simply postponed; the main reason could be a numerical error present in even a deterministic evaluation of the response functions of $C_{ijkl}^{(eff)}$. The homogenizing stresses (first-order corrector in the EET formula) are constant within the fiber and exhibit large variations in the matrix area. So, a randomization of the matrix properties should increase those fluctuations a few times and decisively affect the graphs analyzed here. The second reason, of almost the same importance, is that the sensitivity gradients of $C_{ijkl}^{(eff)}$ are essentially smaller with respect to e_2 than to e_1.

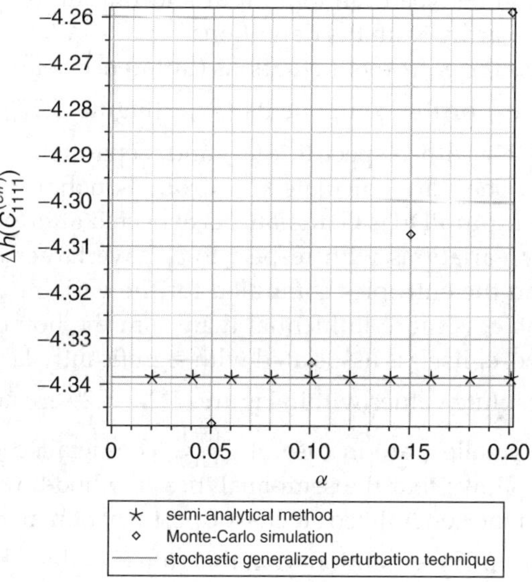

Figure 5.27 Entropy fluctuations of the component $C_{1111}^{(eff)}$ with randomized e_1. Reproduced with permission from John Wiley & Sons Ltd.

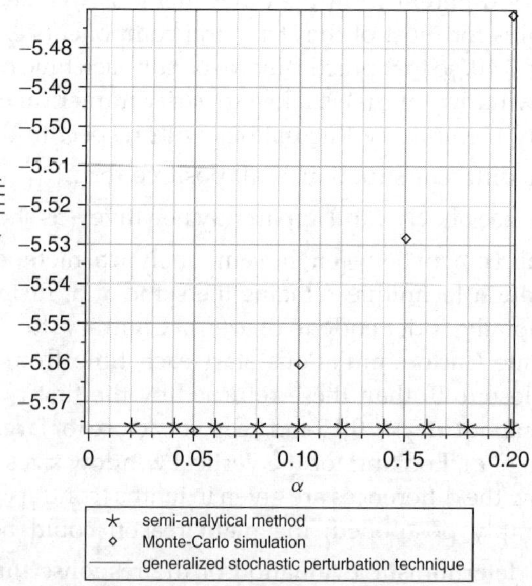

Figure 5.28 Entropy fluctuations of the component $C_{1122}^{(eff)}$ with randomized e_1. Reproduced with permission from John Wiley & Sons Ltd.

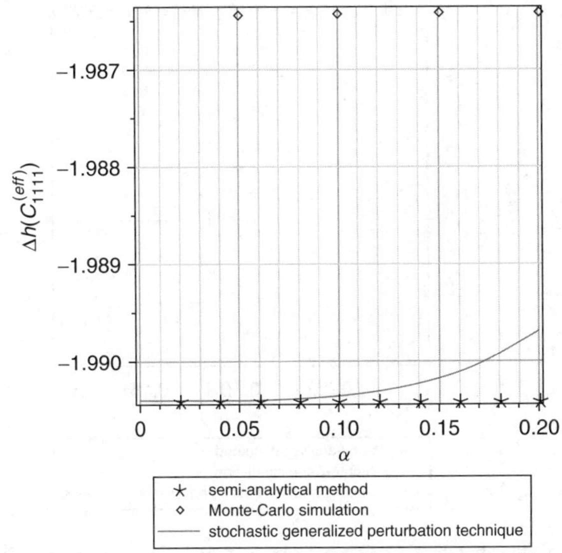

Figure 5.29 Entropy fluctuations of the component $C_{1212}^{(eff)}$ with randomized e_1. Reproduced with permission from John Wiley & Sons Ltd.

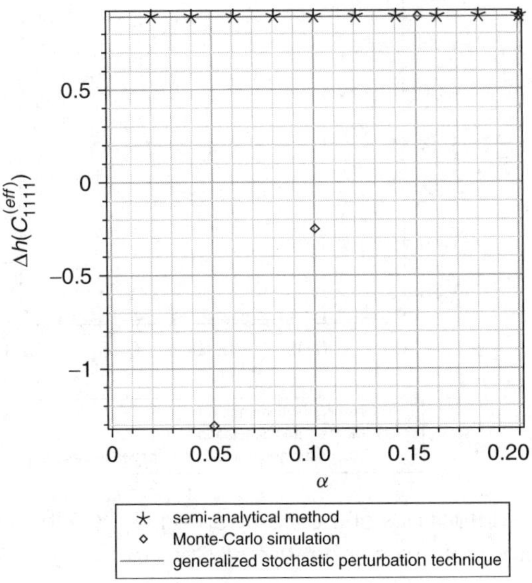

Figure 5.30 Entropy fluctuations of the component $C_{1111}^{(eff)}$ with randomized e_2. Reproduced with permission from John Wiley & Sons Ltd.

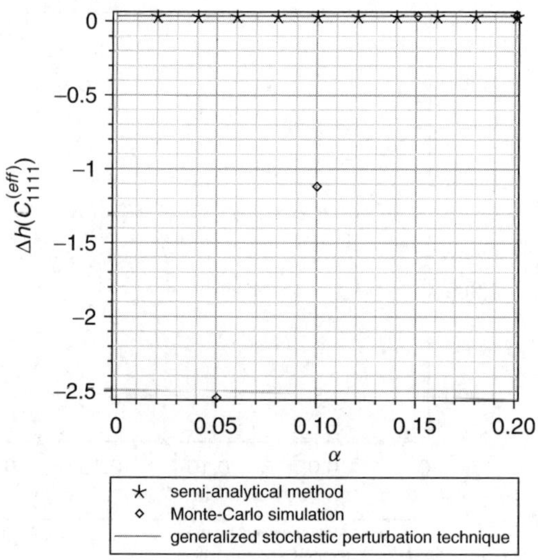

Figure 5.31 Entropy fluctuations of the component $C_{1122}^{(eff)}$ with randomized e_2. Reproduced with permission from John Wiley & Sons Ltd.

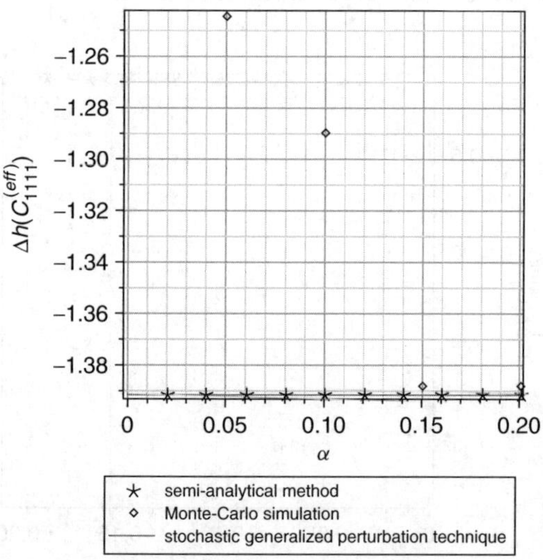

Figure 5.32 Entropy fluctuations of the component $C_{1212}^{(eff)}$ with randomized e_2. Reproduced with permission from John Wiley & Sons Ltd.

Example 5.6: Validation of the stochastic perturbation method in homogenization of 1D composite

A computational illustration is provided for the two-component composite with mean values of elastic parameters $e_1 = 84.0\,\text{GPa}$, $\nu_1 = 0.22$ (for reinforcement) and for the matrix taken as $e_2 = 4.0\,\text{GPa}$, $\nu_2 = 0.34$ (both having the same volume fractions). Now, the Young's modulus of the stronger material is taken as the input random

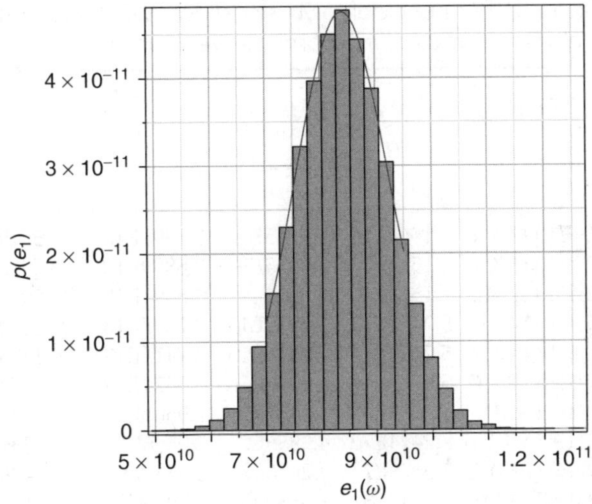

Figure 5.33 Initial histogram and theoretical PDF of the reinforcement Young's modulus. Reproduced with permission from IFTR PAS

variable in numerical simulation, where the standard deviation corresponds to its 10% random dispersion – the theoretical PDF and simulated counterpart (total number of random trials equal to 10^5) are shown in Figure 5.33. The entire analysis is provided using the symbolic environment of the system Maple™, where (i) direct numerical integration following classical definitions of the probability theory, (ii) Monte Carlo simulation and statistical estimation, together with (iii) the response functions and the generalized stochastic perturbation technique equations are implemented [85]. The first strategy needs some attention since general integration in the system Maple™ is unavailable in this case for the unbounded real domain, so that bounded numerical integration is provided within the limits 0 and double expectation. These bounds are determined after some a posteriori error-based analysis, where minimization of the computational domain width is carried out with respect to some a priori given error level (less than a single percent). The results for all homogenized tensor components are collected in Table 5.4 in the form of expected values (Pa), variances (Pa^2), standard deviations (Pa), fourth central probabilistic moments (Pa^4), as well as coefficients of variation, skewness, and concentration (all dimensionless).

The general conclusion is that all the methods return almost the same results – the agreement in the expected values as well as in the coefficients of concentration is perfect, some extremely small differences appear for the second-order characteristics, while the largest differences are noticed in case of skewness. The agreement of these techniques mainly follows the fact that the effective tensor components come from the algebraic transformations during the smearing of the original materials within the RVE, but the homogenization function has an analytical form unlike in most 2D problems, where some small differences caused by the SFEM itself are observed. Quite naturally, the largest variations between the methods are noticed for

Table 5.4 A comparison of analytical (AM), statistical (MC), and perturbation-based (PM) probabilistic characteristics of the homogenized tensor

Probabilistic parameters	$C_{1111}=C_{2222}$	C_{3333}	$C_{1133}=C_{3311}=C_{3322}=C_{2233}$	$C_{1122}=C_{2211}$	$C_{1212}=C_{2121}$	$C_{1313}=C_{3131}=C_{2323}=C_{3232}$
$E[X]$	2.9929E10 (AM)	1.1564E10	6.9545E9	3.8391E10	3.5919E10	5.7197E9
	2.9927E10 (MC)	1.1564E10	6.9545E9	3.8389E10	3.5916E10	5.7197E9
	2.9929E10 (PM)	1.1564E10	6.9546E9	3.8391E10	3.5919E10	5.7198E9
$\mathrm{Var}(X)$	5.5343E18 (AM)	5.2451E15	1.8970E15	8.7387E18	1.1852E19	6.1053E14
	5.5260E18 (MC)	5.2453E15	1.8971E15	8.7414E18	1.1855E19	6.1054E14
	5.5211E18 (PM)	5.0237E15	1.8170E15	8.7324E18	1.1852E19	5.8366E14
$\sigma(X)$	2.3504E9 (AM)	7.2423E7	4.3555E7	2.9561E9	3.4426E9	2.4709E7
	2.3507E9 (MC)	7.2424E7	4.3556E7	2.9566E9	3.4431E9	2.4709E7
	2.3497E9 (PM)	7.0878E7	4.2626E7	2.9551E9	3.4426E9	2.4159E7
$\mu_4(X)$	9.1634E37 (AM)	1.0343E32	1.3530E31	2.2936E38	4.2138E38	1.4152E30
	9.1348E37 (MC)	1.0260E32	1.3421E31	2.2864E38	4.2008E38	1.4035E30
	9.1451E37 (PM)	9.0917E31	1.1893E31	2.2878E38	4.2138E38	1.2350E30
$\alpha(X)$	0.0785 (AM)	0.0063	0.0063	0.0770	0.0958	0.0043
	0.0785 (MC)	0.0063	0.0063	0.0770	0.0959	0.0043
	0.0785 (PM)	0.0061	0.0061	0.0770	0.0958	0.0042
$\beta(X)$	−0.0066 (AM)	−0.6053	−0.6053	−0.0085	9.0630E−9	−0.6193
	−0.0057 (MC)	−0.6000	−0.6000	−0.0075	8.6663E−4	−0.6139
	−0.0094 (PM)	−0.8259	−0.8259	−0.0120	2.1499E−8	−0.8416
$\kappa(X)$	3.0026 (AM)	3.7596	3.7596	3.0034	3.0000	3.7967
	2.9915 (MC)	3.7292	3.7292	2.9921	2.9889	3.7652
	3.0000 (PM)	3.6024	3.6024	3.0001	3.0000	3.6254

the homogenized tensor components indexed with "3," since the random variable is smeared in this direction together with the deterministic quantity – the Young's modulus of the weaker material. This tensor also shows some probabilistic damping since the output coefficients of variation are generally smaller than for the input random variable. Also, generally one can conclude that quite independently from the numerical strategy, the effective tensor appears to be a Gaussian random variable, since its components have higher-order characteristics typical for this distribution. The generalized stochastic perturbation method seems to be efficient but some further implementation needs to be provided to eliminate small discrepancies in third-order characteristic numerical determination.

Example 5.7: Semi-analytical probabilistic determination of the homogenized tensor

The main issue in this example is to verify the influence of random dispersion in composite constituent's material parameters on the first four probabilistic characteristics of the homogenized tensor according to the semi-analytical method. It consists of determination of the response functions for the homogenized tensor w.r.t. the given

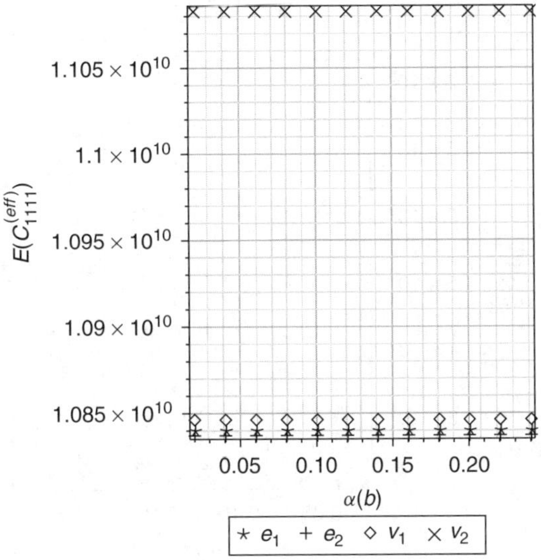

Figure 5.34 The expected values of $C_{1111}^{(eff)}$. Reproduced with permission from John Wiley & Sons Ltd.

random input variable and further symbolic recovery of probabilistic moments from Equations (1.1), (1.2) and (1.4). We assume that the resulting tensor's components are all distributed according to the Gaussian PDF given in Equation (1.3). It is assumed that the Young's moduli as well as the Poisson ratios are independent Gaussian random variables with a priori given expectations; these parameters are randomized independently. Their coefficients of variation belong each time to the interval $\alpha \in [0.0, 0.25]$ – the lower bounds correspond to a deterministic problem, where an upper bound was taken before as the extreme random dispersion. Generally, we would like to verify how random dispersion of the given input variable influences the dispersion of the homogenized tensor as well as its remaining basic probabilistic characteristics. The results of this analysis – the expectations $E\left[C_{ijkl}^{(eff)}\right]$, standard deviations $\sigma\left(C_{ijkl}^{(eff)}\right)$, asymmetry $\beta\left(C_{ijkl}^{(eff)}\right)$, and flatness $\gamma\left(C_{ijkl}^{(eff)}\right)$ coefficients – are contained in Figures 5.34–5.37 for $C_{1111}^{(eff)}$, Figures 5.38–5.41 for $C_{1122}^{(eff)}$, and Figures 5.42–5.45 for $C_{1212}^{(eff)}$. The horizontal axes of these graphs include the input coefficient of variation of the following four separate input random variables: e_1 – Young's modulus of the fiber, e_2 – Young's modulus of the matrix, ν_1 – Poisson ratio of the fiber, ν_2 – Poisson ratio of the matrix. The first observation, quite consistent with the previous analyses, is that the expected values of all effective tensor components (Figures 5.34, 5.38, and 5.42) are quite insensitive to the input coefficient of variation. Let us mention that this is not the case for a generalized stochastic perturbation technique, where some very small variations may be noticed each time.

The differences between the expected values obtained during randomization of e_1, e_2, ν_1, and ν_2 for $C_{1111}^{(eff)}$, $C_{1122}^{(eff)}$, and $C_{1212}^{(eff)}$ separately are almost negligible. One may

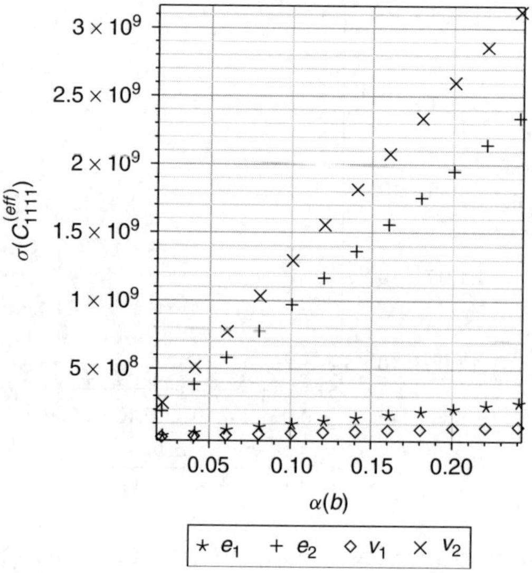

Figure 5.35 The standard deviations of $C_{1111}^{(eff)}$. Reproduced with permission from John Wiley & Sons Ltd.

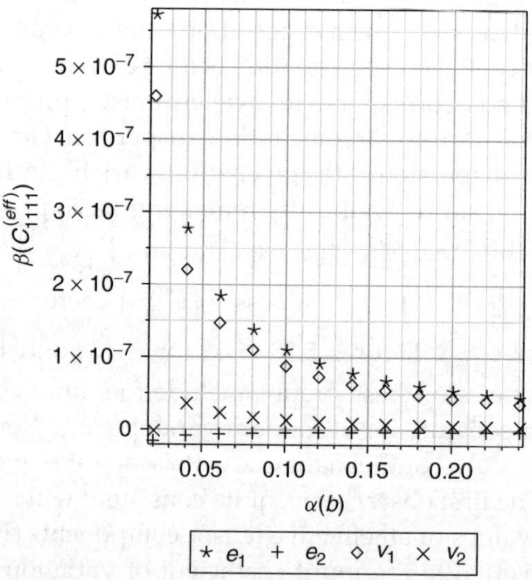

Figure 5.36 The asymmetry coefficient of $C_{1111}^{(eff)}$. Reproduced with permission from John Wiley & Sons Ltd.

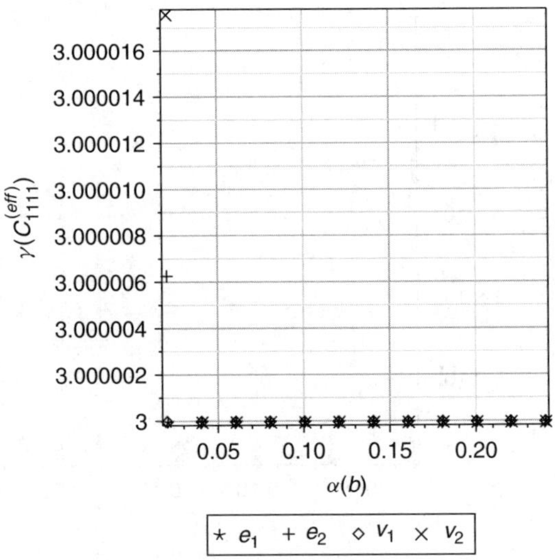

Figure 5.37 The flatness coefficient of $C_{1111}^{(eff)}$. Reproduced with permission from John Wiley & Sons Ltd.

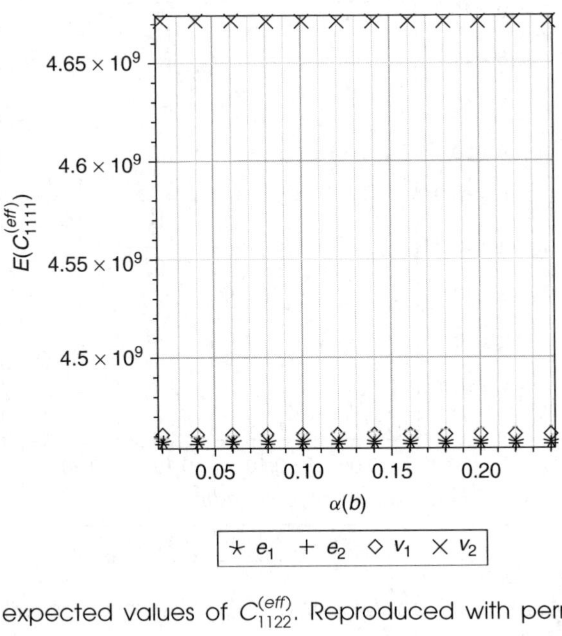

Figure 5.38 The expected values of $C_{1122}^{(eff)}$. Reproduced with permission from John Wiley & Sons Ltd.

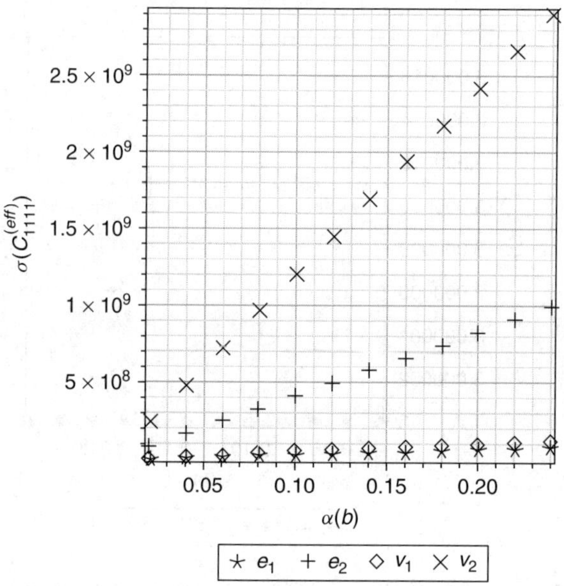

Figure 5.39 The standard deviations of $C_{1122}^{(eff)}$. Reproduced with permission from John Wiley & Sons Ltd.

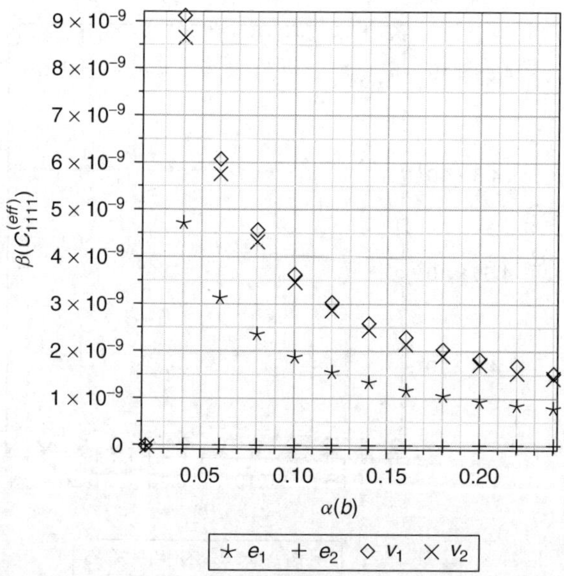

Figure 5.40 The asymmetry coefficient of $C_{1122}^{(eff)}$. Reproduced with permission from John Wiley & Sons Ltd.

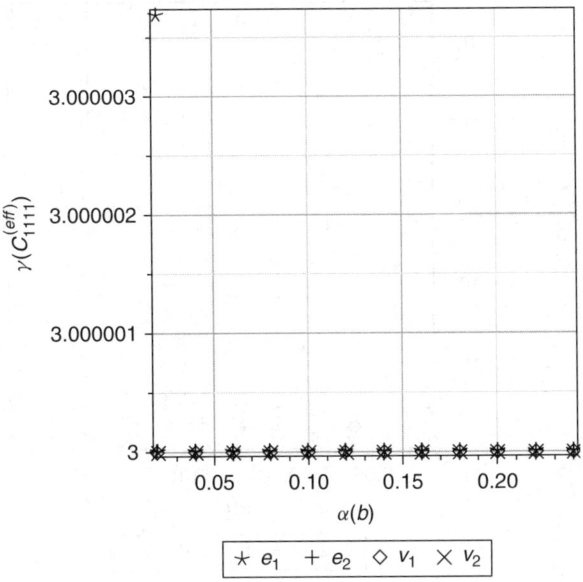

Figure 5.41 The flatness coefficient of $C_{1122}^{(eff)}$. Reproduced with permission from John Wiley & Sons Ltd.

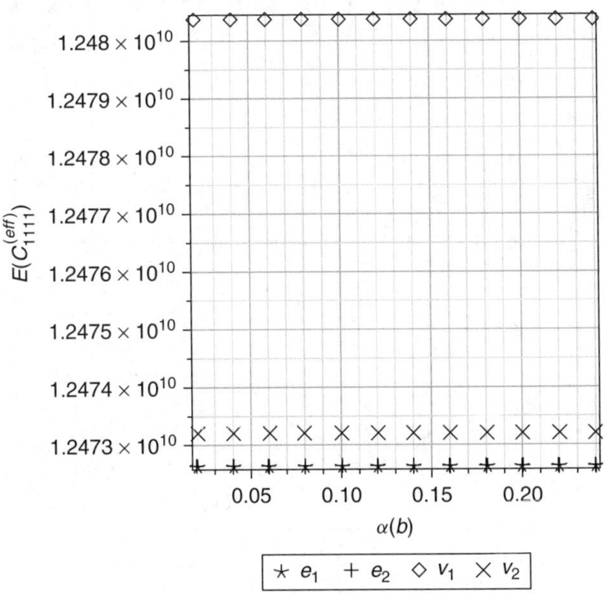

Figure 5.42 The expected values of $C_{1212}^{(eff)}$. Reproduced with permission from John Wiley & Sons Ltd.

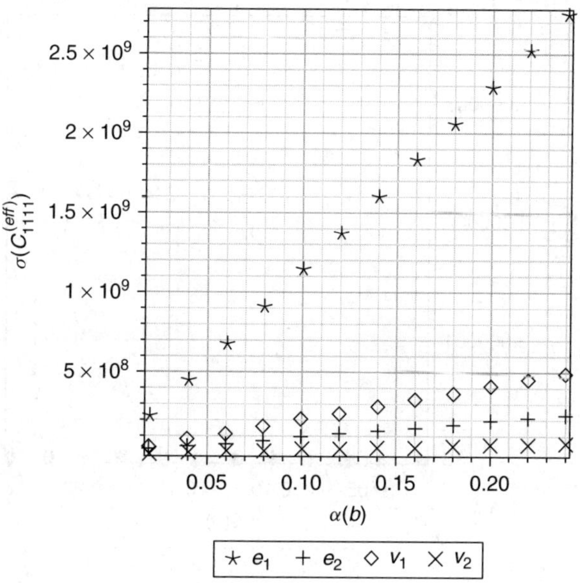

Figure 5.43 The standard deviations of $C_{1212}^{(eff)}$. Reproduced with permission from John Wiley & Sons Ltd.

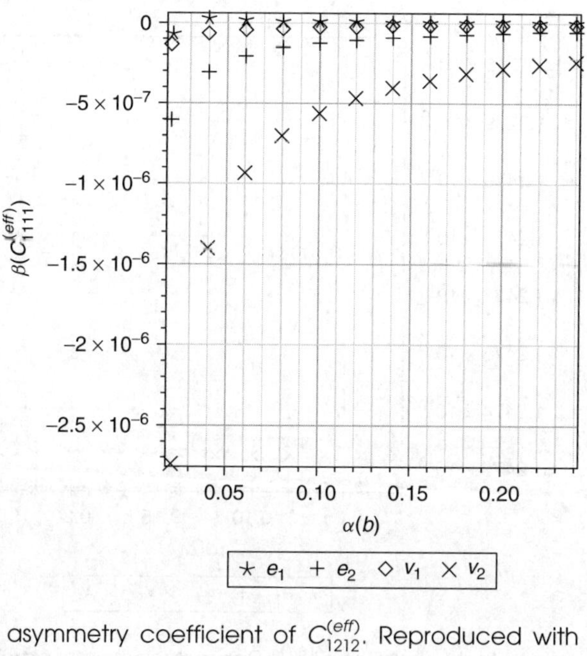

Figure 5.44 The asymmetry coefficient of $C_{1212}^{(eff)}$. Reproduced with permission from John Wiley & Sons Ltd.

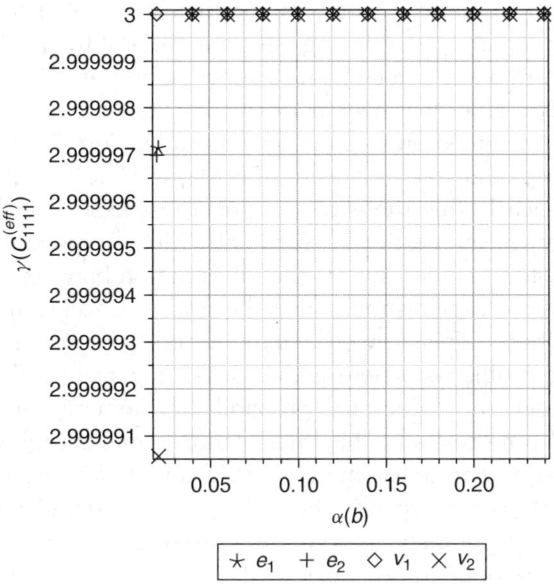

Figure 5.45 The flatness coefficient of $C_{1212}^{(eff)}$. Reproduced with permission from John Wiley & Sons Ltd.

conclude that $E\left[C_{1111}^{(eff)}\right]$ and $E\left[C_{1122}^{(eff)}\right]$ reach a maximum when $\nu_2 = \nu_2(\omega)$ and $E\left[C_{1212}^{(eff)}\right]$ in the case of $\nu_1 = \nu_1(\omega)$. Further, analogously to the previous simulation- and perturbation-based studies, all the interrelations between the output standard deviations of $C_{ijkl}^{(eff)}$ and the input coefficients of variation remain strictly linear (cf. Figures 5.35, 5.39, and 5.43). It is apparent that the random dispersion in matrix elastic characteristics is decisive for $C_{1111}^{(eff)}$ and $C_{1122}^{(eff)}$; the randomness in fiber properties is of marginal importance. This changes in case of $C_{1212}^{(eff)}$, where uncertainty in the fiber characteristics prevails but also a difference between the fibers' Poisson ratio effect and matrix characteristic effect is not so apparent as before. The asymmetry and flatness coefficients of $C_{ijkl}^{(eff)}$ are analyzed and compared with each other to confirm the thesis that these tensor components all have Gaussian distributions; this was first detected thanks to the Monte Carlo simulation-based homogenization method [71]. Now the results are almost the same – all flatness coefficients are clearly equal to 3; some extremely small numerical discrepancies are observed for input coefficient of variation equal to 0. The coefficients of asymmetry do not equal 0 perfectly for the entire parametric variability of the input, but with relatively small error all the output distributions may be treated as fully symmetric. An interesting observation here would be that the larger the input coefficient of variation, the smaller the absolute value of the coefficient β. This asymptotic behavior is observed through the positive values for $C_{1111}^{(eff)}$ and $C_{1122}^{(eff)}$, while for $C_{1212}^{(eff)}$ it is through the very small negative numbers only.

The last group of numerical tests was completely devoted to determination of the computer time consumption during determination of the first four probabilistic moments for the variable X, which is determined as the function of the polynomial expansion order. This study has been performed on a 64-bit version of the Windows 7 operating system installed on a computer with Intel Core *i*5 M430 2.27 GHz with 3.86 GB RAM space available and using the Maple™, v. 13. The results given in Figure 5.46 show clearly that the expansion of up to 15th order has no significant time cost, even for the highest-order moments. Most engineering cases are from elastostatics or elastodynamics, where the interrelation between the random input and the random structural response may be guessed according to the so-called "engineering intuition." This, however, may not be the case for elastoplasticity or general non-linear problems, where higher-order expansions could be important. As one could expect, the higher the probabilistic moment computed, the larger the computational time cost and this interrelation remains highly non-linear. This difference is most apparent for the analysis of 25th order, where the expected value is still determined within a few seconds, the variance needs about 10^3 seconds, while the third- and fourth-order moment computations are so expensive (even more than 10^4 seconds) that they are omitted here for brevity of presentation (one must remember that the CPU time usage is not directly the same as the real computer time necessary to complete the specific task, especially for larger values from this graph). The graph is compared with Figure 5.46 showing analogous CPU time cost, but for the lognormal variable being an argument of the polynomial expansion. It is quite apparent now that the expected values and variances are calculated a few times faster for the lognormal polynomials, whereas the third and higher moments, inversely, need incomparably

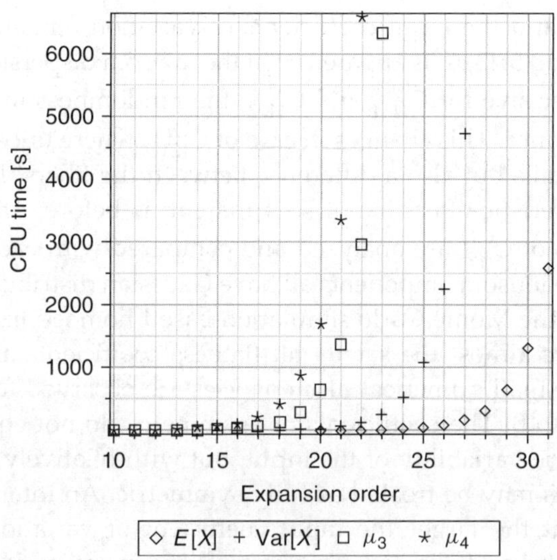

Figure 5.46 Computer cost of the Gaussian polynomial expansion procedure. Reproduced with permission from John Wiley & Sons Ltd.

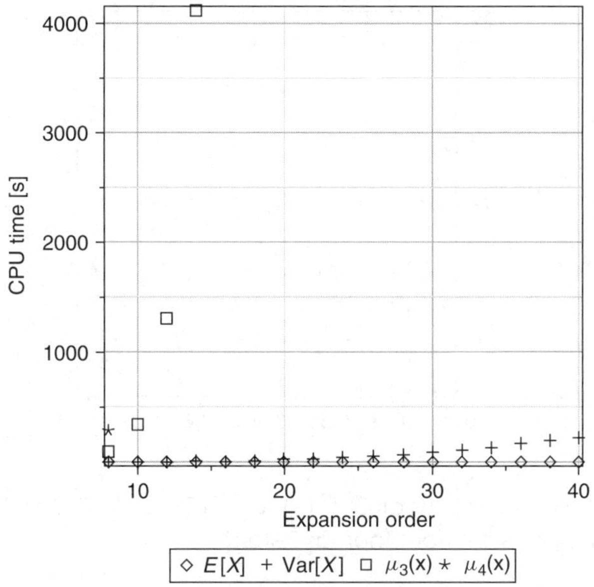

Figure 5.47 Computer cost of the lognormal polynomial expansion procedure. Reproduced with permission from John Wiley & Sons Ltd.

more time-consuming calculations than is necessary for polynomial expansions of Gaussian variables. Since higher than fourth probabilistic moments may not have a direct application for interpretation of the structural response, this computational cost seems to be complete for further structural numerical analyses. Independently from the expansion order chosen for the particular homogenization problem solution, the total computational time cost is always many times smaller than in the Monte Carlo simulation-based analysis.

Example 5.8: Stochastic fluctuations of the round fiber radius in homogenization

According to the engineering evidence [128], degradation of the fibers sometimes results in the uncertain loss of their diameters, so that the main aim of this computational example is to verify quantitatively this effect [80]. Let us consider for this purpose a composite with square periodicity cell and unitary dimensions discretized as in Example 5.4; the fiber has a round cross-section and the expected value of its initial radius is taken as $E[R] = 0.40$. The elastic properties of the glass fiber and epoxy matrix are adopted as follows: Young's moduli $e_1 = 84$ GPa and $e_2 = 4.0$ GPa, Poisson ratios $\nu_1 = 0.22$ for the fiber and $\nu_2 = 0.34$ for the matrix; for brevity, we postpone in this initial study the aging rate of the Young's moduli of the components. The first part of the computations consists of hybrid FEM–symbolic determination of the response functions of the homogenized tensor components with respect to the fiber radius ranging from 0.40 to 0.35 to be used in further stochastic perturbation-based analysis. The results of this procedure are given in Figure 5.48 – a continuous line represents $C^{(eff)}_{1111}(R)$, the dash-dot line is for the component $C^{(eff)}_{1122}(R)$, while the space-dash line

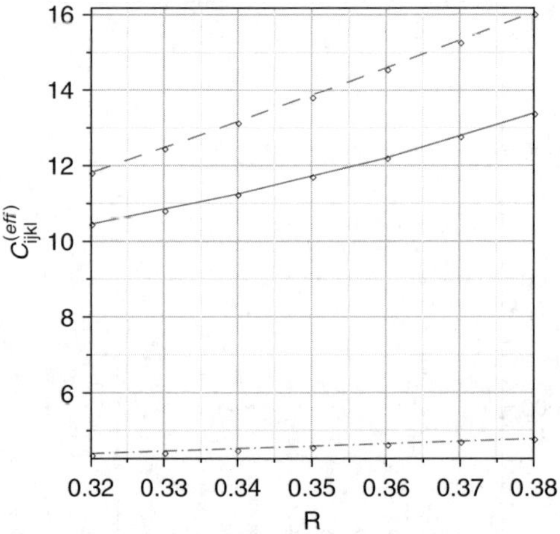

Figure 5.48 The approximations of $C_{1111}^{(eff)}(R)$, $C_{1122}^{(eff)}(R)$, $C_{1212}^{(eff)}(R)$ (in GPa). Reproduced with permission from Thomas Telford

corresponds to $C_{1212}^{(eff)}(R)$. It is apparent that the polynomial interpolations obtained are smooth, continuous, and monotonous within the given range of variability (the vertical axes are given in gigapascals).

Having determined these functions it is possible to compute partial derivatives of the EET components with respect to the fiber radius and, next, to include them in the general relation for probabilistic moments of any order. Assuming finally the specific form of the aging process we can analyze the expected values of the EET components with respect to the additional time changes in expectations and standard deviations of the fiber radius subject to stochastic decay [86]. The results of further numerical experiments are presented in Figure 5.49 for $C_{1111}^{(eff)}(R)$ (expectations (a) and standard deviations (b)), in Figure 5.50 for $C_{1122}^{(eff)}(R)$ (same order as before), and for $C_{1212}^{(eff)}(R)$ (expectations (a) and variances (b) in Figure 5.51). The variability range for the expected values corresponds to its decrease from initial value 0.40 to final value 0.35, while its coefficient of variation ranges from 0.0 to 0.25. Such a presentation assures the maximum generality of the analysis since the continuous spectrum of the aging process is collected here and, furthermore, the aging interdependence with time remains implicit (to finally fit this time scale using a combination of the first two moments into the given level of an environment with aggressive behavior).

All the probabilistic moments given in Figures 5.49–5.51 show the same tendencies – they decrease together with a decrease of both expectations and coefficients of variation of the fiber radius (a maximum is obtained for the upper limits of these variables). Analyzing these figures one may notice that having even a linear decay in expectations for the simplest aging rule, the additional decay in standard deviations (or variances) remains quadratic. Therefore, the grids given on the surfaces presented are not adequate to the specific aging mechanisms. Further, a decrease of the expected values for the fiber from 0.40 to 0.34 corresponds (for most

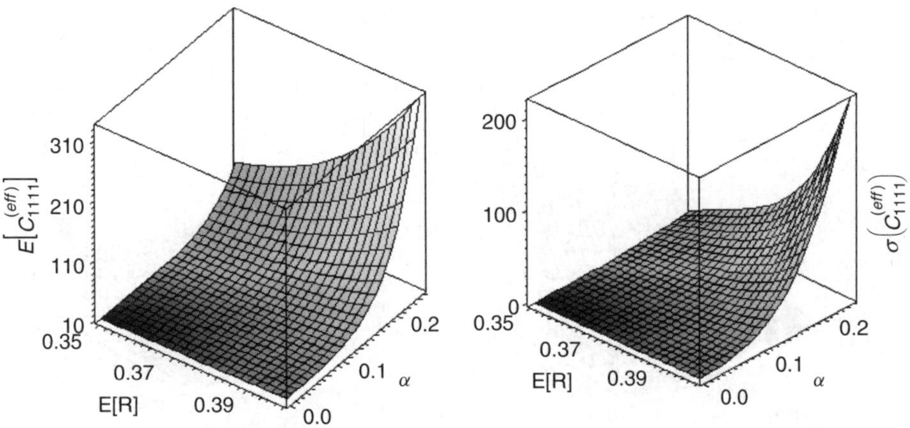

Figure 5.49 The expected values and standard deviations for $C_{1111}^{(eff)}(\omega; t)$ (in GPa and GPa^2). Reproduced with permission from Thomas Telford

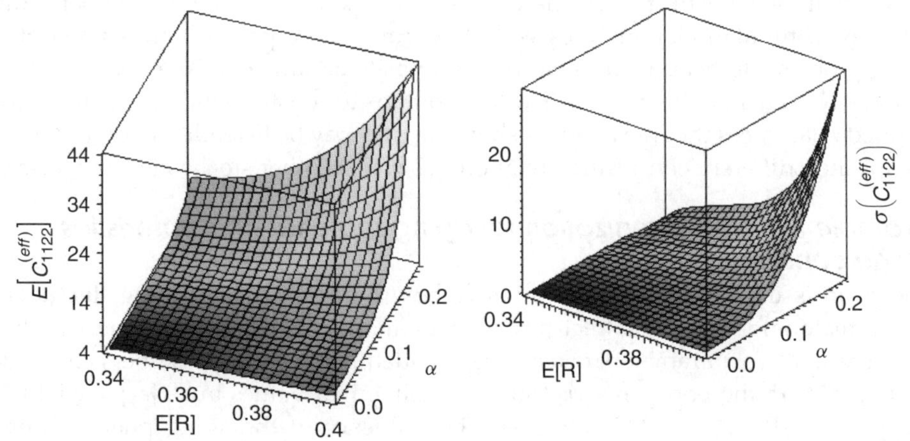

Figure 5.50 The expected values and standard deviations for $C_{1122}^{(eff)}(\omega; t)$ (in GPa and GPa^2). Reproduced with permission from Thomas Telford

engineering fibers) to the period of time from 50 to 100 years rather than few-year intervals only. The second important observation is that the expected values of all homogenized tensor components decrease more than 10 times from their initial values in the case of deterministic analysis (when the parameter is close to 0), while the analogous decrease for the maximum value of α is not so significant. Comparing the influence of the expected value of the fiber radius and its coefficient of variation (for the given variability intervals) it is clear that they are of almost the same importance in the case of final standard deviations (or variances); the coefficient of variation is essentially more important for the changes in the homogenized

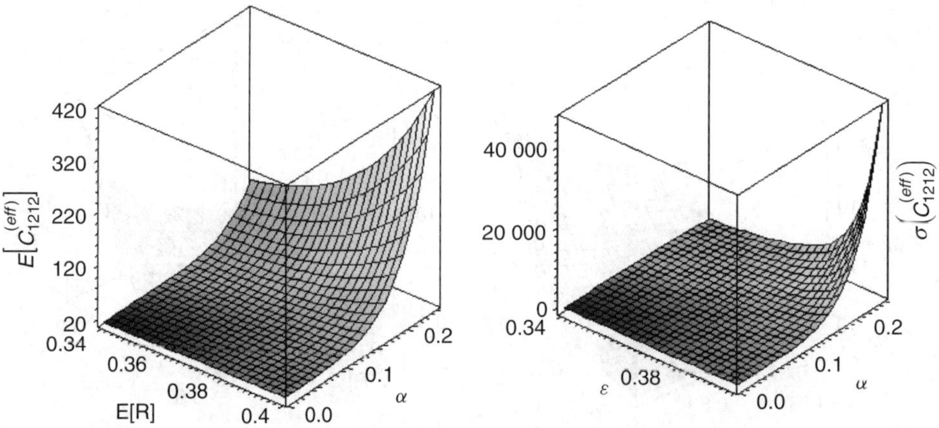

Figure 5.51 The expected values and variances for $C_{1212}^{(eff)}(\omega; t)$ (in GPa and GPa^2 and square gigapascals). Reproduced with permission from Thomas Telford

tensor component expectations than the relevant expected value of the fiber radius. The next computational analyses will allow for a more precise comparison of the aging process effects applied to various composite parameters like geometrical and, separately, material characteristics of the composite constituents. The methodology applied guarantees that the analysis shown above may be extended to other types of composites, different constitutive relations, as well as other stochastic aging rules.

Example 5.9: Homogenization for aging material characteristics of the constituents

The main issue in this study is to verify how much stochastic aging fluctuations of the material properties affect probabilistic moments of the EET time variations [75]. The material parameters taken here are adequate for the glass fibers embedded periodically in the epoxy matrix, so that their initial values are $E[e_1] = 84.0$ GPa, $E[e_2] = 4.0$ GPa, $\nu_1 = 0.22$, $\nu_2 = 0.34$. The influence of the aging phenomenon is simulated numerically in two different scenarios: the first one assumes that the matrix only is subject to a decrease of its Young's modulus according to the equation $e_2 = e_{20} - 0.02\frac{\text{GPa}}{\text{year}} \times t$. The second scenario obeys the aging of both the matrix and the fiber, where a decrease of the reinforcement Young's modulus is described by a quite analogous equation like $e_1 = e_{10} - 0.02\frac{\text{GPa}}{\text{year}} \times t$, where both e_{10} and e_{20} are equal to the expectations given above. The fiber has a round shape, is centrally located in the RVE, and occupies 50% of its area; both components are perfectly connected. The initial determination of the simulation input probabilistic moments together with the post-processing stage consisting of the probabilistic moments of the homogenized tensor time evolution are prepared using Maple™. The results of these computations within the first 20 years of the aging process are demonstrated in Figures 5.52–5.58 (the time parameter is introduced on all horizontal axes, while the vertical axes correspond to the probabilistic moment values). The first two figures show the

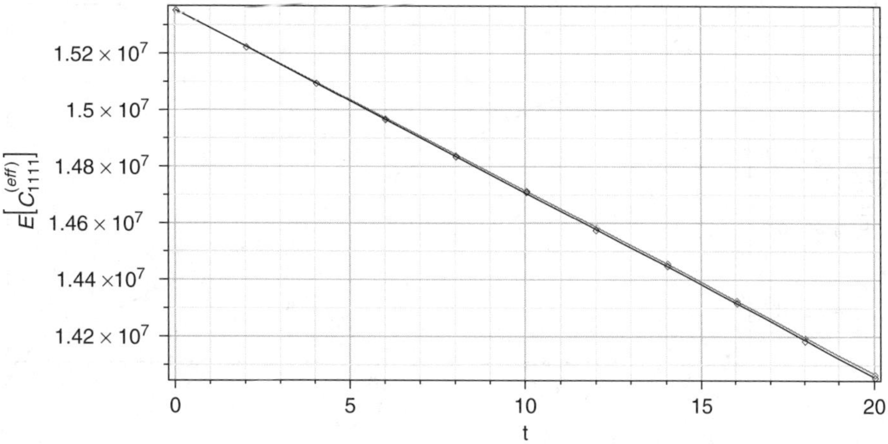

Figure 5.52 Evolution of the expected values for the component $C_{1111}^{(eff)}(\omega; t)$. Reproduced with permission from Begell House

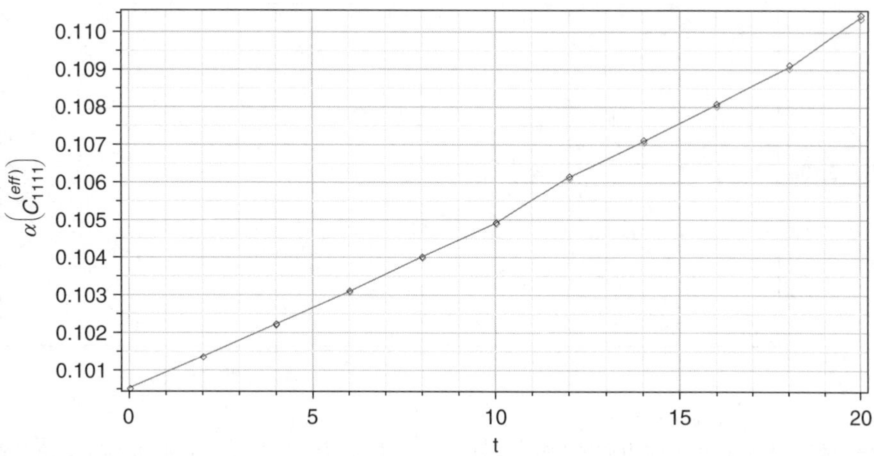

Figure 5.53 Evolution of the coefficients of variation for the component $C_{1111}^{(eff)}(\omega; t)$. Reproduced with permission from Begell House

expected values and the coefficients of variation for the first component of the homogenized tensor, Figures 5.54 and 5.55 show the same parameters for $C_{1122}^{(eff)}(\omega; t)$. The evolution of the PDFs for all components is given in Figures 5.56–5.58, where parts (a) correspond to the first scenario and parts (b) contain the results for the second scenario with aging of both composite components.

As is apparent in the first four figures, the expected values of the first two homogenized tensor components decrease almost linearly with time, where the first scenario given by the higher lines results (lighter line) in a slower decrease of the homogenized expectations. The coefficients of variation of these effective tensor

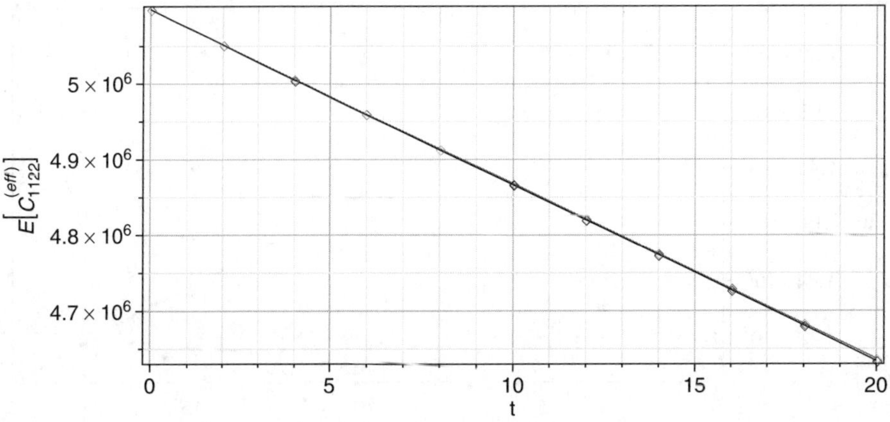

Figure 5.54 Evolution of the expected values for the component $C^{(eff)}_{1122}(\omega; t)$. Reproduced with permission from Begell House

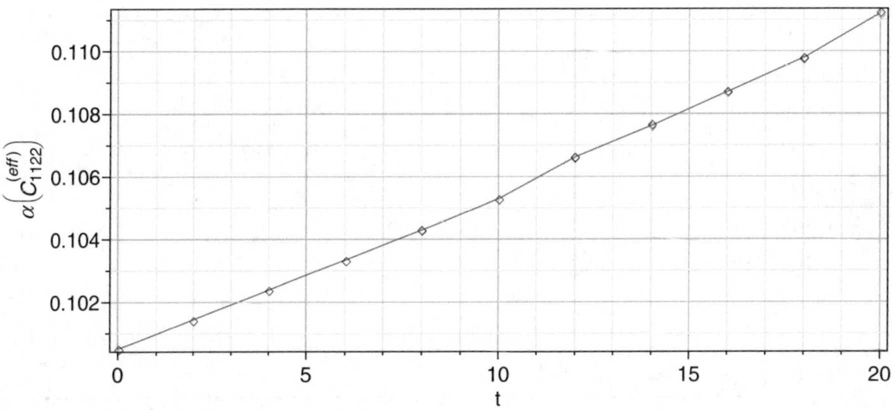

Figure 5.55 Evolution of the coefficients of variation for the component $C^{(eff)}_{1122}(\omega; t)$. Reproduced with permission from Begell House

components behave in a quite opposite way – they increase almost linearly with time in such a way that the continuous line in lighter color representing the first scenario as for the mean values demonstrates a somewhat slower increase in comparison with the darker diamonds reflecting the second aging scenario. Neglecting the scenario character, we observe almost 10% overall decrease of the expectations for the first 20 years of the aging stochastic process and a very similar increase in the ratio of the standard deviation to these expectations.

The time fluctuations of the PDFs (histograms) for the homogenized tensor components are shown in Figures 5.56–5.58, where all possible values of the analyzed effective tensor components are given on the horizontal axes. The essential differences

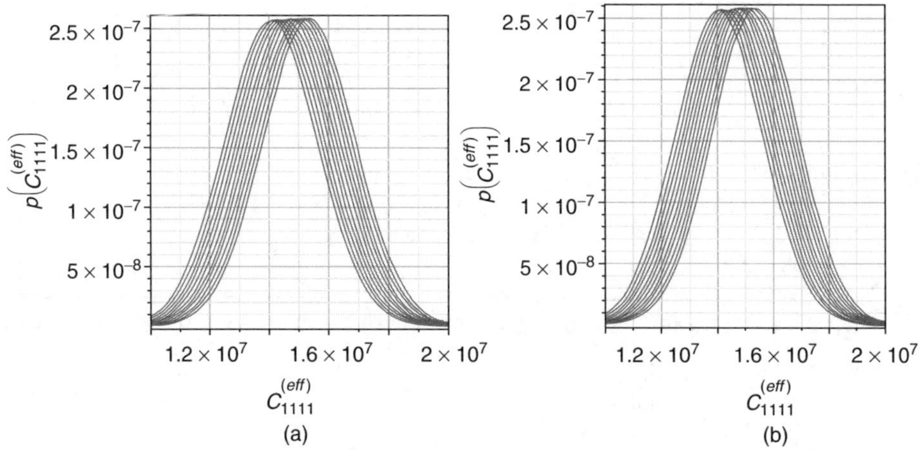

Figure 5.56 Evolution of the density function distributions for the component $C_{1111}^{(eff)}(\omega; t)$: (a) first scenario; (b) second scenario. Reproduced with permission from Begell House

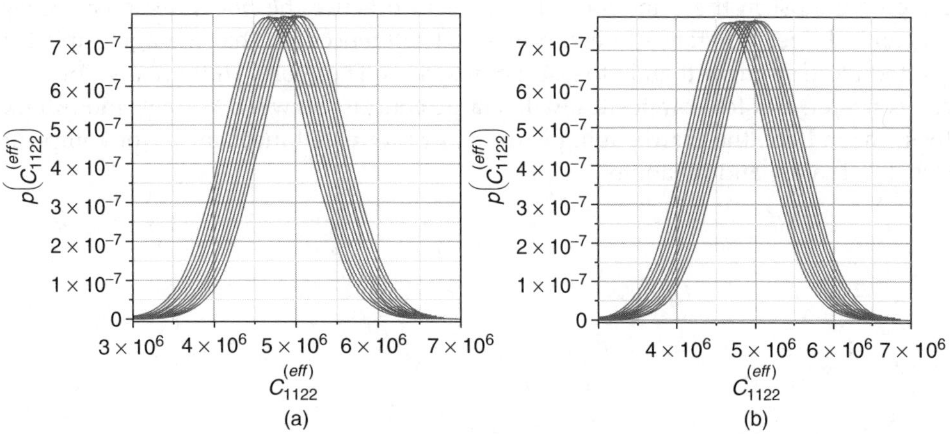

Figure 5.57 Evolution of the density function distributions for the component $C_{1122}^{(eff)}(\omega; t)$: (a) first scenario; (b) second scenario. Reproduced with permission from Begell House

between the first and second scenarios are apparent for the component $C_{1212}^{(eff)}(\omega; t)$ only (the final plot is thinner (a) and thicker (b) in Figure 5.58). These outputs come from the systematic plots of the Gaussian curve corresponding to the homogenized tensor for exploitation time varying from 0 to 20 years with a 2-year increase like before. The first bell-shaped curve is drawn and next, moving from the right to the left of the horizontal axis values, we plot these curves after the next time increment in Figures 5.56–5.58; this evolution in probability densities for various time moments

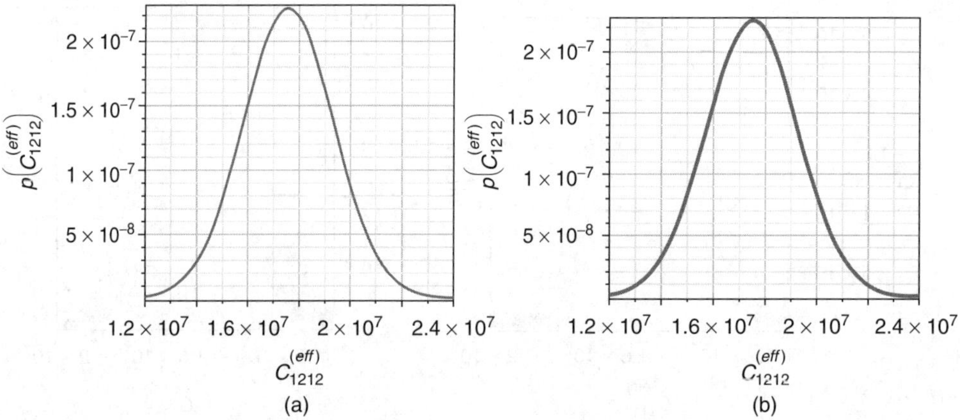

Figure 5.58 Evolution of the density function distributions for the component $C_{1212}^{(eff)}(\omega; t)$: (a) first scenario; (b) second scenario. Reproduced with permission from Begell House

follows of course the additional decrease of the expectations noticed before. It is necessary to point out that the aging laws for the fiber and the matrix have been adopted almost in the same form, however in practice the fibers demonstrate significantly larger resistance to environmental influences, so that the aging must be limited more strongly than in the case of matrices. Therefore, without any doubt we can conclude that for composites with a large contrast between Young's moduli like these taken here, the matrix aging is the most influential mechanism for composite structural safety and reliability.

6

Concluding Remarks

1. The stochastic perturbation method known mainly from its previous first- and second-order realizations has been presented here in its generalized version assuming that practically any order of the common fundamental Taylor expansion is allowable. This was invented to significantly increase the overall accuracy of probabilistic analysis, especially for larger coefficients of variation of input random variables. It was possible due to the response function idea, which is based upon continuous and differentiable function (including higher derivatives also with preferably recursive formulas) between desired structural responses, like a deflection at some point of the mesh or grid and the given random input parameter. They are proposed here as the polynomials of random variables with constant coefficients but may be harmonic or exponential also. This response function is recovered using here simple polynomial interpolation or advanced least squares techniques made on several numerical solutions of the initial boundary value problem, where the deterministic value of the randomized parameter fluctuates in a deterministic sense in the certain neighborhood of its mean value. So, we search for some kind of an analytical solution to this problem in a global or in a local sense using classical discrete numerical techniques. Usually the set of 10–11 experimental input values for randomized parameter has symmetric distribution (most frequently uniform in this interval) with respect to its mean. The main source of computational error in this method is deterministic polynomial interpolation procedure, since further probabilistic analysis is based on classical integral definitions, unlike the direct differentiation procedure applied before, where successive solutions of increasing-order algebraic equations were responsible for this error. Of course, an error following the Taylor expansion with random parameters of the certain order remains common for both methods. Computational implementation's effort and overall time of numerical analysis of the technique proposed is significantly lower than for traditional Monte Carlo simulations with comparable precision, especially for expectations and standard deviations. It does not depend on the choice between the global response determination (a single-state function

The Stochastic Perturbation Method for Computational Mechanics, First Edition. Marcin Kamiński.
© 2013 John Wiley & Sons, Ltd. Published 2013 by John Wiley & Sons, Ltd.

vs. random input) or the local search (state functions at any point of the grid or mesh).

2. The generalized stochastic perturbation method is verified here as suitable for engineering and scientific problems with response functions being lower-order polynomials, while a little bit less efficient for higher-order responses and large input random deviations. Indeed, we have no evidence on highly nonlinear problems with random parameters, particularly state-dependent (relying on a temperature or actual strain, for instance). Especially, linear response functions appearing frequently in elastic systems with randomized Young's modulus always return results extremely close to the Monte Carlo simulation method with large populations. However, we can expect some numerical instabilities for coupled analyses, even with a single random Gaussian parameter, where the final efficiency may depend upon the partial differential equations system being solved. The perturbation method has been verified theoretically and computationally as probabilistically convergent for the polynomial response functions of practically any degree. A choice of optimal degree in symbolic recovery of this response proceeds as minimization within the set of satisfactory interpolations fitting well the discrete data and showing no local oscillations. Further, stochastic perturbation technique should be of a single order higher than the degree of the polynomial response function. Of course, higher orders do not give any extra components as higher-order partial derivatives for random inputs simply vanish; at the same time computational error of lower-order expansions cannot be neglected, especially for third and fourth central moments. It must be underlined that the least-squares method employed to recover the response functions in both global and local formulations in its non-weighted version shows decisively larger variations of especially for higher central probabilistic moments and additional coefficients. The best results are noticed for the weighted LSM, where the spatial distribution of the weights has a similar shape to the Dirac function – it shows a maximum equivalent to the input parameter mean value and a few times smaller weights elsewhere. We can apply a triangular distribution of these weights within the trial interval of random parameter variability also but with worse efficiency.
3. The stochastic perturbation technique displayed in this book is applicable in its RFM version with any commercial and academic software realizing some discrete technique, except for these displayed in this book we can additionally modify the finite volume method [8, 26, 28], the discrete element method, as well as a variety of meshless techniques as well as even molecular dynamics matrix formulations. Further application in reliability analysis (in both FORM and SORM version) depends strongly on whether the structural output verified a posteriori is Gaussian or not; the final values of the reliability index may be computed straightforwardly according to Eurocode 0 while our verification is positive. Recalculation of this index in SORM is not complicated as the limit function shape factor (its curvature) is obtained with additional symbolic double differentiation of the response function with respect to the random input parameter. Unfortunately, the DDM alternative to the RFM needs each time a brand new computer implementation of the discrete

computations code or direct access to the existing well-documented source code of the deterministic original program. The only exception is when symbolic computer algebra is used – then we can make use of the entire implementation of the specific boundary value problems in this environment or build a macro-procedure that reads system matrices from a deterministic procedure, calculates their derivatives for the random input and, after some rearrangement, inserts them once more as an input to the same program. This is possible since algebraic equations for successive orders have essentially modified RHS vectors using lower-order equations, while the LHS system matrix remains always the same.

4. Computational experiments carried out using the SBEM and contained in Chapter 3 show that this technique seems to be the most efficient for analysis of boundary value problems with random shape and/or boundary conditions [16], especially in the RFM version, where small modifications of the few nodes are quite straightforward and sufficient (that can be automatically provided by the computer program itself). Of course, a similar situation will occur when one randomizes the boundary conditions imposed on displacement and/or boundary tractions (stresses) [16]. The SFEM application in this context usually demands some remeshing of the entire computational domain, except for the global mesh consisting of a few macro-elements, where geometrical uncertainty appears within the only one. The SFDM is also not attractive in this situation except for uncertainty in global dimensions, because randomization of boundary waviness and/or some regular holes inside larger domains results each time in a change from regular to the irregular grid. Computational implementation of the DDM version in this context is not automatic in a symbolic environment and must proceed in a way typical for the SFEM. It would be indeed interesting to develop stochastic perturbation-based implementation of the hybrid FBEM [172], which can link the pros of uncertainty analysis coming from both methods or some numerical experiments carried out with complex fundamental solutions typical for electromagnetics [49].
5. It needs to be emphasized that the SFDM application presented in Chapter 4 to the regular grids in Cartesian coordinates is relatively easy in both DDM and RFM versions, because algebraic difference relations may be symbolically partially differentiated with respect to the random input variable successively many times, and then, immediately solved. As the solution procedure for linear algebraic equation systems no restriction on the system matrix dimension and, further, these systems can be automatically generated, even large-scale systems may be modeled in this way. Usually, difference equations may be reorganized algebraically so as to have random input quantities on the RHS only, which speeds up the entire algorithm – the only problem may occur when random fields are considered, like an elastic beam or plate with varying thickness, where some correcting factors need to be inserted even in the deterministic case. Symbolic implementations of difference equations rewritten for polar or cylindrical coordinates are more complicated during implementation of perturbation analysis. The other unresolved problems are – the SFDM for the irregular grids and 3D visualization to show

variations of the basic probabilistic characteristics coming from the SFDM studies (common for all the methods in this book).

6. The homogenization method displayed in Chapter 5 is a very specific area of application of the perturbation technique, where deterministic geometrical perturbation is used to introduce the effective elasticity tensor components and, separately, the stochastic perturbation technique to randomize this problem. Their common usage enables determination of this tensor basic probabilistic characteristics, where material or some geometrical composite parameters are random (not violating periodicity with respect to the single periodicity cell). We used here four different probabilistic techniques – analytical, Monte Carlo simulation, and semi-analytical contrasted with the stochastic perturbation approach – and they give very similar results, where the Gaussian input in material parameters results in the Gaussian homogenized tensor. The homogenization problem solution was an occasion to demonstrate symbolic computations of the probabilistic entropy variations in boundary value problems with uncertain parameters. It is naturally provided for continuous variables, unlike classical applications in information theory, where discrete distributions were analyzed in this context, especially in computer science (data processing and cryptology applications). The chapter devoted to this issue also contains computational analysis of material parameters', fiber radius' as well as interface defects' stochastic aging according to the simple linear decay with two Gaussian coefficients. As one could expect, numerical results confirm that such a decay results in a linear decrease of the homogenized tensor expectations as well as in almost parabolic increase of variances for this tensor components. It could be extended further with exponential degradation of the Young's modulus as well as Poisson ratios and/or exponential increase of the stochastic interface defects appearing between the composite constituents.
7. One of the unexpected opportunities discovered during various implementations of the perturbation-based SFEM is its semi-analytical formulation introduced in the last chapter – to homogenize periodic composite materials by only. It follows directly from the fact that after global or local response function recovery we have polynomial analytical interrelations between random structural output and input variable. It is not necessary at this point to employ a perturbation technique to provide a Taylor expansion with random parameters to derive final probabilistic moments and characteristics of an output. Thanks to the symbolic integration procedures implemented in widely used mathematical software, we can follow traditional integral definitions of these moments. Contrary to the generalized perturbation method, a semi-analytical approach cannot be used automatically – we need to detect the final distribution type first, after some computational experiments, which is indeed relatively easy for the Gaussian variables only. Then we are able to carry out symbolic integration of the moments using the bell-shaped distribution curve. That is why this new computational tool has been used with a deterministic homogenization procedure with the well-documented fact that the resulting tensor is also distributed according to the Gaussian function [71].

8. The generalized stochastic perturbation technique undoubtedly has a lot of potential applications and offers new research directions. Although we presented a single-variable analysis shown consecutively in all numerical examples, a multivariable formulation with at least first-order cross-correlations has been introduced briefly in the first chapter. It is of course possible to extend this method toward many correlated random input variables, however, we need to realize that such an analysis needs precise numerical knowledge about any (order) cross-correlations inserted in the perturbation-based equations. Usually, expected values, standard deviations, and sometimes skewness and kurtosis for uncertain parameters are experimentally established only, which in this case is not satisfactory at all. The only opportunity realistic from an engineering point of view is to analyze uncorrelated random inputs via simple extension of the equations provided here that sometimes (with reach and adequate data sets) may be expanded with first-order covariances. Further, we can usually assume that the dead load of the structure, its main geometrical parameters, as well as the strength and physical properties are usually independent of each other, but some material parameters themselves may and should have some statistical correlation. The second important perspective is an analysis of various probability density functions, like the lognormal, for instance, but also Fisher–Trippett, Weibull, or Poisson, all having many well-documented applications in science and engineering. It is clear that they will need full Taylor expansions with even and odd central probabilistic moments and, further, symbolic integral derivation of higher central probabilistic moments as their recursive formulas most frequently are not available – this will allow for the first time a validation of the generalized stochastic perturbation technique for non-symmetric probability density functions on a large scale. That is why the Readers will find necessary formulas in the Appendix to include derived probabilistic moments for the probability distribution of their choice into their own computer programs. Finally, considering large-scale [144] or multi-scale systems modeling and increasing computer power, one may invent some parallelization procedure for the perturbation-based programs analogous to that worked out before in [110] for instance, and relevant to other stochastic methods. Undoubtedly, the RFM approach should be many times faster than the DDM and applicable also for the commercial programs or the black box type academic software. Another advantage of multiscale analysis is the multi-resolution technique, where random variables and processes may be transferred between neighboring scales using some specially modified wavelet transforms [70, 161]. This can be the basis for the reliability-based structural optimization [127] including certain multi-resolutional features [18, 71].

Appendix

Distribution name	ChiSquare
PDF	$\begin{cases} 0 & t<0 \\ \dfrac{t^{\frac{1}{2}\nu-1}\ \mathrm{e}^{-\frac{1}{2}t}}{2^{\frac{1}{2}\nu}\Gamma\left(\frac{1}{2}\nu\right)} & \textit{otherwise} \end{cases}$
E[b]	ν
σ(b)	$\sqrt{2}\,\sqrt{\nu}$
α(b)	$\dfrac{\sqrt{2}}{\sqrt{\nu}}$
β(b)	$\dfrac{2\sqrt{2}}{\sqrt{\nu}}$
κ(b)	$\dfrac{3\ (\nu+4)}{\nu}-3$
μ_1(b)	0
μ_2(b) = Var(b)	$2\,\nu$
μ_3(b)	$8\,\nu$
μ_4(b)	$12\,\nu\,(\nu+4)$
μ_5(b)	$32\,\nu\ (5\,\nu+12)$
μ_6(b)	$40\,\nu\ \left(52\,\nu+3\,\nu^2+96\right)$
μ_7(b)	$96\,\nu\ \left(480+35\,\nu^2+308\,\nu\right)$
μ_8(b)	$112\,\nu\ \left(5760+4176\,\nu+15\,\nu^3+680\,\nu^2\right)$
μ_9(b)	$256\,\nu\ \left(40320+315\,\nu^3+6608\,\nu^2+32112\,\nu\right)$
μ_{10}(b)	$288\,\nu\ \left(645120+134960\,\nu^2+105\,\nu^4+9800\,\nu^3+554112\,\nu\right)$

The Stochastic Perturbation Method for Computational Mechanics, First Edition. Marcin Kamiński.
© 2013 John Wiley & Sons, Ltd. Published 2013 by John Wiley & Sons, Ltd.

Distribution name	Erlang
PDF	$\begin{cases} 0 & t<0 \\ \dfrac{\left(\frac{t}{b}\right)^{c-1} e^{-\frac{t}{b}}}{b\,\Gamma(c)} & \textit{otherwise} \end{cases}$
E[b]	$b\,c$
σ(b)	$b\sqrt{c}$
α(b)	$\frac{1}{\sqrt{c}}$
β(b)	$\frac{2}{\sqrt{c}}$
κ(b)	$\frac{3\,(2+c)}{c}-3$
μ_1(b)	0
μ_2(b) = Var(b)	$b^2 c$
μ_3(b)	$2\,b^3\,c$
μ_4(b)	$6\,b^4 c + 3\,b^4\,c^2$
μ_5(b)	$24\,b^5\,c + 20\,b^5\,c^2$
μ_6(b)	$120\,b^6\,c + 130\,b^6\,c^2 + 15\,b^6\,c^3$
μ_7(b)	$720\,b^7\,c + 924\,b^7\,c^2 + 210\,b^7\,c^3$
μ_8(b)	$5040\,b^8\,c + 7308\,b^8\,c^2 + 2380\,b^8\,c^3 + 105\,b^8\,c^4$
μ_9(b)	$64224\,b^9\,c^2 + 26432\,b^9\,c^3 + 2520\,b^9\,c^4 + 40320\,b^9\,c$
μ_{10}(b)	$303660\,b^{10}\,c^3 + 44100\,b^{10}\,c^4 + 945\,b^{10}\,c^5 + 362880\,b^{10}\,c + 623376\,b^{10}\,c^2$

Distribution name	Exponential
PDF	$\begin{cases} 0 & t<0 \\ \dfrac{e^{-\frac{t}{b}}}{b} & \textit{otherwise} \end{cases}$
E[b]	b
σ(b)	b
α(b)	1
β(b)	2
κ(b)	6
μ_1(b)	0
μ_2(b) = Var(b)	b^2
μ_3(b)	$2\,b^3$
μ_4(b)	$9\,b^4$
μ_5(b)	$44\,b^5$
μ_6(b)	$265\,b^6$
μ_7(b)	$1854\,b^7$
μ_8(b)	$14833\,b^8$
μ_9(b)	$133496\,b^9$
μ_{10}(b)	$1334961\,b^{10}$

Distribution name	Gamma
PDF	$\begin{cases} 0 & t < 0 \\ \dfrac{\left(\frac{t}{b}\right)^{c-1} e^{-\frac{t}{b}}}{b\,\Gamma(c)} & \textit{otherwise} \end{cases}$
E[b]	$b\,c$
σ(b)	$b\sqrt{c}$
α(b)	$\dfrac{1}{\sqrt{c}}$
β(b)	$\dfrac{2}{\sqrt{c}}$
κ(b)	$\dfrac{3\,(2+c)}{c} - 3$
μ_1(b)	0
μ_2(b) = Var(b)	$b^2\,c$
μ_3(b)	$2\,b^3\,c$
μ_4(b)	$6\,b^4\,c + 3\,b^4\,c^2$
μ_5(b)	$24\,b^5\,c + 20\,b^5\,c^2$
μ_6(b)	$120\,b^6\,c + 130\,b^6\,c^2 + 15\,b^5\,c^3$
μ_7(b)	$720\,b^7\,c + 924\,b^7\,c^2 + 210\,b^7\,c^3$
μ_8(b)	$5040\,b^8\,c + 7308\,b^8\,c^2 + 2380\,b^8\,c^3 + 105\,b^8\,c^4$
μ_9(b)	$64224\,b^9\,c^2 + 26432\,b^9\,c^3 + 2520\,b^9\,c^4 + 40320\,b^9\,c$
μ_{10}(b)	$303660\,b^{10}\,c^3 + 44100\,b^{10}\,c^4 + 945\,b^{10}\,c^5 + 362880\,b^{10}\,c + 623376\,b^{10}\,c^2$

Distribution name	Poisson
PDF	$\sum_{k=0}^{\infty} \dfrac{\lambda^k e^{-\lambda}\,\mathrm{Dirac}\,(t-k)}{k!}$
E[b]	λ
σ(b)	$\sqrt{\lambda}$
α(b)	$\dfrac{1}{\sqrt{\lambda}}$
β(b)	$\dfrac{1}{\sqrt{\lambda}}$
κ(b)	$\dfrac{1}{\lambda}$
μ_1(b)	0
μ_2(b) = Var(b)	λ
μ_3(b)	λ
μ_4(b)	$3\,\lambda^2 + \lambda$
μ_5(b)	$10\,\lambda^2 + \lambda$
μ_6(b)	$15\,\lambda^3 + 25\,\lambda^2 + \lambda$
μ_7(b)	$105\,\lambda^3 + 56\,\lambda^2 + \lambda$

$\mu_8(b)$	$105\lambda^4 + 490\lambda^3 + 119\lambda^2 + \lambda$
$\mu_9(b)$	$1260\lambda^4 + 1918\lambda^3 + 246\lambda^2 + \lambda$
$\mu_{10}(b)$	$945\lambda^5 + 9450\lambda^4 + 6825\lambda^3 + 501\lambda^2 + \lambda$

Distribution name	Inverse Gaussian
PDF	$\begin{cases} 0 & t < 0 \\ \frac{1}{2}\sqrt{2}\sqrt{\frac{\lambda}{\pi t^3}}\, e^{-\frac{1}{2}\frac{\lambda(t-\mu)^2}{\mu^2 t}} & \textit{otherwise} \end{cases}$
E[b]	μ
$\sigma(b)$	$\mu\sqrt{\frac{\mu}{\lambda}}$
$\alpha(b)$	$\sqrt{\frac{\mu}{\lambda}}$
$\beta(b)$	$\frac{3\mu^2}{\lambda^2\left(\frac{\mu}{\lambda}\right)^{3/2}}$
$\kappa(b)$	$\frac{3(\lambda+5\mu)}{\lambda} - 3$
$\mu_1(b)$	0
$\mu_2(b) = \mathrm{Var}(b)$	$\frac{\mu^3}{\lambda}$
$\mu_3(b)$	$\frac{3\mu^5}{\lambda^2}$
$\mu_4(b)$	$\frac{3\mu^6(\lambda+5\mu)}{\lambda^3}$
$\mu_5(b)$	$\frac{15\mu^8(2\lambda+7\mu)}{\lambda^4}$
$\mu_6(b)$	$\frac{15\mu^9(21\lambda\mu + 63\mu^2 + \lambda^2)}{\lambda^5}$
$\mu_7(b)$	$\frac{315\mu^{11}(33\mu^2 + \lambda^2 + 12\lambda\mu)}{\lambda^6}$
$\mu_8(b)$	$\frac{105\mu^{12}(495\lambda\mu^2 + \lambda^3 + 54\mu\lambda^2 + 1287\mu^3)}{\lambda^7}$
$\mu_9(b)$	$\frac{945\mu^{14}(2145\mu^3 + 858\lambda\mu^2 + 110\mu\lambda^2 + 4\lambda^3)}{\lambda^8}$
$\mu_{10}(b)$	$\frac{945\mu^{15}(36465\mu^4 + 15015\lambda\mu^3 + 110\lambda^3\mu + 2145\lambda^2\mu^2 + \lambda^4)}{\lambda^9}$

Distribution name	Laplace
PDF	$\frac{1}{2}\frac{e^{-\frac{\lvert -t+a\rvert}{b}}}{b}$
E[b]	a
σ(b)	$\sqrt{2}\,b$
α(b)	$\frac{\sqrt{2}\,b}{a}$
β(b)	0
κ(b)	3
μ_1(b)	0
μ_2(b) = Var(b)	$2\,b^2$
μ_3(b)	0
μ_4(b)	$24\,b^4$
μ_5(b)	0
μ_6(b)	$720\,b^6$
μ_7(b)	0
μ_8(b)	$40320\,b^8$
μ_9(b)	0
μ_{10}(b)	$3628800\,b^{10}$

Distribution name	Log-Normal
PDF	$\begin{cases} 0, t < 0 \\ \frac{\sqrt{2}}{2t\,\sigma\sqrt{\pi}} exp\left(-\frac{(\ln(t)-\mu)^2}{2\sigma^2}\right) & otherwise \end{cases}$
E[b]	$e^{\mu+\frac{1}{2}\sigma^2}$
σ(b)	$\sqrt{e^{2\mu+\sigma^2}\left(e^{\sigma^2}-1\right)}$
α(b)	$\frac{\sqrt{e^{2\mu+\sigma^2}\left(e^{\sigma^2}-1\right)}}{e^{\mu+\frac{1}{2}\sigma^2}}$
β(b)	$\frac{e^{3\mu+\frac{9}{2}\sigma^2}-3\,e^{3\mu+\frac{5}{2}\sigma^2}+2\,e^{\frac{3}{2}\sigma^2+3\mu}}{\left(e^{2\mu+\sigma^2}\left(e^{\sigma^2}-1\right)\right)^{3/2}}$
κ(b)	$\frac{e^{4\mu+8\sigma^2}-4\,e^{4\mu+5\sigma^2}+6\,e^{4\mu+3\sigma^2}-3\,e^{2\sigma^2+4\mu}}{\left(e^{2\mu+\sigma^2}\right)^2\left(e^{\sigma^2}-1\right)^2}-3$
μ_1(b)	0
μ_2(b) = Var(b)	$e^{2\mu+2\sigma^2}-e^{2\mu+\sigma^2}$

$\mu_3(b)$	$e^{3\mu+\frac{9}{2}\sigma^2} - 3e^{3\mu+\frac{5}{2}\sigma^2} + 2e^{\frac{3}{2}\sigma^2+3\mu}$
$\mu_4(b)$	$e^{4\mu+8\sigma^2} - 4e^{4\mu+5\sigma^2} + 6e^{4\mu+3\sigma^2} - 3e^{2\sigma^2+4\mu}$
$\mu_5(b)$	$e^{5\mu+\frac{25}{2}\sigma^2} - 5e^{5\mu+\frac{17}{2}\sigma^2} + 10e^{5\mu+\frac{11}{2}\sigma^2} - 10e^{5\mu+\frac{7}{2}\sigma^2} + 4e^{\frac{5}{2}\sigma^2+5\mu}$
$\mu_6(b)$	$e^{6\mu+18\sigma^2} - 6e^{6\mu+13\sigma^2} + 15e^{6\mu+9\sigma^2} - 20e^{6\mu+6\sigma^2} + 15e^{6\mu+4\sigma^2} - 5e^{3\sigma^2+6\mu}$
$\mu_7(b)$	$e^{7\mu+\frac{49}{2}\sigma^2} + 6e^{\frac{7}{2}\sigma^2+7\mu} - 7e^{7\mu+\frac{37}{2}\sigma^2} + 21e^{7\mu+\frac{27}{2}\sigma^2} - 35e^{7\mu+\frac{19}{2}\sigma^2}$ $+35e^{7\mu+\frac{13}{2}\sigma^2} - 21e^{7\mu+\frac{9}{2}\sigma^2}$
$\mu_8(b)$	$e^{8\mu+32\sigma^2} - 7e^{4\sigma^2+8\mu} - 8e^{8\mu+25\sigma^2} + 28e^{8\mu+19\sigma^2} - 56e^{8\mu+14\sigma^2}$ $+70e^{8\mu+10\sigma^2} - 56e^{8\mu+7\sigma^2} + 28e^{8\mu+5\sigma^2}$
$\mu_9(b)$	$e^{9\mu+\frac{81}{2}\sigma^2} - 36e^{9\mu+\frac{11}{2}\sigma^2} - 9e^{9\mu+\frac{65}{2}\sigma^2} + 8e^{\frac{9}{2}\sigma^2+9\mu} + 36e^{9\mu+\frac{51}{2}\sigma^2}$ $-84e^{9\mu+\frac{39}{2}\sigma^2} + 126e^{9\mu+\frac{29}{2}\sigma^2} - 126e^{9\mu+\frac{21}{2}\sigma^2} + 84e^{9\mu+\frac{15}{2}\sigma^2}$
$\mu_{10}(b)$	$e^{10\mu+50\sigma^2} - 120e^{10\mu+8\sigma^2} - 10e^{10\mu+41\sigma^2} + 45e^{10\mu+6\sigma^2} + 45e^{10\mu+33\sigma^2}$ $-9e^{5\sigma^2+10\mu} - 120e^{10\mu+26\sigma^2} + 210e^{10\mu+20\sigma^2} - 252e^{10\mu+15\sigma^2} + 210e^{10\mu+11\sigma^2}$

Distribution name	Maxwell
PDF	$\begin{cases} 0 & t < 0 \\ \dfrac{\sqrt{2}\sqrt{\frac{1}{\pi}}\, t^2 e^{-\frac{1}{2}\frac{t^2}{\alpha^2}}}{\alpha^3} & otherwise \end{cases}$
E[b]	$\dfrac{2\sqrt{2}\alpha}{\sqrt{\pi}}$
$\sigma(b)$	$\alpha\sqrt{\dfrac{3\pi-8}{\pi}}$
$\alpha(b)$	$\dfrac{1}{4}\sqrt{\dfrac{3\pi-8}{\pi}}\sqrt{2}\sqrt{\pi}$
$\beta(b)$	$-\dfrac{2\sqrt{2}(5\pi-16)}{\pi^{3/2}\left(\dfrac{3\pi-8}{\pi}\right)^{3/2}}$
$\kappa(b)$	$\dfrac{16\pi-192+15\pi^2}{(3\pi-8)^2} - 3$

$\mu_1(b)$	0
$\mu_2(b) = \mathrm{Var}(b)$	$\frac{\alpha^2\,(3\pi - 8)}{\pi}$
$\mu_3(b)$	$-\frac{2\sqrt{2}\alpha^3\,(5\pi - 16)}{\pi^{3/2}}$
$\mu_4(b)$	$\frac{\alpha^4\,(16\pi - 192 + 15\pi^2)}{\pi^2}$
$\mu_5(b)$	$-\frac{2\,\alpha^5\sqrt{2}\,(-256 + 51\pi^2 - 80\pi)}{\pi^{5/2}}$
$\mu_6(b)$	$\frac{\alpha^6\,(648\pi^2 - 2240\pi - 2560 + 105\pi^3)}{\pi^3}$
$\mu_7(b)$	$-\frac{2\sqrt{2}\,\alpha^7\,(-3072 + 543\pi^3 + 168\pi^2 - 4928\pi)}{\pi^{7/2}}$
$\mu_8(b)$	$\frac{\alpha^8\,(-18816\pi^2 - 28672 - 71680\pi + 11232\pi^3 + 945\pi^4)}{\pi^4}$
$\mu_9(b)$	$-\frac{2\sqrt{2}\,\alpha^9\,(-32768 + 6585\pi^4 + 15264\pi^3 - 72576\pi^2 - 116736\pi)}{\pi^{9/2}}$
$\mu_{10}(b)$	$\frac{3\alpha^{10}\,(-494592\pi^2 + 62200\pi^4 - 471040\pi - 21120\pi^3 - 98304 + 3465\pi^5)}{\pi^5}$

Distribution name	Gaussian
PDF	$\frac{1}{2}\,\frac{\sqrt{2}\,e^{-\frac{1}{2}\frac{(t-\mu)^2}{\sigma^2}}}{\sqrt{\pi}\,\sigma}$
E[b]	μ
$\sigma(b)$	σ
$\alpha(b)$	$\frac{\sigma}{\mu}$
$\beta(b)$	0
$\kappa(b)$	0
$\mu_1(b)$	0
$\mu_2(b) = \mathrm{Var}(b)$	σ^2
$\mu_3(b)$	0
$\mu_4(b)$	$3\,\sigma^4$
$\mu_5(b)$	0
$\mu_6(b)$	$15\,\sigma^6$
$\mu_7(b)$	0
$\mu_8(b)$	$105\,\sigma^8$
$\mu_9(b)$	0
$\mu_{10}(b)$	$945\,\sigma^{10}$

Distribution name	Power
PDF	$\begin{cases} 0 & t < 0 \\ \dfrac{c\,t^{c-1}}{b^c} & t \le b \\ 0 & otherwise \end{cases}$
E[b]	$\dfrac{b\,c}{c+1}$
σ(b)	$\dfrac{b\sqrt{\dfrac{c}{2+c}}}{c+1}$
α(b)	$\dfrac{\sqrt{\dfrac{c}{2+c}}}{c}$
β(b)	$-\dfrac{2c\,(c-1)}{\left(\dfrac{c}{2+c}\right)^{3/2}(5c+6+c^2)}$
κ(b)	$\dfrac{3\,(2+c)\,(-c+3c^2+2)}{(c^2+7c+12)\,c}-3$
μ_1(b)	0
μ_2(b) = Var(b)	$\dfrac{b^2\,c}{4c^2+c^3+5c+2}$
μ_3(b)	$-\dfrac{2b^3\,c\,(c-1)}{8c^4+24c^3+c^5+34c^2+23c+6}$
μ_4(b)	$\dfrac{3b^4\,c\,(-c+3c^2+2)}{13c^6+68c^5+186c^4+c^7+289c^3+257c^2+122c+24}$
μ_5(b)	$-\dfrac{4b^5c\,(-6+c+11c^3-6c^2)}{151c^7+659c^6+19c^8+1745c^5+c^9+2921c^4+3109c^3+2041c^2+754c+120}$
μ_6(b)	$(5b^6c\,(19c^2+2c+53c^4-26c^3+24))/(7264c^7+19090c^6+1830c^8$ $+290c^9+26c^{10}+c^{11}+34026c^5+41330c^4+33695c^3+17644c^2+5364c+720)$
μ_7(b)	$-(6b^7c\,(-120-155c^4+85c^3-85c^2-34c+309c^5))/(5040+237594c^7$ $+43308c+447168c^6+382381c^3+576514c^4+34c^{12}+505c^{11}+4332c^{10}$ $+23934c^9+90012c^8+c^{13}+167140c^2+602517c^5)$
μ_8(b)	$(7b^8c)\,(720+625c^4-345c^3+324c+496c^2-1059c^5+2119c^6))/(40320$ $+8220681c^4+4606616c^3+819c^{13}+43c^{14}+c^{15}+391824c+6527781c^7$ $+9149c^{12}+66962c^{11}+340074c^{10}+1239174c^9+3305610c^8+1731932c^2$ $+10391143c^5+9576511c^6)$

$\mu_9(b)$	$-(8b^9c(-5040-2988c+4858c^5+1603c^3-3556c^2+16687c^7-8344c^6-3220c^4))/(362880+50736447c^9+17726c^{14}+198320710c^6+164906282c^7+53c^{16}+c^{17}+121784221c^4+180333713c^5+59170688c^3+3929616c+1092035c^{12}+5242572c^{11}+18758016c^{10}+165823c^{13}+1258c^{15}+104604811c^8+19545948c^2)$
$\mu_{10}(b)$	$(9b^{10}c(29996c^2+148329c^8+43274c^6+40320+20321c^4+28944c-7756c^3-27664c^5-74164c^7))/(3628800+43287840c+309500343c^{11}+239048136c^2+3058337922c^8+4010924343c^7+4088662164c^6+373389c^{15}+3093348c^{14}+18913187c^{13}+1822923673c^9+810641924c^3+1888265726c^4+3202134249c^5+850285888c^{10}+87346658c^{12}+64c^{18}+c^{19}+1851c^{17}+32094c^{16})$

Distribution name	Rayleigh
PDF	$\begin{cases} 0 & t < 0 \\ \dfrac{t\,e^{-\frac{1}{2}\frac{t^2}{b^2}}}{b^2} & \textit{otherwise} \end{cases}$
E[b]	$\frac{1}{2}b\sqrt{2}\sqrt{\pi}$
$\sigma(b)$	$\sqrt{2-\frac{1}{2}\pi}\;b$
$\alpha(b)$	$\dfrac{\sqrt{2-\frac{1}{2}\pi}\;\sqrt{2}}{\sqrt{\pi}}$
$\beta(b)$	$\dfrac{1}{2}\,\dfrac{\sqrt{2}\;\sqrt{\pi}\;(-3+\pi)}{\left(2-\frac{1}{2}\pi\right)^{3/2}}$
$\kappa(b)$	$-\dfrac{-32+3\pi^2}{(-4+\pi)^2}-3$
$\mu_1(b)$	0
$\mu_2(b)=\mathrm{Var}(b)$	$-\frac{1}{2}b^2\pi+2b^2$
$\mu_3(b)$	$-\frac{3}{2}b^3\sqrt{2}\sqrt{\pi}+\frac{1}{2}b^3\sqrt{2}\pi^{3/2}$
$\mu_4(b)$	$8b^4-\frac{3}{4}b^4\pi^2$
$\mu_5(b)$	$\frac{1}{2}b^5\sqrt{2}\pi^{5/2}-\frac{25}{2}b^5\sqrt{2}\sqrt{\pi}+\frac{5}{2}b^5\sqrt{2}\pi^{3/2}$
$\mu_6(b)$	$-\frac{5}{8}b^6\pi^3-\frac{15}{2}b^6\pi^2+48b^6+15b^6\pi$
$\mu_7(b)$	$-\frac{231}{2}b^7\sqrt{2}\sqrt{\pi}+\frac{35}{4}b^7\sqrt{2}\pi^{3/2}+\frac{63}{8}b^7\sqrt{2}\pi^{5/2}+\frac{3}{8}b^7\sqrt{2}\pi^{7/2}$
$\mu_8(b)$	$252b^8\pi+384b^8-14b^8\pi^3-\frac{7}{16}b^8\pi^4-70b^8\pi^2$
$\mu_9(b)$	$-\frac{2511}{2}b^9\sqrt{2}\sqrt{\pi}-63b^9\sqrt{2}\pi^{3/2}+\frac{1}{4}b^9\sqrt{2}\pi^{9/2}+\frac{441}{4}b^9\sqrt{2}\pi^{5/2}+\frac{45}{4}b^9\sqrt{2}\pi^{7/2}$
$\mu_{10}(b)$	$3840b^{10}-\frac{135}{8}b^{10}\pi^4+3915b^{10}\pi-630b^{10}\pi^2-\frac{525}{2}b^{10}\pi^3-\frac{9}{32}b^{10}\pi^5$

Distribution name	Triangular
PDF	$\begin{cases} 0 & t < a \\ \dfrac{2\,(t-a)}{(b-a)\,(c-a)} & t \le c \\ \dfrac{2\,(b-t)}{(b-a)\,(-c+b)} & t \le b \\ 0 & otherwise \end{cases}$
E[b]	$\frac{1}{3}a + \frac{1}{3}b + \frac{1}{3}c$
σ(b)	$\frac{1}{6}\sqrt{2a^2 + 2b^2 + 2c^2 - 2ab - 2ac - 2bc}$
α(b)	$\dfrac{\sqrt{2\left(a^2 + b^2 + c^2 - ab - bc - ac\right)}}{2\,(a+b+c)}$
β(b)	$\frac{4}{5}\dfrac{2b^3 - 3b^2c - 3ab^2 + 12abc - 3a^2b - 3bc^2 + 2a^3 - 3ac^2 + 2c^3 - 3a^2c}{\left(2a^2 + 2b^2 + 2c^2 - 2ab - 2ac - 2bc\right)^{3/2}}$
κ(b)	$-\frac{3}{5}$
μ_1(b)	0
μ_2(b) = Var(b)	$\frac{1}{18}a^2 + \frac{1}{18}b^2 + \frac{1}{18}c^2 - \frac{1}{18}ab - \frac{1}{18}ac - \frac{1}{18}bc$
μ_3(b)	$\frac{1}{135}b^3 - \frac{1}{90}b^2c - \frac{1}{90}ab^2 + \frac{2}{45}abc - \frac{1}{90}a^2b - \frac{1}{90}bc^2 + \frac{1}{135}a^3 + \frac{1}{135}c^3 - \frac{1}{90}ac^2 - \frac{1}{90}a^2c$
μ_4(b)	$\frac{1}{135}b^4 - \frac{2}{135}ab^3 - \frac{2}{135}b^3c + \frac{1}{45}a^2b^2 + \frac{1}{45}b^2c^2 - \frac{2}{135}bc^3 - \frac{2}{135}a^3b + \frac{1}{45}a^2c^2 - \frac{2}{135}ac^3 + \frac{1}{135}c^4 + \frac{1}{135}a^4 - \frac{2}{135}a^3c$
μ_5(b)	$\frac{4}{1701}b^5 - \frac{10}{1701}ab^4 - \frac{10}{1701}b^4c + \frac{4}{1701}a^2b^3 + \frac{4}{1701}b^3c^2 + \frac{32}{1701}ab^3c - \frac{8}{567}a^2b^2c - \frac{8}{567}ab^2c^2 + \frac{4}{1701}a^3b^2 + \frac{4}{1701}b^2c^3 + \frac{32}{1701}abc^3 - \frac{10}{1701}bc^4 - \frac{10}{1701}a^4b + \frac{32}{1701}a^3bc - \frac{8}{567}a^2bc^2 + \frac{4}{1701}a^5 + \frac{4}{1701}a^3c^2 - \frac{10}{1701}ac^4 + \frac{4}{1701}a^2c^3 + \frac{4}{1701}c^5 - \frac{10}{1701}a^4c$

$\mu_6(b)$	$\frac{31}{20412}c^6+\frac{31}{20412}a^6-\frac{7}{972}ab^3c^2-\frac{7}{972}ab^2c^3+\frac{7}{972}abc^4+\frac{7}{972}a^4bc$ $-\frac{7}{972}a^3b^2c-\frac{7}{972}a^3bc^2-\frac{7}{972}a^2b^3c+\frac{7}{324}a^2b^2c^2-\frac{7}{972}a^2bc^3$ $-\frac{31}{6804}b^5c+\frac{7}{972}ab^4c-\frac{163}{20412}b^3c^3+\frac{53}{6804}b^2c^4-\frac{31}{6804}bc^5-\frac{31}{6804}a^5b$ $-\frac{31}{6804}a^5c+\frac{53}{6804}a^4b^2+\frac{53}{6804}a^4c^2-\frac{163}{20412}a^3b^3-\frac{163}{20412}a^3c^3+\frac{53}{6804}a^2b^4$ $+\frac{53}{6804}a^2c^4-\frac{31}{6804}ab^5-\frac{31}{6804}ac^5+\frac{31}{20412}b^6+\frac{53}{6804}b^4c^2$
$\mu_7(b)$	$\frac{1}{1458}c^7-\frac{7}{2916}ab^6-\frac{7}{2916}a^6b+\frac{1}{324}a^5b^2-\frac{5}{2916}a^4b^3-\frac{7}{2916}a^6c$ $+\frac{1}{324}a^5c^2-\frac{5}{2916}a^4c^3-\frac{5}{2916}a^3c^4-\frac{5}{2916}b^4a^3+\frac{1}{324}b^5a^2+\frac{1}{324}a^2c^5$ $-\frac{7}{2916}ac^6-\frac{5}{2916}b^4c^3-\frac{5}{2916}b^3c^4+\frac{1}{324}b^2c^5-\frac{7}{2916}bc^6-\frac{5}{486}b^4a^2c$ $+\frac{2}{243}a^5bc-\frac{5}{486}a^4b^2c-\frac{5}{486}a^4bc^2-\frac{5}{486}a^2c^4b+\frac{10}{729}ab^3c^3$ $-\frac{5}{486}ab^2c^4+\frac{2}{243}abc^5+\frac{2}{243}ab^5c-\frac{5}{486}ab^4c^2+\frac{10}{729}a^3b^3c+\frac{1}{324}b^5c^2$ $+\frac{10}{729}a^3c^3b+\frac{1}{1458}a^7-\frac{7}{2916}b^6c+\frac{1}{1458}b^7$
$\mu_8(b)$	$-\frac{79}{10935}a^2b^5c+\frac{158}{10935}a^2b^4c^2+\frac{158}{32805}ab^6c-\frac{79}{10935}ab^5c^2$ $+\frac{79}{32805}b^4c^3a-\frac{316}{32805}b^3c^3a^2+\frac{79}{32805}b^3c^4a-\frac{79}{10935}a^5b^2c+\frac{79}{32805}a^4b^3c$ $+\frac{158}{10935}a^4b^2c^2-\frac{316}{32805}a^3b^3c^2-\frac{316}{32805}a^3b^2c^3-\frac{79}{10935}a^5bc^2+\frac{158}{32805}c^6ab$ $+\frac{158}{32805}a^6bc+\frac{79}{32805}a^4bc^3+\frac{79}{32805}a^3bc^4+\frac{158}{10935}a^2b^2c^4+\frac{13}{32805}a^8$ $+\frac{13}{32805}b^8+\frac{139}{32805}a^4b^4-\frac{127}{32805}a^3b^5+\frac{103}{32805}a^2b^6-\frac{52}{32805}ab^7$ $-\frac{127}{32805}b^5c^3+\frac{139}{32805}b^4c^4-\frac{127}{32805}b^3c^5+\frac{139}{32805}c^4a^4-\frac{127}{32805}c^5a^3$ $+\frac{103}{32805}c^6a^2-\frac{52}{32805}c^7a+\frac{103}{32805}c^6b^2-\frac{52}{32805}c^7b-\frac{52}{32805}a^7b$ $+\frac{103}{32805}a^6b^2-\frac{127}{32805}a^5b^3-\frac{52}{32805}a^7c+\frac{103}{32805}a^6c^2-\frac{127}{32805}a^5c^3$ $+\frac{103}{32805}b^6c^2+\frac{13}{32805}c^8-\frac{79}{10935}a^2bc^5-\frac{79}{10935}ab^2c^5+\frac{79}{32805}a^3b^4c$ $-\frac{52}{32805}b^7c$

$\mu_9(\mathrm{b})$

$$
\begin{aligned}
&\frac{224}{1082565}\,b^9-\frac{112}{120285}\,a\,b^8-\frac{112}{120285}\,a\,c^8+\frac{218}{120285}\,b^7\,a^2+\frac{218}{120285}\,c^7\,a^2\\
&-\frac{112}{120285}\,a^8\,b-\frac{112}{120285}\,a^8\,c-\frac{721}{360855}\,a^3\,b^6-\frac{721}{360855}\,a^3\,c^6+\frac{97}{120285}\,a^4\,b^5\\
&+\frac{97}{120285}\,a^4\,c^5+\frac{97}{120285}\,a^5\,b^4+\frac{97}{120285}\,a^5\,c^4+\frac{218}{120285}\,a^7\,b^2+\frac{218}{120285}\,a^7\,c^2\\
&-\frac{721}{360855}\,a^6\,b^3-\frac{721}{360855}\,a^6\,c^3-\frac{112}{120285}\,b\,c^8+\frac{218}{120285}\,c^7\,b^2-\frac{721}{360855}\,b^3\,c^6\\
&+\frac{97}{120285}\,b^4\,c^5+\frac{97}{120285}\,b^5\,c^4+\frac{1054}{120285}\,a^3\,b^5\,c-\frac{161}{24057}\,a\,b^6\,c^2+\frac{1054}{120285}\,a\,b^5\,c^3\\
&+\frac{92}{24057}\,a\,c^7 b-\frac{161}{24057}\,a\,b^2\,c^6+\frac{1054}{120285}\,a\,b^3\,c^5-\frac{194}{24057}\,a\,b^4\,c^4-\frac{139}{24057}\,a^3\,b^4\,c^2\\
&+\frac{1112}{72171}\,a^3\,b^3\,c^3+\frac{92}{24057}\,a^7\,b\,c-\frac{161}{24057}\,a^6\,b^2\,c+\frac{278}{40095}\,a^2\,b^2\,c^5-\frac{161}{24057}\,a^2\,b\,c^6\\
&-\frac{139}{24057}\,a^2\,b^3\,c^4-\frac{161}{24057}\,a^2\,b^6\,c+\frac{278}{40095}\,a^2\,b^5\,c^2-\frac{139}{24057}\,a^2\,b^4\,c^3\\
&-\frac{161}{24057}\,a^6\,b\,c^2+\frac{278}{40095}\,a^5\,b^2\,c^2+\frac{1054}{120285}\,a^5\,b\,c^3+\frac{1054}{120285}\,a^5\,b^3\,c\\
&-\frac{194}{24057}\,a^4\,b^4\,c-\frac{139}{24057}\,a^4\,b^3\,c^2-\frac{139}{24057}\,a^4\,b^2\,c^3-\frac{194}{24057}\,a^4 b\,c^4\\
&-\frac{139}{24057}\,a^3\,b^2\,c^4+\frac{1054}{120285}\,a^3\,b\,c^5-\frac{112}{120285}\,b^8 c+\frac{224}{1082565}\,a^9-\frac{721}{360855}\,b^6 c^3\\
&+\frac{224}{1082565}\,c^9+\frac{92}{24057}\,a\,b^7\,c+\frac{218}{120285}\,b^7\,c^2
\end{aligned}
$$

$\mu_{10}(\mathrm{b})$

$$
\begin{aligned}
&-\frac{85}{144342}\,b^9\,c+\frac{11}{4374}\,a^8\,b\,c+\frac{67}{48114}\,a^8\,b^2-\frac{49}{24057}\,a^7\,b^3+\frac{37}{16038}\,a^6\,b^4\\
&-\frac{109}{48114}\,a^5\,b^5+\frac{37}{16038}\,a^4 b^6-\frac{85}{144342}\,a^9 b-\frac{49}{24057}\,a^3 b^7+\frac{67}{48114}\,a^8\,c^2\\
&-\frac{49}{24057}\,a^7\,c^3+\frac{37}{16038}\,a^6\,c^4-\frac{109}{48114}\,a^5\,c^5+\frac{37}{16038}\,a^4 c^6-\frac{85}{144342}\,a^9 c\\
&-\frac{49}{24057}\,a^3\,c^7-\frac{49}{24057}\,b^3 c^7-\frac{85}{144342}\,b\,c^9-\frac{85}{144342}\,a\,c^9+\frac{37}{16038}\,b^6\,c^4\\
&-\frac{109}{48114}\,b^5\,c^5+\frac{37}{16038}\,b^4 c^6+\frac{67}{48114}\,a^2 b^8+\frac{67}{48114}\,a^2 c^8+\frac{67}{48114}\,b^2\,c^8\\
&-\frac{85}{144342}\,a\,b^9+\frac{17}{144342}\,a^{10}+\frac{67}{48114}\,b^8\,c^2+\frac{17}{144342}\,b^{10}-\frac{11}{2187}\,a^7\,b\,c^2\\
&+\frac{11}{2187}\,a^6\,b\,c^3-\frac{11}{4374}\,a^5\,b\,c^4-\frac{11}{4374}\,a^4\,b^5\,c+\frac{55}{4374}\,a^4\,b^4\,c^2+\frac{55}{4374}\,a^4\,b^2\,c^4\\
&+\frac{11}{2187}\,a\,c^6\,b^3-\frac{11}{2187}\,a\,c^7\,b^2+\frac{11}{4374}\,a\,c^8\,b+\frac{22}{2187}\,a^2\,c^6\,b^2-\frac{11}{2187}\,a^2\,c^7\,b\\
&+\frac{11}{4374}\,a\,b^8\,c-\frac{11}{2187}\,a\,b^7\,c^2+\frac{11}{2187}\,a\,b^6\,c^3-\frac{11}{4374}\,a\,b^5\,c^4-\frac{11}{4374}\,a\,b^4\,c^5
\end{aligned}
$$

	$-\frac{11}{2187}a^7b^2c+\frac{11}{2187}a^6b^3c+\frac{22}{2187}a^6b^2c^2-\frac{11}{4374}a^5b^4c-\frac{22}{2187}a^5b^2c^3$ $-\frac{22}{2187}a^5b^3c^2-\frac{11}{4374}a^4bc^5+\frac{11}{2187}a^3b^6c-\frac{22}{2187}a^3b^5c^2-\frac{22}{2187}a^3b^2c^5$ $+\frac{11}{2187}a^3c^6b-\frac{11}{2187}a^2b^7c+\frac{22}{2187}a^2b^6c^2-\frac{22}{2187}a^2b^5c^3+\frac{55}{4374}a^2b^4c^4$ $-\frac{22}{2187}a^2b^3c^5-\frac{49}{24057}b^7c^3+\frac{17}{144342}c^{10}$

Distribution name	Uniform
PDF	$\begin{cases} 0 & t < a \\ \frac{1}{b-a} & t < b \\ 0 & \textit{otherwise} \end{cases}$
E[b]	$\frac{1}{2}a+\frac{1}{2}b$
σ(b)	$\frac{1}{6}\sqrt{3}\,(b-a)$
α(b)	$\frac{\sqrt{3}\,(b-a)}{3\,(a+b)}$
β(b)	0
κ(b)	$-\frac{6}{5}$
μ_1(b)	0
μ_2(b) = Var(b)	$\frac{1}{12}a^2-\frac{1}{6}ab+\frac{1}{12}b^2$
μ_3(b)	0
μ_4(b)	$\frac{1}{80}a^4-\frac{1}{20}a^3b+\frac{3}{40}a^2b^2-\frac{1}{20}ab^3+\frac{1}{80}b^4$
μ_5(b)	0
μ_6(b)	$\frac{1}{448}a^6-\frac{3}{224}a^5b+\frac{15}{448}a^4b^2-\frac{5}{112}a^3b^3+\frac{15}{448}a^2b^4-\frac{3}{224}ab^5$ $+\frac{1}{448}b^6$
μ_7(b)	0
μ_8(b)	$\frac{1}{2304}a^8-\frac{1}{288}a^7b+\frac{7}{576}a^6b^2-\frac{7}{288}a^5b^3+\frac{35}{1152}a^4b^4-\frac{7}{288}a^3b^5$ $+\frac{7}{576}a^2b^6-\frac{1}{288}ab^7+\frac{1}{2304}b^8$
μ_9(b)	0

$\mu_{10}(b)$	$\frac{1}{11264}a^{10} - \frac{5}{5632}a^9 b + \frac{45}{11264}a^8 b^2 - \frac{15}{1408}a^7 b^3 + \frac{105}{5632}a^6 b^4 - \frac{63}{2816}a^5 b^5 + \frac{105}{5632}a^4 b^6 - \frac{15}{1408}a^3 b^7 + \frac{45}{11264}a^2 b^8 - \frac{5}{5632}a b^9 + \frac{1}{11264}b^{10}$

Distribution name	Weibull
PDF	$\begin{cases} 0 & t < 0 \\ \frac{c\,t^{c-1} e^{-\left(\frac{t}{b}\right)^c}}{b^c} & \textit{otherwise} \end{cases}$
E[b]	$b\,\Gamma\left(\frac{c+1}{c}\right)$
$\sigma(b)$	$\sqrt{\Gamma\left(\frac{2+c}{c}\right) - \Gamma\left(\frac{1+c}{c}\right)^2}\; b$
$\alpha(b)$	$\frac{\sqrt{\Gamma\left(\frac{2+c}{c}\right) - \Gamma\left(\frac{c+1}{c}\right)^2}}{\Gamma\left(\frac{c+1}{c}\right)}$
$\beta(b)$	$\frac{-3\Gamma\left(\frac{c+1}{c}\right)\Gamma\left(\frac{2+c}{c}\right) + \Gamma\left(\frac{3+c}{c}\right) + 2\Gamma\left(\frac{c+1}{c}\right)^3}{\left(\Gamma\left(\frac{2+c}{c}\right) - \Gamma\left(\frac{c+1}{c}\right)^2\right)^{3/2}}$
$\kappa(b)$	$\frac{6\Gamma\left(\frac{c+1}{c}\right)^2\Gamma\left(\frac{2+c}{c}\right) - 4\Gamma\left(\frac{c+1}{c}\right)\Gamma\left(\frac{3+c}{c}\right) + \Gamma\left(\frac{4+c}{c}\right) - 3\Gamma\left(\frac{c+1}{c}\right)^4}{\left(\Gamma\left(\frac{2+c}{c}\right) - \Gamma\left(\frac{c+1}{c}\right)^2\right)^2} - 3$
$\mu_1(b)$	0
$\mu_2(b)$ = Var(b)	$b^2\left(\Gamma\left(\frac{2+c}{c}\right) - \Gamma\left(\frac{c+1}{c}\right)^2\right)$
$\mu_3(b)$	$b^3\left(-3\Gamma\left(\frac{c+1}{c}\right)\Gamma\left(\frac{2+c}{c}\right) + \Gamma\left(\frac{3+c}{c}\right) + 2\Gamma\left(\frac{c+1}{c}\right)^3\right)$
$\mu_4(b)$	$b^4\left(6\Gamma\left(\frac{c+1}{c}\right)^2\Gamma\left(\frac{2+c}{c}\right) - 4\Gamma\left(\frac{c+1}{c}\right)\Gamma\left(\frac{3+c}{c}\right) + \Gamma\left(\frac{4+c}{c}\right) - 3\Gamma\left(\frac{c+1}{c}\right)^4\right)$

$\mu_5(\mathrm{b})$	$b^5\left(-10\,\Gamma\left(\frac{c+1}{c}\right)^3\Gamma\left(\frac{2+c}{c}\right)+10\,\Gamma\left(\frac{c+1}{c}\right)^2\Gamma\left(\frac{3+c}{c}\right)\right.$ $\left.-5\,\Gamma\left(\frac{c+1}{c}\right)\Gamma\left(\frac{4+c}{c}\right)+\Gamma\left(\frac{5+c}{c}\right)+4\,\Gamma\left(\frac{c+1}{c}\right)^5\right)$
$\mu_6(\mathrm{b})$	$-b^6\left(-15\,\Gamma\left(\frac{c+1}{c}\right)^4\Gamma\left(\frac{2+c}{c}\right)+20\,\Gamma\left(\frac{c+1}{c}\right)^3\Gamma\left(\frac{3+c}{c}\right)\right.$ $-15\,\Gamma\left(\frac{c+1}{c}\right)^2\Gamma\left(\frac{4+c}{c}\right)+6\,\Gamma\left(\frac{c+1}{c}\right)\Gamma\left(\frac{5+c}{c}\right)-\Gamma\left(\frac{6+c}{c}\right)$ $\left.+5\,\Gamma\left(\frac{c+1}{c}\right)^6\right)$
$\mu_7(\mathrm{b})$	$-b^7\left(21\,\Gamma\left(\frac{c+1}{c}\right)^5\Gamma\left(\frac{2+c}{c}\right)-35\,\Gamma\left(\frac{c+1}{c}\right)^4\Gamma\left(\frac{3+c}{c}\right)\right.$ $+35\,\Gamma\left(\frac{c+1}{c}\right)^3\Gamma\left(\frac{4+c}{c}\right)-21\,\Gamma\left(\frac{c+1}{c}\right)^2\Gamma\left(\frac{5+c}{c}\right)+7\,\Gamma\left(\frac{c+1}{c}\right)\Gamma\left(\frac{6+c}{c}\right)$ $\left.-\Gamma\left(\frac{7+c}{c}\right)-6\,\Gamma\left(\frac{c+1}{c}\right)^7\right)$
$\mu_8(\mathrm{b})$	$-b^8\left(-28\,\Gamma\left(\frac{c+1}{c}\right)^6\Gamma\left(\frac{2+c}{c}\right)+56\,\Gamma\left(\frac{c+1}{c}\right)^5\Gamma\left(\frac{3+c}{c}\right)\right.$ $-70\,\Gamma\left(\frac{c+1}{c}\right)^4\Gamma\left(\frac{4+c}{c}\right)+56\,\Gamma\left(\frac{c+1}{c}\right)^3\Gamma\left(\frac{5+c}{c}\right)$ $-28\,\Gamma\left(\frac{c+1}{c}\right)^2\Gamma\left(\frac{6+c}{c}\right)+8\,\Gamma\left(\frac{c+1}{c}\right)\Gamma\left(\frac{7+c}{c}\right)-\Gamma\left(\frac{8+c}{c}\right)$ $\left.+7\,\Gamma\left(\frac{c+1}{c}\right)^8\right)$
$\mu_9(\mathrm{b})$	$-b^9\left(36\,\Gamma\left(\frac{c+1}{c}\right)^7\Gamma\left(\frac{2+c}{c}\right)-84\,\Gamma\left(\frac{c+1}{c}\right)^6\Gamma\left(\frac{3+c}{c}\right)\right.$ $+126\,\Gamma\left(\frac{c+1}{c}\right)^5\Gamma\left(\frac{4+c}{c}\right)-126\,\Gamma\left(\frac{c+1}{c}\right)^4\Gamma\left(\frac{5+c}{c}\right)$ $+84\,\Gamma\left(\frac{c+1}{c}\right)^3\Gamma\left(\frac{6+c}{c}\right)-36\,\Gamma\left(\frac{c+1}{c}\right)^2\Gamma\left(\frac{7+c}{c}\right)+9\,\Gamma\left(\frac{c+1}{c}\right)\Gamma\left(\frac{8+c}{c}\right)$ $\left.-\Gamma\left(\frac{9+c}{c}\right)-8\,\Gamma\left(\frac{c+1}{c}\right)^9\right)$

$\mu_{10}(b)$	$-b^{10}\left(-45\,\Gamma\left(\frac{c+1}{c}\right)^{8}\Gamma\left(\frac{2+c}{c}\right)+120\,\Gamma\left(\frac{c+1}{c}\right)^{7}\Gamma\left(\frac{3+c}{c}\right)\right.$ $-210\,\Gamma\left(\frac{c+1}{c}\right)^{6}\Gamma\left(\frac{4+c}{c}\right)+252\,\Gamma\left(\frac{c+1}{c}\right)^{5}\Gamma\left(\frac{5+c}{c}\right)$ $-210\,\Gamma\left(\frac{c+1}{c}\right)^{4}\Gamma\left(\frac{6+c}{c}\right)+120\,\Gamma\left(\frac{c+1}{c}\right)^{3}\Gamma\left(\frac{7+c}{c}\right)$ $-45\,\Gamma\left(\frac{c+1}{c}\right)^{2}\Gamma\left(\frac{8+c}{c}\right)+10\,\Gamma\left(\frac{c+1}{c}\right)\Gamma\left(\frac{9+c}{c}\right)-\Gamma\left(\frac{10+c}{c}\right)$ $\left.+9\,\Gamma\left(\frac{c+1}{c}\right)^{10}\right)$

References

[1] Ahammed, M. and Melchers, R.E. (2009) A convenient approach for estimating time-dependent structural reliability in the load space. *Probabilistic Engineering Mechanics*, **24**(3), 467–472.

[2] Babuška, I., Tempone, R., and Zouraris, G.E. (2005) Solving elliptic boundary value problems with uncertain coefficients by the finite element method: the stochastic formulation. *Computer Methods in Applied Mechanics and Engineering*, **194**, 1251–1294.

[3] Bathe, K.J. (1996) *Finite Element Procedures*, Prentice-Hall, Englewood Cliffs, NJ.

[4] Beer, G. (2001) *Programming the Boundary Element Method. An Introduction for Engineers*, Wiley-VCH Verlag GmbH, New York.

[5] Bendat, J.S. and Piersol, A.G. (1971) *Random Data: Analysis and Measurement Procedures*, Wiley-VCH Verlag GmbH, New York.

[6] Bensoussan, A., Lions, J.L., and Papanicolaou, G. (1978) *Asymptotic Analysis for Periodic Structures*, North-Holland, Amsterdam.

[7] Bhangale, R.K. and Ganesan, N. (2006) Static analysis of simply supported functionally graded and layered magneto-electro-elastic plates. *International Journal of Solids and Structures*, **43**, 3230–3253.

[8] Bijelonja, I., Demirdzic, I., and Muzaferija, S. (2006) A finite volume method for incompressible linear elasticity. *Computer Methods in Applied Mechanics and Engineering*, **195**, 6378–6390.

[9] Björck, A. (1996) *Numerical Methods for Least Squares Problems*, SIAM, Philadelphia.

[10] Brebbia, C.A. and Dominguez, J. (1996) *Boundary Elements. An Introductory Course*, Computational Mechanics Publications, Southampton.

[11] Breitung, K., Casciati, F., and Faravelli, L. (1997) A random field description for stochastic boundary elements. *Engineering Analysis with Boundary Elements*, **19**(3), 223–229.

[12] Burczyński, T. (1995) Boundary element method in stochastic shape design sensitivity analysis and identification of uncertain elastic solids. *Engineering Analysis with Boundary Elements*, **15**(2), 151–160.

[13] Carlsaw, H.S. and Jaeger, J.C. (1986) *Conduction of Heat in Solids*, Clarendon Press, Oxford.

[14] Castaneda, P.P. and Suquet, P.M. (1998) Nonlinear composites, in *Advances in Applied Mechanics*, Vol. **34** (eds E. Van Der Giessen and T.Y. Wu), Academic Press, New York.

[15] Chaboche, J.L. (2008) A review of some plasticity and viscoplasticity constitutive theories. *International Journal of Plasticity*, **24**, 1642–1693.

The Stochastic Perturbation Method for Computational Mechanics, First Edition. Marcin Kamiński.
© 2013 John Wiley & Sons, Ltd. Published 2013 by John Wiley & Sons, Ltd.

[16] Chantasiriwan, S. (2005) Solutions of partial differential equations with random Dirichlet boundary conditions by multiquadric collocation method. *Engineering Analysis with Boundary Elements*, **29**(12), 1124–1129.

[17] Char, B.W. *et al.* (1992) *First Leaves: A Tutorial Introduction to Maple V*, Springer-Verlag, Berlin.

[18] Chellappa, S., Diaz, A.R., and Bendsoe, M. (2004) Layout optimization of structures with finite-sized features using multiresolution analysis. *Structural and Multidisciplinary Optimization*, **26**, 77–91.

[19] Cheng, H.D. and Lafe, O.E. (1991) Boundary element solution for stochastic groundwater flow: random boundary conditions and recharge. *Water Resources*, **27**(2), 231–242.

[20] Chou, Y.F. (1991) *Applications of Discrete Functional Analysis to the Finite Difference Method*, Elsevier, Amsterdam.

[21] Christensen, R.M. (1979) *Mechanics of Composite Materials*, Wiley-VCH Verlag GmbH, New York.

[22] Clayton, J.D. and Chung, P.W. (2006) An atomistic-to-continuum framework for non-linear crystal mechanics based on asymptotic homogenization. *Journal of Mechanics and Physics of Solids*, **54**, 1604–1639.

[23] Collatz, L. (1966) *The Numerical Treatment of Differential Equations*, 3rd edn, Springer-Verlag, New York.

[24] Cruse, T.A. (1988) *Boundary Elements Analysis in Computational Fracture Mechanics*, Kluwer Academic Publishers, Dordrecht.

[25] Cruz, M.E. and Patera, A.T. (1995) A parallel Monte-Carlo finite element procedure for the analysis of multicomponent media. *International Journal for Numerical Methods in Engineering*, **38**, 1087–1121.

[26] Cueto-Felgueroso, L., Colominas, I., Nogueira, X., and Casteleiro, M. (2007) Finite volume solvers and moving least-squares approximations for the compressible Navier–Stokes equations on unstructured grids. *Computer Methods in Applied Mechanics and Engineering*, **196**, 4712–4736.

[27] Dasgupta, G., Papusha, A.N., and Malsch, E. (2001) First order stochasticity in boundary geometry: a computer algebra BE environment. *Engineering Analysis with Boundary Elements*, **25**(9), 741–751.

[28] Durany, J., Pereira, J., and Varas, F. (2006) A cell-vertex finite volume method for thermohydrodynamic problems in lubrication theory. *Computer Methods in Applied Mechanics and Engineering*, **195**, 5949–5961.

[29] El Harrouni, K., Ouazar, D., Wrobel, L.C., and Cheng, A.H.D. (1997) Uncertainty analysis of groundwater flow with DRBEM. *Engineering Analysis with Boundary Elements*, **19**(3), 217–221.

[30] Elishakoff, I. (1995) Random vibration of structures: a personal perspective. *Applied Mechanics Review*, **48**(12), 809–825.

[31] Elishakoff, I. and Ren, Y.J. (1995) Some exact solutions for the bending of beams with spatially stochastic stiffness. *International Journal of Solids and Structures*, **32**(16), 2315–2327.

[32] Emery, A.F. (2004) Solving stochastic heat transfer problems. *Engineering Analysis with Boundary Elements*, **28**(3), 279–291.

[33] Falsone, G. (2005) An extension of the Kazakov relationship for non-Gaussian random variables and its use in the non-linear stochastic dynamics. *Probabilistic Engineering Mechanics*, **20**, 45–56.

[34] Feller, W. (1965) *An Introduction to Probability Theory and its Applications*, Wiley-VCH Verlag GmbH, New York.

[35] Fish, J. and Ghouali, A. (2001) Multiscale analytical sensitivity analysis for composite materials. *International Journal for Numerical Methods in Engineering*, **50**, 1501–1520.

[36] Florence, C. and Sab, K. (2006) A rigorous homogenization method for the determination of the overall ultimate strength of periodic discrete media and an application to general hexagonal lattices of beams. *European Journal of Mechanics A/Solids*, **25**, 72–97.

[37] Friebel, C., Doghri, I., and Legat, V. (2006) General mean-field homogenization schemes for viscoelastic composites containing multiple phases of coated inclusions. *International Journal of Solids and Structures*, **43**, 2513–2541.

[38] Ghanem, R.G. and Spanos, P.D. (1991) *Stochastic Finite Elements. A Spectral Approach*, Springer-Verlag, Berlin.

[39] Ghanem, R.G. and Spanos, P.D. (1997) Spectral techniques for stochastic finite elements. *Archives of Computer Methods in Engineering*, **4**(1), 63–100.

[40] Gitman, I.M., Gitman, M.B., and Askes, H. (2006) Quantification of stochastically stable representative volumes for random heterogeneous media. *Archives of Applied Mechanics*, **75**, 79–92.

[41] Grigoriu, M. (1999) Stochastic mechanics. *International Journal of Solids and Structures*, **37**, 197–214.

[42] Guedri, M., Lima, A.M.G., Bouhaddi, N., and Rade, D.A. (2010) Robust design of viscoelastic structures based on stochastic finite element models. *Mechanical Systems Signal Processing*, **24**(1), 59–77.

[43] Gutierrez, M.A. and de Borst, R. (1999) Numerical analysis of localization using a viscoplastic regularization: influence of stochastic material defects. *International Journal for Numerical Methods in Engineering*, **44**, 1823–1841.

[44] Haftka, R.T. and Gürdal, Z. (1992) *Elements of Structural Optimization*, Kluwer Academic Publishers, Dordrecht.

[45] Harth, T., Schwan, S., Lehn, J., and Kollmann, F.G. (2004) Identification of material parameters for inelastic constitutive models: statistical analysis and design of experiments. *International Journal of Plasticity*, **20**, 1403–1440.

[46] Hien, T.D. and Kleiber, M. (1997) Stochastic finite element modelling in linear transient heat transfer. *Computer Methods in Applied Mechanics and Engineering*, **144**, 111–124.

[47] Hoffman, J.D. (1992) *Numerical Methods for Engineers and Scientists*, McGraw-Hill, New York.

[48] Honda, R. (2005) Stochastic BEM with spectral approach in elastostatic and elastodynamic problems with geometrical uncertainty. *Engineering Analysis with Boundary Elements*, **29**(5), 415–427.

[49] Hromadka, T. and Lai, Ch. (1986) *The Complex Variable Boundary Element Method in Engineering Analysis*, Springer-Verlag, New York.

[50] Huang, B., Li, Q.S., Tuan, A.Y., and Zhu, H. (2009) Recursive approach for random response analysis using non-orthogonal polynomial expansions. *Computational Mechanics*, **44**, 309–320.

[51] Hughes, T.J.R. (2000) *The Finite Element Method – Linear Static and Dynamic Finite Element Analysis*, Dover Publications, New York.

[52] Hurtado, J.E. (2002) Reanalysis of linear and nonlinear structures using iterated Shanks transformation. *Computer Methods in Applied Mechanics and Engineering*, **191**, 4215–4229.

[53] Hurtado, J.E. and Barbat, A.H. (1998) Monte-Carlo techniques in computational stochastic mechanics. *Archives of Computer Methods in Engineering*, **5**(1), 3–30.

[54] Idiart, M.I., Moulinec, H., Ponte Castañeda, P., and Suquet, P. (2006) Macroscopic behavior and field fluctuations in viscoplastic composites: second-order estimates versus full-field simulations. *Journal of Mechanics and Physics of Solids*, **54**, 1029–1063.

[55] Jefferson, G., Garmestani, H., Tannenbaum, R., Gokhale, A., and Tadd, E. (2005) Two-point probability distribution function analysis of co-polymer nano-composites. *International Journal of Plasticity*, **21**, 185–198.

[56] Jeulin, D. and Ostoja-Starzewski, M. (eds) (2001) *Mechanics of Random and Multiscale Structures*, CISM Courses and Lectures, Vol. **430**, Springer-Verlag, NewYork.

[57] Jianting, Z. and Satish, M.G. (1997) A boundary element method for stochastic flow problems in a semiconfined aquifier with random boundary conditions. *Engineering Analysis with Boundary Elements*, **19**(3), 199–208.

[58] Kalamkarov, A.L., and Kolpakov, A.G. (1997) *Analysis, Design and Optimization of Composite Structures*, Wiley.

[59] Kamiński, M. (1996) Homogenization in elastic random media. *Computer Assisted Mechanics and Engineering Science*, **3**(1), 9–21.

[60] Kamiński, M. (1999) Boundary element method homogenization of the periodic linear elastic composites. *Engineering Analysis with Boundary Elements*, **23**(10), 815–823.

[61] Kamiński, M. (1999) Stochastic second order perturbation BEM formulation. *Engineering Analysis with Boundary Elements*, **2**, 123–130.

[62] Kamiński, M. (2001) On stochastic microstructural finite elements. *International Journal of Fracture*, **110**(1), L3–L8.

[63] Kamiński, M. (2001) Stochastic hybrid stress-based finite element method. *Hybrid Methods in Engineering*, **3**(1), 25–51.

[64] Kamiński, M. (2001) Stochastic perturbation approach to the stress-based finite element method. *International Journal of Solids and Structures*, **38**(21), 3831–3852.

[65] Kamiński, M. (2001) Stochastic problem of the viscous incompressible fluid flow with heat transfer. *Zeitschrift für Angewandte Mathematik und Mechanik*, **81**(12), 827–837.

[66] Kamiński, M. (2001) The stochastic second order perturbation technique in the finite difference method. *Communications in Numerical Methods in Engineering*, **17**(9), 613–622.

[67] Kamiński, M. (2002) On probabilistic viscous incompressible flow of some composite fluids. *Computational Mechanics*, **28**(6), 505–515.

[68] Kamiński, M. (2002) Stochastic perturbation approach to engineering structures vibrations by the finite difference method. *International Journal of Sound and Vibration*, **251**(4), 651–670.

[69] Kamiński, M. (2002) Stochastic problem of fiber-reinforced composite with interface defects. *Engineering Computations: International Journal of Computer-Aided Engineering*, **19**(7), 854–868.

[70] Kamiński, M. (2004) Stochastic perturbation approach to wavelet-based multiresolutional analysis. *Numerical Linear Algebra with Engineering Applications*, **11**(4), 355–370.

[71] Kamiński, M. (2005) *Computational Mechanics of Composite Materials. Sensitivity, Randomness and Multiscale Behaviour*, Springer-Verlag, New York.

[72] Kamiński, M. (2005) Multiscale homogenization of *n*-component composites with semi-elliptical stochastic interface defects. *International Journal of Solids and Structures*, **42**(11&12), 3751–3590.

[73] Kamiński, M. (2007) Application of the generalized perturbation-based stochastic boundary element method to elastostatics. *Engineering Analysis with Boundary Elements*, **31**(6), 514–527.

[74] Kamiński, M. (2007) Generalized perturbation-based stochastic finite element method in elastostatics. *Computers and Structures*, **85**(10), 586–594.

[75] Kamiński, M. (2008) Homogenization of the fiber-reinforced composites under stochastic ageing processes. *International Journal of Multiscale Computational Engineering*, **6**(4), 361–370.

[76] Kamiński, M. (2009) A generalized stochastic finite difference method. *Journal of Theoretical and Applied Mechanics*, **47**(4), 957–975.

[77] Kamiński, M. (2009) Sensitivity and randomness in homogenization of periodic fiber-reinforced composites via the response function method. *International Journal of Solids and Structures*, **46**(3&4), 923–937.

[78] Kamiński, M. (2009) Stochastic interface defects in composite materials subjected to the aging processes. *International Journal of Multiscale Computational Engineering*, **7**(4), 371–380.

[79] Kamiński, M. (2010) An introduction to the plasticity problems using the generalized stochastic perturbation technique. *Computational Mechanics*, **45**(4), 349–361.

[80] Kamiński, M. (2010) Effective elastic parameters of the periodic fibrous composites with random fibers according to the perturbation approach with response function. *Communications in Numerical Methods in Engineering*, **26**(21), 1548–1558.

[81] Kamiński, M. (2010) Generalized stochastic perturbation technique in engineering computations. *International Journal of Mathematical and Computer Modelling*, **51**(3&4), 272–285.

[82] Kamiński, M. (2010) Potential problems with random parameters by the generalized perturbation-based stochastic finite element method. *Computers and Structures*, **88**, 437–445.

[83] Kamiński, M. (2010) Structural design sensitivity using boundary elements and polynomial response function. *Structural and Multidisciplinary Optimization*, **41**, 107–116.

[84] Kamiński, M. (2011) Homogenization of fiber-reinforced composites with random properties using the least squares response function approach. *International Journal of Multiscale Computational Engineering*, **9**(3), 257–270.

[85] Kamiński, M. (2011) Homogenization of fiber-reinforced composites with random properties using the weighted least squares response function approach. *Archives of Mechanics*, **63**(5&6), 479–505.

[86] Kamiński, M. (2011) Homogenization of fibrous composites under stochastic ageing. *ICE Proceedings Engineering and Computational Mechanics*, **164**(3), 155–162.

[87] Kamiński, M. (2011) Least squares stochastic boundary element method. *Engineering Analysis with Boundary Elements*, **35**(5), 776–784.

[88] Kamiński, M. (2011) On semi-analytical probabilistic finite element method for homogenization of the periodic fiber-reinforced composites. *International Journal for Numerical Methods in Engineering*, **86**(9), 1144–1162.

[89] Kamiński, M. (2011) Probabilistic analysis of transient problems by the least squares stochastic perturbation-based finite element method. *CMES: Computer Modelling in Engineering and Sciences*, **80**(1), 113–140.

[90] Kamiński, M. (2011) Steady state reaction–diffusion problems with random parameters using the generalized stochastic finite difference method. *Journal of Applied Computer Science*, **19**(2), 31–45.

[91] Kamiński, M. (2011) Structural sensitivity analysis in nonlinear and transient problems using the local response function technique. *Structural and Multidisciplinary Optimization*, **43**(2), 261–274.

[92] Kamiński, M. (2012) Probabilistic entropy in homogenization of the periodic fiber-reinforced composites with random elastic parameters. *International Journal for Numerical Methods in Engineering*, **90**(8), 939–954.

[93] Kamiński, M. (2012) Stochastic boundary element method analysis of the interface defects in composite materials. *Composite Structures*, **94**, 394–402.

[94] Kamiński, M. and Carey, G.F. (2005) Stochastic perturbation-based finite element approach to fluid flow problems. *International Journal of Numerical Methods in Heat and Fluid Flow*, **15**(7), 671–697.

[95] Kamiński, M. and Corigliano, A. (2012) Sensitivity, probabilistic and stochastic analysis of the thermo-piezoelectric phenomena in solids by the stochastic perturbation technique. *Meccanica*, **47**, 877–891.

[96] Kamiński, M. and Hien, T.D. (1999) Stochastic finite element modeling of transient heat transfer in layered composites. *International Communications in Heat and Mass Transfer*, **26**(6), 791–800.

[97] Kamiński, M. and Hien, T.D. (1999) Stochastic structural defects in heat conduction problems for composite materials. *Archives of Mechanics*, **51**(3&4), 269–288.

[98] Kamiński, M. and Kleiber, M. (2000) Numerical homogenization of n-component composites including stochastic interface defects. *International Journal for Numerical Methods in Engineering*, **47**, 1001–1027.

[99] Kamiński, M. and Kleiber, M. (2000) Perturbation based stochastic finite element method for homogenization of two-phase elastic composites. *Computers and Structures*, **78**(6), 811–826.

[100] Kamiński, M. and Leśniak, M. (2012) Homogenization of metallic fiber-reinforced composites under stochastic ageing. *Composite Structures*, **94**, 386–393.

[101] Kamiński, M. and Schrefler, B.A. (2000) Probabilistic effective characteristics of cables for superconducting coils. *Computer Methods in Applied Mechanics and Engineering*, **188**(1–3), 1–16.

[102] Kamiński, M. and Spanos, P.D. (2001) Stochastic second order perturbation approach to finite difference method in vibration analysis. *International Journal of Applied Mechanics and Engineering*, **2**(6), 369–393.

[103] Kamiński, M. and Świta, P. (2011) Generalized stochastic finite element method in elastic stability problems. *Computers and Structures*, **89**, 1241–1252.

[104] Kamiński, M. and Szafran, J. (2009) Eigenvalue analysis for high telecommunication towers with lognormal stiffness by the response function method and SFEM. *Computer Assisted Mechanics and Engineering Science*, **16**, 279–290.

[105] Kamiński, M. and Szafran, J. (2009) Random eigenvibrations of elastic structures by the response function method and the generalized stochastic perturbation technique. *Archives of Civil and Mechanical Engineering*, **9**(4), 5–32.

[106] Kamiński, M. and Szafran, J. (2012) Stochastic analysis and reliability of the steel telecommunication towers. *CMES: Computer Modelling in Engineering and Sciences*, **83**(2), 143–168.

[107] Kanit, T. (2006) Apparent and effective physical properties of heterogeneous materials: representativity of samples of two materials from food industry. *Computer Methods in Applied Mechanics and Engineering*, **195**, 3960–3982.

[108] Karakostas, C.Z. and Manolis, G.D. (1998) A stochastic boundary solution applied to groundwater flow. *Engineering Analysis with Boundary Elements*, **21**(1), 9–21.

[109] Karakostas, C.Z. and Manolis, G.D. (2002) Dynamic response of tunnels in stochastic soils by the boundary element method. *Engineering Analysis with Boundary Elements*, **26**(8), 667–680.

[110] Keese, A. and Matthies, H.G. (2005) Hierarchical parallelisation for the solution of stochastic finite element equations. *Computers and Structures*, **83**, 1033–1047.

[111] Kleiber, M. (ed.) (1998) *Handbook of Computational Solid Mechanics*, Springer-Verlag, Berlin.

[112] Kleiber, M. and Hien, T.D. (1992) *The Stochastic Finite Element Method*, Wiley-VCH Verlag GmbH, New York.

[113] Krishnamoorthy, C.S. (1994) *Finite Element Analysis*, 2nd edn, McGraw-Hill, New York.

[114] Kumar, A. and Chakraborty, D. (2009) Effective properties of thermo-electro-mechanically coupled piezoelectric fiber reinforced composites. *International Journal of Materials and Design*, **30**(4), 1216–1222.

[115] Kunz, K.S. and Luebbers, R.J. (1993) *The Finite Difference Time Domain Method for Electromagnetics*, CRC Press, Boca Raton, FL.

[116] Lepage, D. (2006) Stochastic finite element method for the modeling of thermoelastic damping in micro-resonators. PhD dissertation, University of Liege.

[117] Lin, Y.K. and Cai, G.Q. (1995) *Probabilistic Structural Dynamics*, McGraw-Hill, New York.

[118] Liu, W.K., Belytschko, T., and Mani, A. (1986) Random field finite elements. *International Journal for Numerical Methods in Engineering*, **23**, 1831–1845.

[119] Luciano, R. and Willis, J.R. (2006) Hashin–Shtrikman based FE analysis of the elastic behaviour of finite random composite bodies. *International Journal of Fracture*, **137**, 261–273.

[120] Lux, J., Ahmadi, A., Gobbé, C., and Delisée, C. (2006) Macroscopic thermal properties of real fibrous materials: volume averaging method and 3D image analysis. *International Journal of Heat and Mass Transfer*, **49**, 1958–1973.

[121] Ma, W. and Hu, G. (2006) Influence of fiber's shape and size on overall elastoplastic property for micropolar composites. *International Journal of Solids and Structures*, **43**, 3025–3043.

[122] Manolis, G.D. and Karakostas, C.Z. (2003) A Green's function method to SH-wave motion in a random continuum. *Engineering Analysis with Boundary Elements*, **27**(2), 93–100.

[123] Manolis, G.D. and Shaw, R.P. (1995) Wave motions in stochastic heterogeneous media: a Green's function approach. *Engineering Analysis with Boundary Elements*, **15**(3), 225–234.

[124] Manolis, G.D. and Shaw, R.P. (1996) Boundary integral formulation for 2D and 3D thermal problems exhibiting a linearly varying stochastic conductivity. *Computational Mechanics*, **17**, 406–417.

[125] Markovic, D. and Ibrahimbegovic, A. (2006) Complementary energy based FE modeling of coupled elasto-plastic and damage behavior for continuum microstructure computations. *Computer Methods in Applied Mechanics and Engineering*, **195**, 5077–5093.

[126] Matache, A.M., Babuska, I., and Schwab, C. (2000) Generalized p-FEM in homogenization. *Numerische Mathematik*, **86**, 319–375.

[127] Maute, K., Weickum, G., and Eldred, M. (2009) A reduced-order stochastic finite element approach for design optimization under uncertainty. *Structural Safety*, **31**(6), 450–459.

[128] Melchers, R.E. (2002) *Structural Reliability Analysis and Prediction*, John Wiley & Sons, New York.

[129] Milton, G. (2002) *The Theory of Composites*, Cambridge University Press, Cambridge.

[130] Minkowycz, W.J., Sparrow, E.H., Schneider, G.E., and Fletcher, R.H. (1988) *Handbook of Numerical Heat Transfer*, Wiley-Interscience, New York.

[131] Mitchell, A.R. and Griffiths, D.F. (1980) *The Finite Difference Method in Partial Differential Equations*, Wiley-VCH Verlag GmbH, New York.

[132] Möller, B. and Beer, M. (2004) *Fuzzy Randomness – Uncertainty in Civil Engineering and Computational Mechanics*, Springer-Verlag, Berlin.

[133] Mura, T. (1982) *Micromechanics of Defects in Solids*, Sijthoff and Noordhoff, Netherlands.

[134] Naji, A., Cheng, A.H.D., and Ouazar, D. (1999) BEM solutions of stochastic seawater intrusion problems. *Engineering Analysis with Boundary Elements*, **23**(7), 529–537.

[135] Nayfeh, A. (1973) *Perturbation Method*, Wiley-VCH Verlag GmbH, New York.

[136] Noor, A.K. and Shah, R.S. (1993) Effective thermoelastic and thermal properties of unidirectional fiber-reinforced composites and their sensitivity coefficients. *Composite Structures*, **26**, 7–23.

[137] Nowacki, W. (1975) Thermodiffusion in solids. *Journal of Theoretical and Applied Mechanics*, **13**(2), 143–158.

[138] Oden, J.T. (1972) *Finite Elements of Nonlinear Continua*, McGraw-Hill, New York.

[139] Oden, J.T., Babuška, I., Nobile, F., Feng, Y., and Tempone, R. (2005) Theory and methodology for estimation and control of errors due to modeling, approximation and uncertainty. *Computer Methods in Applied Mechanics and Engineering*, **194**, 195–204.

[140] Ostoja-Starzewski, M. (2005) Scale effects in plasticity of random media: status and challenges. *International Journal of Plasticity*, **21**, 1119–1160.

[141] Owen, D.R.J. and Hinton, E. (1980) *Finite Elements in Plasticity – Theory and Practice*, Pineridge Press, Swansea.

[142] Papadopoulos, V. and Papadrakakis, M. (2005) The effect of material and thickness variability on the buckling load of shells with random initial imperfections. *Computer Methods in Applied Mechanics and Engineering*, **194**, 1405–1426.

[143] Papadopoulos, V., Papadrakakis, M., and Deodatis, G. (2006) Analysis of mean and mean square response of general linear stochastic finite element systems. *Computer Methods in Applied Mechanics and Engineering*, **195**, 5454–5471.

[144] Papadrakakis, M. (ed.) (1993) *Solving Large-Scale Problems in Mechanics. The Development and Application of Computational Solution Methods*, Wiley-VCH Verlag GmbH, New York.

[145] Pepper, D.W. and Heinrich, J.C. (1992) *The Finite Element Method*, Computational and Physical Processes in Mechanics and Thermal Sciences, Hemisphere, New York.

[146] Pietrzak, J., Rakowski, G. and Wrzesniowski, K. (1986), *Matrix Analysis of Structures (in Polish)*, Polish Scientific Publishers, Warsaw.

[147] Rahman, S. (2008) A polynomial dimensional decomposition for stochastic computing. *International Journal for Numerical Methods in Engineering*, **76**, 2091–2116.

[148] Rohan, E., Cimrman, R., and Lukes, V. (2006) Numerical modeling and homogenized constitutive law of large deforming fluid saturated heterogeneous solids. *Computers and Structures*, **84**, 1095–1114.

[149] Roy, R.V. and Grilli, S.T. (1997) Probabilistic analysis of flow in random porous media by stochastic boundary elements. *Engineering Analysis with Boundary Elements*, **19**(3), 239–255.

[150] Samaey, G., Kevrikidis, I.G., and Roose, D. (2006) Patch dynamics with buffers for homogenization problems. *Journal of Computational Physics*, **213**, 264–287.

[151] Sanchez-Palencia, E. (1980) *Non-Homogeneous Media and Vibration Theory*, Lecture Notes in Physics, Vol. **127**, Springer-Verlag, Berlin.

[152] Schäfer, M. (2006) *Computational Engineering – Introduction to Numerical Methods*, Springer-Verlag, Berlin.

[153] Schenk, C.A., Pradlwarter, H.J., and Schüeller, G.I. (2005) Non-stationary response of large, non-linear finite element systems under stochastic loading. *Computers and Structures*, **83**, 1086–1102.

[154] Schüeller, G.I. (2007) On the treatment of uncertainties in structural mechanics and analysis. *Computers and Structures*, **85**, 235–243.

[155] Shannon, C.E. (1948) A mathematical theory of communication. *The Bell System Technical Journal*, **27**, 379–423.

[156] Shannon, C.E. (1948) A mathematical theory of communication. *The Bell System Technical Journal*, **27**, 623–656.

[157] Sigmund, O. (1994) Materials with prescribed constitutive parameters: an inverse homogenization problem. *International Journal of Solids and Structures*, **31**, 2313–2329.

[158] Sobczyk, K. (1991) *Stochastic Differential Equations. Applications in Physics, Engineering and Mechanics*, Kluwer Academic Publishers, Dordrecht.

[159] Soedel, W. (1993) *Vibrations of Shells and Plates*, Marcel Dekker, New York.

[160] Song, Y.S. and Youn, J.R. (2006) Evaluation of effective thermal conductivity for carbon nanotube/polymer composites using control volume finite element method. *Carbon*, **44**, 710–717.

[161] Spanos, P.D., Tezcan, J., and Tratskas, P. (2005) Stochastic processes evolutionary spectrum estimation via harmonic wavelets. *Computer Methods in Applied Mechanics and Engineering*, **194**, 1367–1383.

[162] Steeves, C.A. and Fleck, N.A. (2006) In-plane properties of composite laminates with through-thickness pin reinforcement. *International Journal of Solids and Structures*, **43**, 3197–3212.

[163] Taflove, A. (1998) *Advances in Computational Electrodynamics. The Finite-Difference Time-Domain Method*, Artech House, London.

[164] To, C.W.S. and Kiu, M.L. (1994) Random responses of discretized beams and plates by the stochastic central difference method with time co-ordinate transformation. *Computers and Structures*, **53**(3), 727–738.

[165] Vanmarcke, E. (1983) *Random Fields. Analysis and Synthesis*, MIT Press, Cambridge, MA.

[166] Van Noortwijk, J.M. and Frangopol, D.M. (2004) Two probabilistic life-cycle maintenance models for deteriorating civil infrastructures. *Probabilistic Engineering Mechanics*, **19**, 345–359.

[167] Wah, T. and Calcote, L.R. (1970) *Structural Analysis by Finite Difference Calculus*, Van Nostrand Reinhold, New York.

[168] Wan, X. and Karniadakis, G.E. (2009) Solving elliptic problems with non-Gaussian spatially-dependent random coefficients. *Computer Methods in Applied Mechanics and Engineering*, **198**(21–26), 1985–1995.

[169] Wang, H.F. and Anderson, M.P. (1995) *Introduction to Groundwater Modeling: Finite Difference and Finite Element Methods*, Academic Press, New York.

[170] Wang, W.X., Luo, D., Takao, Y., and Kakimoto, K. (2006) New solution method for homogenization analysis and its application to the prediction of macroscopic elastic constants of materials with periodic microstructures. *Computers and Structures*, **84**, 991–1001.

[171] Witteveen, J.A.S. and Bijl, H. (2009) Effect of randomness on multifrequency aeroelastic responses resolved by unsteady adaptive stochastic finite elements. *Journal of Computational Physics*, **228**(18), 7025–7045.

[172] Wolf, J.P. (2003) *The Scaled Boundary Finite Element Method*, Wiley-VCH Verlag GmbH, New York.

[173] Xiu, D. (2007) Efficient collocational approach for parametric uncertainty analysis. *Communications in Computational Physics*, **2**(2), 293–309.

[174] Xiu, D. and Karniadakis, G.E. (2003) A new stochastic approach to transient heat conduction modeling with uncertainty. *International Journal of Heat and Mass Transfer*, **46**, 4681–4693.

[175] Xiu-Li, G. and Melchers, R.E. (2002) Effect of response surface parameter variation on structural reliability estimates. *Structural Safety*, **23**, 429–444.

[176] Xu, X.F. and Graham-Brady, L. (2005) A stochastic computational method for evaluation of global and local behavior of random elastic media. *Computer Methods in Applied Mechanics and Engineering*, **194**, 4362–4385.

[177] Zaiser, M. and Aifantis, E.C. (2006) Randomness and slip avalanches in gradient plasticity. *International Journal of Plasticity*, **22**, 1432–1455.

[178] Zhang, H., Zhang, S., Guo, X., and Bi, J. (2005) Multiple spatial and temporal scales method for numerical simulation of non-classical heat conduction problems: one dimensional case. *International Journal of Solids and Structures*, **42**, 877–899.

[179] Zhang, H.W., Zhang, S., Bi, J., and Schrefler, B.A. (2007) Thermo-mechanical analysis of periodic multiphase materials by a multiscale asymptotic homogenization approach. *International Journal for Numerical Methods in Engineering*, **69**(1), 87–113.

[180] Zienkiewicz, O.C. and Taylor, R.L. (2005) *Finite Element Method for Solid and Structural Mechanics*, 6th edn, Elsevier, Amsterdam.

Index

The Stochastic Perturbation Method for Computational Mechanics, First Edition. Marcin Kamiński.
© 2013 John Wiley & Sons, Ltd. Published 2013 by John Wiley & Sons, Ltd.

Mechanical